FERTILIZERS
AND SOIL AMENDMENTS

FERTILIZERS AND SOIL AMENDMENTS

Roy Hunter Follett

Professor of Agronomy
Extension Soils Specialist
Colorado State University

Larry S. Murphy

Great Plains Director
Potash and Phosphate Institute
Former Professor of Agronomy
Kansas State University

Roy L. Donahue

Soil Scientist and Agronomist
American Registry of Certified Professionals
in Agronomy, Crops, and Soils
Professor Emeritus of Soil Science
Michigan State University

Prentice-Hall, Inc., Englewood Cliffs, New Jersey 07632

Library of Congress Cataloging in Publication Data

Follett, Roy H.
   Fertilizers and soil amendments.

   Includes bibliographies and index.
   1. Fertilizers and manures. 2. Soil amendments.
3. Plants—Nutrition. I. Murphy, Larry S., joint
author. II. Donahue, Roy Luther, joint
author. III. Title.
S633.F62   631.8   80-25799
ISBN 0-13-314336-8

© 1981 by Prentice-Hall, Inc., Englewood Cliffs, N.J. 07632

Printed in the United States of America

10 9 8 7 6 5 4

Editorial/production supervision by Leslie I. Nadell
Interior design by Ian M. List and Leslie I. Nadell
Cover design by Edsal Enterprises
Manufacturing buyer: John Hall

PRENTICE-HALL INTERNATIONAL, INC., *London*
PRENTICE-HALL OF AUSTRALIA PTY. LIMITED, *Sydney*
PRENTICE-HALL OF CANADA, LTD., *Toronto*
PRENTICE-HALL OF INDIA PRIVATE LIMITED, *New Delhi*
PRENTICE-HALL OF JAPAN, INC., *Tokyo*
PRENTICE-HALL OF SOUTHEAST ASIA PTE. LTD., *Singapore*
WHITEHALL BOOKS LIMITED, *Wellington, New Zealand*

This book is dedicated to

the fertilizer manufacturers, dealers, and researchers and to the farmers who have developed such a mutually beneficial fertilizer delivery system that it is the envy of the world.

Contents

Preface

In this book the authors are presenting current knowledge in an ever-changing science and technology of fertilizers and soil amendments. Data and method of presentation have come from research, extension, and classroom teaching activities of the authors and from fertilizer scientists and technologists in the United States and around the world.

Although much research in the field of fertilizers and soil amendments has been accomplished, instructors of the subject have been handicapped by a lack of suitable textbooks written for students, instructors, and those in the fertilizer business. Unless the teacher is closely associated with soil fertility research and extension and has access to numerous current publications, the preparation of a comprehensive and up-to-date course in fertilizers and plant nutrition is difficult and time-consuming. Fertilizers and soil amendments, when used as indicated in this book, are estimated to increase yields by 40% in the United States and worldwide. This translates into 40% more food, feed, and fiber for people and their animals.

Food to feed the increasing population is a current problem facing not only the developing nations but the entire populace of the world. World demand for food and fiber has increased tremendously since 1960. With it there has been a corresponding increase in the production and consumption of fertilizers. The increase of manufactured fertilizers in the western world, particularly in North America, constitutes an essential factor in the rising food, fiber, and shelter supply and in the progressively high standard of living in these areas.

The authors wish to acknowledge the assistance of many individuals, government agencies, and industries who have furnished illustrations, published reports, and much valuable information in the compilation of this book. To these people, too numerous to list here, we are especially grateful.

Heartfelt thanks are extended to each of our wives, Barbara, Sandy, and Lola, respectively, who not only have been patient with us but have encouraged us to complete the task of writing this book.

We extend our thanks to the following secretaries for their excellent assistance in the typing of the manuscript: Kathleen Williams, Extension Agronomy Secretary, Kansas State University; Mary Hughes, Secretary, DPRA, Manhattan, Kansas; Carol Whitney, Agronomy Secretary, Kansas State University; and Theora Correll, Typist and Artist, Forsyth, Missouri.

The authors would appreciate suggestions from the readers on innovations for the next editions.

Dr. Roy H. Follett, Extension Specialist, Soil Fertility and Management, Department of Agronomy, Kansas State University, Manhattan, Kansas 66506.

Dr. Larry S. Murphy, Great Plains Director, Potash and Phosphate Institute, 200 Research Drive, Manhattan, Kansas 66502.

Dr. Roy L. Donahue, Route 1, Box 169, Forsyth, Missouri 65653.

PEER REVIEWERS

Outstanding scientists and educators were selected to review the first edition of *Fertilizers and Soil Amendments*. These reviewers made many valuable suggestions for improving the book and the authors responded and gratefully acknowledge their contributions. In order to assure that each reviewer was not overloaded with the entire manuscript, each was asked to review a certain chapter dealing with subjects closely aligned to his interests and specialty.

Dr. D. C. Adriano, Associate Professor of Ecology, Savannah River Ecology Laboratory, Drawer E, Aiken, SC 29801.

Dr. James D. Beaton, Western Director, Potash & Phosphate Institute, 12304 Lake Moraine Rise, S. E., Calgary, Alta., Canada T2J 3Z2.

Dr. William F. Bennett, Dean of Resident Instruction, Texas Technological University, Lubbock, TX 79409.

Dr. Roscoe Ellis, Jr., Professor of Soils, Department of Agronomy, Kansas State University, Manhattan, KS 66506.

Dr. Ronald F. Follett, National Program Staff, USDA–ARS, Bldg. 005 ARC-WEST, Beltsville, MD 20705.

Dr. Ardell D. Halvorson, Northern Plains Soils & Water Research Center, USDA–ARS, Box 1109, Sidney, MT 59270.

Dr. J. Benton Jones, Jr., Division Chairman and Department Head, Department of Horticulture, The University of Georgia, Athens, GA 30602.

Dr. Bernard D. Knezek, Professor of Soil Science, Department of Crop and Soil Sciences, Michigan State University, East Lansing, MI 48824.

Dr. Robert M. Koch, Sr., President, National Limestone Association, Suite 501, 3251 Old Lee Highway, Fairfax, VA 22030.

Dr. John F. Marten, Staff Economist, Farm Journal, 5430 Hillside Lane, West Lafayette, IN 47906.

Dr. John J. Mortvedt, Soil Chemist, Soils and Fertilizer, Research Branch, Tennessee Valley Authority, Muscle Shoals, AL 35660.

Dr. Werner L. Nelson, Senior Vice President, Potash & Phosphate Institute, 402 Northwestern Ave., West Lafayette, IN 47906.

Dr. L. Fred Welch, Professor, Soil Fertility, Department of Agronomy, University of Illinois, Urbana, IL 61455.

Dr. William C. White, Vice President, The Fertilizer Institute, 1015 18th Street, N. W., Washington, D. C. 20036.

Dr. Ronald D. Young, Chemical Engineer, Division of Chemical Development, Tennessee Valley Authority, Muscle Shoals, AL 35660.

Roy Hunter Follett
Fort Collins, Colorado

Larry S. Murphy
Manhattan, Kansas

Roy L. Donahue
Forsyth, Missouri

CHAPTER ONE

Fertilizers
and Plant Nutrition

1.1 NUTRITION AND FERTILIZATION

All living things grow and reproduce in response to an interaction of dynamic factors in their ambient ecological environment. Maximum yield results only when plants compete successfully for the essentials of life. Essentials for optimum biological efficiency include these positive factors:

1. Favorable air and soil temperatures.
2. Optimum available soil water and soil air.
3. Adequate light as a source of energy for green plants to carry on photosynthesis.
4. Essential elements for adequate nutrition of plants available when needed and supplied in balanced proportions.

When the soil elements essential for efficient plant nutrition and economic production are low in availability or are not in balance, chemical fertilizers and soil amendments are required. For example, when available nitrogen is low, nitrogen fertilizers are applied to enhance plant growth. Furthermore, when the soil is so acid that many soil nutrients such as phosphorus are not readily available or aluminum is present in toxic quantities, lime is applied. The efficient use of fertilizers and lime supplements the nutrient-supplying capacity of soil minerals and soil organic matter and decreases specific toxicities to achieve optimum agronomic and economic plant nutrition and production (Fig. 1–1).

1.1.1 Plant Nutrition

In nature, soil, water, and air contain 88 elements; 16 more have been made by scientists, totaling 104. Most of these may be absorbed by higher plants but so far only 16 (some scientists claim 19) have been considered essential. These 16, their source, and the most common chemical forms in which they are absorbed are as follows:

From air and water, plants absorb hydrogen (H^+ and HOH), oxygen (O_2, OH^-, CO_3^{2-}, SO_4^{2-}), and carbon (CO_2).

The primary nutrients from soils and fertilizers are nitrogen (NH_4^+ and NO_3^-), phosphorus ($H_2PO_4^-$ and HPO_4^{2-}), and potassium (K^+).

The secondary nutrients absorbed by plants from soils and fertilizers are magnesium (Mg^{2+}), calcium (Ca^{2+}), and sulfur (SO_4^{2-}).

The micronutrients absorbed by plants from soils and fertilizers are chlorine (Cl^-), copper (Cu^{2+}), boron ($H_2BO_3^-$ and $B(OH)_4^-$), iron (Fe^{2+} and Fe^{3+}), manganese (Mn^{2+}), molybdenum (MoO_4^{2-}), and zinc (Zn^{2+}).

Plants use these 16 elements and heat, and light to manufacture foods, starting with hexose sugar, as illustrated by this simple diagram of photosynthesis:

$$6\ CO_2 + 12\ H_2O \xrightarrow[\text{Chlorophyll in plant}]{\substack{\text{Light/heat} \\ \text{energy from sun}}} C_6H_{12}O_6 + 6\ O_2$$

$$\underset{\text{From air}}{6\ CO_2} \quad \underset{\text{From soil}}{12\ H_2O} \qquad \underset{\text{Hexose sugar}}{C_6H_{12}O_6} \quad \underset{\text{Oxygen}}{6\ O_2}$$

Simple hexose sugars are then transformed to sucrose ($C_{12}H_{22}O_{11}$), more complex sugars, starch, cellulose, and pectin. Likewise, amino acids, fatty acids, glycerol [$C_3H_5(OH)_3$], true fats (triglycerides), waxes, and phospholipids either are produced directly or are synthesized from products of photosynthesis. About 90% of all dry matter of green plants is manufactured during photosynthesis.

1.1.2 Mechanisms of Nutrient Uptake

The plant obtains its carbon from the carbon dioxide of the air, oxygen from atmospheric and soil air, and hydrogen from soil water. The other 13 essential nutrients are absorbed in ionic form primarily through plant roots growing in the soil. Nutrient uptake includes three mechanisms: mass flow, diffusion, and root interception.

(a)

(b)

(c)

Fig. 1-1. Human survival and well-being depend on luxuriant plant growth. Plants need plentiful sunshine, water, and essential elements. (a) Sunshine is adequate here but *water* is a limiting factor for growth of food crops in western Africa; for this reason the family is migrating. (b) *Water* is plentiful in Puerto Rico but plant nutrients from the natural soil are inadequate and unbalanced for banana production. Frequent and massive applications of chemical *fertilizers* are used to enhance economic production. (c) *Water* for wheat is usually plentiful in the midwestern United States but the plant-nutrient-supplying power of the soil is often too low and unbalanced. This wheat field has just been topdressed with a solution of urea-ammonium nitrate fertilizer to obtain maximum economic yields [Courtesies: (a) Purcell/United Nations/USAID; (b) Puerto Rico Department of Agriculture; (c) Tryco Mfg. Co.]

1. *Mass flow.* The movement toward roots of plant nutrients dissolved in the soil water. Essential elements in the soil solution around the plant roots are absorbed as cations and anions.

2. *Diffusion.* The movement of nutrient ions from the soil solution into root cells by *passive* or *active* absorption without mass flow of water. In *passive* absorption, each nutrient cation or anion moves independently from an area of high concentration of that nutrient to an area of low concentration of the same nutrient. In *active* absorption, ions move into cells against a concentration gradient using energy from respiration.

3. *Root interception.* The extension (growth) of plant roots into new soil areas where untapped supplies of nutrients are available. Roots then absorb nutrients as in mass flow and/or diffusion.

For the proper nutrition and growth of corn, for example, the relative amounts of selected nutrients absorbed by the three mechanisms of nutrient uptake are as follows:[1]

Nitrogen, sulfur—mostly mass flow.

Phosphorus, potassium—mostly diffusion.

Calcium, magnesium, molybdenum—mass flow and root interception.

1.1.3 Balanced Nutrition Through Fertilization

Because all plant and animal life is dependent on food produced by plants through photosynthesis, *balanced* nutrition through fertilization of plants is of universal, crucial concern.

The term *balanced nutrition through fertilization* is easy to define but difficult to practice. It means the continuous availability to plants of all 16 essential elements, with none in excess and none in deficient amounts (Fig. 1–2). Balanced nutrition of plants in relation to animals can also mean applying fertilizers to increase the nutritional value of the forage or feed. For example, phosphorus fertilizer added to alfalfa may increase the percentage of phosphorus in the forage to 0.3% phosphorus, an acceptable level for dairy cows. Oat grain, however, may contain 0.3% phosphorus without the addition of phosphorus fertilizer (Fig. 1–3).

Plants vary in their requirements from one cultivar to another, among species, and within the same plant cultivar at different physiological stages. For example, in mixture with bromegrass, alfalfa has a higher potassium requirement than bromegrass but bromegrass absorbs potassium more readily than alfalfa. Such mixtures require more than usual potassium fertilizer. Also, the amount of boron needed for maximum production of sugar beets may be

[1] Roy L. Donahue, Raymond W. Miller, and John C. Shickluna, *Soils: An Introduction to Soils and Plant Growth,* 4th ed., Prentice-Hall, Inc., Englewood Cliffs, N.J., 1977, pp. 125, 126.

Fig. 1-2. The plant nutrients in the soil down the middle of this lowland rice field in eastern India are not in balance even though up to 101 kg/ha (90 lb/A) of nitrogen (N) and up to 67 kg/ha (60 lb/A) of potassium (K) were applied on some of the plots. *Reason:* The first limiting factor is phosphorus (P), which tests only 1.8 kg/ha (1.65 lb/A) of P on the surface soil and 0.7 kg/ha (0.6 lb/A) of P on the 15- to 30-cm (6- to 12-in.) depth of soil. In this example, a severe imbalance of nutrients (low phosphorus in the soil) resulted in complete crop failure of rice. (Courtesy Roy L. Donahue.)

toxic to soybeans. Balanced fertilization for corn means making readily available (as percentage of total growing-season needs):

25 days after emergence: 2% of the phosphorus, 4% of the nitrogen, and 6% of the potassium.

50 days after emergence: 30% of the phosphorus, 40% of the nitrogen, and 55% of the potassium.

75 days after emergence: 62% of the phosphorus, 70% of the nitrogen, and 90% of the potassium.

Some agriculturists say that balanced fertilization means to add as fertilizers *all* the nutrients removed by a crop. This hypothesis was stated first by von Liebig in Germany about 1840. It is true today in nutrient culture solutions and almost true on deep sands and highly leached Oxisols of the humid tropics. This concept of plant-soil relationships, however, completely ignores the nutrient-supplying capacity of the soil. This capacity is deter-

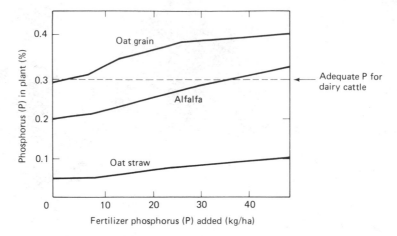

Fig. 1–3. "Balanced nutrition through fertilization" could mean increasing the percentage of phosphorus (P) in forage/fodder/grain by fertilizers to achieve adequate phosphorus (0.3% P) intake when fed to dairy cattle. Under these experimental conditions, oat grain had adequate phosphorus with no phosphorus fertilizer, but alfalfa was adequate only after an application of 30 kg/ha (27 lb/A) of phosphorus. (Courtesy U.S. Plant, Soil, and Nutrition Laboratory, Ithaca, N.Y.)

mined by soil and plant analyses correlated with field response in each contrasting soil taxonomic-anthropogenic-agronomic area.

1.2 SOIL-FERTILIZER-PLANT RELATIONSHIPS

The soil supplies 13 of the 16 elements required for the nutrition of higher plants. These essential elements must be available continuously in balanced proportions. Plants also require mechanical support, sunshine, the right amount of soil water, plentiful oxygen surrounding plant roots, and freedom from plant toxins.

Furthermore, the physical condition of the soil must be favorable to support the plant and to hold and release simultaneously adequate oxygen and water for optimum plant growth. Substances toxic to plants, such as an excess of soluble salts, may be present in the soil as a result of its formation in arid regions. This leaves only sunshine as an essential for the growth of higher plants that is not soil-related.

The kind of soil existing at any one time and place is determined by five factors of soil formation: parent material, climate, topography, biosphere, and time.

It is assumed that all soils with the identical five factors of soil formation would have the same taxonomy and be mapped as the same soil. It can also be assumed that, barring drastic influences by people, plants growing on the same soil will respond to fertilizers in a similar way. Exceptions to this generalization include soils in old fence rows; soils in one field that was treated differently when it was managed as two fields; areas where fertilizers and/or lime were applied unevenly; areas where old roads, homesteads, and barn lots used to be; and areas leveled, eroded, or otherwise drastically disturbed.

The preceding logic is the best technique yet devised for predicting the fertilizer response of crops on soils where such plants have never been grown before. Phrased differently, soil taxonomy and soil surveys are a viable technique for the successful transfer of field data and other technology from known soil-plant response and predicting the response on similar soils where the same plant had never been grown.

1.2.1 Soil Taxonomy and Transfer of Technology

The latest United States system of soil taxonomy, adaptable worldwide, was started in 1951, became official in 1965, and was published in one document in 1975.[2]

Throughout the years of development of the system, the soil scientists used this question to guide them: "Do these soil groupings permit us to make more precise predictions of soil behavior?" Soil behavior includes plant response to the application of fertilizers and soil amendments.

The United States system of soil taxonomy comprises 10 soil orders, 47 suborders, 185 great groups (225 worldwide), 970 subgroups, 4500 soil families, and 10,500 soil series. Phases include texture of the surface soil, slope, stoniness, and degree of erosion. An example of this system of soil taxonomy as applied to a particular soil profile is displayed in Fig. 1–4.

Soil taxonomy is the scientific basis for differentiation among soils on soil maps. The use of soil maps permits a more precise transfer of technology, including plant response to fertilizer and lime. The 10 soil orders and their extent in the United States and in the world are detailed in Table 1–1.[3]

1.2.2 Soil Maps and Rational Fertilizer Use

Soil mapping in the United States was started in 1899 to answer practical agronomic questions relating to soil as a medium for plant growth. Today soil mapping has the same primary objective, even though much more is known about soil taxonomy, soil fertility, plant nutrition, and erosion in relation to the rational use of fertilizers and soil amendments to increase plant growth.

About half of the land area of the United States has been mapped with

[2] Soil Survey Staff, "Soil Taxonomy: A Basic System of Soil Classification for Making and Interpreting Soil Surveys," USDA-Soil Conservation Service, *Agriculture Handbook No. 436,* Dec., 1975.

[3] For an expanded explanation, see Roy L. Donahue, Raymond W. Miller, and John C. Shickluna, *Soils: An Introduction to Soils and Plant Growth,* 4th ed., Prentice-Hall, Inc., N.J., 1977, pp. 408–462.

DEPTH (cm)

HORIZON

A

20

40   B

60

C

Fig. 1-4. This Ultisol is mapped as
Ruston fine sandy loam and is clas-
sified in the fine-loamy, siliceous,
thermic family of Typic Paleudults.
The complete taxonomy follows:
Order—Ultisols; Suborder—Udults;
Great Group — Paleudults; Sub-
group—Typic Paleudults; Family—
Typic Paleudult, fine-loamy, sili-
ceous, thermic; Series — Ruston;
Phase—fine, sandy loam, 3 to 8%
slopes, slightly eroded. (Location:
Louisiana.) (Courtesy USDA—Soil
Conservation Service.)

sufficient detail so that the maps are useful in developing a more precise fertilizer and lime recommendation. The map units consist of compound symbols
that identify the soil series, the range of slopes, and the erosion classes. When
soil-testing laboratories have this information sent to them along with each
soil sample, the chemical analyses can be interpreted into more economical
fertilizer and lime recommendations. For example, with the *same* chemical soil
test:

1. An eroded phase of a soil series should be terraced and/or strip-cropped
   before fertilizers and lime can be used efficiently. Applying fertilizers that
   are carried by sheet erosion to eutrophy surface waters is an agronomic,
   economic, and ecological loss.
2. When the soil series name designates a soil profile with a strongly acid B
   and C horizon, *more* fertilizer and lime are recommended.
3. When the soil series name designates a well-drained soil in temperate
   regions, minimum tillage can be recommended. A poorly drained soil in
   cool areas will not usually warm fast enough in the spring to germinate
   seeds unless it is tilled.

However, it must never be assumed that a taxonomic unit on a soil map is
identical to a soil fertility unit. A high correlation did exist on virgin soils but

Table 1-1. Areas of Soil Orders of the United States and the World
and Their Percentages of Total Land Area

Soil Order	United States			World		
	Square Miles	Square Kilometers	Percentage of Total Land Area	Square Miles	Square Kilometers	Percentage of Total Land Area
Alfisols	478,645	1,237,202	13.51	6,630,200	17,138,200	13.19
Aridisols	411,860	1,064,576	11.63	9,433,600	24,384,000	18.76
Entisols	282,140	729,276	7.96	4,162,400	10,759,200	8.28
Histosols	18,600	48,077	0.53	452,500	1,169,600	0.90
Inceptisols	642,050	1,659,571	18.15	4,464,000	11,538,400	8.88
Mollisols	890,200	2,300,989	25.14	4,310,900	11,143,200	8.57
Oxisols	500	1,292	0.01	4,277,900	11,057,600	8.51
Spodosols	171,620	443,603	4.84	2,150,000	5,557,600	4.28
Ultisols	451,620	1,167,347	12.76	2,794,100	7,222,400	5.56
Vertisols	35,125	90,791	0.99	905,300	2,340,000	1.80
Miscellaneous	158,560	409,846	4.48	10,697,100	27,650,000	21.27
Totals	3,540,920	9,152,570	100.00	50,278,000	129,960,200	100.00

SOURCE: Soil Conservation Service. For the United States: "Soil Map of the United States," 1967, scale—1:7,500,000. For the world: "Soil Map of the World," 1971, scale—1:50,000,000.

the use of fertilizer and lime on crops and pastures has reduced the degree of correlation.

1.3 FOOD AND FERTILIZER OPTIONS

The Food and Agriculture Organization of the United Nations reports that world food production per person in recent years has varied from about a 1.75% increase per year for all developed countries to 0.3% for all developing countries, averaging 0.8% worldwide.[4] The food production increase and population increase must be adequate and in balance to provide enough food for everyone. What are the facts?

In the developed countries the population increase has averaged 1%, compared with a food increase of 1.75%. *Result:* surplus food. Population growth in the developing countries has averaged 2.4%, eight times the food increase of 0.3%. *Result:* The developing countries must obtain surplus food from the developed countries or starve. On a world basis, the population growth has averaged 2% and the food production increase, about 1%. *Result:* Worldwide we must try to decrease population growth and aim to double the rate of increase of food production during the next 25 years (Fig. 1-5).

[4] "Population, Food Supply, and Agricultural Development." In *The State of Food and Agriculture, 1974,* Chapter 3, Food and Agriculture Organization of the United Nations, Rome, Italy, 1975.

(a)

(b)

(c)

(d)

Fig. 1–5. Worldwide acreages of the principal food and feed grains are in this order, high to low: wheat, rice, grain sorghum, and corn (maize). (a) An improved variety of wheat raised at high elevations in Equador, South America. (b) Fertilizing rice in Indonesia. (c) With a slingshot, this man scares birds and monkeys away from the ripening grain sorghum in central India. (d) Corn (maize) in Outer Mongolia. [Courtesies: (a) F. Botts for FAO; (b) United Nations; (c) Roy L. Donahue; (d) N. G. Ipatenko for FAO.]

A decrease in population growth seems next to impossible, but an increase in food production is possible by increasing the total hectares (acres) of arable land and/or increasing yields per hectare (acre), primarily with an increase in use of fertilizer.

An estimate was made of the hectares (acres) available in the world for cultivation of food crops that *are not now cultivated*. This means forested and grassed areas where precipitation is adequate for crop production, as well as arid areas with a potential for irrigation, as follows:

Region	Million Hectares
South America	608
Africa	567
North America	243
Asia	122
Australia and New Zealand	122
U.S.S.R.	122
Europe	40
Total	1824

NOTE: Hectares $\times$ 2.471 = acres.

About 1 billion hectares (2.47 billion acres) are in the tropics where the population is increasing faster than food production.[5]

It *cannot* be assumed that these potentially arable soils are ideal for crop production without expensive inputs. Most of them will require very expensive land clearing, road building, wild animal control, disease and insect control, weed control, irrigation water, soil amendments, and chemical fertilizers. Sometimes the success of new land developments depends on the health and happiness of resettled farm families.

Most professional people who have studied the "more hectares or more yield per hectare" option have concluded the following:

1. The option is specific to each sovereign country or part of the country.
2. The greatest need for more food is in the densely populated countries where most arable land is already being used for food production. The principal exception is most of South America where many countries have more hectares of soils to clear and cultivate. However, South America has more than 10% of its land surface classified in the soil order of Oxisols, the least fertile of all 10 world soil orders (see Section 1.2).
3. Increasing yields per hectare (acre) in most countries is the more feasible of the two alternatives because it is cheaper, and the greatest need is in countries where population pressure is greatest and there are almost no new hectares (acres) to clear.

[5] United Nations' World Food Conference, "The World Food Problem–Proposals for National and International Actions," published for the World Food Conference, Rome, Italy, 1974.

Whether the option is for more hectares (acres) or greater yields per hectare (or both), more lime on strongly acid soils, more acidifying and sodium-replacing salts on sodic soils, more organic mulches, and especially more chemical fertilizers will be necessary to enhance yields of food crops.

For immediate results, chemical fertilizers are the best single agricultural input for increasing crop yields per hectare (acre). In the developing countries the increase in use of fertilizer has exceeded 10% a year and is expected to continue at this rate in the future. With good management, about 4.5 kg (10 lb) of food grains can be expected for each kilogram (pound) of fertilizer used. *Enlightened people estimate that half of the increase in yield of food crops in developing countries can be attributed to the use of fertilizers.*[6]

1.3.1 Plant Scientists' Contributions to Options

Plant scientists are now breeding and selecting plants with greater potential response to fertilizers as well as with greater tolerance for adverse climates and soils. Examples follow:

1. Plants with hybrid vigor and greater yield potential such as corn, grain sorghum, and many vegetable crops.
2. Plants with improved nutritional quality such as more essential amino acids in corn grain, e.g., lysine.
3. Plants that are a new species as a result of crossing two other plants. The best example is *triticale*, resulting from a cross between wheat and rye. Triticale has the plant vigor of rye and grain similar to wheat. Unlike wheat, however, it is resistant to the many races of rust fungi. Wheats are also being bred that are resistant to one or more races of wheat rust.
4. Plants have been selected and bred to tolerate cold weather, hot weather, wet soils, dry soils, high humidity, low humidity, saline soils, sodic soils, compacted soils, acid soils, alkaline soils, allic (high aluminum) soils, low-nutrient soils, organic soils, and high-nutrient (especially high nitrogen) soils.
5. Certain plants have been developed to resist some nematode species, specific insects, specific diseases, and some rodents.
6. The most noted of the principal food plants that have been bred and selected to stand (not lodge) and to respond to high-nitrogen fertilization are the dwarf wheats developed by the International Maize and Wheat Improvement Center (CIMMYT) in Mexico and the dwarf rices bred in the Philippines by the International Rice Research Institute (IRRI). These dwarfs are also insensitive to length of day.
7. Forage plants and grains have been bred and selected to contain a better

[6] John A. Hannah, "Fertilizer—A Key to Solution of World Food Problem," *Fertilizer News,* Jan., 1978, pp. 14–17. Fertilizer Assn. of India, New Delhi, India.

balance of nutrients when fed to livestock. For example, there is a great genetic variation in the calcium and phosphorus contents of corn grain, alfalfa hay, and peanut hay. Potassium is likely to be in excess in some forages, thus decreasing plant uptake of magnesium and leading to grass tetany (hypomagnesemia) in cattle. Some cultivars of wheat and barley, as well as specific cultivars of alfalfa, orchardgrass, and timothy may contain adequate magnesium for cattle. Other nutrients that usually vary in concentration from one cultivar to another include sodium, chlorine, copper, zinc, iron, manganese, iodine, molybdenum, selenium, and sulfur.[7]

8. A specific example of plant breeding programs to increase plant efficiency and yield is cited from the University of California. A new wheat selection, UC 44-111, has been bred for a *low* nitrate concentration in the leaves; whereas, the Anza cultivar has a *high* nitrate concentration. Both cultivars have high yields and Anza accounted for approximately half of the wheat planted in California in 1976. UC 44-111 is a new cultivar. Less nitrogen fertilizer will be needed per hectare (acre) as the UC 44-111 cultivar replaces Anza.[8]

9. In Nigeria (western Africa), the International Institute of Tropical Agriculture is testing lowland rice cultivars under *low* levels of nitrogen, phosphorus, potassium, sulfur, and zinc. Surprisingly, two rice cultivars produced relatively high yields when all of these nutrients were low. Furthermore, five other cultivars produced well under low levels of all three major nutrients: nitrogen + phosphorus + potassium.[9]

10. The International Rice Research Institute near Manila, Philippines, evaluated yields of wetland (paddy) rice cultivars with varying levels of nitrogen fertilizer during both the dry seasons and the wet seasons. During the *dry seasons,* a cultivar, BG 90-2, from Sri Lanka (formerly Ceylon) produced 76% as much rough (unhulled) rice with *no nitrogen* fertilizer as compared with 60 kg/ha of nitrogen. The same comparison during the wet season resulted in 80% as much yield.[10]

Wetland rice (paddy) on the no-nitrogen-fertilizer plots utilized nitrogen from residual sources and from atmospheric N_2 fixation by blue-green algae. Traditionally, wetland rice is grown in 5 to 10 cm (2 to 4 in.) of water.

---

[7] R. R. Hill, Jr. and S. B. Guss, "Genetic Variability for Mineral Concentration in Plants Related to Mineral Requirements of Cattle," *Crop Science,* **16,** Sept.-Oct., 1976, pp. 680-684.

[8] Calvin O. Qualset, John D. Prato, and Herbert E. Vogt, "Breeding Success with Spring Wheat Germplasm." In *California Agriculture,* **31,** No. 9, Summer, 1977, pp. 26, 27.

[9] K. Alluri and I. W. Buddenhagen, "Evaluation of Rice Cultivars for Their Response to Limiting Nutrients." In *International Rice Research Newsletter, 3/77,* June, 1977, p. 2, IRRI, Manila, Philippines.

[10] International Rice Research Institute, 1976. Annual Report for 1975. Los Baños, Philippines, p. 77.

1.3.2 Soil Scientists' Contributions to Options

No one has recorded in history when soil science was generally accepted as a discrete scientific discipline; neither is anyone ever likely to do so, for soil science is a synthesis of sciences. Soil scientists include soil chemists, soil physicists, soil microbiologists, soil taxonomists, soil surveyors, soil conservationists, and many other disciplines. Furthermore, some of the contributions to soil science and fertilizer technology have been made by geologists, plant nutritionists, plant physiologists, crop scientists, agronomists, chemical engineers, and agricultural engineers. Soil scientists, however, may properly claim a major share of the credit for the efficient use of fertilizers and for soil amendments such as lime, animal manures, and sewage sludges, and gypsum and sulfur to reclaim sodic soils.

Early foundations of soil science were laid by scientists searching for the "principle of vegetation," as exemplified by Van Helmont in 1652 with his willow tree, and by Jethro Tull's experiments on tillage in 1731. J. B. Boussingault in 1841 advanced the quantitative aspects of the science of *field* research by measuring and weighing all inputs and outputs.

Outstanding among the early "soil scientists" was Justus von Liebig, a German *laboratory* chemist who published in 1840 *Chemistry and Its Application to Agriculture and Physiology*. His principal contributions to "soil science" included his theses, as follows:

1. Carbon for plant nutrition came from CO_2 of the air and not from humus.
2. Phosphates were necessary for seed production.
3. Potassium was essential for the development of grasses and cereals.
4. By analyzing plant ash he could formulate a correct fertilizer for the same crop when it was planted again.
5. "The law of the minimum," which means that plant growth is retarded by the essential element in least relative amount.

Partly as a result of von Liebig's controversial stand against so many current concepts, in 1843, J. B. Lawes and J. H. Gilbert established the Rothamsted Experiment Station near London, England, to *field* test hypotheses on soil-fertilizer-plant relationships. Excellent field and laboratory research has continued there to this day.

In the United States, the oldest field research plots were established by George E. Morrow in 1876, at the University of Illinois in Urbana. Three of the original 10 plots are continuing today. Research on these field plots has demonstrated conclusively that even on "productive" soil crop yields decrease without fertilizers, lime, and organic residues.

Now, every state in the United States has many laboratories and field stations where research is conducted on fertilizers and soil amendments. Most of these facilities were established by the respective states and are being assisted by the United States Department of Agriculture, the National Science Foundation, private foundations, and many private companies and associations.

As viewed from the state agricultural universities, principal current contributions of soil scientists consist of the following:

1. Establishing soil-testing services to aid farmers and other land managers in selecting the most efficient fertilizer and soil amendment options.
2. Cooperating with the Tennessee Valley Authority and private fertilizer companies in field testing new fertilizer formulations.
3. Cooperating with others such as crop scientists, horticulturists, livestock specialists, farmers, ranchers, and municipalities in finding the most efficient use of new fertilizer formulations, lime, animal manures, and sewage sludges on specific soil series in all climatic regions of the United States.
4. Cooperating with the U.S. Agency for International Development in the State Department, the Consultative Group on International Agriculture, the 14 International Agricultural Research Centers, the United Nations Development Program, the Food and Agriculture Organization of the United Nations, and the International Bank for Reconstruction and Development (World Bank) in research and extension development of food, feed, and fiber throughout the world (Fig. 1-6).

Fig. 1-6. Efficient uses of nitrogen, phosphorus, and potassium are being dramatized with words and music by a group of college students of agriculture before an audience of village farm families in central India. This work was conducted under contract with the U.S. Agency for International Development. (Courtesy Frank Shuman, Univeristy of Illinois.)

1.4 ALTERNATIVES TO CHEMICAL FERTILIZERS

There are many alternatives and substitutes for chemical fertilizers. Some are rational substitutes, others are rational partial substitutes, whereas a third group are nonrational alternatives.

The most popular rational substitutes for chemical fertilizers are animal manures, green-manure crops, sewage sludges, composts, and crop residues. These should be used to their agronomic, economic, and ecological maximum. Partial substitutes include microbes that, under ideal environments, are capable of fixing atmospheric N_2. Included among these microbes are *Rhizobium, Azotobacter, Clostridium, Spirillum,* and blue-green algae. The term *partial substitute* is used because seldom do any of these microbes fix enough nitrogen for optimum yields of the following crop. Chemical nitrogen fertilizers must be added to supplement the nitrogen fixed. Nonrational substitutes for chemical fertilizers include those substances promoted to perform "miracles." The mystic ingredients usually have no agronomic value and are changed too rapidly for the respective state fertilizer control agency to prosecute successfully.[11,12,13,14,15,16]

1.5 INTERNATIONAL FERTILIZER DEVELOPMENT CENTER

In April of 1974 the then Secretary of State, Henry Kissinger, offered the assistance of the United States to the world through the United Nations General Assembly in the establishment of an international fertilizer development center. By October of the same year such a nonprofit corporation was established adjacent to, but not affiliated with, the National Fertilizer Development Center operated by the Tennessee Valley Authority at Muscle Shoals, Ala.

Activities of the International Fertilizer Development Center provide the following worldwide technical services:

1. Increasing the efficiency of nitrogen fertilizers in food production, especially in the tropics.

[11] Martin Alexander, *Introduction to Soil Microbiology,* 2nd ed., John Wiley & Sons, New York, 1977.

[12] Michael A. Cole, "Blue-green Algae a Fertilizer?" *Crops and Soils Magazine,* 30, No. 3, Dec., 1977, pp. 7–9. American Society of Agronomy, Madison, Wis.

[13] Carl S. Hoveland (ed.), "Biological N Fixation in Forage-Livestock Systems," American Society of Agronomy, Madison, Wis., 1976.

[14] J. W. Johnson, L. F. Welch, and L. T. Kurtz, "Soybeans' Role in Nitrogen Balance," Illinois Research, University of Illinois, Summer, 1974, pp. 6, 7.

[15] W. Lockeretz, R. Klepper, B. Commoner, M. Gertler, S. Fast, and D. O'Leary, "Organic and Conventional Crop Production in the Corn Belt: A Comparison of Economic Performance and Energy Use for Selected Farms," Report No. CBNS-AE-7, Center for the Biology of Natural Systems. St. Louis, Mo., Washington University, 1976.

[16] "Opportunities for Energy Savings in Crop Production," Congressional Research Service, Library of Congress. U.S. House of Representatives Committee on Science and Technology, 95th Congress, Dec., 1977, pp. 23–36.

2. Testing the efficiency of foreign sources of phosphorus fertilizers.
3. Providing the technology for more efficient fertilizer formulations.
4. Offering expertise for increasing the accuracy of methods of forecasting fertilizer supply and demand, country by country, around the world.
5. Assisting nations to adopt the most efficient technologies for manufacturing fertilizers.
6. Offering all aspects of technical assistance on fertilizer technology and its efficient use, on a consulting basis.
7. Supplying information on sources of spare parts for fertilizer plants.
8. Giving technical training in fertilizer plant operation.
9. Providing a world fertilizer information media (Fig. 1–7).

1.6 WORLD FERTILIZER CONSUMPTION TO THE YEAR 2000[17]

The United Nations sponsored a panel of world fertilizer production experts in Vienna, Austria, on November 16–18, 1976, and asked them to predict fertilizer consumption to the year 2000. This was done by projection of fertilizer trends from 1955 to 1974. The results are displayed in Fig. 1–8.

Other predictions included the following:

1. A 3.7-fold (370%) increase in world consumption from 1974 [75.8 million mt (83.6 million t)] to 278.5 million mt (307 million t) in the year 2000, of $N + P_2O_5 + K_2O$.
2. In developing countries, the $N:P_2O_5:K_2O$ ratio of consumption from 1974 to 2000 is predicted to be about 4:2:1, respectively. By contrast, in the developed countries, the 1974 ratio of 1.4:1.0:1.0 is expected to be 2.2:1.0:1.1 by the year 2000, of $N:P_2O_5:K_2O$, respectively.
3. The developing countries in 1974 produced only 64% of their consumption—the remaining 36% was imported from developed countries.
4. Nitrogen fertilizers around the world to the year 2000 will be mostly urea, followed by ammonium nitrate. Anhydrous ammonia and fluid nitrogen fertilizers are expected to supply about 10% of world total nitrogen (N) for direct application but to be used mostly in the United States. Again, on a world basis by the year 2000, 10 to 15% more of the nitrogen will be supplied by ammonium phosphates.
5. Phosphorus fertilizer materials are expected to rank in world consumption in this order: diammonium phosphate > monoammonium phosphate > triple superphosphate > nitrophosphates.
6. Potassium will continue to be applied mostly as potassium chloride.
7. Mixed, granulated, or bulk-blended materials are predicted to be mostly urea + diammonium phosphate + muriate of potash.

[17] Principal source: United Nations Industrial Development Organization. "Draft World-Wide Study of the Fertilizer Industry, 1975–2000," International Centre for Industrial Studies. Vienna, Austria, Nov. 16–18, 1976.

(a)

(b)

(c)

(d)

Fig. 1-7. The International Fertilizer Development Center at Muscle Shoals, Ala., works to improve fertilizer technology, to test new fertilizers, and to train people around the world to operate fertilizer plants and to use fertilizers efficiently. (a) A modern phosphorus fertilizer plant in Florida owned and operated by the International Minerals and Chemical Corporation of Chicago, Ill. (b) A phosphorus fertilizer plant in India, Coromandel Fertilizer Company, partly owned by the International Minerals and Chemical Corporation. (c) A retail outlet in India of the Coromandel Fertilizer Company. (d) Nitrogen and phosphorus fertilizers in India increased the yield of rough (paddy) rice by 93% over no fertilizer.

18

(e)

(f)

(g)

(h)

Fig. 1-7 (cont.). (e) A complete (nitrogen + phosphorus + potassium) fertilizer manufactured in Brazil. (f) An agricultural extension worker shows a farmer in Afghanistan the new pelleted, concentrated fertilizer for use on wheat. (g) A concentrated phosphorus fertilizer plant in Tunisia, northern Africa. (h) A fertilizer plant in Senegal, western Africa. [Courtesies: (a) and (b) International Minerals and Chemical Corporation; (c) Roy L. Donahue; (d) Frank Shuman, University of Illinois; (e) J. R. Monato for International Finance Corporation (IFC), a unit of the International Bank for Reconstruction and Development (World Bank); (f) K. Muldoon for International Development Association: (g) R. Shaw for IFC; (h) P. R. Johnson for IFC.]

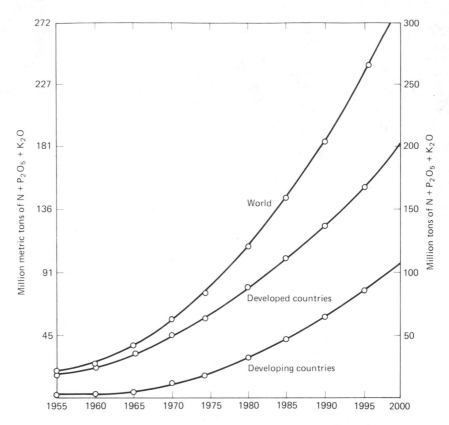

Fig. 1-8. World fertilizer consumption, 1955–1974 (actual) and 1975–2000 estimated). (Source: United Nations Industrial Development Organization, UNIDO/ICIS-22, Nov. 10, 1976, Vienna, Austria.)

SUMMARY

When the soil does not supply the essential nutrients in balanced proportions and in adequate quantities for optimum plant nutrition, growth, and reproduction, fertilizers should be applied. Essential soil nutrients, however, are only a part of the total essential plant requirements. Also necessary for optimum plant response is the presence of favorable temperature, oxygen, carbon dioxide, water, light, and mechanical support, and the absence of insects, diseases, and toxins.

The amounts and kinds of fertilizers needed for maximum economic yields can be approximated by a soil test that has been calibrated with field-plot research. Where the soil taxonomy, soil test, and field-plot research on plant response to fertilizers are correlated, more accurate predictions on plant-fertilizer response can be made.

Alternatives to chemical nitrogen fertilizers are animal manures, green-manure crops, sewage effluents, crop residues, and biological (microbial) nitrogen (N_2) fixation.

The International Fertilizer Development Center at Muscle Shoals, Ala., was established to help all countries in the world to improve their manufacturing, storing, and efficient use of chemical fertilizers for greater food production and faster country development.

World consumption of fertilizers is expected to increase at an accelerating rate to the year 2000.

PROBLEMS

1. Criticize this statement: "All the world needs to be self-sufficient in food is to have low-cost and adequate fertilizer."

2. Explain "balanced nutrition and fertilization of plants."

3. What scientific relationships exist among soil taxonomy, soil survey, field-plot research, soil testing, and lime and fertilizer recommendations?

4. Enumerate several alternatives to the use of chemical fertilizers.

5. Explain the primary services offered by the International Fertilizer Development Center.

REFERENCES

Allaway, W. H., "The Effect of Soils and Fertilizers on Human and Animal Nutrition," Agriculture Information Bulletin No. 378, 1975. USDA—Agricultural Research Service, USDA—Soil Conservation Service, and Cornell University Agricultural Experiment Station. Supt. of Documents, Washington, D.C.

Beeson, Kenneth C. and Gennard Matrone, *The Soil Factor in Nutrition: Animal and Human.* Marcel Dekker, Inc., New York, 1976.

Bornemisza, Elemer and Alfredo Alvarado (Ed.), *Soil Management in Tropical America.* Feb. 10–14, 1974. Published 1975 by North Carolina State University, Raleigh, N.C.

Epstein, Emanuel, *Mineral Nutrition of Plants.* John Wiley & Sons, New York, 1971.

Evans, L. T., "The Natural History of Crop Yield," *American Scientist,* 68: 388–397, July-August, 1980.

Juo, A. S. R.,"Interaction and Balance of Macro- and Micronutrients in Highly Weathered Soils in the Humid Tropics." In *Micronutrients in Agriculture,* FAO/India/Norway Seminar, Sept. 17–21, 1979, New Delhi, India.

Loehr, Raymond C. (Ed.), *Food, Fertilizer, and Agricultural Residues.* Proceedings of aConference at Ithaca, N.Y., 1977. Ann Arbor Science, Ann Arbor, Mich., 1977.

National Academy of Sciences. *Nitrates: An Environmental Assessment.* Washington, D.C., 1978.

National Academy of Sciences. *Soils of the Humid Tropics.* Washington, D.C., 1972.

National Academy of Sciences. *Supporting Papers: World Food and Nutrition Study, Vol. II.* Study Team 4, Resources for Agriculture and Study Team 5, Weather and Climate, 1977. Report of Subgroup C, "Fertilizers," pp. 145–193.

Plant Adaptation to Mineral Stress in Problem Soils. Proceedings of a Workshop Held at the National Agricultural Library, Beltsville, Md., Nov. 22, 23, 1976. Cornell University, Ithaca, N.Y., 1977.

Sanchez, P. A. and S. W. Buol, "Soils of the Tropics and the World Food Crisis," *Science,* **188,** May 9, 1975, 598–603.

"Soils for Management of Organic Wastes and Waste Waters," Soil Science Society of America, American Society of Agronomy, and Crop Science Society of America. Madison, Wis., 1977.

"Suggested Fertilizer-Related Policies for Governments and International Agencies," Technical Bulletin IFDC-T-10, Aug., 1977. International Fertilizer Development Center, Muscle Shoals, Ala.

U.S. Department of Agriculture. "Improving Soils with Organic Wastes," 1978.

Wharton, Clifton R., Jr., "Food, the Hidden Crisis," *Science,* **208,** June 27, 1980, No. 4451, p. 1.

CHAPTER TWO

Nitrogen Fertilizers

2.1 INTRODUCTION

Most of the nitrogen exists in the earth and its atmosphere as the inert gas N_2, which represents about 78% of the earth's atmosphere. In that form, however, nitrogen cannot be utilized directly by higher plants. This chapter serves to outline the processes involving conversion of elemental nitrogen into fixed forms that can be utilized by plants by the activities of symbiotic bacteria such as genus *Rhizobium* and nonsymbiotic nitrogen-fixing organisms, and also by the techniques that have been devised industrially to convert elemental nitrogen into fixed forms.

In addition, the various reactions of nitrogen in the soil and the atmosphere are examined along with nitrogen nutrition in plants. Overall relationships of these factors are coordinated in Fig. 2-1.

2.2 SOIL REACTIONS—CHEMISTRY

Most of the nitrogen that exists in the soil is in organic matter in a very large number of compounds normally found in plants. If we were to analyze the soil for its organic nitrogen compounds, we would find some proteins, amino acids, amides, amines, alkaloids, nucleic acids, nucleotides, and many others. Nitrogen in organic form is unavailable for plant use. It is available

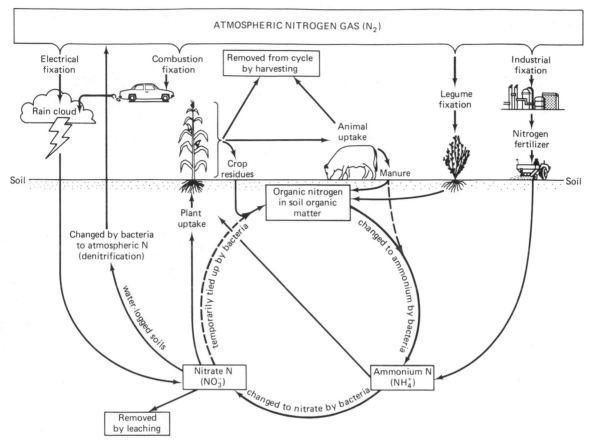

Fig. 2–1. The nitrogen cycle. (Source: W. E. Fenster, "Nitrate Pollution of Surface and Ground Waters and Its Relationship to Fertilizer Nitrogen." In *Soils, Soil Management, and Fertilizer Monographs,* University of Minnesota, Special Report 24, Jan., 1973, pp. 25–27.

once it has been converted to inorganic forms as bacteria decompose organic compounds.

Nitrogen in these organic compounds exists in the form of amino (NH_2^-) groups that represent the same oxidation state (-3) as represented by ammonia (NH_3) and ammonium (NH_4^+). One can calculate the oxidation state of elements such as the nitrogen in this case by remembering that oxygen normally has a -2 oxidation state and hydrogen a $+1$ oxidation state. By determining the net charge on an ion and the number of hydrogen (H^+) or oxygen (O^{2-}) atoms that are combined with it, one can determine the charge on the atom under consideration—that charge being synonymous with oxidation state. In the case of the amino group, which has a -1 net charge, and two hydrogens with a $+1$ charge each, by the process of algebraic addition it's obvious that the charge on the nitrogen is -3.

2.2.1 Symbiotic Nitrogen Fixation

An important contribution to nitrogen nutrition of all plants is made by certain species, principally the legumes that enter into a symbiotic relationship with certain soil bacteria. In this relationship, the bacteria inhabit nodules on host plant roots and are provided with photosynthate by the host plant. The bacteria subsequently convert elemental nitrogen (N_2) into a form usable by the host plant. The contribution of this form of nitrogen to plant nutrition, in general, is quite substantial. Quantities of nitrogen fixed in any symbiotic relationship between bacteria such as *Rhizobia* and a legume such as alfalfa vary with the strain of the organism, the condition of the host plant, and various environmental conditions. Nitrogen fixation by this technique has been estimated as high as 500 kg/ha of nitrogen (448 lb/A) by a crop of clover in New Zealand. However, climatic conditions and soil conditions must be optimum for such high amounts of nitrogen to be fixed, and amounts are normally much less (Table 2-1). In developing countries that may not have nitrogen fertilizer readily available, this mechanism of nitrogen fixation and addition to the plant environment becomes highly important.

In considering symbiotic nitrogen fixation, the type of bacteria that may produce nitrogen in the nodules on roots of one crop may not necessarily be adapted to a second host plant. For instance, the inoculum that is normally applied to soybeans prior to seeding to ensure adequate numbers of *Rhizobium* for root nodulation and nitrogen fixation is not the same strain that would be utilized in inoculating alfalfa. Purchased inoculum indicates on the label the plants for which the bacteria are effective.

Environmental conditions that tend to favor the fixation of nitrogen by the symbiotic process include a relatively high soil pH (6.5 or higher), adequate soil moisture, a good supply of available calcium, warm temperatures, and a low supply of available inorganic soil nitrogen. Bacteria that can enter into this symbiotic relationship with host plants have a high requirement for calcium, thus the positive correlation of high pH and available calcium to their activity.

Temperature affects the activity of any biological process so cooling the

Table 2-1. Estimated Annual Fixation of Atmospheric Nitrogen by Legumes

Legume	Nitrogen Fixed (kg/ha)	Legume	Nitrogen Fixed (kg/ha)
Alfalfa	220	Cowpeas	100
Ladino clover	200	Lespedezas (annual)	95
Sweetclover	130	Vetch	90
Red clover	125	Peas, garden	80
Kudzu	120	Winterpea	60
White clover	115	Peanuts	47
Soybeans	110	Beans, edible	45

soil temperature to near freezing effectively reduces nitrogen fixation to near zero. The temperature effect would also reduce the amount of photosynthate provided by the host plant. Inadequate numbers of bacteria in the soil obviously would reduce fixation and for that reason these organisms are normally added to the seed of legumes at least when they have not been grown in a given field for a few years.

Referring back to inorganic nitrogen supplies, researchers have frequently reported that applications of available nitrogen fertilizer prior to seeding a legume will reduce the number of nodules and greatly reduce the amount of nitrogen fixed by the host plant-bacteria relationship. Apparently, the activity of the organisms is sensitive to the presence of such forms of nitrogen as ammonium and nitrate. An indication that inorganic nitrogen reduces nitrogen fixation and numbers of nodules on soybean plants is given in Tables 2-2 and 2-3, respectively.

2.2.2 Nonsymbiotic Nitrogen Fixation

Fixation of elemental nitrogen by bacteria other than those inhabiting the nodules of legumes and other plants has recently received increased attention. Nitrogen fixation by certain organisms that exist free of any host plant relationship has been indicated by studies in soil microbiology. The organisms that are involved in such nitrogen fixation include blue-green algae and cer-

Table 2-2. Effect of Ammonium on Nitrogen Fixation in Gravel Culture

NH_4 Addition (mg N/pot)	Nitrogen in Plants from		Portion of Nitrogen from Air (Symbiotic) (%)
	Fertilizer (mg/pot)	Air (Symbiotic) (mg/pot)	
Soybeans			
0	0	1639	100
80	68	1692	95
320	252	2243	89
560	464	2185	82
800	648	2423	79
Ladino Clover			
0	0	188	100
80	63	234	75
320	282	159	35
560	527	98	15
800	609	82	12

SOURCE: H. F. Allos and W. V. Bartholomew, "Replacement of Symbiotic Fixation by Available Nitrogen," *Soil Sci.*, 87 (1959), 61 by permission of the American Society of Agronomy

Table 2–3. Effects of Nitrogen Carriers and Time of Nitrogen Application on the Nodulation of Dryland Soybeans

N Rate (kg/ha)	N Carrier	Time N Application	Nodules/ Five Plants	Summary: Mean Values Nodules/Five Plants	
				For Nitrogen Carriers	
0	—	—	445	NH_3	272
168	NH_3	Preplant	144	Urea	294
168	NH_3	Split[b]	281	UAN	251
168	NH_3	Side-dress	391	$Ca(NO_3)_2$	294
168	Urea	Preplant	141	SCU-30	226
168	Urea	Split	253	$LSD_{.05}$ N carriers	NS
168	Urea	Side-dress	488	*For Time of N Application*	
168	N solution	Preplant	125	Preplant	134
168	N solution	Split	182	Split	251
168	N solution	Side-dress	447	Side-dress	416
168	$Ca(NO_3)_2$	Preplant	142	$LSD_{.05}$	56
168	$Ca(NO_3)_2$	Split	295	Time application	
168	$Ca(NO_3)_2$	Side-dress	445		
168	SCU-30[a]	Preplant	118		
168	SCU-30	Split	247		
168	SCU-30	Side-dress	313		
$LSD_{.05}$ Treatment			122		

[a] Sulfur-coated urea.
[b] Half preplant, half side-dressed.
SOURCE: L. J. Meyer et al., "Effects of Nitrogen Carriers and Time of Nitrogen Application on the Yield of Dryland Soybeans," Kansas Fertilizer Research Report of Progress-224, 1974, pp. 146–147, 157.

tain free-living bacteria such as *Azotobacter, Clostridium,* and *Rhodospirillum. Rhodospirillum* is a photosynthetic organism, whereas *Clostridium* and *Azotobacter* are saprophytic organisms of anaerobic and aerobic characteristics, respectively.

Photosynthetic organisms including blue-green algae and *Rhodospirillum* are thought to make only minor contributions to the nitrogen in upland agricultural soils because of the requirement of sunlight in the physiological activity of these organisms. They may play a more major role, however, in soil that is quite wet such as that used for rice culture.

Some microorganisms obtain their energy from the oxidation of organic carbon, and they may contribute more nitrogen to the environment than the photosynthetic organisms. Some microbiologists have estimated that nitrogen fixed by some of these organisms may be in the 20 to 30 kg/ha range annually, but most researchers feel that the value is much lower, possibly around 7 kg/ha of nitrogen. The amounts of nitrogen that have been fixed by free-living organisms in one study are presented in Table 2–4.

Table 2–4. Nonsymbiotic Nitrogen Fixation under Variable Conditions

Treatment	Soil	Ammonia (meq/g soil) Final	Ammonia (meq/g soil) Fixed	Nitrate (meq/g soil) Final	Nitrate (meq/g soil) Fixed	Total Nitrogen Fixed (meq/g soil)	Total Nitrogen Fixed (kg/ha, 15 cm deep)
10 g Yolo soil + 780 mg sucrose	Davis	0.66	0.055	0.75	0.028	0.0123	0.261
Growing lawn, photosyn- thesizing	Davis	1.24	—	0.88	0.0004	0.0004	0.012
	Berkeley	1.78	0.004	1.39	0.001	0.005	0.157
Inverted lawn, decaying	Davis	1.49	0.18	5.44	0.26	0.44	13.77
	Berkeley	3.79	0.087	7.46	0.047	0.134	4.20
Soil with grass removed	Davis	1.07	0.001	9.99	0.007	0.008	0.24
	Berkeley	0.67	0.001	0.46	0.002	0.003	0.095

SOURCE: C. C. Delwiche and J. Wijler, "Non-Symbiotic Nitrogen Fixation in Soil," *Plant and Soil*, 7(2):113, 1956.

2.2.3 Atmospheric Sources of Fixed Inorganic Nitrogen

Electrical discharges in the atmosphere (lightning) result in the conversion of some elemental nitrogen (N_2) to oxides of nitrogen such as nitric oxide (NO) and nitrogen dioxide (NO_2). The electrical arc of lightning produces extremely high temperatures, which in the presence of oxygen and elemental nitrogen result in the production of nitric oxide. This gas further combines with oxygen to produce nitrogen dioxide, which upon being absorbed in water produces nitric acid. Equations for these reactions are:

$$N_2 + O_2 \xrightarrow{\text{Arc}} 2\ NO$$
$$NO + \tfrac{1}{2}\ O_2 \longrightarrow NO_2$$
$$3\ NO_2 + H_2O \longrightarrow 2\ HNO_3 + NO$$

Emissions of nitrogen oxides to the atmosphere from industrial sources and the exhausts of automobiles also provide a source of nitrogen that eventually enters the soil as nitric acid in rainfall.

Another source of nitrogen in the atmosphere that has become increasingly common with industrialization is ammonia gas. This gas is soluble in water and is readily washed out of the atmosphere with rainfall. It is also absorbed directly by surface water bodies such as lakes and also adsorbed directly from the air by soil colloids. Studies in industrialized areas have indicated that 60 to 70 kg/ha of nitrogen (54 to 62 lb/A) could be supplied to soil as ammonia from industrial contamination of the atmosphere.

The contribution of atmospheric nitrogen of all forms to the soil is much less, however, than the amounts previously quoted. Considering the low industrialization of much of the area of the United States and the world, average contributions of these sources of nitrogen to soil nitrogen supplies are probably less than 10 kg/ha of nitrogen (9 lb/A) per year.

2.2.4 Aminization and Ammonification

Whenever organic residues return to the soil, the nitrogen in these plant and animal wastes is slowly returned to the inorganic form by the activity of heterotrophic organisms. These organisms utilize organic carbon compounds as an energy source and are involved in the decay and simplification of organic wastes.

In the process of decomposing organic residues in the soil, nitrogen-containing compounds such as proteins and amino acids are simplified by hydrolytic, reductive, and oxidative microbial activities. Hydrolytic activities involve enzyme systems from the heterotrophic bacteria and saprophytic organisms of all types that, by the addition of water to the compound, separate a complex molecule into simpler forms. Oxidative reactions result in production of carbon dioxide during simplification processes. Hydrolytic and oxidative reactions continue to simplify the nitrogen-containing compounds until in the end a mixture of organic acids, carbon dioxide, and ammonium ions result from the decomposition of such molecules as amino acids. A suggested sequence for such a scheme is presented in Fig. 2-2.

Aminization and subsequently ammonification have the unique characteristic among soil-nitrogen reactions of not involving a change in the oxidation state of nitrogen. Nitrogen of the amino groups (NH_2^-) characteristic of proteins and amino acids has a -3 oxidation state, which is identical to that of the end product of ammonification, the ammonium ion (NH_4^+). Although other portions of the nitrogen-containing compounds may actually undergo oxidation, there is no change in the oxidation status of the nitrogen.

Another example of ammonification is the hydrolysis of urea represented

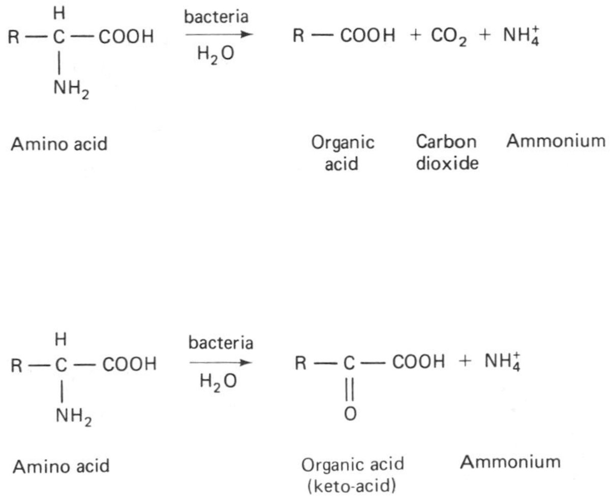

$$
\underset{\text{Amino acid}}{R - \overset{\overset{\text{H}}{|}}{\underset{\underset{\text{NH}_2}{|}}{C}} - COOH} \quad \xrightarrow[\text{H}_2\text{O}]{\text{bacteria}} \quad \underset{\substack{\text{Organic} \\ \text{acid}}}{R - COOH} + \underset{\substack{\text{Carbon} \\ \text{dioxide}}}{CO_2} + \underset{\text{Ammonium}}{NH_4^+}
$$

Fig. 2-2. Generalized mineralization reactions for soil organic nitrogen compounds.

$$
\underset{\text{Amino acid}}{R - \overset{\overset{\text{H}}{|}}{\underset{\underset{\text{NH}_2}{|}}{C}} - COOH} \quad \xrightarrow[\text{H}_2\text{O}]{\text{bacteria}} \quad \underset{\substack{\text{Organic acid} \\ \text{(keto-acid)}}}{R - \overset{}{\underset{\underset{\text{O}}{\|}}{C}} - COOH} + \underset{\text{Ammonium}}{NH_4^+}
$$

1. $$CO(NH_2)_2 + 2H_2O \xrightarrow[\substack{(3)\ \text{other enzymes} \\ (4)\ \text{inorganic catalysts}}]{\substack{(1)\ \text{urease} \\ (2)\ \text{micro-organisms}}} (NH_4)_2CO_3$$

Urea   Water   Ammonium carbonate

Urea in the presence of urease and possibly other enzymes combines with water to form ammonium carbonate.

2. $$(NH_4)_2CO_3 + H_2O \longrightarrow 2NH_4^+ + H_2O + CO_2$$

Ammonium carbonate   Water   Ammonium ion   Water   Carbon dioxide

Fig. 2–3. Urea hydrolysis reactions in the soil are a form of ammonification.

Ammonium carbonate further hydrolizes to form the ammonium ion which immediatly becomes chemically attached to the soil colloids.

in Fig. 2–3. Urea is the simplest of the organic forms of nitrogen and is a component of animal and human wastes. This compound is acted upon by the enzyme urease, which is present in many lower and higher plants. In the presence of water and this enzyme system, urea is rapidly changed into carbon dioxide, ammonium ions, and hydroxyl ions (OH^-). The presence of hydroxyl ions is one of the characteristics associated with ammonification, particularly urea hydrolysis. If one measured the pH of the soil near a hydrolyzing urea prill (particle), one would find that the pH would rise quickly due to the formation of ammonia and hydroxyl ions. The rapidity with which soil pH changes in the vicinity of a urea prill is given in Fig. 2–4.

Many factors affect the rapidity of ammonification reactions. Let's consider urea hydrolysis as a case in point. Temperature, numbers of soil organisms, water, and amounts of plant residue present in the soil all govern the rate of this reaction. Certain higher plants such as soybeans seem to impose a higher urease activity to the soil and subsequently may speed the hydrolysis of urea placed in soil previously cropped to soybeans.

Urease also has the peculiar characteristic of being able to exist as an enzyme apart from the presence of any live bacterial or fungal cells. Much of the enzyme is produced in the soil by bacterial activity, but research has indicated that soil treated with volatile antiseptics still hydrolyzes urea (Fig. 2–5). Researchers have suggested that this enzyme exists in the soil as a clay-protein complex or as a lignin-protein complex. In any event, the soil has the ability to hydrolyze urea rapidly under most conditions. One noted exception, however, is in Alaska where cold soil conditions so restrict urea hydrolysis that urea is not recommended as a fertilizer.

2.2.5 Nitrification

Ammonium ions produced by decomposition of organic materials in soil (ammonification) are immediately subjected to other reactions. One of the most common fates of the ammonium produced by the ammonification process is to enter into the biological process known as *nitrification*. Other fates of the ammonium ions produced include direct absorption by higher plants,

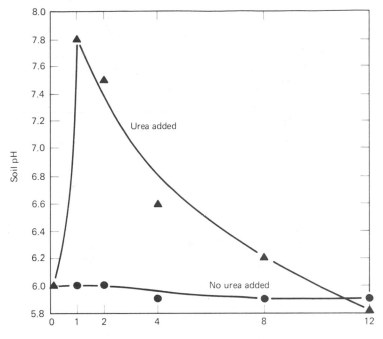

Fig. 2–4. Increased soil pH produced by the hydrolysis of urea in Weldon silt loam. (Source: L. S. Murphy, unpublished data.)

possible fixation by certain clay minerals, and direct utilization by hetero-trophic organisms that are engaged in decomposition of carbonaceous residues.

Nitrification is a two-step process in which the ammonia or ammonium is converted to nitrite (NO_2^-) and finally to nitrate (NO_3^-). These biological oxida-tions involve the loss of electrons by the nitrogen atoms of the ammonium ion and result in its conversion from the -3 oxidation state of ammonia or am-monium to the $+3$ state (nitrite) and on to the $+5$ state (nitrate). This represents a net loss of eight electrons in going from ammonium to nitrate. That oxidation process results in the generation of energy which can be util-ized by the autotrophic organisms *Nitrosomonas* and *Nitrobacter* which carry out this two-step process.

The first reaction of nitrification is mediated by the organism *Nitroso-monas*, which has the characteristics of being tolerant of high concentrations of ammonium ions and operates at an optimum pH ranging from about 8.6 to 8.8. That organism mediates the reaction listed below:

$$\underset{\text{Ammonium}}{NH_4^+} + \tfrac{3}{2} O_2 \longrightarrow \underset{\text{Nitrite}}{NO_2^-} + H_2O + 2\,H^+$$

One of the most intriguing aspects of the first step of the nitrification process is the production of a form of nitrogen (nitrite) that can be toxic to some plants.

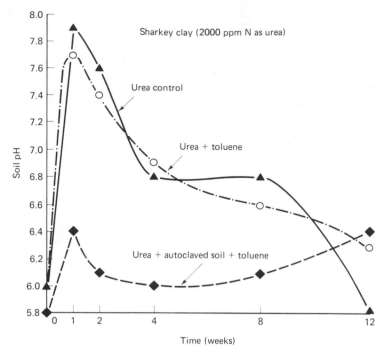

Fig. 2–5. Effects of soil treatments with a volatile antiseptic (toluene) and heat on urea hydrolysis as indicated by soil pH. Heat significantly affected hydrolysis but the toluene had little effect. (Source: L. S. Murphy.)

Equally important is the generation of two protons (hydrogen ions, H^+) with oxidation of each ammonium ion. This process results in acidification of soil upon continued applications of ammoniacal sources of nitrogen. The production of hydrogen ions is the basis for recommendations to maintain liming programs with use of ammoniacal nitrogen sources. The acidifying effects of continued applications of ammonium nitrogen on surface soil pH is indicated in Table 2–5.

The second step in the nitrification process involves conversion of nitrite to nitrate, which is carried out primarily by a second group of obligate autotrophic bacteria of the genus *Nitrobacter*. The reaction representing this process is as follows:

$$2\,NO_2^- + O_2 \longrightarrow 2\,NO_3^-$$

Nitrite   Nitrate

Note that in this reaction no protons (H^+) are produced as was the case in nitrite production. Note also that both reactions require molecular oxygen. The requirement of the organisms for molecular oxygen implies that a factor governing the rate of conversion of ammonium to nitrate is the oxygen supply or aeration of the soil.

Table 2-5. Effects of 20 Years of Nitrogen Fertilization
on Soil pH—Bromegrass

Depth (cm)	Annual N Rate (kg/ha)		
	0	112	224
0–15	5.9	5.3	4.6
15–30	5.7	5.9	5.7
30–45	5.8	5.8	5.8
45–60	5.9	5.9	5.9

SOURCE: Dr. C. E. Owensby, Kansas State University. Unpublished data.

2.2.6 Environmental Factors Controlling Nitrification

The rate of conversion of ammonium to nitrate is affected by several soil environmental conditions, including (1) oxygen supply, (2) population of nitrifying organisms, (3) soil pH, (4) temperature, (5) soil moisture, (6) amount of ammonium ions in the soil, and (7) man-made nitrification inhibitors.

Soil Oxygen Supply

As indicated in the preceding reactions, the bacteria carrying out nitrification require elemental oxygen. Relationships between the oxygen supply in the soil atmosphere and rates of conversion of ammonium to nitrate have been demonstrated repeatedly and are shown in Fig. 2-6.

As the concentration in the soil atmosphere approaches that of the atmosphere above the soil (about 20%), conversion of ammonium to nitrate approaches a maximum. Rapid conversion rates, however, have been recorded at much lower concentrations, around 10% oxygen, in the soil atmosphere.

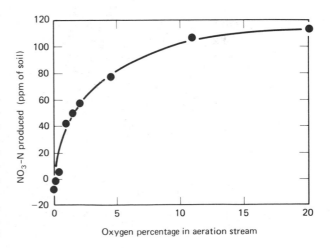

Fig. 2-6. Nitrification of ammonium nitrogen in a loam soil as affected by the percent oxygen in the soil atmosphere (air). Note that nitrification rates are sharply reduced when oxygen in the soil atmosphere drops below 10%, about half that of the normal atmosphere. (Source: John Box, "Nitrogen and Crop Production," Texas Agric. Exp. Sta. Bul. MP-737, 1974, p. 23.)

Conditions that favor the diffusion of gases into and out of the soil promote nitrification. Good soil structure and/or good tilth facilitates the exchange of gases between the soil and the aboveground atmosphere.

Population of Nitrifying Organisms

The number of nitrifying organisms in the soil varies with different soil chemical conditions. It is unlikely that a situation would be encountered where the soil would be totally devoid of these organisms, but factors such as pH and organic matter do have an impact upon the numbers of organisms present. Sabey et al.[1] demonstrated the effects of additions of nitrifying bacteria on production of nitrate in an Iowa soil with an initial pH of 8.3. As the number of nitrifying organisms added to the soil was increased, the nitrate production increased rapidly (Fig. 2-7).

The population of nitrifying organisms in the soil responds to the presence of an adequate supply of ammonium. Thus, the total amount of nitrification that might occur in the soil would probably not be affected by the initial number of organisms, but rate of conversion of ammonium to nitrate could be affected over a short time span. Note in Fig. 2-7 that after about 2 weeks nitrification was occurring even where no additional bacteria were added (check treatment).

Soil pH

Alexander[2] has suggested that pH is chief among the environmental factors influencing the rate of nitrification. Very acid soil depresses the numbers of nitrifying organisms present. Typically, the rate of nitrification declines markedly below pH 6 but does not decline to an essentially negligible rate until below pH 5. This is not to say that nitrates will never occur in soils with a pH more acid than pH 5 but the *rate* of nitrification is extremely slow. An optimum rate of nitrification has been suggested to occur around pH 8.3 to 8.5. Part of the problem associated with low pH effects on nitrification is that such soils may have relatively small supplies of available calcium and phosphorus. Nitrifying bacteria are known to have a high requirement for calcium and thus the relationship may exist between nutrition of the organisms and soil pH.

Due to the microbial sensitivity to acidity and calcium, nitrification in acid soils is markedly enhanced by liming (see Chapter 8).

[1] B. R. Sabey, L. R. Frederick, and W. V. Bartholomew, "The Formation of Nitrate from Ammonium Nitrogen in Soils. III. Influence of Temperature and Initial Population of Nitrifying Organisms on the Maximum Rate and Delay Period," *Soil Sci. Soc. Amer. Proc.*, 23:462, 1959.

[2] M. Alexander, *Introduction to Soil Microbiology*, New York: John Wiley & Sons, 1977, pp. 252–53.

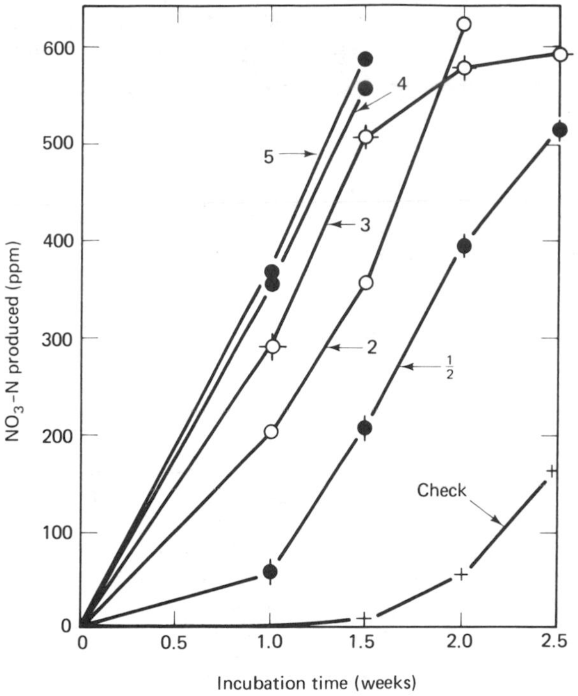

Fig. 2-7. Effects of addition of nitrifying organisms on nitrate production in a silt loam subsoil, pH 8.3. Curve labels indicate milliliters of added inoculum of nitrifying organisms. (Source: B. R. Sabey et al., "The Formation of Nitrate from Ammonium Nitrogen in Soils. III. Influence of Temperature and Initial Population of Nitrifying Organisms on the Maximum Rate and Delay Period," *Soil Sci. Soc. Amer. Proc.,* 23:462, 1959 by permission of the American Society of Agronomy.)

Temperature

Relationships between nitrification and soil temperature have been investigated frequently for many years. The various investigations have indicated that below 5°C (41°F) and above 40°C (104°F) nitrification is very slow. Evidence suggests that nitrate may continue to be formed down to the freezing point despite the fact that the rate drops to a very low level at temperatures below 5°C (41°F). This latter fact has been utilized as a recommendation for fall applications of ammonium fertilizers far in advance of crop needs.

Delaying ammonium application until the soil temperature approaches 5°C at a depth of 10 cm (4 in.) is considered a means of slowing conversion to nitrate. Ammonium ions are adsorbed by the clay and humus and leach less rapidly than nitrate.

Increasing the temperature from the lower levels produces a more rapid oxidation of ammonium. The range varies with different types of organisms because of physiological differences. The optimum is considered to lie somewhere between 30–35°C (86–95°F).

Temperature should not be considered as a total determinant in considering the efficiency of fall applications of nitrogen due to concern for leaching. If fall and winter precipitation are not enough to cause leaching, even though nitrification may occur at low soil temperatures, nitrogen loss is not par-

Table 2-6. Late Summer, Fall, and Spring N Application Effects on Winter Wheat Yield—Oklahoma, Silt Loam Soils

N (kg/ha)	N Source	Time N Application	Yield (kg/ha)
0	—	—	1100 a[a]
34	Urea	Pretillage	2200 b
34	Ammonium nitrate	Pretillage	2250 b
34	Ammonium sulfate	Pretillage	1900 b
67	Urea	Pretillage	2550 c
67	Ammonium nitrate	Pretillage	2550 c
67	Ammonium sulfate	Pretillage	2500 c
34	Urea	Fall	2150 b
34	Ammonium nitrate	Fall	2350 b
34	Ammonium sulfate	Fall	2250 b
67	Urea	Fall	2600 c٬
67	Ammonium nitrate	Fall	2750 c
67	Ammonium sulfate	Fall	2650 c
34	Urea	Spring	2330 b
34	Ammonium nitrate	Spring	2350 b
34	Ammonium sulfate	Spring	2400 b
67	Urea	Spring	2800 c
67	Ammonium nitrate	Spring	2840 c
67	Ammonium sulfate	Spring	2550 c

[a] Numbers followed by the same letter are not significantly different.

SOURCE: Dr. B. B. Tucker, Oklahoma State University.

ticularly a problem. This fact exists in many of the wheat-producing areas of the Great Plains where fall applications of nitrogen have been quite effective despite nitrification during the early fall periods (Table 2-6) (see Chapter 7).

Soil Moisture

Nitrifying bacteria are more sensitive to excess soil water than they are to extremely dry conditions. Although some water is necessary for bacterial activity, excesses of water reduce the soil oxygen supply. That relationship to nitrification has already been discussed. Filling the soil pores with water eliminates oxygen and slows nitrification. As the soil moisture tension declines to zero, indicating saturation, nitrate-nitrogen accumulation in soils has been demonstrated to drop to zero. That fact is demonstrated in work by Parker et al.[3] reported in Fig. 2-8.

Concentration of Ammonium Ions

The activity of nitrifying bacteria is strongly related to the supply of ammonium ions. Numbers of bacteria tend to increase with the presence of large amounts of ammonium ions although initial populations of nitrifying organ-

[3] D. T. Parker and W. E. Larson, "Nitrification as Affected by Temperature and Moisture Content of Mulched Soils," *Soil Sci. Soc. Amer. Proc.*, 26:238, 1963.

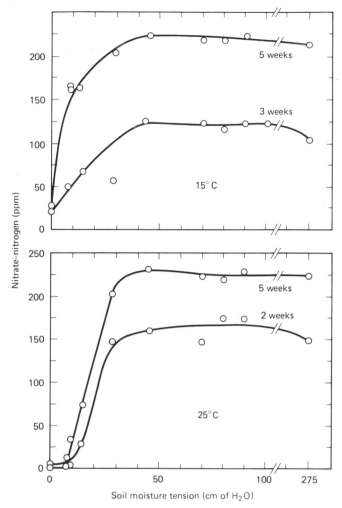

Fig. 2–8. Effects of moisture tension and temperature on nitrate accumulation in a Grundy silt loam following application of 100 ppm NH_4^+–N. (Source: D. T. Parker and W. E. Larson, "Nitrification as Affected by Temperature and Moisture Content of Mulched Soils," *Soil Sci. Soc. Amer. Proc.,* 26:238, 1963 by permission of the American Society of Agronomy.)

isms may have been low. Obviously, no nitrification can occur in a soil devoid of ammonium because this serves as a substrate for the reaction.

The supply of ammonium ions to nitrifying bacteria is closely related to the carbon and nitrogen ratio in the soil. When the carbon:nitrogen ratio is extremely wide (in the vicinity of 60:1) any ammonium released from organic matter (ammonification) is absorbed immediately by the heterotrophic bacterial population that is in the process of decomposing the carbonaceous residue. When the carbon:nitrogen ratio drops to a range of 20 or 25:1, a net release of nitrogen occurs from ammonification and thus ammonium ions are available for nitrification. Therefore, the rate of conversion of nitrogen to nitrate in such a soil would be strictly dependent on the carbon:nitrogen ratios of readily oxidizable carbonaceous residues present. These factors do not imply that the nitrifying organisms in the soil are not present but rather that they are inactive from a lack of oxidizable substrate. Possibly, the numbers of nitrifying organisms could decline due to a shortage of an oxidizable inorganic substrate but their numbers would rapidly increase again as ammonification produces a net ammonium surplus (see Chapter 10).

Nitrification Inhibitors

Another factor involved in the control of nitrogen transformation in soils is the use of nitrification inhibitors. Certain materials are toxic to nitrifying organisms and temporarily inhibit the production of nitrate. One such compound is 2-chloro-6-(trichloromethyl)-pyridine marketed as *N-SERVE* (nitrapyrin) by Dow Chemical Company. This compound is toxic to the organism *Nitrosomonas,* which mediates the conversion of ammonium to nitrite.

Research has been conducted on this compound in the United States for approximately 17 years. Results indicate that the material slows the conversion of ammonium to nitrate and improves the recovery of ammonium nitrogen under conditions where leaching or denitrification occur. Application rates for the compound range from around 0.2 to 0.6 kg/ha of active ingredient. Formulations are available that either are soluble in anhydrous ammonia or nitrogen solutions or are used for coatings on solid nitrogen fertilizers, although the latter method of application is much less common.

The effectiveness of nitrification inhibitors such as nitrapyrin is dependent on conditions conducive to nitrogen losses, leaching, and denitrification. Therefore, primary use would be relegated to areas receiving large amounts of precipitation or supplemental irrigation water and having a coarse texture such as irrigated sands. Some evidence of the effects of this compound on growth of corn plants on a sandy soil can be determined from Fig. 2–9. Other nitrification inhibitors that have been studied and utilized in other parts of the world include potassium azide (KN_3). This compound has been produced in Japan and has been utilized rather widely in that country with some success on a number of crops. Other nitrification inhibitors are also under study.

Fig. 2–9. (a) Anhydrous ammonia plus the nitrification inhibitor nitrapyrin (N-SERVE) produced better nitrogen response early in the growing season than did calcium nitrate on this sandy alluvial soil. (b) Fall (early winter) applications of ammonia plus nitrapyrin also showed early season advantage over spring ammonia alone. (Source: L. S. Murphy.)

2.2.7 Denitrification

A series of soil nitrogen reactions that are essentially the reverse of nitrification are classified as denitrification. These reactions represent a net reduction in the oxidation state of nitrogen from the nitrate or nitrite forms to more reduced compounds, particularly the gases nitric oxide (NO), nitrous oxide (N_2O), and elemental nitrogen (N_2). The primary characteristic of this series of reactions is that the end products represent volatile gases that may be lost from the soil. They also represent forms of nitrogen that are generally unavailable to plants.

Denitrification is a microbial process. Bacteria that are capable of carrying out this process of nitrogen reduction are heterotrophic in nature; i.e., they are engaged in the process of decomposing carbonaceous residues and obtain their energy and carbon supply from organic compound oxidation. Denitrifying bacteria are generally aerobic, and they utilize oxygen as a terminal electron acceptor (see *NOTE*) in the process of carrying sugars through the respiration-glycolysis process in their cells. However, in the absence of oxygen, they are able to utilize the nitrogen atom in nitrate as this terminal electron acceptor. This does not imply that these organisms have the ability to obtain their oxygen from nitrates and nitrites, for that oxygen is already in a reduced form.

In addition to these heterotrophic organisms, several chemoautotrophic organisms are capable of reducing nitrate to molecular nitrogen. These chemoautotrophs are involved in the process of oxidizing other organic substrates in the soil such as sulfur. In this process then they are able to use either oxygen or nitrate as a terminal electron acceptor in a fashion similar to that described earlier for the heterotrophic organisms. Under such conditions the nitrate is converted to gaseous nitrogen. A proposed scheme for denitrification by either heterotrophic or chemoautotrophic organisms follows:

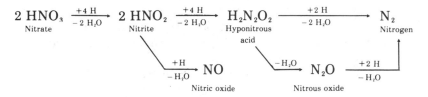

Note that in this series of reactions, ammonium does not appear as an end product of denitrification. This is a point worth remembering because bacteria by a different set of enzymes are able to reduce nitrate or nitrite to ammonium. However, this process is involved in their physiological utilization of

---

NOTE: Terminal electron acceptors are atoms or compounds that serve as a sink for electrons given off in the process of oxidation of carbonaceous compounds in the plant. Such acceptors are chemically reduced in the process of electron acceptance. Molecular oxygen (O_2) is the usual terminal electron acceptor in the physiology of aerobic bacteria.

nitrogen and of the conversion of nitrate or nitrite into a form of nitrogen that can be readily incorporated into amino acids and proteins. The same mechanism of nitrogen utilization also exists in the higher plants.

Considering the microbial reactions with nitrate, three pathways are possible: (1) Nitrate may be reduced to ammonium in the process of converting nitrate-nitrogen into protein nitrogen. (2) Incomplete reduction of nitrate may occur with the production and accumulation of nitrite in the system, an incomplete expression of denitrification. (3) A complete reduction of nitrate may occur to give gaseous compounds including nitric oxide, nitrous oxide, and elemental nitrogen.

Environmental influences on denitrification are essentially the same as those affecting nitrification. These environmental influences include oxygen supply, population of denitrifying organisms, soil pH, temperature, soil moisture, oxidizable organic matter, and amount of nitrate present in the soil.

Soil moisture exerts a primary influence on the rate of denitrification in that saturation of the soil pores with water eliminates oxygen, which is required by the heterotrophic and chemoautotrophic organisms active in the soil.

Oxygen availability is another critical environmental factor in denitrification as outlined earlier. The amount of oxygen present in soil can influence the rate of denitification in two separate ways. First, sufficient oxygen is necessary for the initial formation of nitrite and nitrate that are essential for the denitrification process. As an expression of this importance, consider the situation if ammoniacal nitrogen were added to a flooded rice soil where oxygen is essentially devoid. Lack of oxygen would not produce nitrite or nitrate and subsequently no denitrification could occur. The other role of oxygen is in relation to the decomposition of oxidizable carbon by the heterotrophic organisms. In the case of a lack of oxygen as discussed earlier, nitrate or nitrite would serve as a terminal electron acceptor.

Rate of denitrification is also affected by the amount of readily oxidizable carbonaceous material in the soil. Easily decomposed carbonaceous material such as sugars or organic acids are quickly oxidized and stimulate greater demand for oxygen and, in lieu of oxygen, stimulate greater demand for nitrate as the terminal electron acceptor.

Denitrification is dramatically affected by temperature. The processes proceed very slowly at temperatures slightly above freezing but increase rapidly with rising temperatures. Optimum temperatures for the reaction are in the range of about 25° to 30°C (77° to 86°F). Denitrification can occur at temperatures possibly as high as 60°C (140°F) but in this range the rate declines due to detrimental effects of heat on the bacterial systems. Denitrification may be important at low soil temperatures in an economic sense. At temperatures in the vicinity of 10°C (50°F), plant utilization of nitrate is very low. However, at this temperature, denitrification can still proceed at an appreciable rate (Fig. 2–10).

Soil acidity (pH) also exerts an influence on denitrification both in terms of the amount of nitrogen volatilized as the various gaseous products and also

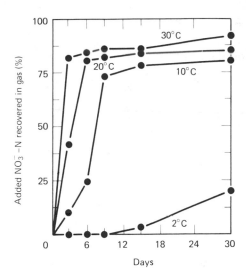

Fig. 2-10. Effects of temperature on denitrification of nitrate in a soil receiving an oxidizable carbon source, glucose. (Adapted from J. M. Bremner and K. Shaw, *J. Agric. Sci.*, 51:40, 1958. Used with permission, Cambridge University Press.)

in terms of the major nitrogen products of the reaction. Denitrifying bacteria are sensitive to high concentrations of hydrogen ions and acid soil and, therefore, tend to promote lower amounts of nitrogen loss under acid conditions than at pH's in the alkaline range. Wijler et al.[4] published relationships of soil pH to total nitrogen loss and to the relative percentage of nitrous oxide, nitric oxide, and elemental nitrogen. Considering these pH effects, the evolution of nitric oxide is quite pronounced at soil pH's 6 to 6.5. Nitric oxide (NO) only appears in significant quantities when the pH is quite low, 4.9–5.6. As the pH increases, nitric oxide may be evolved first even in the pH range of 7.3–8.4, but eventually elemental nitrogen becomes the dominant nitrogen product as pH increases and time proceeds.

2.2.8 Ammonium Fixation

Besides nitrification and absorption by plants, another possible fate of ammonium nitrogen in the soil is reaction with certain silicate clays, particularly those with an expanding lattice. Certain clays such as montmorillonite, vermiculite, and illite react with ammonium, trapping the ions between the layers of the clay platelets. This reaction, known commonly as *fixation*, generally occurs to a greater extent in subsoils than surface soils.

The exact mechanism of ammonium fixation is similar to that of potassium fixation, which is discussed in Chapter 4. Loss of cations within the structure of the clay itself allows ammonium to satisfy exposed negative charges. Such ammonium is considered to be trapped essentially physically and is only slowly available to nitrification and to plant absorption. Apparently ammonium fixed in this mechanism can be replaced by large amounts of cations (Ca^{2+}, Mg^{2+}, H^+), which tend to expand the lattice of the clay. Addi-

[4] J. Wijler and C. C. Delwiche, "Investigations on the Denitrifying Process in Soil," *Plant and Soil*, 5:155, 1954.

tion of significant amounts of cations that tend to contract the clay lattice tend to render the fixed ammonium even more unavailable. Such cations include cesium (Cs^+), rubidium (Rb^+), and potassium (K^+).

Wisconsin researchers (Walsh et al.[5]) studied the amounts of native fixed ammonium and fixation of ammonium added to several soils under various types of conditions. The results tended to show larger amounts of ammonium fixed in the subsoils but all values reported were less than 1 milliequivalent/ 100 g of soil. The Wisconsin researchers also demonstrated that drying the clay to oven-dryness tended to produce more ammonium fixation, which is an expression of collapse of expandable clay lattice due to loss of water. Freezing also tended to enhance ammonium fixation.

Although the phenomenon of ammonium fixation is largely associated with mineral soils, some researchers have reported apparent fixation of ammonium by organic soils. Ammonium fixation was directly related to percentage of carbon in the organic matter, and the suggestion was made that the mechanism of fixation revolved around hydroxyl groups present in the organic matter.

In general, the significance of ammonium fixation in applied agriculture is not generally considered to be great. Under certain soil conditions it may represent a fate of nitrogen that is measured in terms of lowered availability during the first cropping season following ammonium nitrogen application.

2.2.9 Ammonia Volatilization

When ammonia or ammonium salts are added to a soil, one of the common questions that is raised concerning the fate of this material is the possibility of volatilization of ammonia to the atmosphere. Whenever ammonia or ammonium is added to the soil, an equilibrium is established between these two forms of nitrogen and water. That reaction is:

$$NH_3 + H_2O \rightleftharpoons NH_4^+ + OH^-$$

This equilibrium is affected by many factors. If the equilibrium is established in an alkaline soil containing relatively high concentrations of hydroxyl ions in solution, the equilibrium shifts to the left and ammonia gas is formed at the expense of ammonium ions. Loss of water from the system also tends to shift the equilibrium to the left. That loss could be brought about by high temperatures and rapid desiccation of the soil containing the ammonia-ammonium system. Any factor that shifts the equilibrium toward greater amounts of ammonia may increase nitrogen losses by ammonia volatilization.

Microenvironmental conditions can also affect this equilibrium. For instance, in the vicinity of a hydrolyzing urea particle, the soil pH rises rapidly due to the equation just discussed (see Fig. 2–4). The presence of hydroxyl

[5] L. M. Walsh and J. T. Murdock, "Native Fixed Ammonium and Fixation of Applied Ammonium in Several Wisconsin Soils," *Soil Sci.*, 89:183, 1960.

ions in the vicinity of the urea particle causes increased pH. Subsequently this drives the reaction back to the left forming free ammonia, and ammonia volatilization may occur if the reaction is taking place near the soil surface.

The importance of water in retention of ammonia cannot be overemphasized. Water is important in converting ammonia into ammonium, which is subsequently chemically adsorbed by soil colloids and held tightly against leaching. On the other hand, it is possible for ammonia injected into the soil to react directly with protons (reserve acidity) either on the soil colloids or in the soil solution and to be converted to ammonium ions that are subsequently chemically adsorbed. In any event, the final fate of the ammonia added to soil is of concern to persons who depend on this nitrogen source to supply a following crop adequately.

Several investigations have indicated large losses of nitrogen in the ammonia form from surface applications of urea and other nitrogen compounds. Urea has been particularly subjected to criticism from the standpoint of possible volatilization resulting from its hydrolysis at the soil surface. In addition to the factors just mentioned, a low cation exchange capacity can be related to significant amounts of ammonia volatilization due to insufficient exchange capacity to adsorb ammonium ions from the ammonia-water-ammonium system.

Some indication of the magnitude of nitrogen losses (not all ammonia) under laboratory conditions from surface applications of urea and other nitrogen sources are indicated in Table 2–7. However, under field conditions, ammonia volatilization, although difficult to measure, has not been high. Losses from Kansas and Texas studies of surface nitrogen applications (Tables 2–8 and 2–9) have not been nearly so large. Laboratory conditions tend to promote unusually high ammonia volatilization losses due to high rates of airflow and desiccation (Table 2–10).

Losses of ammonia from field applications of anhydrous ammonia have long been a subject for discussion. Stanley and Smith[6] demonstrated that greater depths of anhydrous ammonia application reduced ammonia volatilization losses when the gas could be sealed in the soil (Fig. 2–11). Other work at the University of Missouri has demonstrated that volatilization losses of ammonia under field conditions was higher when soils were wet and sealing of the gas in the soil was more difficult. Greater losses apparently occur under wet conditions than under dry conditions. In addition to problems with sealing the slit made in wet soils by the applicator knife, movement of capillary water to the soil surface in the process of desiccation can carry ammonium ions with it that upon loss of water by evaporation convert to ammonia and are volatilized to the atmosphere.

Studies in Kansas with field losses of ammonia from undercutting blade applications of anhydrous ammonia have demonstrated quite low losses as long as proper depth of application was maintained (Table 2–11). When am-

---

[6] F. A. Stanley and G. E. Smith, "Proper Application Improves Values of NH_3," *Agric. Ammonia News*, April–June, 1955.

Table 2–7. Changes in Total Nitrogen of Different Soils After Treatment with 550 ppm of N (Incubated for 3 Months at Room Temperature, Optimum Soil Moisture)

	Soil Type			Means for Nitrogen Carriers
Nitrogen Carrier	Weldon Silt Loam	Sharkey Clay	Sharpsburg Clay Loam	
	Loss or Gain in Total Nitrogen as Percentage of Treatment			
Sodium nitrate	− 35.7	− 50.3	+17.5	− 22.8
Ammonium nitrate	− 33.0	− 15.9	− 11.7	− 20.2
Ammonium nitrate and straw	− 16.5	− 44.6	+10.5	− 16.9
Ammonium nitrate and molybdenum	− 15.6	− 52.1	− 16.5	− 28.1
Ammonium sulfate	− 2.0	− 28.1	− 12.4	− 14.2
Monoammonium phosphate	− 8.8	− 31.2	− 24.5	− 21.5
Diammonium phosphate	− 5.3	− 10.4	− 15.8	− 10.5
Aqua ammonia	− 47.8	− 61.2	− 17.3	− 42.1
Urea	− 81.2	− 41.3	− 29.3	− 50.6
Urea and straw	− 53.0	− 50.6	− 18.5	− 40.7
Alfalfa hay (ground)	+ 2.4	− 33.9	− 21.7	− 17.7
Means for soil types	− 28.0	− 38.8	− 11.2	

NOTES:

L.S.D.$_{.05}$ for comparison of means for nitrogen carriers within soil types = 13.2.
L.S.D.$_{.05}$ for comparison of means for soil types = 3.7.
L.S.D.$_{.05}$ for comparison of means for nitrogen carriers = 7.6.

SOURCE: G. H. Wagner and G. E. Smith, "Recovery of Fertilizer Nitrogen from Soils," Missouri Agric. Exp. Sta. Bul. 738, 1960, p. 28.

Table 2–8. Ammonia Volatilization from Surface Nitrogen Applications

	Percentage Applied N Lost as NH_3		
N Source	Wheat	Tall Fescue	Bromegrass
Sulfur-coated urea	1.33	2.75	1.75
Urea-ammonium sulfate	2.08	6.83	2.50
Urea	3.00	5.83	2.00
28% N solution	1.08	2.08	2.50
Ammonium nitrate	0.67	1.33	0.67

NOTE: N rate was 134 kg/ha of N (120 lb/A) for all three crops.

SOURCE: Dr. D. A. Whitney, Kansas State University. Unpublished data.

Table 2–9. Ammonia Loss from Surface Applications of Several Ammonium Salts (Houston Black Clay, pH 7.6)

N Source	Percentage Applied N Lost as NH_3	
	24 hr	100 hr
Ammonium fluoride (NH_4F)	59	68
Ammonium sulfate ($(NH_4)_2SO_4$)	40	55
Diammonium phosphate ($(NH_4)_2HPO_4$)	37	52
Ammonium chloride (NH_4Cl)	8	18
Ammonium nitrate (NH_4NO_3)	8	18
Ammonium iodide (NH_4I)	6	15

NOTE: N rate: 550 kg/ha of N.

SOURCE: L. B. Fenn and D. E. Kissel, "Ammonia Volatilization from Surface Applications of Ammonium Compounds on Calcareous Soils: I. General Theory," *Soil Sci. Soc. Amer. Proc.*, 37:855, 1973 by permission of the American Society of Agronomy.

Table 2–10. Soil pH Effects on Cumulative Losses of Nitrogen as Ammonia from Surface Urea Applications (112 kg/ha of N)

Dickson Silt Loam Initial Soil pH	Percentage Added N Volatilized as NH_3 after 10 Days
5.0	9
5.5	13
6.0	19
6.5	27
7.0	38
7.5	52

SOURCE: J. W. Ernst and H. F. Massey, "The Effects of Several Factors on Volatilization of Ammonia Formed from Urea in the Soil," *Soil Sci. Soc. Amer. Proc.*, 24:87, 1960 by permission of the American Society of Agronomy.

monia was applied at a shallow depth of 5 cm (2 in.) with spacings between points of ammonia release as great as 100 cm (40 in.), ammonia volatilization was greater than when a 10-cm (4-in.) depth of application was used with 40-cm (16-in.) spacings between points of release.

Other research has shown some competition between ammonia and water for adsorption sites on the colloids themselves. Under slightly acid conditions, researchers at the Tennessee Valley Authority demonstrated that greater ammonia retention was obtained when ammonia was injected into oven dried soil than when injected into soil at field capacity.

Plant residues on the soil surface can affect ammonia volatilization losses, especially from urea or urea-containing fertilizers. Residues on the surface may keep applied nitrogen from contacting the soil where adsorption reactions would bind ammonia or ammonium ions to soil colloids. Residues apparently contain a significant amount of urease activity and urea hydrolysis

Table 2-11. Ammonia Losses from Application with an Undercutting Blade Were Very Small on a Silt Loam Soil in Kansas. Losses from Sand Were Controlled by Greater Depth of Application and Narrower Spacing Between Points of Ammonia Release

Treatment			
Depth (cm)	(kg/ha N)	Spacing (cm)	% NH$_3$ Lost
		Silt Loam	
5	112	102	0
5	224	102	Trace
		Sand	
5	224	15	6.5
5	224	102	6.4
10	224	102	7.2
10	224	15	0.8

SOURCE: C. L. Swart and L. S. Murphy, unpublished data.

Table 2-12. Nitrogen Fertilizer Source Affects on Dryland Grain Sorghum, 1965-72 [a]

Nitrogen Carrier	Average Yield (kg/ha)		
	1971–1972 5 Sites	1969–1972 12 Sites	1965–1972 29 Sites
No nitrogen	3575	3199	3199
Ammonium nitrate (34-0-0)	3951	3512	3700
Urea (45-0-0)	3951	3575	3638
Liquid nitrogen (28-0-0)	3951	3512	3575
Anhydrous ammonia (82-0-0)	4139	3575	—
Ammonium sulfate (21-0-0)	3951	—	—
Difference required for significance .05	251	188	188

[a] 67 kg/ha of N on sites from 1964 to 1968. 45 kg/ha of N on sites from 1969 to 1972; silt loam soils.

SOURCE: C. A. Thompson, "Fertilizing Dryland Grain Sorghum," Kans. Agric. Exp. Sta. Bul. 579, 1974, 20 pp.

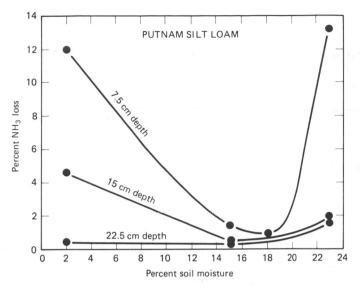

Fig. 2-11. Loss of ammonia from Putnam silt loam when anhydrous ammonia was applied at different depths and varying soil moisture contents. Nitrogen was applied at a rate of 112 kg/ha (100 lb/A) of nitrogen, 102 cm between points of release. (Source: F. A. Stanley and G. E. Smith, "Proper Application Improves Value of NH₃," *Agricultural Ammonia News,* April-June, 1955. Used with permission.)

may proceed rapidly. Lack of sufficient moisture to wash the nitrogen onto the soil surface can combine with large amounts of plant residue and high temperature to produce significant ammonia volatilization losses. Application of nitrogen fertilizer during cool weather and tillage will help control the problem.

In summary, ammonia volatilization losses may be significant under conditions of poor soil tilth, excessively wet conditions at time of ammonia application, or surface applications of some nitrogen fertilizers. Basically, ammonia losses from surface applications of liquid or solid fertilizers can be prevented by placing the nitrogen materials beneath the soil surface or by incorporation shortly after application. This recommendation is particularly effective on soils that have a high temperature at the time of application. Surface application of solid nitrogen materials does not necessarily imply poor performance, however. When soil conditions are cold, volatilization of ammonia is lower. Good yield responses from various nitrogen materials have tended to confirm that ammonia volatilization losses are low (Table 2–12).

2.3 NITROGEN IN PLANT NUTRITION

Nitrogen is one of the essential elements in plant nutrition and one nutrient that is most frequently in short supply in cultivated soils around the world. The availability of nitrogen as an essential element in plant nutrition was established in the 19th century. Its role in plant nutrition has been recognized to be connected to the production of vigorous vegetative growth and associated with a dark green leaf color. The association with the coloration of plants is through the fact that nitrogen is one of two essential elements absorbed from the soil that is an important constituent of the chlorophyll molecule. The other element is magnesium. Lack of adequate nitrogen subsequently produces plants that are lighter green in color due to a smaller amount of chlorophyll.

As indicated in the preceding sections, most of the soil nitrogen is associated with the organic fraction of the soil (see Chapter 10). In order to be utilized by plants, this nitrogen must be released in ammonium form and subsequently undergo nitrification reactions. Plants assimilate both ammonium and nitrate with the latter being preferentially absorbed by most mature plants. Certain plants grown under flooded soil conditions, however, tend to absorb ammonium preferentially. Rice would be an example of this latter class of plants.

Regardless of the form in which nitrogen is absorbed, it is an essential constituent of amino acids, proteins, nucleotides, nucleic acids, amines, and amides. One characteristic of all of these nitrogen compounds is that the nitrogen contained in them is in the same oxidation state, -3. Because it was indicated earlier that many plants preferentially absorb nitrate, which is the most common inorganic form of nitrogen in most soils, it is important to consider how the oxidation state of nitrate nitrogen ($+5$) is converted to the -3 oxidation state of ammonium nitrogen.

2.3.1 Nitrate Reduction

Plant absorption of anions such as nitrate is thought to be affected by metabolic functions including respiration and photosynthesis. The concentration of nitrate-nitrogen within plant root cells is normally higher than that in the soil solution. For nitrate in the soil solution to move into the zone of higher concentration in the plants, energy must be expended. This energy is produced from photosynthesis and subsequent metabolism of sugars by the respiration process in cells. The exact theory of anion absorption is beyond the scope of this discussion, but a carrier mechanism has been proposed to transport ions across the membranes of the living cells.

When nitrate ions reach the plant root, they are either transported to other portions of the plant or metabolized at that location. Metabolism of nitrate begins with the reduction of nitrate to nitrite. That reaction is mediated by an enzyme, nitrate reductase, which requires molybdenum (Mo) and a reduced coenzyme for completion of the reaction. The reduced coenzyme is a source of electrons for the reduction of nitrate to nitrite and is obtained through the light reactions in plants or metabolism of carbohydrates. Molybdenum serves as a link in the electron transfer mechanism between the reduced coenzyme and the nitrate ion.

There is some disagreement on whether this process proceeds by a second step of nitrite reduction to hydroxylamine or whether nitrate is reduced directly to ammonium. If the second step of nitrite to hydroxylamine does exist, some evidence suggests that an enzyme system similar to nitrate reductase is active and involves iron and copper in its activity. A third and final step has been proposed for this mechanism, the reduction of hydroxylamine to ammonium. A summary of this scheme for nitrate reduction follows:

$$NO_3^- + NADPH \xrightarrow[Mo^{6+}]{\text{Nitrate reductase}} NO_2^- + NADP$$

Nitrate   Reduced coenzyme   Nitrite   Oxidized coenzyme

$$NO_2^- + NADPH \xrightarrow[\text{Fe, Cu?}]{\text{Nitrite reductase?}} NH_2OH + NADP$$

Hydroxylamine

$$NH_2OH + NADPH \xrightarrow{\text{Hydroxylamine reductase}} NH_4^+ + NADP$$

Ammonium

Regardless of the intermediate reactions in the process of nitrate reduction to ammonium, the initial step in the formation of proteins has been completed. Nitrogen has been converted to the -3 oxidation state and is next reacted with carbon skeletons or organic acids (Fig. 2-12) to form amino acids such as glutamic acid, which is elaborated or metabolized into over 100 different amino acids. More than 20 of the different amino acids are then joined together through peptide linkages to form proteins. The order in which amino acids are linked together is controlled by the genetics of the plant. Subsequently, environmental factors, genetics, and the amount of nitrogen supplied to the plant affect the amount of protein produced and stored in grain.

Referring back to the process of nitrate reduction, there are several environmental factors that may affect the rate of conversion of nitrate to ammonium. Included in this array of environmental factors are supplies of nitrogen, phosphorus, potassium, sulfur, copper, iron, and molybdenum; light intensity; moisture stress; temperature; stage of plant maturity; genetic variations; and a possible role of certain endogenous (developing within the cell) antimetabolites in plant physiology.

Referring to the effects of these various factors, the activity of nitrate reductase in the plant seems to be dependent on the amount of nitrate-nitrogen supplied to the plant. As more nitrate is supplied, the plant develops an increasing ability to reduce nitrate to ammonium.

Phosphorus exerts an influence on nitrate reduction through its role in the coenzymes necessary for the activity of the nitrate reductase and other enzyme systems. The coenzymes NADH and NADPH and the formation of sugar phosphates in carbohydrate metabolism are particularly important in nitrate reduction.

Potassium may exert an influence on nitrate reduction and nitrogen metabolism in general by controlling the amount of organic acids produced for protein synthesis and also in terms of a controlling influence in sugar metabolism. Insufficient potassium tends to depress the formation of organic acids resulting in an accumulation of nitrate in the plants due to a buildup of

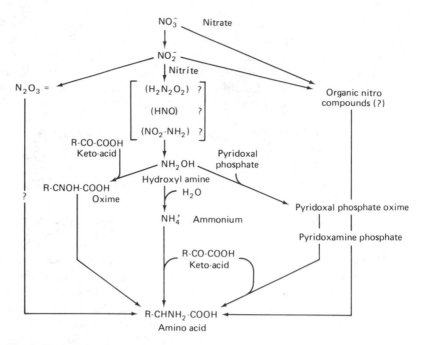

Fig. 2-12. Schematic diagram of possible assimilatory pathways of nitrate-nitrogen and its reduction products.

end products of nitrate reduction. Lower rates of carbohydrate metabolism result in smaller amounts of reduced coenzymes available for nitrate reduction.

The role of sulfur in nitrate reduction probably relates to its effect on protein and enzyme formation. All proteins contain sulfur, thus a reduction in the sulfur content of the plant could lead to a lower enzymatic activity of nitrate reductase.

The role of molybdenum has already been noted as a portion of the nitrate reductase electron transfer mechanism. Other micronutrients implicated in nitrogen metabolism are iron and copper, both probably related to the activity of cytochromes (intracellular enzyme systems).

Light intensity is directly related to the ability of the plant to reduce nitrate nitrogen to the ammonium form. Light reactions are responsible for the production of reduced coenzymes transferring electrons from the photolysis of water to reactions such as nitrate reduction. Nitrate accumulation in plants may occur when they are grown under greatly reduced light intensity due to cloud cover or natural shade.

Moisture stress also tends to affect nitrogen metabolism. Drought-stressed plants have frequently been noted to accumulate large amounts of nitrate nitrogen. This phenomenon may be due to relatively high concentrations of nitrate nitrogen in the soil solution due to desiccation of the soil.

Associated with moisture stress is the problem of high temperature. High temperatures are frequently related to poor nitrate utilization and high nitrate accumulation in the plants and may be associated with some heat-stress damage to enzyme systems.

Stage of maturity of plants also affects their ability to reduce nitrate nitrogen. The nitrate reductase enzyme system tends to become more active as plants mature and subsequently older plants have a greater ability to reduce nitrate nitrogen than very young plants. Plant genera and varietal differences even with the same genus also affect nitrate-nitrogen metabolism. Research with several species has indicated that the ability to metabolize nitrate nitrogen to ammonium is genetically linked. This has also been demonstrated in terms of nitrate accumulation in plants of different genetic backgrounds when grown under the same environmental conditions. In general, corn, sorghum, and oats are nitrate accumulators, whereas legumes and some perennial grasses are nonaccumulators.

The presence of endogenous antimetabolites in plants has been speculated as affecting nitrate reduction, although this cannot be well documented. Certain types of sorghum contain high concentrations of cyanogenic glycosides that upon hydrolysis yield cyanide (CN^-). Cyanide upsets the metabolism of enzymes in plants and animals that are dependent on the presence of heavy metals. It is unlikely that cyanide could influence just nitrate reductase rather than all other oxidative-reductive enzymes, but it is interesting to note that nitrate and cyanide frequently accumulate simultaneously in certain genera (*Sorghum*) under identical environmental conditions.

In addition to its role in the formation of proteins and subsequently its role in enzyme systems in cells, nitrogen also has a very important function in chlorophyll, as outlined earlier. The chlorophyll molecule is comprised essentially of a central magnesium atom around which are coordinated four pyrrole rings each of which contains one nitrogen and four carbon atoms. The role of chlorophyll in capturing light energy in plants has been understood for some time and this important role of nitrogen in chlorophyll can be related back to the process of nitrogen metabolism. Low amounts of chlorophyll result in a low photolysis of water and subsequently small amounts of reduced coenzymes are available for nitrate reduction, illustrated as follows:

$$H_2O + NADP \xrightarrow[\text{Chloroplast}]{\overset{\text{Photolysis of water}}{\text{light}}} \tfrac{1}{2} O_2 + NADPH + H^+$$

Oxidized coenzyme   Reduced coenzyme

Because nitrogen is a primary component of chlorophyll and because chlorophyll provides energy necessary for carbohydrate production, then nitrogen is directly related to carbohydrate production and utilization. In the latter role, carbohydrate utilization, nitrogen exerts its influence in its role in proteins or enzymes. Insufficient supplies of nitrogen result in carbohydrate deposition in vegetative cells that causes cell thickening. Adequate nitrogen supplies on the other hand tend to favor growth, proteins are formed from the manufactured carbohydrates, and the nutritional value of the plants for animal consumption is improved. Subsequently, carbohydrate deposition in the cell walls decreases and a more succulent plant results.

To carry succulence one step further, however, it should be remembered that excessive succulence can result in some crops from excessive nitrogen fertilizer applications. Too much nitrogen can lead to too little deposition of carbohydrate in the cell walls (cellulose) and subsequently the plants are weaker and less able to withstand severe environmental conditions such as severe wind and rain. As a result of excessive succulence, plants are not only more subject to breakage (lodging) but also are more subject to the entrance of disease organisms due to the thin cell walls and easier penetration of the cells. Potassium can help to offset the effects of excessive nitrogen (Chapter 4). Excessive nitrogen fertilization can also exert a detrimental effect on the quality of certain crops that are grown for their carbohydrate content, particularly sugar beets and sugarcane. Very strict control of nitrogen fertilization dramatically increases sugar yields from these crops.

2.4 NITRATE ACCUMULATION

Factors that affect nitrate reduction in plants are outlined in Section 2.3.1. The effects of these environmental conditions in producing poor nitrate reduction and subsequent accumulation of nitrate in the plant suggest problems for animals that might consume plants containing large amounts of nitrate nitrogen.

If the nitrogen supply to the plants is adequate and if excessive nitrogen supplies are responsible for nitrate accumulation in plants, the plant is not detrimentally affected by the presence of large amounts of nitrate nitrogen. However, animals and humans consuming these plant materials would be subjected to metabolic disruptions produced by the presence of this form of nonprotein nitrogen.

Physiologists, physicians, and veterinarians are generally agreed on one of the factors in nitrate toxicity to animals and humans. When large amounts of nitrate nitrogen are ingested by a ruminant animal such as a cow, the bacteria within the rumen reduce nitrate to nitrite in a manner strictly analogous to that discussed in Section 2.3.1. The nitrite produced is subsequently absorbed through the intestine walls and into the bloodstream. Some of these same types of bacteria also exist in the intestines of monogastric animals including humans.

In the bloodstream, nitrite, which is also a strong oxidizing agent, converts the ferrous iron (Fe^{2+}) in the center of the hemoglobin molecule to ferric iron (Fe^{3+}) greatly diminishing the ability of blood to transport oxygen to the sites in the animal's body requiring it for normal metabolic activity. As a result of this phenomenon, animals exhibit symptoms of oxygen starvation with bluish discoloration of mucous membranes and difficulty in breathing (*methemoglobinemia*). Severe cases of nitrate toxicity will result in a rapid decline in the animal's condition, possible abortion of pregnant animals, a drop in milk production, a decline in rate of gain, and eventually death due to a deficiency of oxygen in the blood. The effects of nitrate ingestion on an animal are strictly dependent on the condition of the animal and distinctly different effects may be produced in different animals.

Animals that are subjected to environmental stress such as cold weather, pregnancy, or poor rations generally will tend to be more subject to nitrate toxicity than would those on a high carbohydrate ration with protection from a harsh environment. Very young animals, including human infants, are particularly subject to nitrate toxicity.

Other possible metabolic effects of nitrate in animals include a decline in vitamin A stored in the liver and interference with normal metabolic activity of the thyroid gland. It is not within the scope of this discussion to cover all points of animal physiology, but it should be established that problems with nitrogen metabolism in plants can be readily transmitted to animals that consume the forage.

Some disagreement exists between animal physiologists and veterinarians in the interpretation of nitrate accumulations in forage plants and animal rations. Garner has provided an interpretive evaluation of nitrate nitrogen concentrations in animal rations and listed predicted animal responses. Those points are noted in Table 2–13. Monogastric (one-stomach) animals can also be subjected to the effects of high nitrate in their rations. Nitrate and nitrite concentrations in water are additive to that in rations and compound the toxicity problems.

Table 2-13. Level of Nitrate-Nitrogen in Ration and Expected Animal Response

NO_3-N in Ration (%)	Animal Response
0.00–0.07	Normal if on adequate rations.
0.07–0.14	Milk production drop. Slow at first, increasing after the sixth to eighth week. Typical vitamin A deficiency symptoms in the sixth to eighth week.
0.14–0.21	Milk production loss within 4–5 days. Reproduction difficulty over the period fed and may extend over several weeks after the removal of the feed.
0.21–up	Death; usually several head and suddenly.
Conversion factors for other methods of reporting by laboratories	% KNO_3 × .6132 = % NO_3^- % KNO_3/1.63 = % NO_3^- % KNO_3 × .1385 = % N % KNO_3/7.22 = % N

SOURCE: G. B. Garner, "Learn to Live with Nitrates," Missouri Agric. Exp. Sta. Bul. 708, 1958.

2.5 NITROGEN DEFICIENCY SYMPTOMS

Plants deficient in nitrogen are generally stunted and chlorotic (yellow) in appearance. This chlorosis usually appears first on the lower leaves, the upper leaves remaining green. This is because nitrogen in the lower, older leaves is translocated to the new growing areas of the plant when insufficient supplies are being provided through the roots. This characteristic classifies nitrogen as a "mobile" element.

As cases of nitrogen deficiency become more advanced, the leaves become more chlorotic and eventually die. In grasses such as corn, sorghum, sugarcane, and forage grasses the deficiency symptom may appear as a yellowing in the leaf tip that develops down the center midrib of the leaf. Eventually, with very severe deficiencies the entire leaf becomes chlorotic and eventually necrotic (dead). This chlorosis is usually termed *firing* and is depicted in Fig. 2–13.

Another general deficiency symptom aside from the chlorosis of the leaves is an overall stunting of the plants. Note in Figs. 2–13 and 2–14 that plants that have received insufficient nitrogen are quite chlorotic and much smaller than plants that have received adequate amounts of nitrogen.

Dicotyledonous (broadleafed) plants do not show quite the same effect that nitrogen deficiency produces on grasses. Cotton, for instance, has a pale

Fig. 2–13. Nitrogen deficiency in irrigated corn being produced on a soil (sand) with a low cation exchange capacity and low organic matter content (0.6%). Plants on the right received 224 kg/ha (200 lb/A) of nitrogen as anhydrous ammonia applied parallel to the row immediately prior to planting. (Courtesy L. S. Murphy.)

Fig. 2-14. Fall applications of nitrogen (preplant) for winter wheat on this sandy site were much less effective than comparable applications topdressed in early spring. Excessive soil moisture from heavy rains in the fall contributed to leaching of applied nitrogen. Fall applications of nitrogen in the Great Plains of the United States have been very effective, however, on medium- and fine-textured soils. (Courtesy L. S. Murphy.)

yellowish green color of its leaves, which are considerably reduced in size. Under extremely low levels of soil nitrogen, the symptom may be noted even in the cotyledons or seed leaves. Other broadleafed plants respond with a similar chlorosis; again it appears on the older leaves first.

Legumes do not show nitrogen deficiency unless root nodulation is poor. Along with the development of the deficiency symptoms just described, nitrogen concentrations in the plant tissue may also be reduced. Leaf content of nitrogen has been positively correlated with nitrogen supplies to the plant, and although it can be modified by other environmental conditions, it is a good indication of the adequacy of the nitrogen supply to the crop (Table 2–14). Tissue analysis can also be used as an indication of the protein level of the grain, for low levels of available nitrogen to the plant are also transformed into low grain protein levels of such crops as corn, wheat, sorghum, barley, and oats.

Table 2-14. Relationship of Yield, Leaf N Concentrations and Grain Protein in Winter Wheat Grown on Sandy Soil and Subjected to Excessive Rainfall—Kansas

N (kg/ha)	N Carrier	Time N Application	Yield (kg/ha)	Grain Protein (%)	Leaf N (%)
0	—	—	1915	9.4	2.30
33	SCU-30[a]	Fall	2144	9.6	2.27
67	SCU-30	Fall	2144	9.8	2.58
101	SCU-30	Fall	2688	11.0	2.63
33	Urea	Fall	2426	10.1	2.25
67	Urea	Fall	2083	9.9	2.31
101	Urea	Fall	2365	9.9	2.48
33	NH_4NO_3	Fall	2070	9.8	2.65
67	NH_4NO_3	Fall	2130	9.6	2.16
101	NH_4NO_3	Fall	2627	9.3	2.39
33	SCU-30	Spring	2587	9.8	2.50
67	SCU-30	Spring	2856	10.8	2.76
101	SCU-30	Spring	3273	12.3	3.32
33	Urea	Spring	3266	11.0	2.73
67	Urea	Spring	3454	13.1	3.23
101	Urea	Spring	3541	14.2	3.71
33	NH_4NO_3	Spring	2943	10.4	2.82
67	NH_4NO_3	Spring	3367	12.2	3.22
101	NH_4NO_3	Spring	3220	13.8	3.45

[a] Sulfur-coated urea, 30% of N dissolved in first 7 days.

SOURCE: P. J. Gallagher et al., "Evaluation of Sulfur-Coated Urea as a Nitrogen Source for Winter Wheat," Kansas Fertilizer Research Report of Progress-202, 1973, p. 62.

2.6 NITROGEN FERTILIZERS

2.6.1 Anhydrous Ammonia

The production of most nitrogen fertilizers in the world is dependent on the synthesis of anhydrous ammonia (NH_3). Commercial production of this compound is based on the discoveries of two German chemists, Fritz Haber and Carl Bosch, in 1908. Their studies of the laws of equilibrium chemistry and the subsequent development of a suitable catalyst for the combination of nitrogen and hydrogen under pressure led to development of the processes that are now used all over the world. The first small ammonia synthesis unit was placed in operation in 1911.

Synthesis of ammonia requires elemental nitrogen from the atmosphere and hydrogen from either fossil fuels or water. The synthesis of hydrogen for ammonia production is one of the costly steps in the manufacturing process and one of the reasons why nitrogen synthesis has been so dependent on the availability of energy. Originally hydrogen synthesis was based on the water-gas process, which involves steam and coke and produces a mixture of gases including a workable concentration of hydrogen. Later, this inefficient process was largely discarded for natural gas (methane), which can be converted into carbon dioxide and hydrogen through the use of special reactors, heat, and catalysts.

The basic reactions for production of hydrogen from natural gas have been termed *steam re-forming* of hydrocarbons. A generalized reaction for stream re-forming is presented below:

$$CH_4 + H_2O \rightleftharpoons CO + 3\,H_2$$
$$\underset{\text{Methane}}{} \qquad \underset{\substack{\text{Carbon} \\ \text{monoxide}}}{} \quad \underset{\text{Hydrogen}}{}$$

$$CH_4 + 2\,H_2O \rightleftharpoons CO_2 + 4\,H_2$$
$$\underset{\substack{\text{Carbon} \\ \text{dioxide}}}{}$$

$$2\,CH_4 + O_2 \rightleftharpoons 2\,CO + 4\,H_2$$

A second process follows steam re-forming and increases the efficiency of the process by producing more hydrogen from water. This process is known as *shift conversion* and results in purification and enrichment of the hydrogen-containing gas stream (synthesis gas):

$$CO + H_2O \rightleftharpoons CO_2 + H_2$$

That reaction also produces carbon dioxide, which is a feedstock for the synthesis of urea. The shift conversion reaction is essential to remove carbon monoxide (CO) from the gas stream to avoid poisoning the ammonia synthesis catalyst.

Increasing energy costs have made fertilizer manufacturers examine other processes for production of hydrogen including gasification of coal and a possible return to an improved water-gas method of hydrogen production. In any event, the costs of ammonia synthesis will continue to increase significantly due to the cost of hydrogen production. Other hydrogen sources may eventually be developed including hydrogen from electrolysis of water. This process will require large amounts of low-cost electricity. Hydrogen gas (H_2) could also be used as a fuel for many other domestic and industrial purposes.

Ammonia synthesis requires heat, pressure, and a special catalyst. The reaction known as the Haber-Bosch reaction is outlined here:

$$3\,H_2 + N_2 \underset{\substack{\text{200 atm pressure} \quad \text{heat}}}{\overset{\text{Fe}_3\text{O}_4 \text{ catalyst}}{\rightleftharpoons}} 2\,NH_3$$

The resulting ammonia gas is essentially devoid of water and contains 82% nitrogen. Other characteristics of anhydrous ammonia are presented in Table 2–15. Anhydrous ammonia production in the United States alone totaled 15,656,000 metric tons (17,222,000 short tons) in 1979.

In the United States, Australia, and certain parts of western Europe, anhydrous ammonia is utilized for direct soil application. Further discussion of this follows in Chapter 7. In most other countries of the world, anhydrous ammonia production is utilized entirely for production of other fertilizers and other industrial uses, but very little is consumed by direct application. Direct application of ammonia in the United States accounts for 40% of nitrogen fertilizer applications.

2.6.2 Ammonium Nitrate

Ammonium nitrate (NH_4NO_3) represented the first solid nitrogen product that was produced on a major industrial scale. After World War II, industrial plants that had been designed to produce ammonium nitrate for blasting agents or nitric acid for munitions were converted to ammonium nitrate production for fertilizer purposes. These older plants employed the ammonia technology discussed earlier but also involved a process for conversion of ammonia into nitric acid. The process of converting ammonia into nitric acid involves oxidation of the ammonia gas by oxygen in the presence of a platinum-rhodium catalyst (90% Pt–10% Rh) at pressures lower than that used for ammonia synthesis. Conditions within the reactors themselves are a function of the design. Still, the basic chemistry of nitric acid production, which is known as the Ostwald process, involves the following reactions:

$$4\,NH_3 + 5\,O_2 \xrightarrow{\text{Pt-Rh}} 4\,NO + 6\,H_2O$$
Ammonia   Oxygen   Nitric oxide

$$2\,NO + O_2 \longrightarrow 2\,NO_2$$
Nitrogen dioxide

$$3\,NO_2 + H_2O \longrightarrow 2\,HNO_3 + NO$$
Nitric acid

Descriptively, ammonia is oxidized to nitric oxide (NO), which is subsequently oxidized to nitrogen dioxide (NO_2). Nitrogen dioxide is scrubbed through an aqueous nitric acid solution, which increases the concentration of that solution by producing nitric acid in a reaction with water.

The nitric acid produced is concentrated and then reacted with ammonia in a simple acid-base reaction to produce an aqueous solution of ammonium nitrate:

$$NH_3 + HNO_3 + H_2O \longrightarrow NH_4NO_3 + H_2O$$
Ammonium nitrate

If the ammonium nitrate solution is maintained at a relatively low concentration, the product may be used for the formulation of nitrogen solutions. However, for the production of solid ammonium nitrate, the solution must be concentrated to a very high level by removal of water and then converted to a solid by use of a prilling tower or a rotary drum granulator. In the case of the prilling tower, the concentrated ammonium nitrate solution (95–96% NH_4NO_3)

Table 2-15. Physical Properties of Anhydrous Ammonia

Molecular Weight		17.03
Density of Liquid Ammonia		
−17.8°C (0°F)	0.664 g/ml	(41.34 lb/ft³)
10°C (50°F)	0.617 g/ml	(39.00 lb/ft³)
37.8°C (100°F)	0.584 g/ml	(36.40 lb/ft³)
65.5°C (150°F)	0.535 g/ml	(33.39 lb/ft³)
Boiling Point, °C at 1 atm		−33.4 (−28°F)
Freezing Point, °C at 1 atm		−77.7 (−107.9°F)
Critical Temperature, °C		132.4 (271.4°F)
Critical Pressure		111.5 atm
Vapor Pressure		
−17.8°C (0°F)	1.10 kg/cm²	(15.7 lb/in.²)
10°C (50°F)	5.25 kg/cm²	(74.5 lb/in.²)
37.8°C (100°F)	13.90 kg/cm²	(197.2 lb/in.²)
65.5°C (150°F)	20.65 kg/cm²	(293.1 lb/in.²)
Solubility in Water at 1 atm		
0°C (32°F) wt % NH₃		42.8
20°C (68°F) wt % NH₃		33.1
40°C (104°F) wt % NH₃		23.4
60°C (140°F) wt % NH₃		14.1

is sprayed from nozzles in an upward air stream and allowed to cool and solidify as it falls to the bottom of the shaft approximately 60 m (200 ft) tall.

The particles or prills are further dried, coated with diatomaceous earth, screened, and stored. Coating with diatomaceous earth is merely a mechanism of preventing water from entering into the particle. Absorption of water by the particles can lead to caking and severe handling problems. For that reason, coated ammonium nitrate stored in bulk is subjected to air conditioning in large warehouses to lower the humidity of the atmosphere. Bagged materials are less subject to moisture absorption from the atmosphere. Characteristics of the solid ammonium nitrate are presented in Table 2-16.

Ammonium nitrate, although having excellent agronomic capabilities as a nitrogen source for plants and providing equal amounts of nitrogen from ammonium and nitrate nitrogen, does have some specific characteristics that can make the material dangerous to handle. Contamination with organic materials

Table 2-16. Properties of Ammonium Nitrate

Color	White crystalline
Molecular weight	80.04
Density (25°C, 77°F)	1.725 g/cm³
Melting point	169.6°C (337°F)
Boiling point	210°C (410°F)
Solubility	118 g/100 ml at 0°C (32°F)
	871 g/100 ml at 100°C (212°F)
% N (pure)	35.0
% N (fertilizer grade)	33.5–34

such as carbon black, diesel fuel, or ground plant material provides compounds that are easily oxidized in the presence of ammonium nitrate. Ammonium nitrate is classified by governmental agencies as a hazardous product due to the fact that it is a strong oxidizing agent. The presence of carbon, high heat, high temperature, and pressures in the vicinity of 35 kg/cm^2 (500 lb/in.^{2}) can cause ammonium nitrate detonation. The Texas City, Texas, disaster of 1947 was produced by a fire and an explosion in the hold of a ship containing ammonium nitrate coated with Carbowax. In the early days of ammonium nitrate manufacturing immediately following World War II, this petroleum product (Carbowax) was used in lieu of diatomaceous earth to reduce moisture absorption by solid ammonium nitrate.

This characteristic of ammonium nitrate is utilized commercially as a blasting agent and several hundred thousand tons of ammonium nitrate are used annually in operations requiring movement of large amounts of earth and rocks. The strong oxidizing tendency of ammonium nitrate is reason enough to protect this material in storage from contamination with carbonaceous residue. Storage in heavy-walled, sealed compartments should also be avoided due to the possibility of high pressure in the case of high temperature and fire. In Europe, calcium carbonate (lime) is added to ammonium nitrate to improve handling safety. This lower analysis mixture (20.5% N) is called ammonium nitrate of lime (ANL).

For years, ammonium nitrate has reigned supreme among nitrogen carriers in the world for direct application. Production tonnage in the United States in 1979 represented 6,774,000 metric tons (7,452,000 short tons). Over the last several years, however, a decline in ammonium nitrate usage has been noted with the rapid rise in urea production. Ammonium nitrate production today is subjected to severe environmental restraints, both from the release of nitrogen oxides (NO and NO$_2$) to the atmosphere and also from the dust produced in the prilling and coating process. New ammonium nitrate plant construction in the United States is rare and future use of this material for direct application will probably decline. This should not be interpreted as an indication that ammonium nitrate will not be manufactured because it is a necessary component of nitrogen solutions.

2.6.3 Urea

Although calcium cyanamide (CaCN$_2$) was produced early in the 20th century and does contain carbon, it is no longer in common use. Today urea is the single organic nitrogen fertilizer compound produced synthetically. The existence of urea, CO(NH$_2$)$_2$, in nature has been recognized for over 200 years but commercial synthesis of this material was not achieved until the 20th century.

Synthesis of urea for both fertilizers and an animal feed additive involves the use of anhydrous ammonia and carbon dioxide generated in the production of anhydrous ammonia. In the synthesis of urea, ammonia and carbon dioxide are reacted under pressure to form ammonium carbamate. This material decomposes into ammonia and one molecule of urea. The ammonia is recycled to be combined with more carbon dioxide and the urea is drawn off as a liquid. The reactions for this process are indicated as follows:

$$2\,NH_3 + CO_2 \longrightarrow H_2N - \overset{\displaystyle O}{\overset{\|}{C}} - O - NH_4$$

Ammonia   Carbon dioxide   Ammonium carbamate

$$H_2N - \overset{\displaystyle O}{\overset{\|}{C}} - O - NH_4 \longrightarrow H_2N - \overset{\displaystyle O}{\overset{\|}{C}} - NH_2 + H_2O$$

Urea (liquid)

After liquid urea has been produced, it is concentrated to a specific level for use in the synthesis of urea-ammonium nitrate solutions or for the production of solid urea. The production of solid urea is a critical process, particularly the concentration of the urea-containing solution to a level high enough to produce solid material. This concentration process, if carried out at excessively high temperatures, results in the formation of a compound known as *biuret.*

Biuret, a condensation product of two molecules of urea,

$$2\,H_2N - \overset{\displaystyle O}{\overset{\|}{C}} - NH_2 \longrightarrow H_2N - \overset{\displaystyle O}{\overset{\|}{C}} - \overset{\displaystyle H}{\overset{|}{N}} - \overset{\displaystyle O}{\overset{\|}{C}} - NH_2 + NH_3$$

Urea   Biuret

is toxic to many plants and was present in relatively high concentrations in some of the first solid urea produced in the United States. Strict attention to temperatures and the use of vacuum processes in concentrating urea liquors has led to very low levels of biuret in today's materials. Practically speaking, biuret content of urea-containing solutions and solid urea products is low enough to provide no hazard for plants. Direct foliar applications of urea solutions, however, has led to specifications by that portion of the fertilizer industry for no more than 0.2% biuret by weight in the solution.

Production of solid urea following concentration of the urea liquor usually involves the use of rotary drum granulators similar to the ones involved in ammonium nitrate production. The concentrated urea liquor is sprayed on a rotating drum of fine urea particles and particle size increases as the solution condenses on the surfaces. The solid material produced is dried, screened, and stored. To reduce hygroscopicity some manufacturers use a coating agent such as diatomaceous earth.

Storage of urea does not involve as much problem with moisture absorption from the atmosphere as does storage of ammonium nitrate. Reduced storage costs are a contributing factor to lower manufacturing costs for solid urea.

Basically, urea has emerged as the future solid nitrogen fertilizer around the world. Elimination of the expensive process of nitric acid production, the environmental problems associated with nitric acid production, and the lower

Table 2-17. Properties of Urea

Color	White
Molecular weight	60.06
Density, crystalline	1.32 g/cm³
Density, fertilizer	0.67 g/cm³ (42 lb/ft³)
Solubility	78 g/100 ml at 5°C (41°F)
Melting point	132.7°C (271°F)
Boiling point	Decomposes at atmospheric pressure
% N (pure)	46.67
% N (fertilizer grade)	45–46

storage and transportation costs for urea all combine to produce a cheaper product per unit of nitrogen. Characteristics of solid urea are presented in Table 2-17. United States urea production in 1979 amounted to 5,467,000 metric tons (6,014,000 short tons) with about 80% of this being utilized as fertilizer. Agronomic considerations of urea use are discussed at length in Chapter 7. Refer to Section 2.2.4 for detailed soil reactions of urea.

2.6.4 Nitrogen Solutions

Nitrogen solutions have rapidly increased in popularity in the last 20 years (Table 2-18). Two major classes of nitrogen solutions exist. This classification is based upon the presence or absence of free ammonia or a positive vapor pressure of ammonia in these aqueous solutions. Solutions that contain free ammonia are classified as *pressure solutions,* whereas those without free ammonia are classified as *nonpressure.* Today, nonpressure solutions are utilized in largest quantities worldwide. These nonpressure solutions are comprised primarily of ammonium nitrate and urea but can contain other compounds such as ammonium sulfate and calcium nitrate. Nonpressure solutions are adapted to soil surface and subsurface applications. Pressure solutions, on the other hand, must be applied to the soil or irrigation water in a manner similar to anhydrous ammonia due to the positive vapor pressure of ammonia. Pressure solutions always include ammonia and may include ammonium nitrate, urea, ammonium sulfate, or calcium nitrate.

Generally speaking, pressure solutions have higher nitrogen concentrations than nonpressure solutions. However, the principal exception to this generalization in the United States is aqua ammonia, a pressure solution that contains approximately 20% nitrogen. On the other hand, nonpressure solutions in most common use today contain from 28 to 32% nitrogen and the higher nitrogen concentration of these solutions dictates their greater popularity. They are also more adaptable to different types of application, including application in irrigation water, combinations with herbicides, and direct surface applications, either alone or in combination with other nutrient-containing solutions.

Nitrogen solutions are described by a particular nomenclature. That

Table 2-18. Consumption of Selected Nitrogen Materials for Direct Application

Fiscal Year	Anhydrous Ammonia	Aqua Ammonia	Nitrogen Solutions	Ammonium Nitrate	Urea	Ammonium Sulfate
			Metric Tons of Nitrogen (N)			
1955	263,942	38,526	34,874	341,198	28,157	94,540
1960	529,022	77,618	177,036	378,050	58,723	97,235
1961	605,667	78,626	265,969	405,986	84,043	105,195
1962	697,659	90,712	337,617	425,930	120,731	106,079
1963	915,238	105,861	428,538	453,980	150,180	133,746
1964	1,043,700	143,210	478,868	504,836	167,570	135,332
1965	1,165,425	149,030	541,690	497,716	176,102	147,134
1966	1,460,793	182,558	647,677	555,186	192,377	150,724
1967	1,794,178	161,068	713,083	645,911	207,229	158,940
1968	2,233,873	148,224	730,083	714,860	221,235	152,096
1969	2,342,210	130,650	763,568	785,265	240,686	142,765
1970	2,585,507	130,707	881,711	866,237	220,689	146,283
1971	2,958,327	137,564	942,270	880,991	248,733	168,532
1972	2,711,158	136,513	917,380	923,225	324,896	174,731
1973	2,692,392	120,829	1,001,705	999,099	397,289	181,037
1974	3,116,065	134,610	1,076,824	965,220	425,394	175,477
1975	2,995,267	130,862	1,083,588	856,255	476,528	152,032
1976	3,659,325	132,890	1,473,007	895,959	662,954	199,782
1977	3,673,025	120,185	1,531,843	851,693	777,585	200,205
1978	3,382,990	105,000	1,447,213	745,778	800,684	171,098
1979	3,639,859	89,603	1,567,145	764,793	879,127	143,496

SOURCE: TVA, Fertilizer Trends–1979. Bul. Y-150, January, 1980.

nomenclature involves an expression of percent nitrogen and percentage by weight ammonia, ammonium nitrate, and urea. A typical nomenclature for a nonpressure nitrogen solution would be as follows: 280 (0–39–31). The first figure represents the percent nitrogen with no decimal. The decimal actually is between the second and third digits and is 28% nitrogen in this case. The first digit(s) inside the parentheses refers to the percent by weight of ammonia (0), the second set of digits to the percent by weight of ammonium nitrate (39), and the third set to the percent of urea (31). From the sum of the percentages of these materials, the amount of water in the material may be determined by difference. In this example, water comprises 30% of the solution. If an additional component is utilized, it would appear as a fourth set of digits inside the parentheses and would be explained in the labeling of the solution. Some examples of nomenclature for both pressure and nonpressure nitrogen solutions are given in Tables 2–19 and 2–20.

Another characteristic of nitrogen solutions is the salting-out temperature. This represents the temperature at which crystals begin to form in the solution due to the decline in solubility of the components with declining temperature. Note in Tables 2–19 and 2–20 that the salting-out temperature increases as the concentration of nitrogen in the solution increases, particu-

Table 2–19. Composition Nomenclature and Properties of Some Pressure Nitrogen Solutions

| Solution Nomenclature[a] | Forms of Nitrogen | | | | Vapor Pressure (Gage) at | | | Salting-Out Temperature | | Specific Gravity |
| | Free NH_3 (% N) | Ammonium (% N) | Nitrate (% N) | Urea (% N) | 32°C | 40°C | 49°C | (°C) | (°F) | |
						(kg/cm²)				
201 (24-0-0)	20.1	—	—	—	0	0	0.5	− 52	− 62	0.913
247 (30-0-0)	24.7	—	—	—	0.3	0.8	1.4	− 81	− 112	0.896
370 (17-67-0)	13.7	11.7	11.7	—	0	0.1	0.5	10	50	1.184
410 (19-58-11)	15.6	10.2	10.2	5.1	0.3	0.7	1.3	− 14	7	1.162
410 (22-65-0)	18.3	11.4	11.4	—	0.4	0.7	1.3	− 6	21	1.138

[a] Digits outside parentheses refer to % N, 20.1 % N in the first solution. First set of digits in parentheses refers to % ammonia (NH_3) by weight, second set to % ammonium nitrate (NH_4NO_3), third set to % urea.

SOURCE: Compiled from fertilizer industry data.

larly for the nonpressure solutions. A good example of the effects of concentration of salts in solution on the salt-out temperature can be obtained by examining the data for 28% nitrogen and 32% nitrogen urea-ammonium nitrate (UAN) solutions (Table 2–20). Salt-out temperature for 28% N UAN (UAN = as Urea Ammonium Nitrate) is −18°C (−1°F), whereas that for 32% N UAN is 0°C (32°F). When salting out occurs, no permanent damage results but the nitrogen concentration of the remaining solution declines significantly. As the solution is warmed and agitated, the salts formed will redissolve.

The production of nitrogen solutions involves no special chemistry, for the three major components utilized have already been discussed. A peculiar chemical quirk does exist in the production of these solutions, however, particularly nonpressure solutions. Solutions of either ammonium nitrate alone or urea alone fail to achieve the nitrogen concentration that results from a mixture of the two materials (see Table 2–20). Solubility of both urea and ammonium nitrate is enhanced by the presence of the other compound. This fact is painfully evident to persons who try to handle both solid ammonium nitrate and solid urea with the same equipment in a humid climate. Even slight contamination of one product with the other results in a hygroscopic mass that cakes easily and renders effective application impossible.

The solubility of ammonium nitrate is 118.3 g/100 ml of water at 0°C (32°F). The solubility of urea, on the other hand, is 78 g/100 ml of water at 5°C (41°F). In the presence of each compound in solution 280 (0–39–31) the solubilities of ammonium nitrate and urea have been changed to 130 g/100 ml and 103 g/100 ml at 0°C (32°F), respectively.

Applications of nitrogen solutions are discussed in detail in Chapter 7. Special purposes, such as foliar fertilization, require the use of straight urea solutions. This is because ammonium ions provided by ammonium nitrate can cause severe leaf burn. Leaf burn can also occur with foliar applications of urea solutions if hydrolysis of urea occurs at or near the leaf surface.

Table 2-20. Composition and Nomenclature of Some Nonpressure Nitrogen Solutions

Solution Nomenclature[a]	Combined Ammonia (% N)	Nitrate (% N)	Urea (% N)	Salting-Out Temperature		Specific Gravity 10.5°C (60°F)
				(°C)	(°F)	
160 (0–46–0)	8.0	8.0	—	−12	11	1.207
170 (0–49–0)	8.5	8.5	—	− 9	16	1.222
180 (0–51–0)	9.0	9.0	—	− 5	23	1.235
190 (0–54–0)	9.5	9.5	—	1	33	1.253
200 (0–57–0)	10.0	10.0	—	5	41	1.264
210 (0–60–0)	10.5	10.5	—	9	49	1.286
228 (0–65–0)	11.4	11.4	—	18	65	1.307
280 (0–39–31)	6.9	6.9	14.3	−18	− 1	1.279
280 (0–40–30)	7.0	7.0	14.0	−18	− 1	1.283
300 (0–42–33)	7.4	7.4	15.3	− 9	15	1.301
320 (0–44–35)	7.8	7.8	16.5	0	32	1.327
320 (0–45–35)	7.9	7.9	16.3	0	32	1.325

[a] Digits outside parentheses refer to % N, 16.0% N in the first solution. First set of digits in parentheses refers to % NH_3 by weight, second set to % ammonium nitrate, third set to % urea.

SOURCE: Compiled from fertilizer industry data.

2.6.5 Ammonium Sulfate

Ammonium sulfate [$(NH_4)_2SO_4$] has a long history as a nitrogen source. This material has been manufactured as a by-product from a multitude of industrial processes. At one time, ammonium sulfate represented the most common source of nitrogen in many sections of the world and still remains an important nitrogen source, particularly in areas producing rice. A major disadvantage to its use is its relatively low nitrogen content (21%) and the strong acid residue that it imparts to the soil.

Ammonium sulfate is a frequent by-product of the steel industry, particularly the coking of coal when some ammonia is present in the coke-oven gas. Spent or used sulfuric acid is another by-product material that can be marketed by conversion into ammonium sulfate by reacting with anhydrous ammonia, aqua ammonia, or waste-gas streams that contain some free ammonia. Waste ammonia is frequently recovered in this manner from the production of synthetic fibers. On the other hand, clean sulfuric acid and unused ammonia can be used in producing ammonium sulfate. Production chemistry is very simple:

$$2\,NH_3 + H_2SO_4 \longrightarrow (NH_4)_2SO_4$$

Whatever the source, ammonium sulfate's physical and chemical characteristics remain the same. These characteristics are set out in Table 2-21. The color of solid ammonium sulfate (unrelated to quality) may vary con-

Table 2-21. Properties of Ammonium Sulfate

Color	Mostly white crystalline, but varies with presence of contaminants
Molecular weight	132.14
Density	1.769 g/cm^3
Melting point	Decomposes at 235°C (455°F)
Solubility	70.6 g/100 ml at 0°C (32°F)
	103.8 g/100 ml at 100°C (212°F)
% N (pure)	21.2
% S (pure)	24.2
% N (fertilizer grade)	21
% S (fertilizer grade)	24

siderably depending on its source. Ammonium sulfate produced from petroleum by-products frequently has a dark color due to the presence of hydrocarbons or coal tars. United States consumption of ammonium sulfate is indicated in Table 2-18.

Residual effects of ammonium sulfate on the soil have long been known to be more acidic than residual effects of other nitrogen sources. Guidelines for the acidity produced by ammoniacal nitrogen materials are outlined in Table 2-22. Application of a kilogram of nitrogen as ammonium sulfate requires approximately 5 kg of effective calcium carbonate to offset the residual acidity both from the nitrification of the ammonium nitrogen and the hydrolysis of ammonium sulfate. Compare these values with those from urea, anhydrous ammonia, and ammonium nitrate. In the case of these materials, 2 kg of effective calcium carbonate must be applied for each kilogram of nitrogen supplied in the ammoniacal form.

Continued use of ammonium sulfate on acid soils for extended periods of time without adequate liming can result in drastically reduced plant growth. The high acidity produced results in the destruction of clay minerals and subsequently the release of large amounts of soluble aluminum that is toxic to plants. In addition, phosphorus fixation may be enhanced due to the presence of large amounts of soluble iron and aluminum that react with phosphorus (see Chapter 3). Some pictorial evidence of the effects of long-term applications of ammonium sulfate without adequate liming is given in Fig. 2-15.

An advantage to the use of ammonium sulfate on sulfur-deficient soils is that this material contains 24% sulfur. Sulfur in ammonium sulfate is highly water soluble and readily available (see Chapter 5).

2.6.6 Sodium Nitrate and Calcium Nitrate

Sodium nitrate ($NaNO_3$, 16.5% N) was the first commercial source of nitrogen utilized in the United States. This material, imported from Chile, was the nitrogen source for early cotton production in southeastern United States.

Table 2–22. Equivalent Acidity or Alkalinity of Nitrogen Fertilizers

Materials	% N	Pure CaCO₃ Required to Neutralize Residual Acidity (kg)		
		Per kg NH_4^+-N	Per kg N	Per 100 kg Material
Ammonium sulfate	21	5.4	5.4	110
Anhydrous ammonia	82	1.8	1.8	148
UAN solution				
280 (0–39–31)	28	1.8	1.4	38
Pressure solution				
410 (19–58–11)	41	1.8	1.4	56
Ammonium nitrate	34	1.8	0.9	31
Urea	46	1.8	1.8	83

	% N	Pure CaCO₃ Equivalent Added to Soil (kg)	
		Per kg N	Per 100 kg Material
Calcium nitrate	15.5	1.4	22
Sodium nitrate	16.0	1.8	29

SOURCE: Averages of theoretical and published analyses.

Sodium nitrate is produced by processing a natural ore mined from high, dry Andean valleys. The source of the nitrate in these deposits is of questionable origin but one of the prominent theories relates to its derivation from fumaroles associated with volcanic activity. The nitrogen produced was eventually nitrified to nitrate and deposited beneath detritus washed down from

(a)

(b)

Fig. 2–15. Heavy continuous applications of ammonium nitrogen (a) produced severe depressions in soil pH. Corn growth was impossible due to large amounts of exchangeable aluminum originating from decomposition of clay minerals. Applications of calcium nitrate (b) produced a net increase in soil pH and produced excellent plant growth. Corn plants in the background in the picture on the left are growing on an area which had been limed according to soil test. (Courtesy Dr. M. L. Vitosh, Michigan State University.)

surrounding slopes. The dry climate of the region has protected the highly soluble nitrate salts. This deposit is unique in the world in terms of its extent and accessibility.

Today, sodium nitrate is produced from the Chilean deposits and synthetically from reactions of nitric acid and sodium carbonate. A major disadvantage to its use lies in the low nitrogen content, the large amount of freight associated with movement of the natural material from Chile, and the sodium carried with the nitrate. It assumes a very minor role in terms of nitrogen fertilizer materials used in the United States today.

Calcium nitrate [$Ca(NO_3)_2$, 15,5% N] was one of the first commercial materials manufactured in the 20th century from synthetically fixed nitrogen. This process, developed in Norway, was dependent on the production of nitric acid from electric arc fixation of elemental nitrogen. In the arc process, elemental nitrogen is united with oxygen to produce nitric oxide. Nitric oxide is subsequently oxidized to nitrogen dioxide in the presence of oxygen. Nitrogen dioxide is absorbed in water to produce nitric acid. Reactions for this process follow:

$$N_2 + O_2 \xrightarrow{\text{Arc}} 2\,NO$$

Elemental nitrogen — Nitric oxide

$$NO + \tfrac{1}{2}\,O_2 \longrightarrow NO_2$$

Nitrogen dioxide

$$3\,NO_2 + H_2O \longrightarrow 2\,HNO_3 + NO$$

Nitric acid

$$CaCO_3 + 2\,HNO_3 \longrightarrow Ca(NO_3)_2 + CO_2 + H_2O$$

Calcium carbonate — Calcium nitrate

The nitric acid produced by this process is reacted with calcium carbonate to produce calcium nitrate. Note the similarity to fixation of nitrogen by lightning in Section 2.2.3.

Today, calcium nitrate is still produced in Norway but is derived from nitric acid produced from anhydrous ammonia. The nature of the end product is identical to that which was originally produced but improvements in coating technology have reduced the high hygroscopicity of this material. The disadvantage of the use of calcium nitrate today lies in the high cost per unit of nitrogen that is largely associated with freight required to move the material from Scandinavia to the United States. Agronomically, this material is an excellent source of nitrogen and in addition to providing soluble nitrate also provides a readily available source of calcium to plants.

A characteristic of both sodium and calcium nitrate is the residual effect that these materials exert on the soil. Instead of producing acid, these two materials impart an alkaline residue to the soil and continued use will tend to raise soil pH. Calcium from calcium nitrate also tends to produce aggregation of soil colloids (desirable), whereas sodium from sodium nitrate can eventually cause colloidal dispersion (undesirable). Dramatic differences have been noted between the long-term use of calcium nitrate compared to the effects of continued use of such materials as ammonium sulfate without liming. Comparisons of the pictures in Fig. 2–15 indicate how plant growth varies where liming is not practiced when acidic, ammoniacal nitrogen sources are used for extended periods.

2.6.7 Slow-Release Nitrogen Fertilizers

Less than optimum rates of recovery by plants of applied nitrogen have been recognized for years. Results of field and lysimeter studies have shown 70 to 50% or less of fertilizer nitrogen applied is recovered by crop plants. Reasons for the poor nitrogen recovery relate to problems with ammonia volatilization, leaching, and denitrification of nitrate. Rapid dissolution of the nitrogen products supplied allows the possibility of rapid uptake by plants. Large amounts of available nitrogen at one period in time may lead to luxury nitrogen consumption by plants, which represents absorption without increased growth.

The problem of slowing nitrogen release from various compounds has been approached by altering the solubility of materials or to develop compounds that have low water solubility. Methods of altering the release of nitrogen from soluble materials has been to coat water-soluble compounds with materials that are less water soluble thus retarding entry of water into the particle and the movement of nitrogen out. Coatings applied to soluble nitrogen materials generally have been of three types:

1. Impermeable coatings with small pores that allow slow entrance of water and slow passage of solubilized nitrogen compounds out of the encapsulated area.
2. Impermeable coatings that require breakage by physical or chemical or biological action before the nutrient is dissolved.
3. Semipermeable coatings through which water diffuses and creates internal pressures sufficient to disrupt the coating.

Sulfur-coated urea (SCU) has been developed in recent years as a product with the characteristics of slow nitrogen release. This material is basically urea with a coating of elemental sulfur including a binding agent, a sealant, and a microbicide (type 2). Elemental sulfur was chosen as a coating agent because of its relatively low cost and ease of handling. Nitrogen release rates can be varied by controlling the thickness of the sulfur coating. The nitrogen content of SCU ranges from about 10 to 37% depending on the thickness of the sulfur coating. Nitrogen release is calculated as the amount of the nitrogen that will solubilize in 7 days. Characteristics of SCU that have been examined by the Tennessee Valley Authority include material ranging from

10% nitrogen solubilized in the first 7 days up to 40% nitrogen solubilized in the first 7 days. These forms of SCU are noted as SCU-10 and SCU-40, respectively.

Agronomic evaluations of SCU have indicated some possible improvement in nitrogen efficiency, particularly with crops with very large demands for nitrogen throughout an extended growing season. Rice, bermudagrass, and sugarcane have been crops where some positive responses to SCU versus conventional urea have occurred.

Uncoated organic compounds of low water solubility have been studied as sources of nutrients for over 50 years. Intensive studies during the last 25 to 30 years have resulted in the production of certain urea-derived compounds that have a characteristic of slow nitrogen solubility in water. Typical of these types of compounds are those produced by the reaction of urea and formaldehyde to produce condensation products known generally as urea-formaldehyde or urea-form.

Urea reacts with the formaldehyde in the presence of a catalyst to form a mixture of compounds that are classified under the general name of urea-form. These materials are white, orderless compounds that contain varying amounts of nitrogen but averaging around 38%. A basic component of urea-forms is methylene-urea polymers varying in chain length and in degree of cross-linking between chains. Basically, as the compound links are extended, the water solubility of these materials declines.

Although the urea-forms have been examined agronomically in the United States, they have not received wide attention except as nitrogen sources for very high-value crops, turf, and ornamentals. More attention has been given to these compounds in Europe and in Japan.

Isobutylidenediurea (IBDU) is a reaction product of urea and isobutyraldehyde. The reaction has been utilized in Germany and Japan to produce a compound containing approximately 32% nitrogen and having a low water solubility (0.1–0.01%). This compound also has the ability to be granulated with soluble nutrient sources of phosphorus and potassium to supply mixed fertilizer formulations with slow nitrogen release.

Crotonylidene diurea (CDU) is another urea-based material that is produced in Germany, Japan, and France from crotonaldehyde and urea. The nitrogen content of this material is about 32%.

Other slow-release nitrogen compounds have been studied or produced in small quantities in some countries. Basically, the characteristics of these materials involve high cost per unit of nitrogen. Use should be considered as being relegated to very high-value crops, turf, and ornamentals. Extensive agronomic use has not been demonstrated to date as being economically feasible in the United States.

2.6.8 Natural Organic Nitrogen Fertilizers

In addition to the manufactured nitrogen sources, nitrogen is available from harvested organic materials, wastes, or crops grown for symbiotic nitrogen fixation. This latter aspect of nitrogen supply to crops was discussed in Section 2.2.1.

Natural nitrogen-containing compounds available for use include bird guano, poultry and animal manures of all types, and municipal sewage sludges. Bird guano (excreta) has one of the longest histories of direct use and contains not only nitrogen but appreciable quantities of phosphorus and other plant nutrients. Supplies of guano are relatively scarce and are not used extensively today.

Use of animal wastes as a source of nutrients for crops has long been recommended because of the efficiency involved in the operation and the fact that the materials being added to the soils impart not only nitrogen but carbon and many other plant nutrients. Analyses of animal wastes vary widely depending on the composition of the rations, the types of animals involved, and the storage techniques utilized in stockpiling manure prior to its application. Some indication of the range in nitrogen concentrations of beef feedlot wastes can be obtained from Table 2-23.

Studies of the availability of nitrogen from animal waste applications have indicated that not all the nitrogen obtained in the wastes is available the first year. From nitrogen contents of around 1% in feedlot wastes, calculations from California, Kansas, and Wisconsin have indicated that about 13 kg of nitrogen is available the first year per metric ton of applied animal waste on a dry matter basis. Again, this is a function of the climate, the original composition of the manure, and storage. Data from Texas have indicated generally higher concentrations of nitrogen in beef feedlot wastes if the material is collected and applied after short storage periods. Ranges of composition of

Table 2-23. Variations in Analyses of Manure from Beef Feedlots

Feedlot	Nutrient Content (% on dry weight basis)					
	N	P	K	Na	Ca	Mg
1	2.05	0.83	2.28	1.13	1.98	0.78
2	3.50	1.00	2.33	—	0.60	0.50
3	1.00	0.40	1.10	0.20	0.80	0.40
4	0.88	0.58	0.98	0.25	0.98	0.43

Average, High, and Low Composition of 40 Manure Samples Taken from a Beef Feedlot Stockpile

	Nutrient Content (% on dry weight basis)					
	N	P	K	Na	Ca	Mg
Average	1.04	0.41	1.09	0.23	0.78	0.40
High	1.59	0.54	1.40	0.32	1.03	0.50
Low	0.64	0.31	0.85	0.15	0.62	0.30

SOURCE: W. L. Powers et al., "Guidelines for Applying Beef Feedlot Manure to Fields," Kansas Coop. Ext. Circular 502, 1973.

several types of animal wastes have been compiled and are presented in Table 2-24. Remember that in attempting to use this table these represent only ranges and the midpoint is not necessarily an average of the composition of these wastes. Each major source of manure should be analyzed before it is used.

Municipal wastes also contain nitrogen and have been utilized extensively all over the world as nitrogen supplies for crops. Highly processed and leached sewage sludge may contain very low amounts of nitrogen, much less than 1%. Nitrogen has been volatilized to the atmosphere or has been leached out into the wastewaters. This is a fate of much of the nitrogen in animal and human wastes, for nitrogen is largely voided in the form of urea.

Seaweed has been utilized as a source of nitrogen under some conditions where it has been harvested for extraction of certain chemicals. Again, composition varies depending on the source, and this practice is less widespread than the use of animal manures for nitrogen. Composition of several naturally occurring materials is given in Table 2-25 (see also Chapter 10).

Table 2-24. Composition of Various Manures and Waste Materials

	Dry Matter (%)	N (%)	K (%)	P (ppm)	Ca (ppm)	Mg (ppm)	S (ppm)	Fe (ppm)	Zn (ppm)	Cu (ppm)	Mn (ppm)	B (ppm)
				All Values Are Reported on a Wet or "As-Received" Basis								
Beef feedlot manure[a]												
High values	38.5	1.08	0.95	3,680	7,027	3,405	—	6,970	69	—	—	—
Low values	24.2	0.44	0.58	2,100	4,231	2,032	—	4,876	36	—	—	—
Mean values	32.4	0.71	0.74	2,830	5,275	2,675	—	6,001	45	—	—	—
Liquid poultry manure[b]	8.1	0.16	0.24	180	624	157	—	0.5	1.6	0.2	1.3	—
Liquid swine manure[a]	0.6	0.01	0.06	20	24	9	—	0.1	2.1	0.1	0.1	—
Liquid swine manure[a]	2.6	0.09	0.07	239	58	178	—	1.2	32.0	1.4	3.7	—
Liquid dairy manure[a]	8.6	0.24	0.19	210	406	248	—	7.8	2.9	0.3	2.1	—
Dairy cattle manure[b]	21	0.56	0.50	1,000	2,800	1,100	500	400	15	5	10	15
Sheep manure[b]	35	1.40	1.00	2,100	5,850	1,850	900	160	25	5	10	10
Swine manure[b]	25	0.50	0.38	1,400	5,700	800	1,350	280	60	5	20	40
Poultry manure[b]												
(w/o litter)	46	1.56	0.35	4,000	37,000	2,900	3,100	465	90	15	90	60
Poultry manure[c]												
High values	28	1.50	0.74	7,100	30,100	2,900	—	2,500	1,600	1.1	80	—
Low values	26	1.00	0.70	6,800	27,900	2,600	—	2,200	1,300	0.9	80	—
Activated sewage sludge[d]	100	5.56	0.36	25,860	13,000	5,600	9,800	56,000	2,500	916	134	33
Lagoon sewage sludge[d]	100	1.71	0.23	18,750	111,470	996	15,000	32,000	2,750	775	440	7

[a] L. S. Murphy, W. L. Powers, and H. L. Manges, unpublished data, Kansas State University, 1971.

[b] E. J. Benne, C. R. Hoglund, E. D. Longnecker, R. L. Cook, Michigan Agric. Exp. Sta. Circ. Bul. 231, 1961.

[c] L. S. Robertson, Department of Crop and Soil Sciences, Michigan State University. Manure utilized in fertilization study with corn (Personal communication).

[d] B. D. Knezek, Department of Crop and Soil Sciences, Michigan State University (Personal communication).

Table 2-25. Primary Nutrients Contained in Organic Fertilizers
(Average Analysis of Fertilizers without Losses from
Leaching or Decomposition)

	Percentage Nitrogen (N)	Percentage Phosphorus (P)	Percentage Potassium (K)
	Dry Weight Basis		
Bulky organic materials			
Alfalfa hay	2.5	0.2	1.7
Alfalfa straw	1.5	0.1	1.3
Bean straw	1.2	0.1	1.0
Cattle manure	2.0	1.0	2.0
Cotton bolls	1.0	0.1	3.3
Grain straw	0.6	0.1	0.9
Hog manure	2.0	0.6	1.5
Horse manure	1.7	0.3	1.5
Olive pomace	1.2	0.4	0.4
Peanut hulls	1.5	0.1	0.7
Peat and muck	2.3	0.2	0.6
Poultry manure	4.3	1.6	1.6
Sawdust and wood shavings	0.2	0.0	0.2
Seaweeds (kelp)	0.6	0.0	1.1
Sheep manure	4.0	0.6	2.9
Timothy hay	1.0	0.1	1.3
Winery pomace	1.6	0.7	0.7
Organic concentrates			
Animal tankage	9.0	4.4	1.3
Bat guano	10.0	2.0	1.7
Bone meal	4.0	10.1	0.0
Castor pomace	6.0	0.8	0.4
Cocoa shell meal	2.5	0.7	2.1
Cottonseed meal	6.0	1.1	2.3
Dried blood	13.0	0.7	0.7
Fish meal	10.0	2.6	0.0
Fish scrap	5.0	1.3	0.0
Garbage tankage	2.5	0.7	2.3
Hoof and horn meal	12.0	0.9	0.0
Sewage sludge	5.0	2.2	0.4
Soybean meal	7.0	0.5	2.3
Steamed bone meal	0.8	13.0	0.0
Tobacco dust and stems	1.5	0.2	4.2
Wood ashes	0.0	0.9	5.0
Wool wastes	7.5	0.0	0.0

SOURCE: Let's Take a Closer Look at Organic Gardening, Ext. Bul. 555, The Ohio
State University, 1973.

2.7 CROP RESPONSES TO NITROGEN

Crop responses to nitrogen are well-known around the world. Responses have been expressed both in terms of higher yields and improved crop quality. Discussion of recommended nitrogen rates is beyond the scope of this publication but examples of nitrogen response effects are discussed.

Some of the most significant advances in nitrogen fertilization of crops have occurred in the last decade, particularly with new rice cultivars developed at the International Rice Research Institute (IRRI). New cultivars, resistant to lodging, were developed that were better able to respond to higher rates of nitrogen fertilization. These new semidwarf, erect, high-tillering cultivars generally yield more than the traditional tall tropical rices even when no fertilizer is added. The yield advantage of the new cultivars increased with increasing rates of applied fertilizer nitrogen (Fig. 2–16) but was evident even with low rates of nitrogen application.[7]

Nitrogen responses in upland (rain-fed) rice have also been significant but have received less popular press attention than flooded rice. DeDatta[8] and co-workers at IRRI have reported significant nitrogen responses to several nitrogen sources.

Wheat yield and grain protein responses to applied nitrogen are documented in Table 2–14. The quality aspect of grains is obviously important in nutrition of man and animals. Baking characteristics of wheat (loaf volume, mixing time, and dough strength) are also directly related to protein content of the grain and subsequently to nitrogen supplied to the plant.[9] Hard red winter wheats and hard red spring wheats in North America, Australia, and Europe are in demand for their high protein contents.

Robinson et al.[10] noted the relationship of low nitrogen supply to low grain protein (yellow berry) in durum wheats (Fig. 2–17). Low grain protein in durum wheat, like hard red winter and hard red spring wheats, results in price penalties in marketing of the grain. Late season nitrogen applications tend to increase grain protein more than preplant applications but the former have less effect on yield.

Nitrogen yield responses in corn and sorghums have frequently been dramatic (Fig. 2–18). These grains also exhibit protein level increases from added nitrogen just as do small grains. Summarized information from Ohio shows substantial increases in grain nitrogen levels (crude protein), nitrogen

[7] S. K. De Datta, F. A. Saladaga, W. N. Obcemea, and T. Yoshida, "Increasing Efficiency of Fertilizer Nitrogen in Flooded Tropical Rice," Proc., FAI-FAO Seminar on Optimizing Agricultural Production under Limited Availability of Fertilizers, New Delhi, 1974, pp. 265–288.

[8] A. Thasanasongchan, B. B. Mabbayad, and S. K. De Datta, "Methods, Sources and Time of Nitrogen Application for Upland Rice," Philippine J. Crop Sci., 1:179, 1975.

[9] F. C. Stickler, A. W. Pauli, and J. A. Johnson, "Relationship Between Grain Protein Percentage and Sedimentation Value for Four Wheat Varieties at Different Levels of Nitrogen Fertilization," Agron. J., 56:592, 1964.

[10] F. E. Robinson, D. Cudney, and W. F. Lehman, "Yellow Berry of Wheat Linked to Protein Content," Calif. Agric., March, 1977, pp. 16–17.

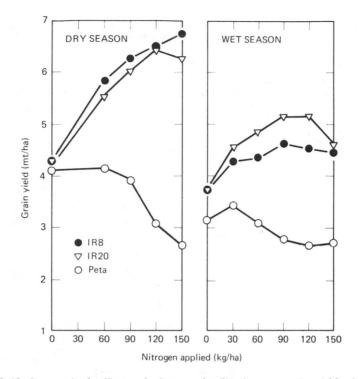

Fig. 2-16. Summarized effects of nitrogen fertilization on grain yield of three cultivars of rice in the Philippines. Note the yield reduction with higher rates of nitrogen when the tall Peta rice was used. IR 8 and IR 20 are short-stemmed rice cultivars. (Source: S. K. De Datta, F. A. Saladaga, W. N. Obcemea, and T. Yoshida, "Increasing Efficiency of Fertilizer Nitrogen in Flooded Tropical Rice," Proc. FAI-FAO Seminar on Optimizing Agricultural Production under Limited Availability of Fertilizers, New Delhi, 1974, pp. 265-88.)

concentration in the fodder and ear leaf nitrogen when increasing rates of nitrogen were added to the soil (Table 2-26). Sorghum yield responses from nitrogen have also been accompanied by higher grain protein percentage.

Yield and water utilization efficiency of many crops including corn,[11] sorghum,[12] and wheat[13] can be increased significantly through application of adequate amounts of nitrogen. Fertilized crops extract more water from the soil profile than do unfertilized crops because of better root growth.

Nitrogen is frequently deficient in growth of trees used for wood products. Biomass production of young loblolly pines was increased 25% by nitrogen

[11] Nathu Singh, R. Singh, H. N. Verma, and Y. Singh, "Yield and Profile Water Depletion Pattern of Rainfed Vijay Maize as Affected by Nitrogen Fertilization," *Indian J. Agron.*, 21:249, 1976.

[12] C. A. Thompson, "Fertilizing Dryland Grain Sorghum," Kan. Agric. Exp. Sta. Bul. 579, 1974, 20 pp.

[13] Ranjodh Singh, A. S. Gill, and H. N. Verma, "Water Use and Yield of Dryland Wheat as Affected by N and P Fertilization in Loamy Sand," *Indian J. Agron.*, 21:254, 1976.

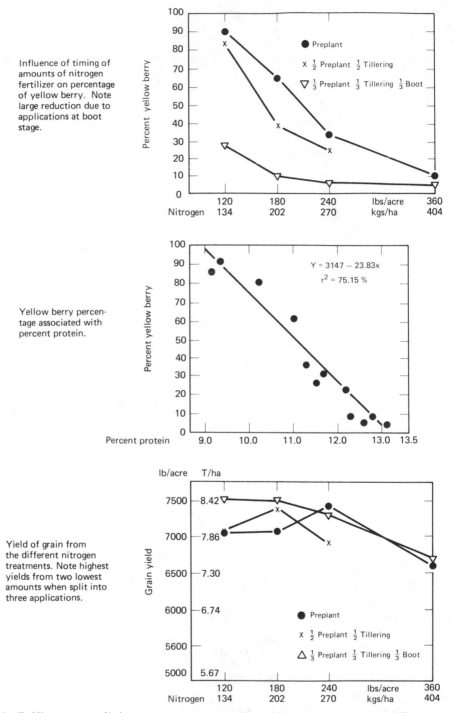

Influence of timing of amounts of nitrogen fertilizer on percentage of yellow berry. Note large reduction due to applications at boot stage.

Yellow berry percentage associated with percent protein.

Yield of grain from the different nitrogen treatments. Note highest yields from two lowest amounts when split into three applications.

Fig. 2-17. Nitrogen supplied to wheat affects not only yield but grain quality as well. Note the relationship of *yellow berry* to rate of nitrogen fertilizer and to grain protein. (Source: F. E. Robinson, D. Cudney, and W. F. Lehman, "Yellow Berry of Wheat Linked to Protein Content," *California Agriculture,* March, 1977, pp. 16–17. Used with permission of the publisher and courtesy of University of California Division of Agricultural Sciences.)

Fig. 2-18. Application of 90 kg/ha (80 lb/A) of nitrogen on right produced a 74% yield increase in forage sorghum at this location in central India. (Courtesy Frank Shuman, University of Illinois.)

Table 2-26. Effect of Nitrogen Fertilizer Rate on Yield and Composition of Continuous Corn (Ohio Summary–7 years at 7 locations)

Nitrogen Treatment (kg/ha N)	Mean Grain[a] Yield (kg/ha)	Fodder[b] Weight (mt/ha)	Nitrogen Content		
			Grain (%)	Fodder (%)	Ear Leaf[c] (%)
0	3889	2.78	1.32	0.52	1.80
56	5018	3.29	1.47	0.63	2.45
112	5833	3.55	1.60	0.72	2.80
168	6209	3.66	1.66	0.80	3.25
224	6272	3.59	1.67	0.85	3.30

[a] Maximum yield obtained was 8530 kg/ha.
[b] Fodder at harvest.
[c] Ear leaf sampled at initial silk.
   NOTE: mt/ha x 0.446 = t/a.

SOURCE: E. W. Stroube, "Nitrogen Fertilizer Requirement for Continuous Corn," Ohio Agric. Exp. Sta. Cir. SFT-17, April 1968.

applications in Mississippi,[14] for example, and the practice of fertilization in forestry is rapidly gaining in popularity in the United States and Canada.

The use of nitrogen in European forestry is much more common than in North America. Mayer-Krapoll[15] reported on practices of fertilization of various species of trees and noted excellent responses to nutrient applications, particularly nitrogen.

2.8 NITROGEN SOURCE COMPARISONS

Agronomic comparisons of various nitrogen fertilizers when applied according to recommendations have indicated slight, if any, differences in their efficiency (Table 2–27). Obviously, conditions can vary for a given location that may make one source of nitrogen superior to others, but those conditions are frequently predictable. Data reported earlier in Table 2–12 are a good example of the comparable performance of nitrogen fertilizers on grain sorghum.

Voss[16] noted that comparisons of three nitrogen sources for corn in Iowa, ammonium nitrate, urea, and 28% nitrogen solution, were about equal in effectiveness when averaged across times of application and nitrogen rates. Differences among nitrogen sources were usually 60 to 200 kg/ha of grain, and therefore within the limits of experimental error.

Comparisons of nitrogen fertilizers for rice have produced some differences in efficiency on occasion. In 1960, Wells[17] reported no differences in urea, ammonium nitrate, ammonium sulfate, ammonium chloride, and 32% nitrogen solution as nitrogen sources for flooded rice in Arkansas when applied 45 and 75 days after planting. De Datta and Magnaye[18] surveyed nitrogen fertilizers for rice and concluded that those supplying ammonium-nitrogen were about equally effective in terms of grain yield response but noted that nitrate sources were unsatisfactory for preplant treatments because of denitrification losses of nitrogen (see Section 2.2.7).

The development of slow-release fertilizers such as sulfur-coated urea (SCU) discussed earlier has indicated some potential advantages for such nitrogen sources for crops such as rice. Sanchez et al.[19] recorded up to 59% greater yield increases from use of SCU versus urea or ammonium sulfate

[14] J. B. Baker et al., "Biomass Production and Nitrogen Recovery after Fertilization of Young Loblolly Pines," *Soil Sci. Soc. Amer. Proc.*, 38:958, 1974.

[15] H. Mayer-Krapoll, "The Use of Commercial Fertilizers—Particularly Nitrogen—in Forestry," Pub. by Allied Chemical Corp., 1956, 111 pp.

[16] R. D. Voss, "Nitrogen Efficiency for Corn," Proc. 1978, Iowa Fertilizer and Ag Chemicals Dealers Conference, Des Moines, Jan. 10–11, 1978.

[17] J. P. Wells, "Sources of Nitrogen for Rice," Arkansas Agri. Exp. Sta. Report 115, 1962, 10 pp.

[18] S. K. De Datta and C. P. Magnaye, "A Survey of the Farms and Sources of Fertilizer Nitrogen for Flooded Rice," *Soils and Fertilizers*, 2:103, 1969.

[19] P. A. Sanchez et al., "Performance of Sulfur-Coated Urea Under Intermittently Flooded Rice Culture in Peru," *Soil Sci. Soc. Amer. Proc.*, 37:789, 1973.

Table 2-27. Sources of Nitrogen Fertilizer, Grade, Chemical, and Physical Form and Recommended Method of Application

Fertilizer	Grade (N–P₂O₅–K₂O)	Physical Form	Method of Recommended Application
Anhydrous ammonia (NH_3)	82-0-0	High-pressure liquid	Must be injected 6–8 in. (15–20 cm) deep in friable, moist soil
Aqua ammonia ($NH_3 + H_2O$)	20-0-0 to 24-0-0	Low-pressure liquids	Must be injected 2–3 in. (5–8 cm) deep in friable, moist soil
Low-pressure N solutions ($NH_4NO_3 + NH_3 + H_2O$)	37-0-0 to 41-0-0	Low-pressure liquids	Must be injected 2–3 in. (5–8 cm) deep in friable, moist soil
Pressureless N solutions ($NH_4NO_3 + urea + H_2O$)	28-0-0 to 32-0-0	Pressureless liquids	Spray on surface or side-dress. Incorporate surface application to prevent volatilization loss of NH_3 from the urea.
Ammonium nitrate (NH_4NO_3)	33.5-0-0 34-0-0	Dry prills, granules	Broadcast or side-dress. Can be left on the soil surface.
Ammonium sulfate ($(NH_4)_2SO_4$)	20-0-0	Dry granules	Broadcast or side-dress. Can be left on the soil surface.
Urea ($NH_2 \cdot CO \cdot NH_2$)	45-0-0	Dry prills, granules	Broadcast or side-dress. Incorporate surface application to prevent volatilization loss of NH_3 from the urea.
Sodium nitrate ($NaNO_3$)	16-0-0	Dry granules	Broadcast or side-dress. Can be left on the soil surface.
Calcium nitrate $Ca(NO_3)_2$	15.5-0-0	Dry granules	Broadcast or side-dress. Can be left on the soil surface.
Ammonium nitrate of lime ($NH_4NO_3 + CaCO_3$)	20.5-0-0	Dry granules	Broadcast or side-dress. Can be left on the soil surface.
Diammonium phosphate ($(NH_4)_2HPO_4$)	18-46-0	Dry granules	Broadcast or apply in the row. Can be left on the soil surface.
Potassium nitrate (KNO_3)	13-0-44	Dry granules	Broadcast or apply in the row. Can be left on the soil surface.
Ammonium phosphate	10-34-0 11-37-0	Pressureless liquids	Spray on surface or side-dress.

SOURCE: L. M. Walsh, "Soil and Applied Nitrogen," University of Wisconsin, Ext. Fact Sheet A 2519, 1973.

when the nitrogen was applied preplant. Topdressed applications also yielded advantages for SCU, 20% higher grain response than topdressed applications of urea or ammonium sulfate. The International Rice Research Institute has proved the agronomic superiority of SCU.[20] Best performance of coated nitrogen sources such as SCU has occurred where water management has increased nitrogen loss through leaching or denitrification. Advantages for such a material must be weighed against higher costs (see Section 2.6.7).

[20] International Rice Research Institute, 1978, Annual Report for 1977. Los Baños, Philippines, pp. 296–301.

SUMMARY

Nitrogen is one of the most frequently limiting factors in crop production. Crop responses to this element have been spectacular and will continue to increase in frequency as cultivation of land continues. Both organic and inorganic sources of supplemental nitrogen are available to agricultural producers. Costs and forms of the supplemental nitrogen dictate which of these sources should be used in a given situation.

Production of supplemental nitrogen by industrial techniques has relied heavily on fossil fuels since its inception. Rising costs of these fuels indicate that the cost of industrially fixed nitrogen will continue to increase. The relationship of industrially fixed nitrogen to fossil fuels is through the requirement for hydrogen for the production of anhydrous ammonia, the precursor of most other nitrogen fertilizers.

Use of organic sources of nitrogen is feasible where these materials are available in sufficient quantities. Nitrogen release for plant absorption from these nitrogen sources is not instantaneous and is dependent on bacterial action in the soil to convert the nitrogen to a usable form. Soil chemistry of nitrogen is strongly tied to soil microbiology and essentially all nitrogen oxidation-reduction reactions in the soil are a result of microbial action.

Plant deficiency symptoms of nitrogen are relatively easy to diagnose due to the relationship of nitrogen to chlorophyll production in the plant. Chlorosis resulting from nitrogen deficiency is not the only symptom, but its pattern is one of the most easily recognized of the nutrient deficiency symptoms.

Crop responses to applied nitrogen are measured not only in terms of more mass of product but also in terms of quality. Wheat grain protein, corn grain protein, sugar content of plants, and digestibility of forages all are tied to an adequate but not excessive nitrogen supply.

PROBLEMS

1. Denitrification represents a mechanism of nitrogen loss from soils. Outline the process of denitrification, the compounds involved, and the end products. Also indicate why denitrification occurs and name the organisms involved.

2. Describe at least two processes or reactions representative of ammonification in the soil.

3. Outline with reactions the process of ammonia synthesis beginning with methane and air. Include raw materials, catalysts, reactor conditions, and end products.

4. Describe the role of soil colloids in ammonia and ammonium retention.

5. A certain nitrogen solution has the following analysis and nomenclature: 182 (18.2% N)(0-0-39)(39% urea by weight). This material is applied as a foliar application for soybeans at the rate of 20 gal/A (182 l/ha). You are told by its manufacturer

that the solution weighs 1.31 kg/*l* (10.9 lb/gal). Assume that 10% of the nitrogen was lost by volatilization during the foliar application and that 30% was utilized by the beans and removed in the harvested beans. The remainder is left for a crop of wheat that will be planted following the beans. Fertilizer recommendations for the 30 bu/A (2016 kg/ha) wheat crop you expect call for 50 lb/A (56 kg/ha) of nitrogen. How much nitrogen from some other fertilizer source would you need? Give your answer in both kilograms per hectare (kg/ha) and pounds per acre (lb/A).

Solution 1:

182 *l* N solution/ha × 1.31 kg/*l* = 238 kg solution/ha
238 kg N solution × 18.2% N = 43.3 kg N/ha applied
(100% N − 10% lost − 30% removed in bean crop) × 43.3 kg N/ha applied =
26 kg N/ha remaining in soil
56 kg N/ha recommended for wheat − 26 kg N remaining = **30 kg N/ha additional
needed for wheat**

Solution 2:

20 gal N solution/A × 10.9 lb/gal = 218 lb solution/A
218 lb N solution x 18.2% N = 39.7 lb N/A applied
(100% N − 10% lost − 30% removed in bean crop) × 39.7 lb N/A applied =
23.8 lb N/A remaining in soil
50 lb N/A recommended for wheat − 23.8 lb N remaining = **26.2 lb N/A additional
needed for wheat**

6. An irrigation well pumps 800 gal (3028 *l*)/min. The farmer will be applying 1.0 in. of water (2.54 cm) to an irrigated quarter section; however only 132 A (55 ha) will actually be irrigated. If he maintains a concentration of 20 ppm nitrogen in the water throughout the irrigation cycle, how much 32% nitrogen solution will be needed for the area? Give answer in both pounds and kilograms of nitrogen solution.

   Solution:

   Water/hectare = 10,000 m²/ha × .0254 m (2.54 cm) = 254 m³ H_2O
   254 m³ x 1000 kg/m³ H_2O = 254,000 kg H_2O/ha
   N concentration in H_2O is 20 ppm
   20 ppm = 20 kg N/1,000,000 kg H_2O

   $$\frac{20 \text{ kg N}}{1,000,000 \text{ kg H}_2\text{O}} = \frac{X \text{ kg N}}{254,000 \text{ kg H}_2\text{O}}$$

   $X = \dfrac{20 \text{ kg N} \times 254,000 \text{ kg H}_2\text{O}}{1,000,000 \text{ kg H}_2\text{O}}$ = 5.08 kg N/ha in 2.54 cm H_2O
   5.08 kg N/ha × 55 ha = 279.4 kg N ÷ 32% N solution = 873 kg N solution
   873 kg N solution x 2.2 lb/kg = **1921 lb N solution**

7. Ammonia damage to seedlings would possibly be reduced by (a) shallower depth of ammonia application, (b) wider spacing between points of ammonia release, (c) narrower spacing between points of ammonia release, (d) applications of ammonia parallel to the direction that rows will run.

8. Which of the following forms of nitrogen is most subject to leaching? (a) ammonium, (b) nitrate, (c) elemental nitrogen, (d) nitric oxide, (e) nitrous oxide.

9. Soil acidity can enter into ammonia retention (adsorption) in the soil. The process occurs by (a) hastening conversion to nitrate, (b) stopping oxidation of ammonium to nitrite, (c) providing protons to convert ammonia to ammonium, (d) depressing hydroxyl dissociation on clay surfaces.

10. Nitrification is generally an acid-producing reaction. The acidification process involves (a) consumption of hydroxyls in formation of nitrite, (b) production of protons in the nitrate production step, (c) production of hydroxyls in the nitrite production step, (d) production of protons in the nitrite production step.

11. Assuming that urea-nitrogen had undergone the process mediated by urease and assuming that the nitrogen had been acted upon by aerobic *Nitrosomonas*, aerobic *Nitrobacter*, and oxygen-stressed heterotrophic organisms, in that order. What would be a likely final form that nitrogen might be found in (a) ammonium, (b) ammonia, (c) nitrate, (d) elemental nitrogen?

REFERENCES

Alexander, M., *Introduction to Soil Microbiology*, 2nd ed., New York: John Wiley & Sons, 1977.

Anderson, O. E., F. C. Boswell, and S. V. Stacy, "Effect of Temperature on Nitrification in Georgia Soils," Georgia Agric. Exp. Sta. Bul. N.S. 130, (1965) 22 pp.

Anonymous, "Fertilizer Urea—Its Efficient Use," *Fertilizer News*, 22(9) (1977) 3–18.

Brage, B. L., W. R. Zich, and L. O. Fine, "The Germination of Small Grain and Corn as Influenced by Urea and Other Nitrogenous Fertilizers," *Soil Sci. Soc. Amer. Proc.*, 24 (1960) 294–296.

Broadbent, F. E., "Mineralization, Immobilization and Nitrification (of Nitrogen)," Proc. Natl. Conf. on Mgt. of Nitrogen in Irr. Agric., Univ. of Calif., May, 1978.

Dalal, R. C., "Comparative Efficiency of Soluble and Controlled-Release Sulfur-Coated Urea Nitrogen for Corn in the Tropics," *Soil Sci. Soc. Amer. Proc.*, 38 (1974) 970–974.

Dalal, R. C., "The Use of Urea and Sulfur-Coated Urea for Corn Production on a Tropical Soil," *Soil Sci. Soc. Amer. Proc.*, 39 (1975) 1004–1005.

Diamond, R. B., "A New Way to Minimize Nitrogen Soil Losses," *Rice Farming*, 9(6) (1975) 6–8.

Engelstad, O. P., and R. D. Hauck, "Urea—Will It Become the Most Popular Nitrogen Carrier?" *Crops and Soils*, (May 1974) 11.

Finney, K. F., J. W. Meyer, H. C. Fryer, and F. W. Smith, "Effect of Spraying Pawnee Wheat With Urea Solution on Yield, Protein Content and Protein Quality," *Agron. J.*, 49 (1957) 341–346.

Gallagher, P. J. et al., "Comparisons of Nitrogen Carriers and Time of Nitrogen Application for Irrigated Corn," Kansas Fert. Res. Rept. of Progress-255, (1975) 96–97, 110.

Gascho, G. J., and G. H. Snyder, "Sulfur-Coated Fertilizers for Sugarcane: I. Plant Response to Sulfur-Coated Urea," *Soil Sci. Soc. Amer. Proc.*, 40 (1976) 119–122.

Jones, U. S., *Fertilizers and Soil Fertility*, Reston Publishing Co., Reston, Va., 1979, 368 pp.

Maples, R., J. L. Keogh, and W. E. Sabbe, "Sulfur-Coated Urea as a Nitrogen Source for Cotton," Arkansas Agric. Exp. Sta. Bul. 807 (1976) 23 pp.

Morey, D. D., M. E. Walker, W. H. Marchant, and R. S. Lowrey, "Small Grain Forage Production and Quality as Influenced by Rates of Nitrogen," Georgia Agric. Exp. Sta. Res. Bul. 70 (1969) 19 pp.

Olson, R. A. et al., "Impact of Residual Mineral N in Soil on Grain Protein Yields of Winter Wheat and Corn," *Agron. J.*, 68 (1976) 769.

Page, A. L., and A. C. Chang, "Nutritional Supplements to Plants: Sewage Sludge. (Origin, Composition and Uses)." In USDA-SEA National Workshop, "Utilization of Wastes on Land: Emphasis on Municipal Sewage," University of Maryland, July 15–17, 1980 and Anaheim, California, August 12–14, 1980.

Powers, W. L. et al., "Guidelines for Land Disposal of Feedlot Lagoon Water," Kansas Coop. Ext. Ser. Cir. 485 (1973).

Rolston, D. E., "Volatile Losses of Nitrogen from Soil," Proc. Natl. Conf. on Mgt. of Nitrogen in Irr. Agric., University of Calif., May, 1978.

Tisdale, S. L., and W. L. Nelson, *Soil Fertility and Fertilizers*, 3rd ed., New York and London: Macmillan Publishing Co., Inc., 1975.

Van Burg, P. F. J., "Nitrogen Fertilizing of Grassland in Spring," Netherlands Nitrogen Tech. Bul. 6 (1968) 45 pp.

CHAPTER THREE

Phosphorus Fertilizers

3.1 INTRODUCTION

Phosphorus, one of the macronutrients, is present in the world's soils in varying quantities but is usually higher in virgin soils particularly in areas of low to moderate rainfall. This element is required by plants in quantities that are approximately one-tenth as great as those of nitrogen and potassium. Some estimates of the ranges in phosphorus concentration in soils are given in Table 3-1. The amount of phosphorus available to plants, however, is not necessarily well correlated with the total phosphorus content of soils.

3.2 FORMS OF SOIL PHOSPHORUS

Soil phosphorus exists in three main categories: (1) inorganic phosphorus compounds, (2) organic phosphorus compounds, and (3) soil solution phosphorus. Various inorganic minerals present in the soil contain phosphorus. A general characteristic of most inorganic phosphorus compounds in either soil or parent material is low water solubility. Phosphorus readily reacts with other soil components such as calcium, magnesium, iron, and aluminum to produce the compounds listed in Table 3-2. Soil reactions of applied phosphorus-producing compounds of limited water solubility is the main reason why only very small percentages of applied phosphorus are utilized by the first crop following application.

Table 3-1. Phosphorus Content of Soils

Soil Location	Total Phosphorus (ppm)	Organic Phosphorus (% of total)
Mollisols, Iowa	613	41.6
Hapludalfs, Iowa	574	37.3
Alfisols (clay pan), Iowa	495	52.7
Surface soils, Arizona	703	36.0
Subsoils, Arizona	125	34.0
Alluvial soils, Oregon	1479	29.4
Loess soils, Kansas	650	—
Coastal plains soils, Georgia	86	—

SOURCES: R. W. Pearson and R. W. Simonson, "Organic Phosphorus in Seven Iowa Profiles: Distribution and Amounts as Compared to Organic Carbon and Nitrogen," *Soil Sci. Soc. Amer. Proc.*, 4:162, 1939 and W. H. Fuller and W. T. McGeorge, "Phosphates in Calcareous Arizona Soils, II. Organic Phosphorus Content," *Soil Sci.*, 71:45, 1951 by permission of the American Society of Agronomy.

Relatively more is known of the nature of the calcium phosphates in soils than of the iron and aluminum phosphates. Fluoroapatite, the most insoluble and unavailable of the calcium phosphates, is a primary mineral and is found in even severely weathered soils. Its resistance to weathering and alteration is an indication of its insolubility under most conditions. More soluble and simpler calcium phosphates quickly revert to more insoluble forms and the former rarely exist in the soil for an extended period.

The iron and aluminum phosphates are a complex group of minerals with many different formulas. These types of compounds tend to dominate in acid soils where they are more stable. All have a low solubility and represent a relatively unavailable form of phosphorus to plants. Conditions conducive to their formation are discussed in Section 3.3.4.

Table 3-2. Some Forms of Inorganic Soil Phosphorus

Compound	Formula
Dicalcium phosphate dihydrate	$CaHPO_4 \cdot 2H_2O$
Tricalcium orthophosphate	$Ca_3(PO_4)_2$
Trimagnesium orthophosphate	$Mg_3(PO_4)_2 \cdot 22H_2O$
Octocalcium phosphate	$Ca_4H(PO_4)_3 \cdot 3H_2O$
Fluoroapatite	$Ca_{10}(PO_4)_6F_2$ or $[Ca_3(PO_4)_2]_3 \cdot CaF_2$
Chloroapatite	$Ca_{10}(PO_4)Cl_2$ or $[Ca_3(PO_4)_2]_3 \cdot CaCl_2$
Hydroxyapatite	$Ca_{10}(PO_4)_6(OH)_2$ or $[Ca_3(PO_4)_2]_3 \cdot Ca(OH)_2$
Ferric hydroxyphosphate	$Fe(H_2O)_3(OH)_2H_2PO_4$
Dufrenite	$Fe_2(OH)_3PO_4$
Vivianite	$Fe_3(PO_4)_2 \cdot 8H_2O$
Aluminum hydroxyphosphate	$Al(H_2O)_3(OH)_2H_2PO_4$
Wavellite	$Al_3(OH)_3(PO_4)_2 \cdot 5H_2O$
Dihydrogen phosphate anion	$H_2PO_4^-$
Monohydrogen phosphate anion	HPO_4^{2-}

Organic forms of phosphorus in the soil are even more varied in composition than the inorganics. Many of the phosphorus compounds in plants are found in the soil at least for variable periods of time. The composition of these compounds changes as the materials undergo decomposition (mineralization). The data available on the organic phosphorus compounds in soils indicate the presence of three major groups: (a) nucleic acids, (b) phytin and phytin derivatives that are calcium-magnesium salts of inositol hexaphosphoric acid, and (c) phospholipids. The latter group is more subject to decomposition and subsequently the first two groups represent most of the soil's organic phosphorus.

Soil solution phosphorus is actually the source of phosphorus for plants. Phosphorus is absorbed by plants largely as the primary and secondary orthophosphate anions ($H_2PO_4^-$ and HPO_4^{2-}) directly from the soil solution. Phosphate anion concentrations in the soil solution are a result of the equilibrium that exists between the soil solution and the various forms of phosphorus. Weathering reactions are responsible for releasing phosphate anions from the low solubility compounds calcium, magnesium, iron, and aluminum phosphates. Plant absorption of phosphorus is related to the concentration of phosphorus anions in soil solution and therefore the equilibrium that exists between the phosphorus present in the soil and the soil solution becomes a controlling factor in plant phosphorus absorption.

Concentrations of phosphorus in the soil solution are very low, and even in a soil at field capacity the amount (mass) of phosphorus in the soil solution at any one time is extemely small. Concentrations of phosphorus in the soil solution frequently range from 0.01 to 0.06 ppm. One can calculate the number of times that this concentration must be renewed if the plant is to absorb the proper amount of phosphorus.

Assume that a crop of alfalfa that requires large amounts of phosphorus produced 22 metric tons per hectare of dry forage (9.8 short tons per acre). This forage contained 0.3% total phosphorus, which represents the amount absorbed and removed that year. Multiplying the total production by the amount of phosphorus in the forage gives the amount removed per hectare per year, 66 kg of phosphorus. If you assume that the alfalfa absorbed the phosphorus from the surface meter of the soil, that there are about 3.9 million kg of moist soil (25% water) per hectare to a depth of 1 m, and that the average concentration of phosphorus in that soil solution is 0.04 ppm, the phosphorus concentration in the soil solution would have to have been renewed a total of 1692 times during one growing season. Remember, that is just the phosphorus removed by alfalfa forage and does not take into account that required to produce the roots of the alfalfa. Obviously, the soil conditions must favor weathering of these minerals plus mineralization of organic phosphorus for plant growth to be optimal. Conditions that influence the release of phosphorus from inorganic and organic forms to the soil solution include pH, temperature, the surface area of the primary inorganic phosphorus minerals exposed to weathering, the amount of soil organic matter, and the soil moisture content.

Soil behavior of phosphorus is quite different from that of nitrogen. As you recall, nitrogen is eventually converted to nitrate in the scheme of soil reactions under aerobic conditions. The nitrate anion is not adsorbed by the soil colloids and subsequently is free to move with the soil solution. Phosphorus, on the other hand, also exists in the soil solution in the anionic form but *is not* free to move in the soil solution because it reacts so readily with other soil components to form less soluble phosphorus compounds.

Phosphorus can be removed from the soil surface by erosion, but leaching is not usually a problem in the management of this element. Most studies have shown that phosphorus migrates in the soil less than 2 cm from the point of application in medium- and fine-textured soils. Hergert and Reuss[1] studied phosphorus movement in coarse-textured soils (sands) in Colorado under center pivot irrigation and found that phosphorus movement up to 18 cm did occur when ammonium polyphosphate was applied in the irrigation water. Rolston and co-workers[2] at the University of California reported that organic phosphate fertilizers moved up to 12 cm in a calcareous Panoche clay loam when applied in irrigation water, whereas inorganic orthophosphates moved only about 2 cm. They attributed the greater movement of glycerophosphate to the requirement for hydrolysis of the organic phosphorus compound before it could react with soil components and be immobilized.

The chemistry of phosphorus in the soil is complex and affects not only the fate of applied fertilizer phosphorus but more important perhaps the absorption of this element by plants. Soil reactions of phosphorus are discussed in detail in Section 3.3.

3.3 SOIL CHEMISTRY AND BIOLOGY OF PHOSPHORUS

3.3.1 Mineralization of Organic Phosphorus

Although the nature and reactions of soil organic phosphorus compounds are not as well understood as some of the inorganic phosphorus chemistry, we do understand that release of the organic phosphorus forms to the inorganic state usually occurs before plant absorption and before reactions with soil calcium, magnesium, iron, or aluminum occur. Phosphorus mineralization reactions in a way are similar to those of nitrogen. The reactions are dependent on the actions of fungi, actinomycetes, and bacteria that decompose carbonaceous residues containing phosphorus compounds (Fig. 3–1).

Mineralization of organic phosphorus is enhanced by pH values that are conducive to general microbial actions. Generally, increasing the soil pH from acid to alkaline enhances phosphorus mineralization because of the greater ac-

[1] G. W. Hergert and J. O. Reuss, "Sprinkler Application of P and Zn Fertilizers," *Agron J.*, 68:5, 1976.

[2] D. E. Rolston, R. S. Rauschkolb, and D. L. Hoffman, "Infiltration of Organic Phosphate Compounds in Soil," *Soil Sci. Soc. Amer. Proc.*, 39:1089, 1975.

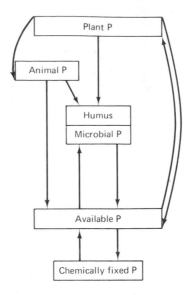

Fig. 3-1. The soil phosphorus cycle involves the activities of plants, animals, and soil microbes. (Source: M. Alexander, *Introduction to Soil Microbiology,* 1961, p. 353 by permission of John Wiley & Sons, Inc.)

tivity of soil bacteria.[3] Rate and magnitude of phosphorus release are both positively related to the quantity of mineralizable phosphorus present. Phosphorus release is generally more rapid in virgin soils than in those that have been under cultivation for some time. Apparently the higher amount of organic phosphorus present in virgin soils leads not only to greater releases of phosphorus but also is associated with mineralization of a higher percentage of the total organic phosphorus. It should be pointed out that from 25 to 90% of the total soil phosphorus is organic. The absolute quantity of organic phosphorus declines with increasing soil depth, which is also correlated to a decrease in organic matter and microbial activity.

Mineralization of phosphorus, like mineralization of nitrogen, is more rapid at higher soil temperatures. Higher rates of biochemical reactions can be expected as the temperature increases. Thompson and Black[4] reported that, below 30°C (86°F), net mineralization of phosphorus responded only slightly to increased temperature but, above 30°C (86°F), the rate of release was markedly affected by higher temperatures.

The organic compounds making up soil organic matter are derived from plants, animals, and microbes. Because these components are directly related to the plants that have grown recently in the soil and because the plants themselves differ in composition, mineralization rates of various residues vary. Pearson, Norman, and Ho[5] studied the rate of mineralization of different plant

[3] L. M. Thompson, C. A. Black, and J. A. Zoellner, "Occurrence and Mineralization of Organic Phosphorus in Soils with Particular Reference to Association with Nitrogen, Carbon and pH." *Soil Sci.,* 77:185, 1954.

[4] L. M. Thompson and C. A. Black, "The Effect of Temperature on the Mineralization of Soil Organic Phosphorus," *Soil Sci. Soc. Amer. Proc.,* 12:323–26, 1947.

[5] R. W. Pearson, A. G. Norman, and C. Ho, "The Mineralization of the Organic Phosphorus of Various Compounds in Soil," *Soil Sci. Soc. Amer. Proc.,* 6:168, 1941.

materials and reported the rate of phosphorus release to be affected by the initial substrate and by time. Apparently, the rate of mineralization of phosphorus can decrease with time as organic acids build up in the media, producing a drop in soil-media pH.

Phytic acid (inositol hexaphosphate), phytin (calcium, magnesium, iron, and aluminum salts of phytic acid), and related compounds are estimated to represent from about 25 to as high as 90% of the soil organic phosphorus. Nucleic acids and possibly phospholipids represent the remainder. Mineralization of these materials occurs at a rate inversely related to their abundance. Phytin and associated compounds are mineralized slowly, particularly in acid soils. Higher pH enhances phosphorus availability from phytin probably due to more microbial activity in the soil. Highly calcareous conditions, however, would tend to lower the solubility of phytin because this material is a calcium or magnesium salt, and large amounts of calcium in the soil would drive the solubility reaction back toward phytin.

Nucleic acids added to the soil experimentally are known to be rapidly mineralized or dephosphorylated. Some controversy surrounds early reports of 20 to 60% of total soil phosphorus in this form. Later studies have suggested that nucleic acid-type phosphorus compounds probably represent from 1 to 10% of the total soil phosphorus.

Some attempts have been made to exploit the process of microbial release of phosphorus, particularly in the USSR. Soviet scientists have inoculated soils with organisms known as phosphobacteria either by direct soil application or as coatings on seeds. They have reported dramatic yield increases and higher phosphorus concentrations in plants. These studies have not been successfully reproduced in other areas of the world, however, and for that reason there is some doubt as to the efficacy of such a phenomenon.

In summary, mineralization of phosphorus is analogous to the mineralization of nitrogen, at least the ammonification process. The oxidation state of phosphorus in the organic materials remains in the $+5$ state during the mineralization process. Similarly, the oxidation state of nitrogen in the amino form (-3) does not change during ammonification. As a general rule, phosphorus release from organic materials responds to the same conditions that favor the release of ammonium nitrogen. Highly significant correlations have been observed between the release of nitrogen and phosphorus from organic materials.

3.3.2 Phosphorus Immobilization

Microbial activity and growth in the soil requires adequate amounts of phosphorus just as it requires adequate amounts of nitrogen, sulfur, and the other elements. Phosphorus is essential for microbial physiology in exactly the same manner as it is for higher plants. For that reason, when a low phosphorus, carbonaceous plant residue is added to the soil, increased microbial activity ensues due to the addition of an oxidizable energy source, but there may be a net decline in the amount of phosphorus available to plants because of the demand of the microbes for phosphorus. Utilization of phosphorus by these organisms converts the phosphorus into forms that are of lower availability to higher plants. The net result is a decline in phosphorus availability in the short run. This phosphorus is eventually released for plant use as the organisms die and the phosphorus reenters the mineralization scheme. Nitrogen, sulfur, and micronutrients all go through this same type of cycle. On the other hand, if the residue had been relatively rich in phosphorus, greater than 0.2% total phosphorus, a net release of phosphorus would have occurred as the organic materials were decomposed. The effect of phosphorus content of plant residues on phosphorus uptake by other plants (ryegrass) is demonstrated in Table 3-3. Stated another way, microbiologists agree that a net mineralization of phosphorus will occur when the carbon to phosphorus ratio in plant residue is less than about 200:1. Net immobilization occurs during initial stages of residue decomposition when the C:P ratio is greater than 300:1 (see Chapter 10).

3.3.3 Hydrolysis of Polyphosphates

Polyphosphates represent a class of phosphorus compounds that may be organic or inorganic in nature. These compounds differ from orthophosphates in the fact that they are longer-chain molecules comprised essentially of several units, at least two, of orthophosphate. Complex polyphosphates exist in cell physiology of higher plants, soil microbes, and animals in the form of energy storage compounds such as adenosine triphosphate (ATP), adenosine diphosphate (ADP), and similar compounds (Fig. 3-2).

Table 3-3. Effects of Phosphorus Content of Plant Residues on Phosphorus Uptake by Ryegrass

P Content of Residue (%)	Yield of Ryegrass (g/pot)	P in Ryegrass (%)	Total P Absorbed (mg/pot)
0.08	1.0	0.08	1.0
0.11	1.1	0.09	1.0
0.14	1.3	0.08	1.0
0.16	1.7	0.08	1.4
0.17	1.9	0.09	1.7
0.19	1.8	0.12	2.2
0.23	1.8	0.18	3.2
0.39	2.0	0.21	4.2
0.42	2.8	0.22	6.2
0.48	2.5	0.27	6.8
0.58	2.7	0.31	8.4
Control soil	1.5	0.21	3.2

SOURCE: W. H. Fuller, D. R. Nielsen, and R. W. Miller, "Some Factors Influencing the Utilization of Phosphorus from Crop Residues," *Soil Sci. Soc. Amer. Proc.*, 20:218, 1956 by permission of the American Society of Agronomy.

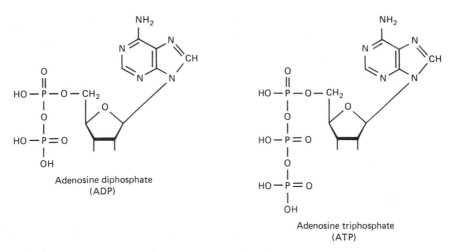

Fig. 3–2. Structural formulas for the two most important energy storage and transfer compounds in plants, adenosine diphosphate (ADP), and adenosine triphosphate (ATP). Energy is stored in the pyrophosphate bonds in these compounds.

Inorganic polyphosphates are regularly introduced into the soil in the form of certain phosphate fertilizers, the nature and properties of which are discussed later in Section 3.6. Basically, these types of compounds are simplified in the soil by reactions that in the strictest sense are a type of mineralization. The simplification process results in splitting the polyphosphate molecules into orthophosphate units (anions). Because this reaction requires water, it is termed *hydrolysis*.

As soon as polyphosphates are introduced into the soil in either the organic or inorganic form, enzyme systems begin the hydrolysis process. These enzyme systems can exist either in microbial cells or as free enzymes in the soil near plant roots. Subbarao[6] and Subbarao et al.[7] examined the agents governing the speed of this reaction and have reported that temperature, pH, and biological activity in the soil including the type of plant roots present are all strongly related to the speed of the reaction. Chemical hydrolysis of polyphosphate is possible but Subbarao and co-workers have reported that biological hydrolysis through the activity of the enzyme pyrophosphatase is much more rapid and more likely to occur. Chemical hydrolysis of polyphosphates would probably take weeks or months, whereas biological hydrolysis would occur in a matter of hours or at best a few days.

The rapidity with which biological hydrolysis of polyphosphates occurs in the soil tends to suggest that the introduction of soluble polyphosphates as

[6] Y. V. Subbarao, "Polyphosphate Hydrolysis and Reaction Products in Soils and Solution," Ph.D. Thesis, Dept. of Agronomy, Kansas State University, 1975.

[7] Y. V. Subbarao, R. Ellis, Jr., G. M. Paulsen, and J. V. Paukstellis, "Kinetics of Pyro- and Tripolyphosphate Hydrolyses in the Presence of Corn and Soybean Roots as Determined by NMR Spectroscopy," *Soil Sci. Soc. Amer. Proc.*, 41:316, 1977.

fertilizer into the soil will result in hydrolysis to the orthophosphate form according to reactions outlined as follows:

$$H_4N\text{-}O\text{-}\overset{\overset{O}{\|}}{\underset{\underset{NH_4}{|}}{P}}\text{-}O\text{-}\overset{\overset{O}{\|}}{\underset{\underset{NH_4}{|}}{P}}\text{-}OH + H_2O \xrightarrow[\text{Enzyme}]{\text{Phosphatase}} H_4N\text{-}O\text{-}\overset{\overset{O}{\|}}{\underset{\underset{NH_4}{|}}{P}}\text{-}OH + H_4N\text{-}O\text{-}\overset{\overset{O}{\|}}{\underset{\underset{OH}{|}}{P}}\text{-}OH$$

Triammonium pyrophosphate   Diammonium phosphate   Monammonium phosphate

The probability of this reaction suggests that uptake of phosphorus by plants from material applied in the polyphosphate form is likely to be in the form of orthophosphate anions.

Studies of the uptake of polyphosphate, specifically the pyrophosphate form of polyphosphate, by Sutton and Larsen[8] produced evidence that the hydrolysis of polyphosphate or pyrophosphate in a sandy loam soil was so rapid that only orthophosphate was taken up by the test crop, ryegrass. Terman and Engelstad[9] suggested that polyphosphates must be hydrolyzed to orthophosphate before phosphorus can be absorbed by the plant. Regardless of whether plants do indeed use a small amount of phosphorus in the polyphosphate form or whether it is all absorbed in the orthophosphate form, there is evidence to suggest that rapid conversion of polyphosphate to orthophosphate does occur in soils under most conditions. Some conditions are described in section 3.6 where polyphosphate hydrolysis is greatly delayed due to reactions with other soil components.

The effects of roots of different plant species on the hydrolysis of polyphosphates in the soil has been studied by Webb,[10] Subbarao and co-workers,[7] and Savant and Racz.[11] Webb reported generally higher pyrophosphatase (phosphatase) activity (the polyphosphate hydrolyzing enzyme) in cereal (monocotyledonous) species than in dicotyledonous species (Fig. 3–3). The activity of this enzyme was significantly higher in leaves than in roots of all species examined. Subbarao and others[7] examined the abilities of corn and soybean to hydrolyze polyphosphates (pyrophosphate and tripolyphosphate). Presence of roots of either species in a growth medium enhanced the rate of polyphosphate hydrolysis. Generally, corn hydrolyzed polyphosphate faster than did soybeans. Savant and Racz attributed pyrophosphate hydrolysis in the presence of plant roots to the phosphatase activity of rhizosphere organisms and roots.

[8] C. D. Sutton and S. Larsen, "Pyrophosphate as a Source of Phosphorus for Plants," *Soil Sci.,* 97:196, 1964.

[9] G. L. Terman and O. P. Engelstad, "Crop Responses to Nitrogen and Phosphorus in Ammonium Polyphosphate," *Comm. Fert. Plant Food Ind.,* 112(6):30, 1966.

[10] B. B. Webb, "Field and Growth Chamber Comparisons of Ortho and Polyphosphate," Ph.D. Thesis, Department of Agronomy, Kansas State University, 1970.

[11] N. K. Savant and G. J. Racz, "Hydrolysis of Sodium Pyrophosphate and Tripolyphosphate by Plant Roots," *Soil Sci.,* 113:18, 1970.

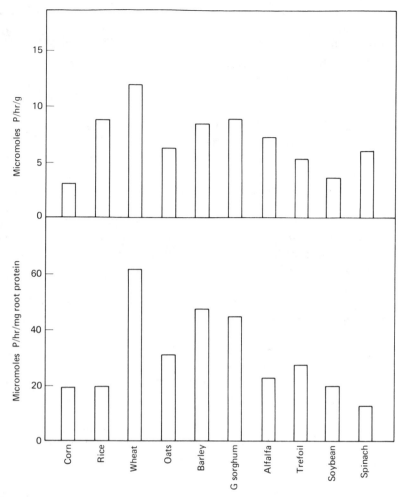

Fig. 3–3. Root pyrophosphatase activity (micromoles P/hr/g material) of several plant species (above). When the activity was expressed per milligram of root protein, similar relationships were recorded (below). (Source: B. B. Webb, "Field and Growth Chamber Comparisons of Ortho and Polyphosphate," Ph.D. Thesis, Department of Agronomy, Kansas State University, 1970.)

3.3.4 Phosphorus Reactions in Acid Soil

Introduction of soluble phosphate compounds or phosphate ions from mineralization of organic matter into an acid soil results in a complex series of reactions, the nature of which is dependent on characteristics of the soil. Generally, the availability of introduced phosphatic fertilizers in an acid soil begins to decline as soon as the fertilizer reaches the soil. The overall process of reactions of soluble calcium phosphates or ammonium phosphates with acid soil components is known as *fixation* or *phosphate retention.*

The end products of the fixation reactions in acid soils are usually iron or aluminum phosphates. In the pH range of 4 to 5, large amounts of soluble iron and aluminum exist either in the soil solution or adsorbed on the soil colloids. The solubility of iron and aluminum phosphates is particularly low and very small quantities of these metals in the soil solution cause precipitation when phosphates are added to the soil.

Highly weathered soils have long been known to be high in these two metals. As an example, highly weathered soils of the southern and southeastern United States and the tropics in general contain high concentrations of soluble iron and aluminum because most of the siliceous minerals have been removed by the weathering process. In the humid southern United States, these soils are frequently acid and rapidly convert added phosphate fertilizers into compounds such as ferric hydroxyphosphate and a suite of similar iron compounds or aluminum analogs (see Table 3.2). All of these materials have the same characteristic of low solubility and low phosphorus availability. When the phosphorus has been incorporated into such compounds, it is "fixed" in terms of its solubility and availability to plants. Long-term weathering will eventually convert the phosphorus back into the available anions in soil solution but that process is slow. Coleman, Thorup, and Jackson[12] studied the abilities of soils to adsorb phosphorus from solution and reported a strong correlation between the amount of exchangeable aluminum and the amount of phosphorus adsorbed by a suspended clay (Fig. 3–4).

Part of the ability of soils to react with phosphorus is due to the presence of large amounts of precipitated iron and aluminum oxides and hydroxides. These precipitates are frequently finely divided and have large effective surface areas providing plenty of surface for reaction with phosphates. These finely divided hydrous oxides have the ability to adsorb phosphate anions directly on their surfaces. This adsorption reaction represents an electrical attachment of the phosphate anion to the surface of the iron or aluminum oxide or hydroxide and produces the same lowering of phosphorus availability as crystallization occurs.

Phosphate anions can also react with silicate clays. Clay minerals high in hydroxyl groups have been suggested to enter into a type of anion exchange with phosphate anions satisfying the charges produced when hydroxyl groups are lost by dissociation or breakage particularly at the edges of the clay particles. Clays likely to enter into this type of anion exchange would be those like kaolinite with a 1:1 lattice ratio, one layer of alumina and one layer of silica in the structure of the clay. Hydroxyls would be associated with the alumina layer. Substitution of phosphate (PO_4^{3-}) for a silicon tetrahedron (group) in the silica layers of the clay may also occur.

Other studies of phosphorus reactions with silicate clays have suggested that calcium-saturated clays possibly form clay-calcium-phosphate linkages, essentially a type of phosphate adsorption although this point is strongly debated. These studies have indicated that clays saturated by sodium are less

[12] N. T. Coleman, J. T. Thorup, and W. A. Jackson, "Phosphate-Sorption Reactions that Involve Exchangeable Al," *Soil Sci.*, 90:1, 1960.

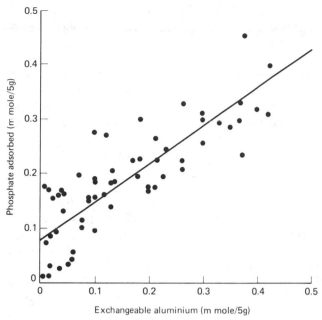

Fig. 3-4. Phosphorus adsorbed by a suspended clay increased dramatically as the amount of exchangeable aluminum increased. (Source: N. T. Coleman, J. T. Thorup, and W. A. Jackson, "Phosphate-Sorption Reactions that Involve Exchangeable Al," *Soil Sci.,* 90:1, 1960. Used with permission.)

likely to adsorb and hold phosphate than clays saturated with calcium because of the greater solubility of sodium phosphates. Ragland et al.[13] demonstrated this phenomenon in some Kentucky soils.

The reactions of phosphate with iron and aluminum in acid soils are partially reversible by the process of liming. The fixation reactions that occur between phosphorus anions and the hydrated iron and aluminum oxides may partially be due to the substitution of phosphate anions for hydroxyls. For instance, if ferric hydroxide [$Fe(OH)_3$)] reacted with the phosphate anion $H_2PO_4^-$, the resultant compound might have the formula $Fe(OH)_2H_2PO_4$, a ferric hydroxyphosphate. Liming supplies hydroxyl ions to the soil solution. Large amounts of hydroxyls could reverse this reaction by replacing the phosphate anion and reverting the iron compound to its original form.

$$\underset{\substack{\text{Ferric}\\\text{hydroxide}}}{Fe(OH)_3} + \underset{\substack{\text{Lime}\\\text{(OH}^-\text{)}}}{H_2PO_4^-} \rightleftharpoons \underset{\substack{\text{Ferric hydroxy}\\\text{phosphate}}}{Fe(OH)_2H_2PO_4} + \underset{\text{Hydroxyl}}{OH^-}$$

Liming then can exert a positive influence on the availability of phosphorus in acid soils through such a replacement reaction. At the same time, liming can at least delay further fixation reactions because the solubility of iron and

[13] J. L. Ragland and W. A. Seavy, "The Effects of Exchangeable Calcium on the Retention and Fixation of Phosphorus by Clay Fractions of Soil," *Soil Sci. Soc. Amer. Proc.,* 21:261, 1957.

aluminum compounds is greatly reduced at higher pH values and less of these two metals are available to react with phosphorus (see Chapter 8).

Up to this point, the discussion of phosphorus reactions in soils has implicated only the orthophosphate anions, $H_2PO_4^-$ and HPO_4^{2-}. Polyphosphate fertilizers can also react directly with components of high iron and aluminum (acid) soils without undergoing the hydrolysis reactions discussed in the preceding section. Polyphosphate reactions in high iron and aluminum soils have been studied but not to the degree of orthophosphates, primarily due to the fact that these compounds (polyphosphates) have not been available as fertilizer materials for such an extended period. In high-iron and/or high-aluminum soils, polyphosphates will probably become fixed and form insoluble materials, but by different products than the reactions of orthophosphates. These insoluble compounds tend to form quickly and the possibility exists that they may represent relatively poor sources of residual phosphorus for subsequent crops. Subbarao[6] examined reactions of ammonium polyphosphate fertilizers with high iron soils and noted the development of iron polyphosphate materials of low solubility.

Some indication of the practical aspects of polyphosphate reactions with high iron soils is borne out in Table 3–4. This study involved applications of

Table 3-4. First-Year and Residual Effects of Phosphorus Applications on Irrigated Grain Sorghum Grown on High Iron Soil

P Applied, 1971 (kg/ha)	Source	1971-First Crop		1972-Second Crop (residual P)	
		Grain Yield (kg/ha)	Leaf P (%)	Grain Yield (kg/ha)	Leaf P (%)
0		2758	0.21	3512	0.18
39	Ammonium orthophosphate (DAP)	5642	0.36	4077	0.21
39	Ammonium polyphosphate (APP)	5015	0.28	3763	0.18
39	Triple superphosphate (TSP)	4703	0.31	4077	0.22
39	Potassium polyphosphate (KPP)	4953	0.25	2968	0.18
LSD$_{.05}$ for P Sources		502	0.07	439	0.02

NOTES:
P rate is mean of 19, 39, and 58 kg/ha of P.
DAP = 18–46–0; APP = 15–62–0; TSP = 0–46–0; KPP = 0–50–40.
Tissue samples, eight-leaf stage, last fully emerged leaf.

SOURCE: J. A. Armbruster, L. S. Murphy, L. J. Meyer, P. J. Gallagher, and D. A. Whitney, "Field and Growth-Chamber Evaluations of Potassium Polyphosphate," *Soil Sci. Soc. Amer. Proc.,* 39:144, 1975 by permission of the American Society of Agronomy. J. A. Armbruster, L. S. Murphy, L. J. Meyer, and P. J. Gallagher, "Residual Effects of Four P Carriers on Corn and Grain Sorghum," Kansas Fert. Res. Report of Progress 194, 1972, p. 127.

ammonium and potassium polyphosphates on a high iron soil in southern Kansas that was cropped to irrigated grain sorghum. Although first-season responses to phosphorus applications were fairly good, analysis of the soil after the first crop indicated a very high percentage of the applied polyphosphate still was unhydrolyzed. This suggested that the polyphosphate had reacted readily to form some type of compound that was resistant to phosphatase hydrolysis. Subsequent cropping without phosphorus application the following season indicated that polyphosphates, both ammonium and potassium, had lower availability than did comparable applications of orthophosphates in the form of diammonium phosphate and triple superphosphate on that high iron soil.

3.3.5 Phosphorus Reactions in Alkaline Soils

A common phenomenon in soils containing appreciable quantities of calcium salts, particularly calcium carbonate, is the process of calcium fixation of phosphorus. Like the fixation reactions occurring in acid soils through the reaction of phosphorus with iron and aluminum, calcium fixation also results in a lowered availability of added phosphorus. Reactions can occur with calcium in the soil solution or with calcium adsorbed on the surfaces of soil colloids or precipitated calcium carbonate. The degree of adsorption of phosphorus on the precipitated calcium carbonate is inversely related to the particle size of the calcium carbonate. Very large particles do not adsorb phosphorus as readily due to the smaller effective surface.

The exact compounds that occur in the soil at the final, equilibrium state of the system are dependent on the amount of calcium present, soil pH, and the presence of other cations such as potassium, magnesium, iron, and aluminum. Reactions of calcium and orthophosphates in particular have been studied for many years and are well documented even if the final products are variable. Lindsay and co-workers[14] as well as Olsen and Flowerday[15] have extensively examined these reactions. Depending on the starting material and the soil conditions, several compounds have been documented as reaction products of orthophosphates in calcareous soil. Some of these are listed in Table 3-5.

Polyphosphates on the other hand react with calcium quite differently due to the different composition of these fertilizer materials. Pyrophosphate, the two-P atom polyphosphate, and one of the main constituents of polyphosphate-containing fertilizers, reacts with calcium, magnesium, potassium, and ammonium to produce a complex series of compounds that have been examined recently by some researchers.

Although the polyphosphate compounds present in liquid or solid am-

[14] W. L. Lindsay, A. W. Frazier, and H. F. Stephenson, "Identification of Reaction Products from Phosphate Fertilizers in Soils," *Soil Sci. Soc. Amer. Proc.,* 26:446, 1962.

[15] S. R. Olsen and A. D. Flowerday, "Fertilizer Phosphorus Interactions in Alkaline Soils," In *Fertilizer Technology and Use,* 2nd ed., edited by R. A. Olson et al., pp. 153–185, Soil Science Society of America, 1971.

Table 3-5. Some Soil Orthophosphate Reaction Products in Alkaline Soil

Compound	Formula
Dicalcium phosphate anhydrite	$CaHPO_4$
Dicalcium phosphate dihydrate	$CaHPO_4 \cdot 2H_2O$
Octocalcium phosphate	$Ca_4H(PO_4)_3 \cdot 3H_2O$
Fluoroapatite	$(Ca_3(PO_4)_2)_3 \cdot CaF_2$
Hydroxyapatite	$(Ca_3(PO_4)_2)_3 \cdot Ca(OH)_2$
Calcium diammonium orthophosphate	$Ca(NH_4)_2HPO_4 \cdot H_2O$
Magnesium ammonium orthophosphate	$MgNH_4PO_4 \cdot 6H_2O$
Trimagnesium orthophosphate	$Mg_3(PO_4)_2 \cdot 22H_2O$

monium polyphosphate fertilizers are readily water soluble and represent excellent sources of phosphorus for plants, their soil reactions where calcium is plentiful can produce relatively insoluble, unavailable forms of phosphorus. Plants subjected to growth under these kinds of conditions but supplied with large amounts of phosphorus may still show strong phosphorus deficiency symptoms.

Subbarao and Ellis[16] recently studied the nature of these compounds and produced some outstanding electron micrographs of reaction products of polyphosphates as they developed on soil surfaces (Fig. 3-5). These micrographs nicely corroborated data that were collected by X-ray diffraction techniques in that the very nature of the crystals themselves was observed and compared to known crystal shapes. Note, for instance, in Fig. 3-5, the crystalline materials growing in an extract of an alkaline soil (Ulysses silt loam). Electron micrographs taken at various times throughout this series of investigations with alkaline soil showed a transition from the original ammonium polyphosphates into insoluble compounds such as calcium ammonium pyrophosphate. By contrast note that the polyphosphate reaction product formed on the surface of a slightly acid soil (Eudora silt loam, pH 6.8) was completely different and was identified as monoammonium phosphate (MAP). Subbarao and Ellis identified many polyphosphate reaction products and several of them are listed in Table 3-6. Monoammonium phosphate (MAP) and diammonium phosphate (DAP) in the same studies produced either no precipitates that could be noted or else the same compounds mentioned earlier.

The effects of calcium reactions of polyphosphates on the availability of applied phosphorus have been observed in both the growth chamber and the field. Subbarao and Ellis[16] noted a lower availability of polyphosphates in the same soil in which calcium ammonium pyrophosphate formation was verified and plants were grown under growth chamber conditions.

[16] Y. V. Subbarao and R. Ellis, Jr., "Reaction Products of Polyphosphates and Orthophosphates with Soils and Influence on Uptake of Phosphorus by Plants," *Soil Sci. Soc. Amer. Proc.*, 39:1085, 1975.

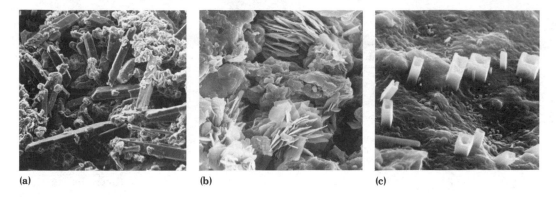

(a)   (b)   (c)

Fig. 3-5. (a) Crystals of calcium ammonium pyrophosphate $Ca(NH_4)_2P_2O_7 \cdot H_2O$ are growing on rod-shaped crystals of triammonium pyrophosphate $(NH_4)_3HP_2O_7 \cdot H_2O$ in the saturated extract of an alkaline soil. Triammonium pyrophosphate was the fertilizer applied. The calcium salt represents a reaction product of lower water solubility. (b) Crystals of calcium ammonium pyrophosphate have formed on the surface of alkaline soil particles. (c) The classic crystal structure of monoammonium phosphate $NH_4H_2PO_4$ which formed on the surface of soil particles in a slightly acid soil when the same ammonium polyphosphate was added. (Source: Y. V. Subbarao and R. Ellis, Jr., "Reaction Products of Polyphosphates and Orthophosphates with Soils and Influence on Uptake of Phosphorus by Plants," *Soil Sci. Soc. Amer. Proc.,* 39:1085, 1975 by permission of the American Society of Agronomy.)

A similar low phosphorus availability in the field from polyphosphate reactions with calcium was observed in Kansas. Data in Table 3-7 demonstrates this effect. An iron study with grain sorghum was established in Scott County, Kansas, on soil that had been leveled for irrigation and in the process cut to a depth of approximately 60 cm (2 ft) exposing the highly calcareous subsoil (pH 8.2). This subsoil, although high in calcium carbonate, was at the same time very low in organic matter, available phosphorus, available iron, and available zinc. In addition to various iron treatments that were applied prior to seeding, phosphorus was applied broadcast preplant as liquid ammonium polyphosphate (10-34-0) at a rate of 58 kg/ha of phosphorus. When the plants emerged, it was obvious that two things were wrong: (1) Iron deficiency did exist but (2) accumulation of anthocyanin pigments indicated a strong possibility of phosphorus deficiency. Analysis of the grain sorghum leaves (Table 3-7) revealed phosphorus concentrations as low as 0.07%, although normal plants in an area that had received manure applications and were outside the most severe area of the exposed subsoil contained phosphorus concentrations as high as 0.40%. Despite the very high rate of phosphorus application, rapid reactions with the soil components had apparently rendered the water-soluble ammonium polyphosphate highly unavailable for plant use. Biological activity in the soil of the exposed subsoil area would have also been quite low due to the removal of the organic matter in the surface soil. This fact could have had a significant effect slowing the rate of hydrolysis of the added polyphosphate and increasing the likelihood of

Table 3–6. Some Polyphosphate Reaction Products Identified in Soil Saturation Extracts and Soils

Reaction Products Formed in Soil Saturation Extracts			
Saturated Solution Source	pH of Saturated Solution	pH of Soil Saturation Extract	P Reaction Products Identified in Extracts
Ulysses Silt Loam			
APP	5.3	5.5	$(NH_4)_3HP_2O_7$ and $Ca(NH_4)_2P_2O_7 \cdot H_2O$
DAP	7.4	7.1	No precipitate
Eudora Silt Loam			
APP	5.3	5.1	$Ca(NH_4)_2P_2O_7 \cdot H_2O$
DAP	7.4	7.1	No precipitate

Reaction Products Formed in Soils		
	P Reaction Products Identified	
P Source	16 Weeks	28 Weeks
Ulysses Silt Loam		
APP	$Ca(NH_4)_2P_2O_7 \cdot H_2O$	$Ca(NH_4)_2P_2O_7 \cdot H_2O$
DAP	$Mg_2KH(PO_4)_2 \cdot 15H_2O$	$Ca_4H(PO_4)_3 \cdot 2\frac{1}{2}H_2O$
Eudora Silt Loam		
APP	$NH_4H_2PO_4$ and $(NH_4)_2H_2P_2O_7$	$Ca_4H(PO_4)_3 \cdot 2\frac{1}{2}H_2O$
DAP	$MgHPO_4 \cdot 3H_2O$	$MgHPO_4 \cdot 7H_2O$

NOTES:
APP = ammonium polyphosphate.
DAP = diammonium phosphate.

SOURCE: Y. V. Subbarao and R. Ellis, Jr., "Reaction Products of Polyphosphates and Orthophosphates With Soils and Influence on Uptake of Phosphorus by Plants," *Soil Sci. Soc. Amer. Proc.*, 39:1085, 1975 by permission of the American Society of Agronomy.

polyphosphate reactions with calcium. More study is needed of the combined effects of biological activity, pH, and calcium carbonate content of the soil on the availability of phosphorus in polyphosphate fertilizers.

Other investigations, however, have not demonstrated poor performance of polyphosphate fertilizer materials on a soil similar to that reported in Table 3–7. Schield and others[17] reported that responses to phosphorus on calcareous soils, low in available phosphorus but moderate in organic matter content,

[17] S. J. Schield, L. S. Murphy, G. M. Herron, and R. E. Gwin, Jr., "Comparative Performance of Polyphosphate Fertilizer for Row Crops," *Comm. Soil Sci. Plant Anal.*, 9:47, 1977.

Table 3-7. Effects of Land Leveling, Soil pH, and Soil Organic Matter Content on Phosphorus Concentrations in Grain Sorghum Following Application of Ammonium Polyphosphate Fertilizer

Area	Soil pH	Soil O. M. (%)	P Applied (kg/ha)	P Concentrations in Leaves (%)
Not leveled	7.7	1.6	58	0.40
Leveled (subsoil exposed)	8.2	0.8	58	0.07

NOTE: P applied as 10-34-0, liquid ammonium polyphosphate.

was essentially identical regardless of whether the phosphorus source was an orthophosphate or a polyphosphate.

3.3.6 Factors Affecting Inorganic Phosphorus Availability

Soil Microorganisms

The principal factors affecting the availability of inorganic phosphorus are soil microorganisms, soil temperature, and soil pH. Microorganisms can increase the availability of inorganic phosphorus compounds resulting from the reactions discussed in the two preceding sections. Genera that have indicated an ability to solubilize calcium phosphates include *Pseudomonas, Penicillium, Aspergillus,* and several others. Activities of these organisms have solubilized phosphorus from compounds such as tricalcium phosphate and apatites. Apparently more phosphorus is released than the organisms consume suggesting that the organisms would have a beneficial effect on the phosphorus supply to higher plants.

Production of organic acids from organic matter seems to be a major pathway of solubilization of insoluble phosphorus compounds. A classic example of the effects of heterotrophic organisms on solubilization of water-insoluble phosphorus compounds is the composting process in which rock phosphate is added to stacked manure. Oxidation of the carbonaceous residues in the manure produces the necessary organic acids that in turn react with the rock phosphate (tricalcium phosphate, apatite). Other composting processes involve the addition of soil, elemental sulfur, and rock phosphate to manure in a compost. Autotrophic organisms that convert ammonium and reduced sulfur compounds into nitrate and sulfate, respectively, can exert an influence on availability of phosphorus due to the acidity produced (nitric and sulfuric acids) in those oxidation reactions.

Insoluble iron phosphates such as strengite ($FePO_4 \cdot 2H_2O$) undergo partial dissolution with the aid of bacteria under reducing conditions (oxygen deficiency) existing in a flooded soil. Patrick et al.[18] reported the results of a study

[18] W. H. Patrick, Jr., S. Gotoh, and B. G. Williams, "Strengite Dissolution in Flooded Soils and Sediments," *Science,* 179:564, 1973.

in which ^{59}Fe-labeled strengite was added to a soil suspension. Their data indicated that the greatest release of phosphorus and iron from strengite occurred under conditions of low oxidation-reduction potential in combination with low pH. Microbial enzyme systems that function in a flooded soil are apparently effective in lowering the activation energy required for strengite reduction. Changing the redox potential alone through the use of hydrogen in Patrick's study was not sufficient to release phosphorus from strengite thus supporting the theory of bacterial enzyme involvement.

Another phosphorus solubilization theory could account for phosphorus release from iron phosphates. Sperber[19] suggested that certain bacteria liberate hydrogen sulfide (H_2S) under reducing conditions and the hydrogen sulfide reacts with iron phosphates to produce ferrous sulfide, liberating the phosphate ions for absorption by plants.

Soil Temperature

Speed of chemical and biological reactions generally increases with increasing temperature. Because biological activity has been known to increase readily with higher temperatures, the connection of bacteria to solubilization of phosphorus compounds discussed in the preceding section would be responsive to higher temperatures. On the other hand, higher mean temperatures tend to hasten soil weathering and increase the soil content of hydrous oxides of iron and aluminum. Greater phosphorus fixation capability exists in those soils and results in a negative correlation of mean temperature to phosphorus availability.

Some studies of soil incubation temperature effects on phosphorus extractability have indicated lower levels of extractable phosphorus with higher incubation temperatures.[20] Higher soil incubation temperatures could produce a greater utilization of phosphorus by bacteria depending on the soils' carbon supply. Assimilation of phosphorus into microbial nucleic acids, phospholipids, or other substances leads to the accumulation of phosphorus compounds not readily extractable.

In summary, higher soil temperatures could exert either a positive or a negative effect on phosphorus availability to plants depending on the chemical characteristics of the soil, the organic matter content, and the microbial activity.

Soil pH

The relationships of pH to availability of iron, aluminum, calcium, and magnesium for reactions with phosphorus were discussed earlier. In most

[19] J. I. Sperber, "Solution of Mineral Phosphates by Soil Bacteria," *Nature*, 180:994, 1957.
[20] A. R. Mack and S. A. Barber, "Influence of Temperature and Moisture on Soil Phosphorus. I. Effect on Soil Phosphorus Fractions," *Soil Sci. Soc. Amer. Proc.*, 24:381, 1960.

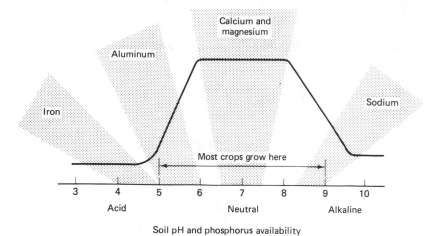

Soil pH and phosphorus availability

Fig. 3-6. The relationships of soil pH and available metals to phosphorus availability. (Source: John Box, "Phosphorus and Crop Production," Texas Coop. Ext. Pub. MP-860, 1976.)

soils, phosphorus availability is highest in the pH range 6.0 to 7.5. Reactions with iron and aluminum are most prevalent at the lower pH values leading to low availability. Calcium-magnesium reactions are more likely at the higher pH values also resulting in lowered phosphorus availability. The relationship of phosphorus availability to soil pH is demonstrated in Fig. 3-6 (see Chapter 8).

3.4 ROLES OF PHOSPHORUS IN PLANTS

Phosphorus, along with nitrogen and potassium, is a major nutrient in plant nutrition. This classification is based primarily on the amount of the element that is found in plants and is not intended to reflect on the relative importance of elements. Phosphorus concentrations are frequently in the range of one-tenth of those of nitrogen and potassium. Actually, phosphorus concentrations in plants are not much larger than some of the other elements, particularly sulfur and magnesium, which are classed as secondary nutrients.

3.4.1 Forms of Phosphorus Absorbed

Phosphorus absorption was mentioned briefly in section 3.3 on soil chemistry and biology. Absorption of this element by plants involves mainly two phosphate anions, the primary orthophosphate anion $H_2PO_4^-$ (also known as the dihydrogen phosphate anion) and the secondary orthophosphate anion HPO_4^{2-} (which is also classed as monohydrogen phosphate). The concentration of these two ions in the soil solution, the source of phosphorus absorbed by

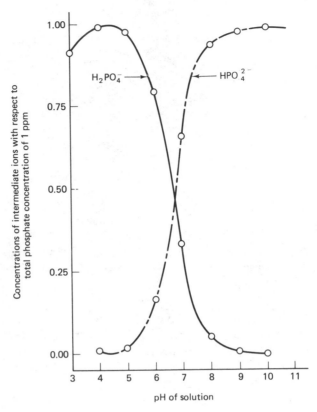

Fig. 3-7. Relationship of pH to the relative concentration of orthophosphate anions in solution. (Source: Arizona Agricultural Experiment Station.)

the plant, is determined by the pH of the soil solution. The relationship of occurrence of these anions to soil solution pH is demonstrated in Fig. 3-7. More acid conditions favor the presence of the less dissociated ion, $H_2PO_4^-$. As the pH of the solution goes up, the phosphate anions tend to act as acid dissociating protons (H^+) and are converted to the HPO_4^{2-} ion.

No more than a few tenths of a part per million of phosphorus in the form of either or both of these anions exists in the soil at any one time. Presence of metals such as iron, aluminum, magnesium, and calcium cause precipitation of phosphate salts as described in Sections 3.3.4 and 3.3.5. Continued renewal of this solution phosphate supply is mandatory for plant growth.

Polyphosphate anions of various types may exist in the soil for short periods. However, most soil scientists consider that absorption of phosphorus in these forms is relatively unlikely due to the rapid hydrolysis of polyphosphates. Organic phosphates of various types are probably not absorbed intact to any great extent because of the size of these molecules and because of the rapid conversions of soluble organic phosphates into inorganic forms by microbial action.

3.4.2 Phosphate Absorption

Plant absorption or uptake of nutrients frequently occurs against a concentration gradient; that is to say, the concentration of a particular element within the root cells is usually higher than that in the soil solution where the ion exists prior to absorption. If root cell membranes were completely permeable, the movement of nutrients by diffusion would tend to be out of the roots into the soil solution because of the lower concentration there. Water (solvent) on the other hand would move into the root until the concentrations of the nutrients (solutes) were equalized in the soil solution and the cell sap.

Root cell membranes are classified as semipermeable, however, so osmosis tends to occur rather than diffusion. In osmosis, water moves toward the area of higher solute or nutrient concentration (into the root) without the movement of nutrients out into the soil solution. Osmosis then can be related to movement of water into the roots but can hardly account for nutrient uptake.

The absorption of solutes by living cells may occasionally follow the laws of diffusion and permeability but a whole group of substances' penetration into living cells cannot be accounted for by the processes of diffusion and osmosis. These substances usually have molecular or ionic weights greater than 50–60 and are usually insoluble in lipids (fats). This would include sugars, many inorganic salts, and ions such as phosphate. Ionic weight, for instance, of $H_2PO_4^-$ is 97.

One theory of absorption against a concentration gradient is dependent on the presence of large, multivalent, protein anions inside the protoplasm of the cell. The large protein anions cannot pass through the cell membrane but inorganic anions and cations can, at least under certain conditions. This combination of organic and inorganic ions and the need to maintain electrical neutrality as some ions move into the cell is governed by a process known as the *Donnan equilibrium*.

Accumulations of ions such as phosphate, however, are much too great to be accounted for by this process. Further, both nutrient anions and cations accumulate simultaneously in the cell. Other types of energy must thus be used to cause such an accumulation according to the laws of thermodynamics. Another source of energy available within the plant is respirational energy developed by the oxidation of carbohydrates. Interestingly, respirational energy has been shown to be proportional to nutrient absorption (Table 3–8). Respirational inhibitors have been shown to interfere with this type of relationship.

Nutrient ions can then be absorbed in two different ways: *active* and *passive* absorption. Active absorption occurs against concentration and Donnan equilibrium gradients upon expenditure of respirational energy. Passive absorption is due to diffusion and follows either simple diffusion or Donnan equilibrium concentration and charge gradients.

In spite of the importance of the nutrient absorption process, there is still controversy over exactly how it occurs. The process seems first to involve an adsorption of the ion on the cell surface. This process of adsorption does not

Table 3-8. Effects of Oxygen on Respiration Rate and Potassium Absorption by Carrot Discs

Atmospheric O_2 (%)	Relative Respiration Rate	Relative K Absorption
2.7	43	22
12.2	78	96
20.8	100	100
43.4	106	117

SOURCE: J. Leavitt, *Plant Physiology*, Englewood Cliffs, N.J.: Prentice-Hall, 1954. p. 56.

require energy and is made possible by the charged nature of the cell surface. Adsorption on the cell surface is also not temperature dependent and occurs at 0°C as well as at normal growing temperatures. Ions adsorbed on the cell surface are transported across the cell membrane into the protoplasm by a system that is linked with respiration possibly through the cytochrome system.

Phosphate absorption then, like the absorption of other nutrient ions, is largely due to active absorption and the expenditure of respirational energy. Respirational reactions in the plant, like any other process in the chemical or biochemical realm, are dependent on temperature as well as other environmental factors. Low temperatures in the soil shortly after germination can have a very depressing effect on phosphorus absorption. Because of the dependency of phosphorus uptake on respiration and because of the relationship of respiration to temperature, plants may develop symptoms of phosphorus deficiency in cold soils even when adequate amounts of phosphorus are available in the soil. Phosphorus absorption is in fact retarded by cold soil conditions and plant analyses during the cold period may show a lower concentration than at a later date when warm weather produces rapid nutrient absorption and rapid plant growth.

3.4.3 Energy Storage and Transfer

Phosphorus exerts a very important role in the plant in energy storage and transfer. This importance is due to the widespread effects of energy transfer and utilization on other metabolic processes.

Phosphorus compounds acting as energy storage and transport "currency" in plants include adenosine diphosphate (ADP) and adenosine triphosphate (ATP). Energy is stored in these compounds in the form of high-energy pyrophosphate bonds between phosphate groups located in a terminal (end) position on the ADP and ATP molecules. Because ATP (see Fig. 3-2) involves three phosphate groups and two pyrophosphate bonds between those groups, more energy is stored and transferred in that compound than in ADP.

The high-energy pyrophosphate bonds are generated in energy-yielding reactions at two separate levels of metabolism. One of these levels is substrate oxidation where hydrogen atoms are transferred to a compound such as phos-phoglyceraldehyde. An example of such an energy-generating reaction would be in the glycolysis process where glucose is broken down. When energy is transferred to phosphoglyceraldehyde, that compound is transformed into a high-energy compound. The chemical rearrangement of the phosphate group into the specific high-energy configuration is termed *phosphorylation.*

A second level of metabolism in which energy is produced and stored in high-energy compounds is along the oxidative chain (cytochrome chain) where hydrogen atoms or electrons are transferred over various carriers to oxygen. Only one high-energy bond can be generated per mole of substrate at the substrate level of phosphorylation discussed in the former paragraph, but ap-proximately 3 moles of high-energy phosphate are formed during the passage of each mole of hydrogen from the oxidative chain to oxygen.

Once the energy has been stored in ADP or ATP, it can be utilized to drive other energy-requiring reactions such as the synthesis of sucrose, starch, and proteins. The process of donating energy to another molecule is also known as phosphorylation. In this reaction ATP is converted back to ADP with a phosphate group left attached to the phosphorylated compound.[21]

Discussion of the role of phosphorus in energy storage and transfer in plants would not be complete without consideration of photosynthesis. Light energy is captured in this process and used to produce sugars that are then used as energy sources for the various energy-requiring reactions in the plant. Sugars are also used for the formation of other plant components including starch, cellulose, organic acids, and proteins.

Phosphorus is involved in the photosynthetic capture of light energy. It was once believed that carbon dioxide received light energy directly, but researchers now know that ATP, the energy currency, is the initial receptor of energy trapped by light striking a chlorophyll molecule.

Formation of ATP through the reaction of ADP with an inorganic phosphate group in the presence of light is termed *photophosphorylation.* The summary reaction of that process is

$$ADP + H_3PO_4 + light \xrightarrow[\text{Cytochromes}]{\text{Coenzyme}} ATP + H_2O$$

3.4.4 Electron Transport

Many processes in the plant involve oxidation-reduction reactions. Compounds known as pyridine nucleotides act as electron or hydrogen trans-porters or carriers in these reactions. Phosphorus is an essential component of these compounds, which are also termed *coenzymes.* Two primary coenzymes are mentioned in the literature: nicotinamide adenine dinucleotide (NAD) and nicotinamide adenine dinucleotide phosphate (NADP). Older literature refers to these materials as DPN and TPN. The oxidized and reduced forms of these compounds are noted by the addition of the letter H to their abbreviations.

[21] P. K. Stumpf, "Phosphate Assimilation in Higher Plants," In *Phosphorus Metabolism,* Vol. II, W. D. McElroy and B. Glass, eds., Johns Hopkins Press, Baltimore, 29 pp.

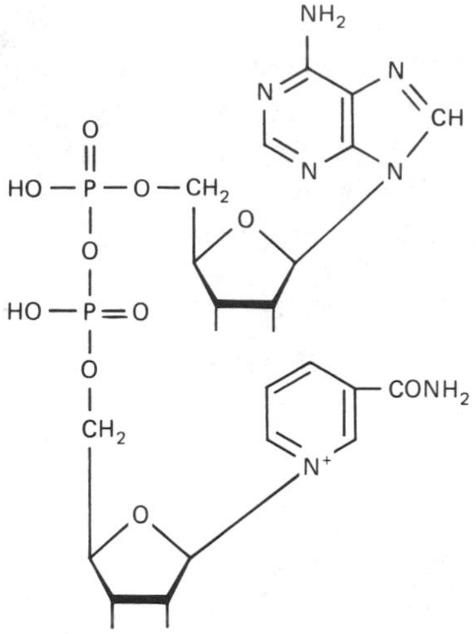

Diphosphopyridine nucleotide (DPN$^+$),
nicotinamide-adenine dinucleotide (NAD$^+$),
or coenzyme I

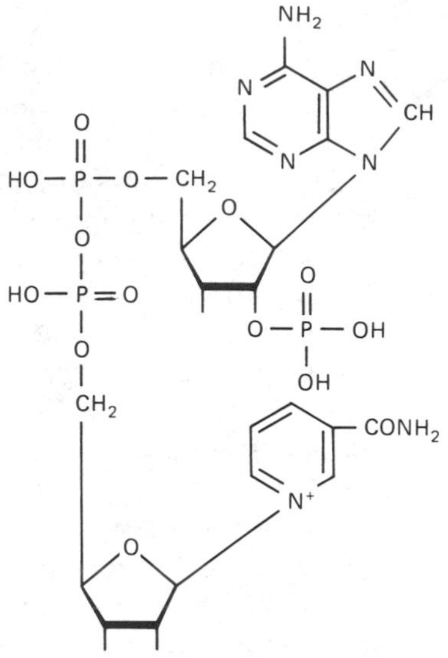

Triphosphopyridine nucleotide (TPN$^+$),
nicotinamide-adenine dinucleotide phosphate
(NADP$^+$), or coenzyme II

Fig. 3-8. Structural formulas of two phosphorus-containing coenzymes involved in electron transfer reactions in plants such as nitrate reduction.

NADH is the reduced form; NAD is the oxidized form of nicotinamide adenine dinucleotide.

These compounds (Fig. 3-8) contain two phosphate groups joined by a pyrophosphate bond. NADP has a third phosphate group separated from the other two. Although these phosphate groups are not directly involved in the chemical reactions as was the case with ADP and ATP, they do affect proper attachment of the coenzymes to the compound being oxidized or reduced. The coenzymes then act to transport electrons or hydrogen between sites of oxidation and reduction reactions. For instance, NADP is reduced in photophosphorylation to NADPH and subsequently transfers electrons to nitrate (NO_3^-) in the reduction of that form of nitrogen to the amino (NH_2^-) form (see Chapter 2). This relation puts phosphorus directly into the nitrogen metabolism scheme. Plants low in phosphorus tend to accumulate nitrate. Similar oxidation-reduction reactions occur in plant metabolism of sulfate and in respiration.

3.4.5 Nutrient Transport

Nutrient absorption against a concentration gradient is discussed in Section 3.4.2. Active absorption requires energy to be expended and the source of that energy is respiration. Energy for the active transport system across the plasma membrane is moved from the point of energy generation to the point of expenditure by the energy currency, ATP. ATP provides energy to the transport mechanism by phosphorylation of a carrier molecule on the inside of the membrane. The carrier travels across the membrane in a cyclic process and attaches to the ion to be brought in through energy provided by ATP. Conditions that limit the production of energy in respiration such as low temperatures and nutrient deficiencies also limit nutrient absorption.

Translocation of nutrients from the roots to the tops has been the subject of much research. Theories of energy requirements for this translocation, aside from the actual movement of the sap in the xylem, suggest a possible energy expenditure in movement from the root cells to the vascular system. Without going into the aspects of those theories, any energy required for such a movement would be supplied by ATP just as in the initial absorption process by the roots.

Movement of other compounds in the plant aside from nutrient ions also involves phosphorus. Street and Lowe[22] reported an example of phosphorus involvement in the movement of sucrose across cell membranes. Their work with tomatoes indicated that sucrose must be phosphorylated by one of the energy currency compounds (ATP) before that uncharged molecule could move across a cell membrane.

3.4.6 Reproduction

Phosphorus is an integral part of the plant reproductive system as a component of the genetic memory system: ribonucleic acid and deoxyribonucleic acid (RNA and DNA, respectively). Specifically, the role of phosphorus in the structure of these two compounds directly or indirectly controls every biochemical reaction occurring in plants.

Phosphorus storage occurs in seeds to prepare them for germination and early growth prior to extensive root development. Phytin, already mentioned as a component of the organic soil phosphorus pool, is the principle storage form of phosphorus in seeds. Very little inorganic phosphorus occurs in seeds, and seedlings must depend on the phosphorus released from phytin to sustain them in the early stages of growth.

3.5 PHOSPHORUS DEFICIENCY SYMPTOMS

Phosphorus deficiency symptoms differ, as expected, in different crops. Generally, a lack of sufficient phosphorus results in a decreased rate of respiration before photosynthesis is slowed. When respiration slows down, sugars start to accumulate in the tissues. As a result of that accumulation, a purple pigment, anthocyanin, develops and gives leaves and lower stems one of the characteristics of phosphorus deficiency. Usually, this pigment develops on the lower leaves because phosphorus tends to be translocated

[22] H. E. Street and J. S. Lowe, "The Carbohydrate Nutrition of Tomato Roots. II. The Mechanism of Sucrose Absorption by Excised Roots," *Ann. Bot.*, 55:307, 1950.

from older leaves to new growth when in short supply. Use of anthocyanin accumulation as a means of determining phosphorus deficiency can be misleading, however. Anthocyanins can also accumulate in seedlings when cold weather slows growth. In such cases, photosynthesis continues but respiration is restricted leading to sugar accumulation that leads to anthocyanin production. Also, not all cultivars of corn and sorghum, for instance, carry the genetic factors for the development of the purple pigment. In such cases, a bronzing effect develops in the same parts of the plants. Small grains also do not show pronounced accumulations of anthocyanins when phosphorus is deficient.

Beside the color effects of phosphorus deficiencies, stunting results in most plants from a lack of sufficient phosphorus. Cotton plants and leaves affected by low phosphorus supplies are stunted but retain their dark green color. Development of secondary branches in cotton also is restricted, which is another expression of the effects of phosphorus on new growth.

Alfalfa and soybeans do not tend to show strong color effects of phosphorus deficiencies but do exhibit stunting (Fig. 3–9). Some reddish discoloration of the stems of both plants can occur when phosphorus is extremely deficient. Seedling development is very restricted in many forage grasses when phosphorus is deficient in the soil. The small seeds of these species do not allow storage of large quantities of phosphorus. Fescue and bromegrass show a cupped appearance of the leaves with phosphorus deficiency as compared to the flatter appearance of normal leaves. Dark green color may accompany phosphorus deficiency in cool-season grasses but also, particularly in established stands, a dark purplish cast may appear down the leaf from the tip. Stems may also show this accumulation of anthocyanins. Despite adequate applications of nitrogen, such grasses usually exhibit very limited growth (stunting) on phosphorus-deficient soils.

In potatoes, phosphorus deficiencies may be reflected more in poor yields than in specific foliage symptoms. Tomatoes and radishes develop anthocyanin accumulation on the underside of their leaves. Tomatoes may also demonstrate a decline in chlorophyll content. Other crops such as celery and carrots develop poor root systems. Cabbage family plants and sweet corn, on the other hand, develop anthocyanins after a general fading out of the chlorophyll in the leaves.

Phosphorus deficiency symptoms in horticultural and forest tree crops is more frequently characterized by slow growth rather than color or foliage effects.

Phosphorus deficiency may frequently be expressed in terms of delayed maturity. Corn, grain sorghum, wheat, tobacco, cotton, and soybeans have exhibited delayed maturity because of phosphorus deficiency. Delayed maturity may also lead to higher grain moisture at the time of harvest. Welch and Boone[23] noted a significant decline in grain moisture of corn with increasing

---

[23] L. F. Welsh and L. V. Boone, "How Fertilizer Affects Field Drying of Corn," *Illinois Research*, 3–4, Spring 1968.

Fig. 3–9. Stunting in alfalfa produced by insufficient soil phosphorus. Plants on the right received 17 kg/ha (15 lb/A) of phosphorus as triple superphosphate. (Courtesy of W. A. Moore and G. A. Raines, Department of Agronomy, Kansas State University.)

rates of phosphorus application (Table 3-9). Keep in mind, however, that many other elements also exert control on crop maturity and this aspect cannot be used as a diagnosis of phosphorus deficiency without other supporting evidence such as soil and plant analyses. (see Chapter 1).

3.6 PHOSPHATIC FERTILIZERS

3.6.1 Organic Phosphorus Sources

Crop residues and animal manures have long been used as sources of phosphorus. The phosphorus concentrations vary in these materials but values for several classes are given in the previous chapter (see Tables 2.23, 2.24, 2.25). Animal manures have been the most common source of organic

Table 3-9. Effect of Phosphorus on Corn Yield and Grain Moisture

P Applied (kg/ha)	1966		1967	
	Moisture (%)	Yield (kg/ha)	Moisture (%)	Yield (kg/ha)
0	31.8	6209	35.4	5833
17	27.8	8216	31.5	7213
35	27.0	8843	29.7	7715
52	26.9	8467	28.2	7965
69	26.5	8718	28.0	8404

SOURCE: L. F. Welch and L. V. Boone, "How Fertilizer Affects Field Drying of Corn," *Illinois Research*, 3–4, Spring 1968.

phosphorus fertilizer but concentrations of phosphorus are low, generally less than 0.25%.

Phosphorus in manures must undergo mineralization prior to being available for plant absorption. Not all the phosphorus contained in animal manure is available the first year due to slow decomposition of some types of organic materials. Large applications of animal manure per acre can provide more than enough phosphorus for crops. Very heavy rates of application over an extended period can increase the available soil test phosphorus to extremely high levels that could be injurious to plant growth due to interference with absorption of micronutrients. Wallingford[24] studied the accumulation of phosphorus in soils from application of large amounts of animal manure and indicated very significant increases in available phosphorus (Fig. 3–10). Similar effects of animal waste applications on phosphorus accumulations in soils have been reported in Texas and California.

Management or animal manure and plant residue applications for phosphorus fertilization dictate the use of soil analyses to determine when accumulations have become significantly high to allow diversion of the manure applications to other land areas. Residual effects of such treatments should be identical to those produced by any other form of phosphorus application, and monitoring by soil analyses is the only technique available to determine when additional applications should be resumed.

Recently some studies with synthesized organic phosphates have been conducted at the University of California at Davis. It was reported that such materials were effective when applied in irrigation water due to a greater degree of movement in the soil than when conventional inorganic sources of phosphorus were utilized. Apparently glycerol phosphate is not readily reactive with soil components and thus moves in the soil for a greater distance. Although this material proved effective, it is not commonly available for grower use and when available it would probably be more expensive than conventional phosphorus sources. Further research and development with this technique may lead to more feasible economics in the future.

[24] G. W. Wallingford, "Effects of Beef-Feedlot Lagoon Water on Soil Chemical Properties and Growth and Composition of Corn Forage," *J. Environ. Qual.*, 3:74, 1974.

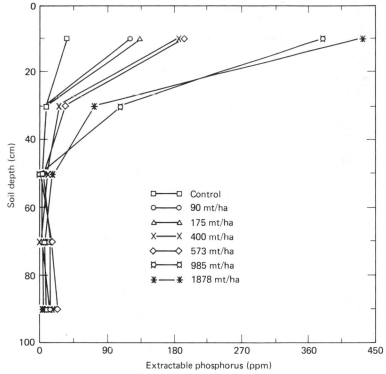

Fig. 3-10. Accumulations of available soil phosphorus were very large from heavy applications of beef feedlot manure in this study conducted in Kansas. Note, however, that only small increases in concentration were noted below 30 cm indicating only slight movement of phosphorus in this silt loam soil. (Source: G. W. Wallingford, "Effects of Solid and Liquid Beef Feedlot Wastes on Soil Characteristics and on Growth and Composition of Corn Forage," Ph.D. Thesis, Department of Agronomy, Kansas State University, 1974.)

3.6.2 Rock Phosphate

The use of rock phosphate as a source of phosphorus for plants has been recognized for over a century. Prior to the direct application of rock phosphate to soil, ground bones were used as a phosphatic fertilizer in Europe. Bones and rock phosphate are similar due to the fact that both contain fluoroapatite. Rock phosphate actually is composed of several different apatite minerals including fluoroapatite, chloroapatite, and hydroxyapatite (see Table 3.2).

Rock phosphate occurs in some parts of the world as a sedimentary deposit. Important deposits and production are located in the countries listed in Table 3-10. In the United States, important production locations are in Florida, North Carolina, Idaho, Wyoming, and Montana.

The geology of phosphate deposits is closely tied with marine geology.

Table 3-10. World Production and Reserves of Phosphate Rock

	Production (1000 metric tons)	Production (%)
North America		
United States	47,256	40.74
Other	279	0.24
South America	754	0.65
Europe		
USSR	24,200	20.86
Other	143	0.12
Africa		
Morocco	17,027	14.68
Tunisia	3,614	3.12
Western Sahara	232	0.02
Other	8,910	7.68
Asia		
China, People's Republic	4,100	3.53
Christmas Island	1,186	1.02
Israel	1,232	1.06
Jordan	1,781	1.54
Vietnam	1,500	1.29
Other	1,687	1.45
Oceania	2,047	1.76
Production, World Total	116,000	100.00

	Total Measured Reserves	Total Identified Reserves
	Millions of Metric Tons	
United States	2,200	8,000
USSR	1,400	3,400
Morocco	18,000	40,000
Jordan	100	300
Australia	—	2,000
All Others	5,300	13,300
World Total	27,000	67,000

SOURCE: W. F. Stowasser, U. S. Bureau of Mines 11, Mineral Commodity Profiles:11. Phosphate. Jan. 1979.

Apatite minerals are widely distributed in igneous rocks but are seldom found in masses of sufficient size for mining. Today, more than 80% of phosphate rock production comes from secondary deposits of marine phosphorite, such as those of the Bone Valley Formation in Florida. Leaching of primary apatite deposits was responsible for moving large amounts of phosphorus into the oceans where the secondary phosphate minerals were deposited after the water became saturated with phosphorus. However, the phosphorus content of the oceans is not uniform. Deep cold waters contain up to 0.3 ppm PO_4^{3-}, whereas warmer, near-surface waters contain as little as 0.01 ppm. Because the solubility of phosphate in upwelling waters is lowered by changes in temperature and pH, phosphate may be precipitated by organic or inorganic processes. Phosphorites, along with shale, chert, and limestone, are depositional products that change in composition as the result of physical, chemical, and biological processes.

The Bone Valley Formation in central Florida, where the largest deposits of phosphate rock occur, was formed 10 to 15 million years ago during the late Miocene or Pliocene Age. The phosphate deposits were first discovered south of Bartow on the Peace River about 1881 by an Army engineer, Capt. Francis LeBaron. Actual mining began later in that decade and by the end of the 19th century as many as 200 phosphate mining companies had been formed.

The known marine phosphorite deposits of central Florida occur in an area roughly 60 by 100 km centered around the town of Mulberry. Phosphatic land pebble deposits are recovered from the lower part of the Bone Valley Formation (Pliocene age) and from the upper part of the underlying Hawthorne Formation (Miocene age). The matrix (ore) ranges in thickness from 0.3 to 11 m and is overlain by 3 to 6 m of sand and clay. Groundwater is close to the surface in these areas. Below the phosphorite deposit is the solid or weathered phosphatic bearing Hawthorne limestone (Fig. 3-11).

Generally, deposits are mined by strip-mining methods similar to that for coal. Regardless of whether the mining occurs in the United States, Morocco, or other countries, techniques are identical. Usually draglines or large shovels are used to remove the overburden from the ore (Fig. 3-12). Depth of the ore bodies varies with the location within a given mining area. Usually, however, the deposits are covered with something less than 20 m of overburden.

The composition of the ore itself varies with location although certain minerals are common to all sedimentary deposits. Concentrations of phosphorus within the ore bodies currently mined ranges from a high of 17% to a low of 7%. Much lower concentrations of phosphorus in rock phosphate do occur but are not economically feasible for mining at present world prices. Such deposits may eventually be opened as the demand for phosphorus increases and the amount available declines.

After the ore is removed from the mine mechanically, it is transported by different techniques to the refinery. Some processors use a slurrying technique in which high-pressure jets of water convert the ore into a slurry that is subsequently pumped to the refinery for dewatering and beneficiation (concentration). Other mines transport the ore to the processing plants by truck or by rail.

When the rock phosphate ore arrives at the plant, it is subjected to processes that remove impurities such as sand and clay. Sand is separated from the phosphate rock by a hydro separation process that involves the use of spiral separators. The spiral nature of this equipment allows the lighter sand

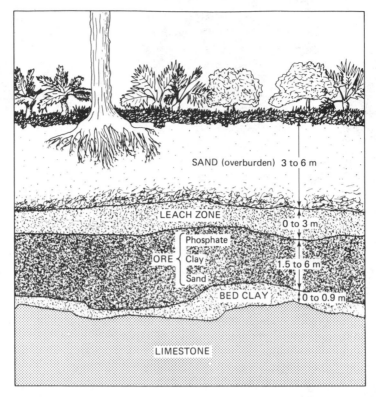

Fig. 3–11. Florida phosphate rock deposits are overlain by sand and clay varying in depth from 3 to 6 m (10–20 ft). Below the phosphate ore is the Hawthorne limestone. (Source: IMC Corporation.)

to be washed over the outer edge of the spiral with the water. The heavier phosphate ore pebbles are concentrated at the inner edge of the spiral and subsequently the phosphorus concentration is increased.

Still remaining in the rock phosphate slurry, however, is clay, another common contaminant of the ore. Clay is partially separated from the phosphate ore by a skimming process in which the rock phosphate pulp is allowed to settle in large tanks and the clay suspension rises to the surface and is skimmed off mechanically. This process works well for separation of the large rock phosphate particles from the clay but significant amounts of very fine phosphate particles remain suspended with the clay.

In the earlier days of processing, this mixture of clay and rock phosphate was considered worthless and was discarded because no process existed for separation of the finely divided phosphate ore and clay. In the past several years, a flotation process has been introduced that is similar to the flotation process used in the separation of sodium and potassium chlorides discussed in Chapter 4. The clay-colloidal phosphate suspension is directed into flotation cells into which an organic reagent and compressed air are introduced along with agitation. The resulting bubbles attract colloidal phosphate particles and

Fig. 3–12. Dragline stripping operation in Florida mining phosphate rock. The dragline operator drops the phosphate ore in this sump where high-pressure water jets slurry the ore that is pumped several miles to the processing plant for beneficiation. (Courtesy IMC Corporation.)

float them to the surface where they are skimmed off and recovered (Fig. 3–13). The reagent is removed by washing and the outflow at the bottom of these flotation cells, which contains primarily clay, is discarded. The beneficiated (concentrated) rock phosphate still has the same chemical composition as far as the phosphorus compounds are concerned. Some free calcium and magnesium carbonates will exist in this material and iron and aluminum phosphates and hydroxides may also be present. As we see in Section 3.6.5, these materials are detrimental to the formation of phosphatic fertilizers. Beneficiated rock phosphate usually contains approximately 13% elemental phosphorus (30% P_2O_5) and is ground, dried, and stored prior to being processed into other phosphatic fertilizers or shipped to consumers.

Beneficiated or unbeneficiated rock phosphate may be used for direct land applications on acid soils with positive effects on plant growth. Obviously, the use of raw rock phosphate or beneficiated rock phosphate would pose a cheaper source of phosphorus than compounds that are produced by additional industrial manipulation. Prior to the development of processes that improve the solubility of phosphate rock, very large tonnages of rock phosphate

Fig. 3-13. Flotation processes are used to recover very small particles of rock phosphate that were lost in the earlier days of phosphate mining. Compressed air, a special reagent and mechanical agitation produce bubbles that adsorb the small particles of rock phosphate and float them out of the cell. (Courtesy IMC Corporation.)

were applied directly to soils in the eastern half of the United States. The insoluble phosphorus compounds in rock phosphate will dissolve only under acid conditions (see Chapter 8). Because soils in the eastern half of the United States are generally acid, this technique was feasible as a source of phosphorus when relatively large amounts of phosphorus were applied. Beneficiated rock phosphate may contain around 1 to 1.5% available phosphorus and thus is not a good material when phosphorus is required by plants immediately, such as in a starter fertilizer. Still, on soils that are quite acid, the material may be practical in terms of improving soil phosphorus availability over a period of years.

Presently, the International Fertilizer Development Center at Muscle Shoals, Alabama, is working on projects involving direct use of rock phosphate on the highly phosphorus-deficient soils of Africa. Direct use of raw rock phosphate in those areas may have an additional benefit, that being a savings of energy due to the elimination of such energy-requiring processes as sulfur production for the manufacture of sulfuric acid used in phosphate fertilizer manufacturing.

The efficacy of the direct use of rock phosphate as a phosphatic fertilizer

Table 3-11. Comparative Performance of Rock Phosphate and Superphosphate as Phosphorus Sources for Crops on Acid Soils—Arkansas

Treatments[a]	Average Annual Yield Increases Versus Control Plots								
	1921-24	1925-28	1929-32	1933-36	1937-40	1941-44	1945-48	1949-51	1921-51
	Corn Grain (kg/ha)								
Superphosphate[b]	659	684	596	263	633	1248	659	1191	753
Rock phosphate[b]	351	426	376	201	721	1154	659	985	659
	Oat Grain (kg/ha)								
Superphosphate	538	545	663	462	957	1079	1541	763	821
Rock phosphate	391	244	236	254	609	817	1258	738	563
	Wheat Grain (kg/ha)								
Superphosphate	558	605	914	1048	1116	1263	1210	1317	995
Rock phosphate	329	363	423	625	571	954	840	1042	632

[a] All plots received N, K, and lime in addition to P.

[b] Superphosphate, 7% P applied at rate of 280 kg/ha material. Rock phosphate, 14% P applied at 280 kg/ha material.

SOURCE: E. O. McLean, D. A. Brown, and C. A. Hawkins, "Comparative Evaluation Studies on Rock and Superphosphate," Ark. Agr. Exp. Sta. Bul. 528, 1952, p. 29.

is indicated in the data presented in Table 3-11. Both crop yield and available soil phosphorus are increased by applications on acid soils. The low solubility of the phosphorus minerals in rock phosphate and the fact that they are largely calcium salts is depressed further by high pH and calcareous soil conditions. Rock phosphate used on such soils is not likely to be an effective form of phosphorus because of the low solubility. However, increasing costs of energy dictate continuing evaluation of rock phosphate.

3.6.3 Normal Superphosphate

Improvement in the effectiveness of rock phosphate as a source of phosphorus for crops was noted in England in the 1840s when bone meal and phosphate rock were acidulated with sulfuric acid. The patent that was issued on this process in 1845 was eventually used to fund the establishment of the Rothamsted Experiment Station in England. The material produced by this process has been known since the 1840s as single superphosphate, normal superphosphate (NSP), or ordinary superphosphate (OSP) in deference to a more concentrated material that is discussed in the following sections.

The process used in the manufacture of normal superphosphate (NSP) involves mixing ground, beneficiated rock phosphate with sulfuric acid of approximately 60 to 72% concentration and allowing the mass to react (cure) for a period of up to 6 weeks. The process chemistry depicted in Fig. 3-14

Normal superphosphate

$$[Ca_3(PO_4)_2]_3 \cdot CaF_2 \ + \ 7\,H_2SO_4 \longrightarrow 3\,Ca(H_2PO_4)_2 \ + \ 7\,CaSO_4 \ + \ 2\,HF$$

Rock phosphate   Sulfuric acid   Monocalcium phosphate   Gypsum   Hydrofluoric acid

Fig. 3–14. Theoretical reactions involved in the production of normal superphosphate.

demonstrates that the fluoroapatite of the rock phosphate is converted to monocalcium phosphate with the simultaneous production of gypsum. This reaction is not instantaneous and the curing process then is required to allow reactions to be completed and for the material to be physically conditioned. Otherwise, the incompletely reacted mass would have a very acid pH and would be of very poor physical characteristics (sticky, wet).

Several different industrial processes for the manufacture of normal superphosphate have been devised. Originally, the process involved mixing the batches of rock phosphate and sulfuric acid together and dropping them into a container known as a "den" that had a removable wall (Fig. 3–15). After a short initial curing period, the block of normal superphosphate was cut out of the den and conveyed to a pile for additional curing. At the end of the

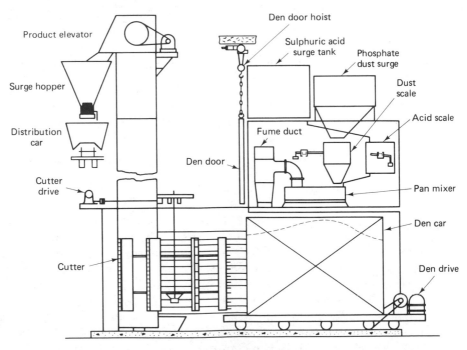

Fig. 3–15. Batch production of normal superphosphate involves the use of a curing den where reactions between the sulfuric acid and the rock phosphate occur. The solid mass of superphosphate is cut out of the den mechanically and sent to a storage pile for completion of the reactions. (Source: Tennessee Valley Authority.)

prolonged curing period, the material was broken up, screened, and bagged or shipped in bulk.

More recently, superphosphate production has involved the use of a continuous process (Fig. 3–16) where the materials react for a relatively short period of time prior to being removed automatically and transported to a curing pile where normal chemical reactions are completed.

The quality of the phosphate rock influences the availability of phosphorus in the resulting normal superphosphate. Large concentrations of calcium carbonate in the rock phosphate require greater consumption of sulfuric acid and increase the gypsum content of the end product, lowering the phosphorus concentration. Normal superphosphate has a phosphorus concentration ranging from 8 to 8.8% P (18–20% P_2O_5).

The low concentration of phosphorus in the material is one of the main disadvantages to its use in the world fertilizer trade today. For many years normal superphosphate was the world's only processed phosphatic fertilizer, but in recent years it has fallen into disfavor because of the high freight costs per unit of phosphorus. Substantial quantities of the material are still used in various parts of the world but much higher concentrations of phosphorus are now possible in other materials and subsequently their use has climbed far above that of normal superphosphate. Currently, one attractive use of the material in fertilizer formulation is for the production of ammoniated mixed fertilizers (ammoniated superphosphates). The advantage of the use of ammonia in the formulation of mixed fertilizer is that this nitrogen source is

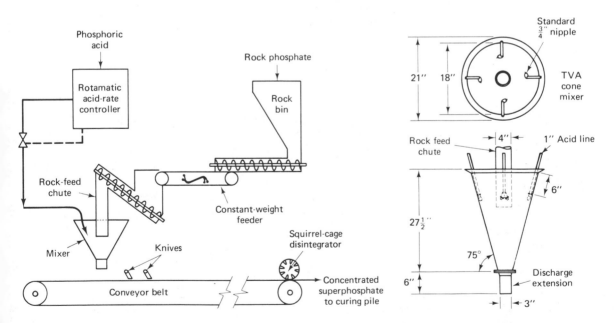

Fig. 3–16. Schematic drawing of the process for the manufacture of triple superphosphate involving continuous flow and use of the TVA cone mixer (inset). (Source: Tennessee Valley Authority.)

Ammoniation of normal superphosphate

$$Ca(H_2PO_4)_2 \quad + \quad NH_3 \quad \longrightarrow \quad CaHPO_4 \quad + \quad NH_4H_2PO_4$$

Monocalcium phosphate   Ammonia   Dicalcium phosphate   Monoammonium phosphate

$$NH_4H_2PO_4 \quad + \quad CaSO_4 \quad + \quad NH_3 \quad \longrightarrow \quad CaHPO_4 \quad + \quad (NH_4)_2SO_4$$

Monoammonium phosphate   Calcium sulfate   Ammonia   Dicalcium phosphate   Ammonium sulfate

$$2\,CaHPO_4 \quad + \quad CaSO_4 \quad + \quad 2\,NH_3 \quad \longrightarrow \quad Ca_3(PO_4)_2 \quad + \quad (NH_4)_2SO_4$$

Dicalcium phosphate   Calcium sulfate   Ammonia   Tricalcium phosphate   Ammonium sulfate

Ammoniation of triple superphosphate

$$Ca(H_2PO_4)_2 \quad + \quad NH_3 \quad \longrightarrow \quad CaHPO_4 \quad + \quad NH_4H_2PO_4$$

Monocalcium phosphate   Ammonia   Dicalcium phosphate   Monoammonium phosphate

$$3\,CaHPO_4 \quad + \quad NH_3 \quad \longrightarrow \quad NH_4H_2PO_4 \quad + \quad Ca_3(PO_4)_2$$

Dicalcium phosphate   Ammonia   Monoammonium phosphate   Tricalcium phosphate

Ammoniation usually expressed in terms of kg NH_3 per 20 kg of P_2O_5. For normal superphosphate, the normal range is 4–6 kg NH_3 per 20 kg P_2O_5; for tripple super-phosphate, 3–4 kg NH_3 per 20 kg P_2O_5.

Fig. 3–17. Ammoniation of superphosphates must be strictly controlled to prevent reversion of the water-soluble phosphorus compounds to less available forms. Excessive ammoniation tends to produce water insoluble tricalcium phosphate.

usually cheaper per unit of nitrogen. Normal superphosphate reacts with more ammonia without problems of ammonia retention than more concentrated superphosphates. The difference is due to the presence of gypsum in normal superphosphates.

Excessive ammoniation of normal superphosphate or triple superphosphate causes reversion of the phosphorus compounds to more insoluble compounds. The reactions outlined in Fig. 3–17 indicate what happens when ammonia is mixed with normal superphosphate. Upon excessive ammoniation, tricalcium phosphate is produced, a compound with very low water solubility. Although the presence of dicalcium phosphate in ammoniated superphosphate is allowable, tricalcium phosphate is undesirable. Subsequently, experience has shown that an ammoniation rate of approximately 3 kg of ammonia per 10 kg of P provided as normal superphosphate is the maximum allowable rate without excessive reversion and lowered phosphorus availability. Ammoniation of triple superphosphate, on the other hand, which also contains monocalcium phosphate as a prime phosphorus material but less gypsum, allows use of only about 2 kg of ammonia per 10 kg of phosphorus. Above that rate, reversion results.

Normal superphosphate has good agronomic qualities and high water solubility, and it provides an excellent source of sulfur in addition to phosphorus. Approximately 14% sulfur is present in most normal superphosphates.

3.6.4 Phosphoric Acid

Orthophosphoric acid (H_3PO_4) is produced from the acidulation of rock phosphate with sulfuric acid. The difference in this reaction and the one that produces normal superphosphate is that a higher concentration of sulfuric acid is used ranging up to 93%. The same beneficiated rock phosphate that might be used to produce normal superphosphate can be treated with sulfuric acid by a continuous or batch process. Due to the higher concentrations of the sulfuric acid the phosphate minerals are solubilized. The resulting impure phosphoric acid contains large amounts of gypsum from reactions of the sulfuric acid with calcium. During the acidulation process in production of both phosphoric acid and normal superphosphate, fluorine present in the phosphate rock is liberated as hydrogen fluoride (HF) unless silica is added to the reacting medium to trap the highly corrosive hydrogen fluoride.

The impure phosphoric acid produced contains gypsum that must be filtered out of the acid to produce a material that is suitable for further industrial uses. Vacuum filtration is used for separation of the gypsum from the phosphoric acid. The waste gypsum ($CaSO_4 \cdot 2H_2O$) is a perfectly good source of sulfur and is utilized as a soil amendment in many areas of sodic soils (see Chapter 9). Some emerging countries such as India also utilize the gypsum produced in this manner for the production of ammonium sulfate by reacting ammonia directly with the gypsum.

The amount of sulfuric acid that must be added to beneficiated rock phosphate to produce the desired phosphoric acid is a function of the quality of the rock phosphate but varies with the presence of contaminants such as calcium carbonate and magnesium carbonate. These two carbonates react with the acid and reduce the amount of fluoroapatite that is converted into H_3PO_4.

As a further step in the process of producing phosphoric acid of approximately 17 to 24% P (40–54% P_2O_5) content, the impure, dilute (13% P) phosphoric acid that has had the gypsum removed is heated in a vacuum to drive off the water.

Phosphoric acid has been and continues to be used directly as a phosphatic fertilizer although problems in handling the material occur because of its corrosive nature. Figure 3–18 shows an applicator currently being used in the regions of western Idaho, eastern Washington, and Oregon for wheat production that allows simultaneous application of anhydrous ammonia and orthophosphoric acid or other liquid source of phosphorus. Anhydrous ammonia and phosphoric acid are handled by two separate lines on each shank. The two compounds are delivered to the same soil zone, however, where they react to produce varying types of ammonium phosphates. Proponents of this type of phosphorus fertilization suggest that benefits arise from the placement of ammonia and phosphorus together due to

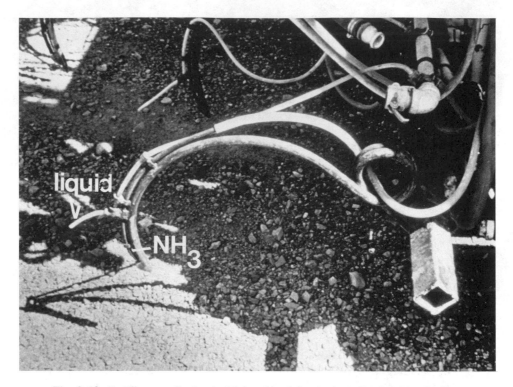

Fig. 3–18. Fertilizer applicator in Idaho rigged for dual application of anhydrous ammonia and phosphoric acid or ammonia and ammonium polyphosphate liquid. The ammonia and liquid lines are separated at the delivery point to prevent the expanding ammonia from freezing the liquid. (Courtesy Cenex, Inc.)

a slowdown of fixation reactions brought on by both the presence of high concentrations of ammonia and high concentrations of phosphorus in a limited amount of soil.

Phosphoric acid produced as indicated in the preceding discussion is known in the fertilizer industry as *wet process acid, green acid, black acid,* or *merchant-grade acid.* The colors are caused by the presence of impurities. Black is due to the presence of traces of carbon, particularly from phosphate deposits in the western part of the United States (Wyoming, Idaho, and Montana). The color of the acid has nothing to do with the chemistry of the phosphorus and the phosphorus contained in such a material is fully as effective as phosphorus produced by the furnace method.

Furnace-grade orthophosphoric acid results from the subjection of rock phosphate to an electric arc furnace that produces elemental phosphorus. Elemental phosphorus is reacted with oxygen to produce phosphorus pentoxide (P_2O_5), which is subsequently absorbed in water to produce phosphoric acid (H_3PO_4). This process produces a pure phosphoric acid that has essentially the same phosphorus content as the wet acid. This form of phosphoric acid is utilized in the formulation of a very limited number of fertilizer materials and is characterized by its high price compared to wet process acid. Despite claims that have been made to the contrary, the properties of the two forms of orthophosphoric acid in terms of soil reaction and availability to plants are identical.

Wet process phosphoric acid is used in much larger amounts for the production of other phosphatic fertilizers than for direct application as a phosphorus source.

3.6.5 Triple Superphosphate

Triple superphosphate (TSP) was developed to increase the phosphorus concentration in superphosphate. This material varies somewhat today in its phosphorus content but ranges from approximately 17 to 21% P (40–46% P_2O_5). The major phosphorus compound in triple superphosphate is monocalcium phosphate, the same phosphorus compound present in largest concentrations in normal superphosphate. The basic difference between these two materials is the absence of gypsum in triple superphosphate.

The production of triple superphosphate involves the treatment of rock phosphate with wet process phosphoric acid. The beneficiated, dried rock phosphate is subjected to phosphoric acid in either a batch or continuous process. Like the reactions producing normal superphosphate, these reactions do not go to completion immediately but must be allowed to react over an extended time period, usually about 30 days. The chemistry of the process can be somewhat complex if one considers all the intermediate reactions. They are summarized as a general reaction that is given in Fig. 3–19. Excessive amounts of magnesium and calcium carbonates in the rock phosphate tend to produce a lower solubility of phosphorus due to some of the fluoroapatite not being solubilized by the reaction with the phosphoric acid.

Triple superphosphate is widely used in the manufacture of other fertilizer grades in dry bulk blends or for direct application. It has a high water solubility, can be ammoniated, and in general is an excellent phosphorus source. Agronomically, there is little difference in the availability of phosphorus from either normal or triple superphosphate. The sulfur content of triple superphosphate is quite low (0–1% S) due to the absence of sulfuric acid in the formulation process. Some small amounts of sulfur may be present from contaminants of sulfuric acid in the phosphoric acid.

Triple superphosphate

$$[Ca_3(PO_4)_2]_3 \cdot CaF_2 \ + \ 12\,H_3PO_4 \ + \ 9\,H_2O \longrightarrow 9\,Ca(H_2PO_4)_2 \ + \ CaF_2$$

Rock phosphate   Phosphoric acid   Water   Monocalcium phosphate   Calcium fluoride

Fig. 3–19. Theoretical reactions involved in the production of triple superphosphate.

3.6.6 Ammonium Phosphates

The principal ammonium phosphates used for chemical fertilizers are *monoammonium phosphate, diammonium phosphate,* and *ammonium polyphosphate.*

Monoammonium Phosphate. Monoammonium phosphate (MAP) has been an important source of phosphorus for plants for many years. This compound is produced by the simple neutralization of wet process phosphoric acid with ammonia or ammonium hydroxide. The simple reaction given in Fig. 3–20 produces a compound of high water solubility and good agronomic properties. Recently, monoammonium phosphate has come back into favor in the formulation of suspension fertilizers because of its production from relatively low-quality phosphoric acids and because it can be ammoniated by utilizing the cheapest source of nitrogen, anhydrous ammonia.

The development of ammonium phosphates was accelerated by research indicating that the presence of the ammonium ion enhanced the plant uptake of the phosphate ion.[25,26] Part of this effect may be due to the process of forced ammonium absorption by plants when monoammonium phosphate is used as a starter material. The presence of ammonium ions in the vicinity of the roots and subsequent absorption of the ammonium ions leads to the expulsion of a hydrogen ion from plant roots maintaining a low pH in the vicinity of the roots. Ammonium ions originating from a phosphorus compound may also tend to enhance the availability of the phosphorus by keeping calcium fixation at a minimum.

Concentrations of nutrients in monoammonium phosphate range from 11 to 13% nitrogen and 21 to 24% P (48–55% P_2O_5). The material mixes well with nitrogen materials and potassium sources, and it may serve as an ingredient of bulk blends.

Consumption of MAP and other phosphorus sources in the United States is indicated in Fig. 3–21.

Diammonium Phosphate. Over a period from approximately 1965 to 1980, the production and use of monoammonium phosphate (MAP) was overshadowed by the production and use of diammonium phosphate (DAP). Diammonium phosphate involves the same acid-base chemistry as the production of monoammonium phosphate. Two moles of ammonia are added per mole of phosphoric acid in the neutralization reaction (see Fig. 3–20).

The primary advantage in the manufacture of diammonium phosphate was in relation to its higher nitrogen content, which ranged from 18 to 21% nitrogen. Phosphorus concentrations normally range from 20 to 23% P (46–53% P_2O_5).

[25] M. H. Miller and A. J. Ohlrogge, "Principles of Nutrient Uptake from Fertilizer Bands. I. Effect of Placement of Nitrogen Fertilizer on the Uptake of Band-Placed Phosphorus at Different Soil Phosphorus Levels," *Agron. J.,* 50:95, 1958.

[26] R. A. Olson and A. F. Drier, "Nitrogen—A Key Factor in Fertilizer Phosphorus Efficiency," *Soil Sci. Soc. Amer. Proc.,* 20:509, 1956.

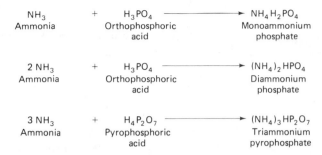

Fig. 3-20. Theoretical reactions involved in the production of monoammonium phosphate, diammonium phosphate, and triammonium pyrophosphate (ammonium polyphosphate) fertilizers. The first two compounds are both orthophosphates.

Characteristically, the pH of a saturated solution of diammonium phosphate (pH 8.0) is higher than that of monoammonium phosphate (pH 4.0) due to the presence of more ammonium in the former material. The fact that a higher pH is associated with the saturated solution of diammonium phosphate gives rise to one of the agronomic characteristics of the material. Application of large quantities of diammonium phosphate in direct seed contact on high pH soils is conducive to the same type of germination damage produced by seed contact of urea or anhydrous ammonia. As discussed in

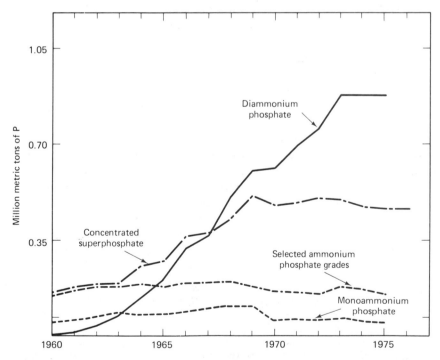

Fig. 3-21. Consumption of several phosphate fertilizer materials in the United States. (Source: USDA, Commercial Fertilizers, Statistical Reporting Service, Annual Reports.)

Chapter 2, the ammonia-water equilibrium under high pH tends to shift to the left forcing the formation of ammonia molecules. Ammonia exerts some toxic influence on plant growth when in direct seed contact and germination damage from high concentrations of DAP have been observed (Allred and Ohlrogge[27]). On acid soils this problem does not seem to be particularly great. Because most starter fertilizer applications placed in direct seed contact do not involve extremely large quantities of material, the germination damage problems with DAP become more academic. In the wheat production areas of the Great Plains of the United States, general recommendations call for a maximum of 100 kg/ha (89 lb/A) of DAP in direct seed contact to avoid germination damage. Under extremely calcareous, high-pH situations, that value could be high. Most seed applications of DAP for corn, grain sorghum, soybeans, and small grains do not involve applications higher than 67 kg/ha (60 lb/A) of material. High concentrations of any soluble salts in direct seed contact introduce the risk of salt injury through plasmolysis.

The water solubility of monoammonium phosphate is lower than that of diammonium phosphate at lower temperatures, 23 g/100 ml of solution at 10°C versus 39 g/100 ml of solution at 10°C, respectively. At higher temperatures, the solubilities become 38 g/100 ml at 40°C for MAP and 47 g/100 ml for DAP. Simultaneously, the ammonia vapor pressure of DAP solutions increases as the temperature rises but the ammonia vapor pressure of MAP solutions is largely unaffected.

These characteristics produce a basis for possible agronomic differences between these two types of phosphate fertilizers when the materials are placed close to the seed. Beaton and Read[28] studied the initial effects of sources of phosphorus on growth of oats on a calcareous soil. Their data indicated that MAP provided more phosphorus for the plants than did DAP. Tucker[29] also reported differences in the effectiveness of phosphorus sources for wheat grown on calcareous soils in Oklahoma 6 weeks after planting (Table 3–12). In both cases, however, placement of the phosphorus was close to the seed. Tucker rated MAP and DAP as equally effective on acid and neutral pH soils. Similar observations (Fig. 3–22) have been made in Kansas where comparisons of several types of phosphorus fertilizers under neutral soil pH conditions showed no differences in the effects of MAP and DAP on wheat plant growth and yield when these materials were placed in direct seed contact at a constant rate of 112 kg/ha of material. Placement of urea-ammonium phosphates in direct seed contact at that location did produce germination damage, however.

[27] S. E. Allred and A. J. Ohlrogge, "Principles of Nutrient Uptake from Fertilizer Bands. VI. Germination and Emergence of Corn as Affected by Ammonia and Ammonium Phosphate," *Agron. J.*, 56:309, 1964.

[28] J. D. Beaton and D. W. L. Read, "Effects of Temperature and Moisture on Phosphorus Uptake from a Calcareous Saskatchewan Soil Treated with Several Pelleted Sources of Phosphorus," *Soil Sci. Soc. Amer. Proc.*, 27:61, 1963.

[29] B. B. Tucker, "Phosphorus Fertilization in Oklahoma," Dept. of Agronomy, Oklahoma State Univ., Mimeographed Report, 1969.

Table 3–12. Effectiveness of Several Phosphorus Sources for Wheat

Type of Phosphate	Analysis (%)			Soil Reaction		
	N	P	K	Acid	Neutral	Calcareous
Ordinary super	0	9	0	1[a]	2	4
Triple super	0	23	0	1	2	4
Diammonium	21	23	0	1	1	3
Diammonium	18	20	0	1	1	3
Monoammonium	11	21	0	1	1	1
Ammonium poly	11	16	0	1	1	2

[a] Relative rating of wheat growth 6 weeks after emergence. The lower the number, the greater the growth.

SOURCE: B. B. Tucker, "Phosphorus Fertilization in Oklahoma," Mimeographed Circular, Dept. of Agronomy, Oklahoma State University, 1969.

Aside from the placement effects that could occur when MAP and DAP are compared, other agronomic effects of the two phosphorus sources are similar. Although these phosphorus sources have historically been used for direct application or in bulk blends, both materials can be utilized in the production of suspension fertilizers. Ammonium orthophosphate solutions have

(b)

(c)

(a)

Fig. 3–22. Different phosphorus fertilizers can exert varying effects on wheat germination and growth when placed in direct seed contact. In this series of pictures, (a) monoammonium phosphate (11–48–0), (b) diammonium phosphate (18–46–0), and (c) urea-ammonium phosphate (34–17–0) were applied in seed contact at a constant rate of 112 kg/ha (100 lb/A). Monoammonium phosphate and diammonium phosphate produced identical effects on this slightly acid soil. The urea in the urea-ammonium phosphate injured germination of the wheat. (Courtesy L. S. Murphy.)

been relatively rare in recent years due to the low phosphorus concentrations in such solutions.

Ammonium Polyphosphates. Another class of ammonium phosphates has developed in the past 20 years, the ammonium polyphosphates. The term *polyphosphate* is a generality that includes several different compounds including ammonium pyrophosphate, ammonium tripolyphosphate, and ammonium tetrapolyphosphate.

Generally, the ammonium polyphosphates are characterized by higher phosphorus concentrations because two molecules of orthophosphate have been condensed through the elimination of water (Fig. 3–23). As a result, the concentration of phosphorus in polyphosphates is higher due to the elimination of water (Table 3–13). The fertilizer industry, particularly the liquid fertilizer segment, has been strongly involved in the use of ammonium polyphosphates in their formulations.

The chemistry of the ammonium polyphosphates also allows them to react with metals such as magnesium, iron, and zinc with a net twofold effect. Magnesium and iron are frequent contaminants of wet process phosphoric

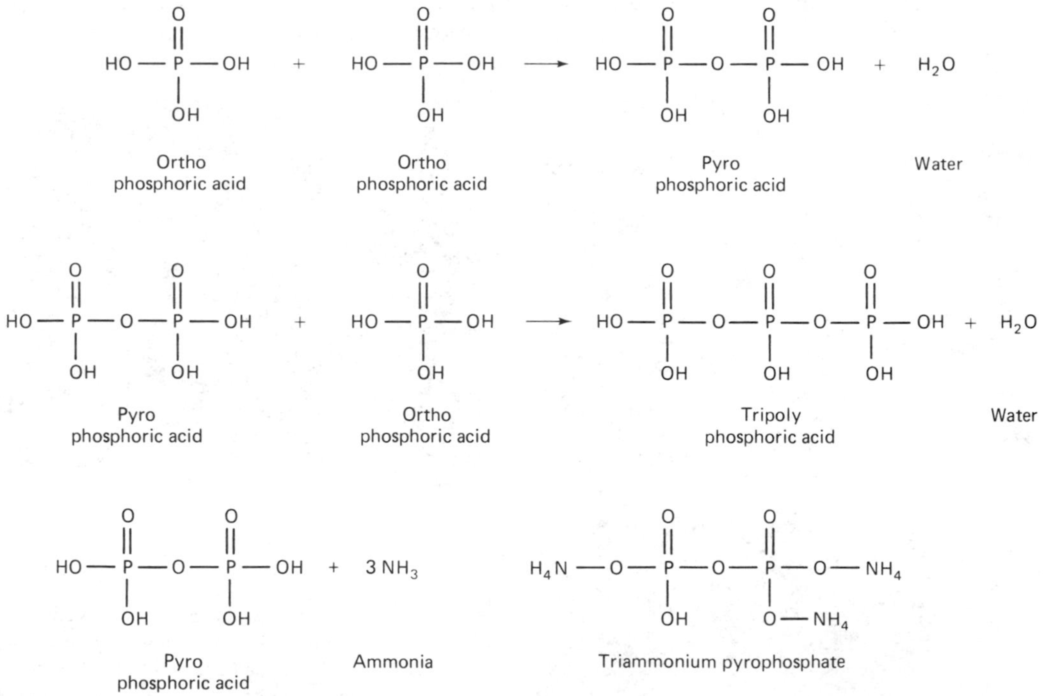

Fig. 3–23. Synthesis reactions for the formation of pyrophosphoric acid (superphosphoric acid). The reaction is a condensation of two molecules of orthophosphoric acid by splitting out one molecule of water. Longer-chain acids (tripolyphosphoric acid) can be formed by further condensation. Ammoniation produces ammonium polyphosphates. (Courtesy L. S. Murphy.)

Table 3-13. Composition of Some Phosphatic Fertilizers

Material	Frequently Used Abbreviations	Analysis (%)				Form of Phosphorus	Percentage Total P Available	Formula of Main P Compound
		N	P_2O_5	K_2O	S			
Rock phosphate	RP	—	25-40	—	—	Orthophosphate	14-65	$[Ca_3(PO_4)_2]_3 \cdot CaF_2$
Rhenania phosphate	—	—	28	—	—	Orthophosphate	97	$Ca_4P_2O_9$
Normal or ordinary superphosphate	NSP or OSP	—	16-22	—	11-12	Orthophosphate	97-100	$Ca(H_2PO_4)_2$
Wet process phosphoric acid	—	—	48-53	—	—	Orthophosphate	100	H_3PO_4
Triple superphosphate	TSP or CSP	—	44-53	—	1-1.5	Orthophosphate	97-100	$Ca(H_2PO_4)_2$
Ammonium phosphates								
Monoammonium phosphate	MAP	11-13	48-62	—	0-2	Orthophosphate	100	$NH_4H_2PO_4$
Diammonium phosphate	DAP	18-21	46-53	—	0-2	Orthophosphate	100	$(NH_4)_2HPO_4$
Ammonium polyphosphate	APP	10-15	35-62	—	—	Mixture of ortho and polyphosphates	100	$(NH_4)_3HP_2O_7 + NH_4H_2PO_4$ + others
Urea-ammonium phosphate	UAP or UAPP	21-34	16-42	—	—	Mixture of ortho and polyphosphates	100	$NH_4H_2PO_4, (NH_4)_3HP_2O_7$
Nitric phosphates	NP	14-29	14-28	0-20	—	Orthophosphate	80-100	$CaHPO_4, NH_4H_2PO_4$
Ammoniated normal superphosphate	—	2-5	14-21	—	9-11	Orthophosphate	97-100	$NH_4H_2PO_4, CaHPO_4$
Ammoniated triple superphosphate	—	4-6	44-53	—	0-1	Orthophosphate	96-100	$CaHPO_4, NH_4H_2PO_4$
Basic slag	—	—	9-18	—	0-1	Orthophosphate	60-90	Variable
Potassium phosphates								
Monopotassium phosphate	—	—	51	35	—	Orthophosphate	100	KH_2PO_4
Dipotassium phosphate	—	—	41	54	—	Orthophosphate	100	K_2HPO_4
Potassium polyphosphate	—	—	51	40	—	Polyphosphate and orthophosphate	100	$K_3HP_2O_7, KH_2PO_4$, others

acid, having been present in the rock phosphate initially. When liquid fertilizers are manufactured from wet process orthophosphoric acid, these two metals, magnesium and iron, lead to precipitation of their phosphates. The resulting sludge has a lowering effect on the phosphorus concentration of the solution and also makes the material extremely difficult to pump and spray. The benefit from the presence of polyphosphates in solutions is the fact that magnesium and iron can be sequestered by being attached to the various polyphosphate species between two adjacent hydroxyl groups. The net effect is to render these materials more soluble and in essence prevent the precipitation that would occur if only orthophosphate was present in the fertilizer solution. An example of the technique involved in sequestering is demonstrated in Fig. 3–24.

The other advantage for the use of polyphosphates through the sequestering reaction is the fact that these materials are able to maintain higher concentrations of certain micronutrient metals in solution without precipitation. One of the most common utilizations of this phenomena is for the incorporation of zinc into liquid fertilizers. Zinc orthophosphates are quite insoluble and tend to precipitate quickly. Research at the Tennessee Valley Authority has determined the concentrations of metals that can be maintained in orthophosphate- and polyphosphate-containing liquid fertilizers. That information is summarized in Table 3–14. Note the much higher zinc concentrations where polyphosphates represent a significant percentage of the phosphate anions in a fertilizer solution.

In terms of fertilizer-manufacturing technology, polyphosphates have the advantages of higher concentrations of phosphorus in solutions and are able to maintain higher concentrations of some micronutrient metal in solution without precipitation.

To be strictly correct in terms of the notation "polyphosphate," it should be recognized that most fertilizers that contain polyphosphates also contain orthophosphates. It is not uncommon to find polyphosphate concentrations in

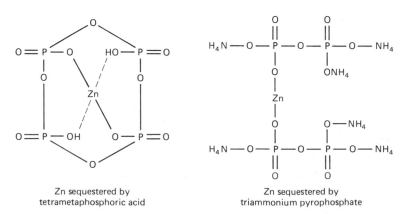

Zn sequestered by
tetrametaphosphoric acid

Zn sequestered by
triammonium pyrophosphate

Fig. 3–24. Sequestering of a metal such as zinc by polyphosphates reduces the possibility of precipitation in fertilizer solutions. (Courtesy L. S. Murphy.)

Table 3–14. Solubilities of Micronutrients in Liquid Ammonium Polyphosphate and Liquid Ammonium Orthophosphate

| Materials Added | In 11-37-0 Ammonium Polyphosphate Composition (wt %) | | | | In 8-24-0 Orthophosphate (wt %) |
	N	P	Micronutrient Element (B, Zn, Mn, Fe, Cu)	Time Stable (days)	Micronutrient Element (B, Zn, Mn, Fe, Cu)
$Na_2B_4O_7 \cdot 10H_2O$	10.2[a]	15.1	0.9	>30	0.9
ZnO	10.7[a]	15.8	2.1	90[b]	0.05
$ZnSO_4$	10.4[a]	15.2	2.3	3[b]	0.06
Mn_3O_4	10.9	16.1	0.2	>30	—
$MnSO_4 \cdot H_2O$	10.9	16.1	<0.08	0	<0.08
$Fe_2(SO_4)_3 \cdot 9H_2O$	10.5[a]	15.5	1.0[a]	>30	0.08
CuO	10.8	16.1	0.7	>30	—
$CuSO_4 \cdot 5H_2O$	10.4[a]	15.4	1.5[a]	>30	0.13

[a] Calculated value, no analysis.
[b] Precipitate identified as form of $ZnNH_4PO_4$.

SOURCE: A. V. Slack, H. M. Potts, and H. B. Shaffer, Jr., "Effect of Polyphosphate Content on Properties and Use of Liquid Fertilizers," *J. Agr. and Food Chem.*, 13:165, 1965.

solutions ranging from 20 to 80% of the total phosphorus present. Just how much polyphosphate may be necessary in a solution is a function of the use, the storage time, and the amount of micronutrients that are carried in the formulations.

The production of polyphosphates employs wet process orthophosphoric acid that is subsequently dehydrated to what is termed *superphosphoric acid.* Characteristics of some phosphoric acids including superphosphoric acid are presented in Table 3–15. Note that as the polyphosphate content becomes higher, the percentage of longer-chain polyphosphates is increased. Also the viscosity of the solution increases as the polyphosphate content increases. Eventually it becomes impractical to produce too high a polyphosphate content because the materials must be handled at extremely high temperatures in order to be fluid.

The reactions involved in the production of ammonium polyphosphates (APP) are similar to those in the production of MAP and DAP (see Fig. 3–20.) Superphosphoric acid is ammoniated to a given N:P ratio by a simple acid-base reaction. The development of what is known as the *pipe reactor* or *T reactor* for the production of ammonium polyphosphates has greatly simplified the production of such fertilizer materials. TVA[30] pioneered the development of the very simple equipment indicated in Fig. 3–25 that permits the produc-

[30] R. S. Meline, R. G. Lee, and W. C. Scott, "Use of a Pipe Reactor in Production of Liquid Fertilizer with Very High Polyphosphate Content," *Fertilizer Solutions*, 16(2):32, 1972.

Table 3-15. Characteristics and Composition of Superphosphoric Acids

%P	(% P$_2$O$_5$)	1	2	3	4	5	6	7[a]	8	9	>9
		Distribution (%) of P in Chains Containing Indicated Number of P Atoms									
23.6	54.3	100									
31.7	73.0	81.44	18.27	0.28	0.00						
32.1	74.0	71.37	27.19	1.37	0.06	0.00					
32.6	75.0	60.23	35.84	3.58	0.32	0.03	0.00				
33.0	76.0	49.00	42.76	7.05	1.03	0.14	0.02	0.00			
33.4	77.0	38.47	46.90	11.49	2.50	0.51	0.10	0.02	0.00		
33.8	78.0	29.21	47.73	16.24	4.91	1.39	0.38	0.10	0.03	0.01	0.00
34.3	79.0	21.52	45.33	20.34	8.11	3.03	1.09	0.38	0.13	0.04	0.02
34.7	80.0	15.46	40.23	22.87	11.55	5.47	2.49	1.10	0.48	0.20	0.15
35.2	81.0	10.94	33.30	23.18	14.35	8.32	4.64	2.51	1.33	0.70	0.73
35.6	82.0	7.71	25.58	21.18	15.59	10.76	7.13	4.59	2.90	1.80	2.77
36.0	83.0	5.46	18.05	17.32	14.77	11.81	9.06	6.77	4.95	3.56	8.27
36.4	84.0	3.87	11.53	12.51	12.06	10.91	9.47	7.99	6.60	5.37	19.69
36.9	85.0	2.67	6.56	7.86	8.37	8.36	8.01	7.47	6.82	6.13	37.75
37.3	86.0	1.68	3.32	4.29	4.92	5.30	5.48	5.51	5.42	5.25	58.83
37.8	87.0	0.82	1.61	2.18	2.64	2.98	3.24	3.43	3.55	3.62	75.93

[a] Orthophosphate contains 1 P atom, pyrophosphate 2, tripolyphosphate 3, etc.

SOURCE: J. D. Fleming, "Polyphosphates Are Revolutionizing Fertilizers. I. What Polyphosphates Are," *Farm Chemicals,* 30, August 1969.

tion of either liquid or solid fertilizers with high polyphosphate contents. By means of the pipe reactor it is possible to produce liquid ammonium polyphosphate containing as high as 80 to 90% of the phosphorus in the polyphosphate form.

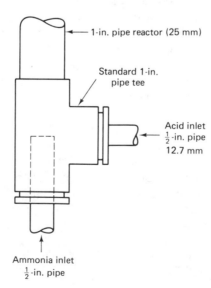

1-in. pipe reactor (25 mm)

Standard 1-in. pipe tee

Acid inlet
$\frac{1}{2}$-in. pipe
12.7 mm

Ammonia inlet
$\frac{1}{2}$-in. pipe

Fig. 3-25. T-reactor (pipe reactor) for mixing of ammonia and superphosphoric acid in the manufacture of ammonium polyphosphate fertilizers. (Source: R. S. Meline, R. G. Lee, and W. C. Scott, "Use of a Pipe Reactor in Production of Liquid Fertilizers with Very High Polyphosphate Content," *Fertilizer Solutions,* 16(2):32–45, 1972. Used with permission.)

TVA has also pioneered in the development of solid ammonium polyphosphates such as 15–62–0 and 12–53–0. Granular ammonium polyphosphate has the same advantages as liquid polyphosphates, namely higher concentrations of phosphorus and subsequently lower freight charges for the movement of phosphorus. This solid product has essentially the same agronomic properties as MAP or DAP and can be used to produce a liquid polyphosphate. In the manufacture of solid APP, the heat of reaction between ammonia and superphosphoric acid is used to produce an essentially anhydrous melt. The melt is granulated with recycled fines in a pug mill and then screened and cooled.

Agronomic characteristics of polyphosphates in general are very good. Basically the materials have about the same characteristics as MAP with little germination damage being produced by placement in direct seed contact even at high soil pH levels. Discussion of agronomic characteristics of phosphorus fertilizers is expanded in Section 3.7.

3.6.7 Nitric Phosphates

Phosphatic fertilizers produced under the designations "nitric phosphate," "nitrophosphate," or "nitrophoska" are normally limited to those materials that involve the acidulation of rock phosphate with nitric acid. Interest in the production of nitric phosphates has been generated by shortages or high costs of sulfur and sulfuric acid. Increasing costs for sulfur recovery by the Frasch method makes the use of nitric acid for acidulation of rock phosphate somewhat more attractive. In all probability, developments of processes for recovery of sulfur from industrial emissions and the price of sulfuric acid from those sources will have a major effect on the geographic development of nitric phosphates in the future.

The first commercially feasible process for the formulation of phosphatic fertilizers using nitric acid to solubilize the phosphorus in rock phosphate originated from patents published in 1930 by Erling Johnsen, the so-called *Odda process*. This process was further developed by Dutch State Mines, Norsk Hydro, Hoechst, and BASF, all European firms. In this process, the acidulate produced by the reaction of rock phosphate and nitric acid (Fig. 3–26) is cooled causing crystallization of calcium nitrate tetrahydrate, a reaction product of the nitric acid and the calcium of the rock phosphate. Calcium nitrate is separated from the acidulate by centrifugation or filtration. The remaining slurry is ammoniated, granulated, and dried producing a nitrogen-phosphorus fertilizer with a typical analysis of 20–20–0. Following ammoniation, potassium can be added as either potassium chloride or potassium sulfate, the latter being favored in some cases due to a tendency of potassium chloride to promote degradation of the nitric phosphate.

Calcium nitrate separated in the first stages of producing is a valuable by-product and can be dehydrated and marketed separately. In Europe, calcium nitrate is also treated with ammonium nitrate to form ammonium nitrate of

Phosphorus Fertilizers

Inital step

$$Ca_{10}F_2(PO_4)_6 + 20\,HNO_3 + 4\,H_3PO_4 \xrightarrow{\;H_2O\;} 10\,H_3PO_4 + 10\,Ca(NO_3)_2 + 2\,HF$$

| Rock phosphate | Nitric acid | Phosphoric acid | Phosphoric acid | Calcium nitrate | Hydro-fluoric acid |

Second step, removal of calcium nitrate

$$10\,H_3PO_4 + 10\,Ca(NO_3)_2 + 2\,HF + 21\,NH_3 \longrightarrow 9\,CaHPO_4 + NH_4H_2PO_4 + 20\,NH_4NO_3 + CaF_2$$

| Phosphoric acid | Calcium nitrate | Hydro-fluoric acid | Ammonia | Dicalcium phosphate | Monoammonium phosphate | Ammonium nitrate | Calcium fluoride |

Fig. 3–26. Reactions involved in the production of nitric phosphates vary with different industrial processes. This process involves use of a mixture of nitric and phosphoric acids to dissolve the rock phosphate. (Source: A. V. Slack, "It's Time to Consider Nitric Phosphates," *Farm Chemicals,* June, 1967, 124. Used with permission.)

lime or calcium ammonium nitrate (see Chapter 2) or ammoniated to produce ammonium nitrate solutions.

The agronomic availability of phosphorus in the final product depends on the combined calcium to phosphorus ratio, which is governed by the amount of calcium removed as calcium nitrate, the characteristics of the rock phosphate, the concentration of the nitric acid, and the amount of nitric acid used. The proportion of water-soluble phosphorus to total phosphorus in the product may be increased by increasing the amount of calcium removed. Commercially, the economic limit seems to be removal of about 70% of the calcium nitrate to yield a product containing about 50% water-soluble phosphorus. If all the calcium nitrate were allowed to remain and simply acidulated, all the phosphorus theoretically would wind up as relatively insoluble dicalcium phosphate.

Modifications of various processes for production of nitric phosphates include addition of a soluble sulfate salt to cause precipitation of calcium sulfate, which may be either removed or left in the product but which results in an increased percentage of water-soluble phosphorus. Still another type of process involves the introduction of phosphoric or sulfuric acid in the extraction. Those processes frequently include ammoniation to convert calcium nitrate to ammonium nitrate.

When phosphoric acid is added in sufficient quantities, the ammoniation portion of the process yields some ammonium phosphate and increases the water-soluble phosphorus content of the final material.

It is generally agreed that phosphate fertilizers should have some degree of water solubility to give them applicability to a wide range of crops and soils. There is not as much agreement on the minimum solubility acceptable, but 40 to 50% seems to be about as near a concensus as can be reached in the United States. Webb and Pesek[31] reported extensively on this property in the

[31] J. R. Webb and J. T. Pesek, "An Evaluation of Phosphorus Fertilizers Varying in Water Solubility. I. Hill Applications for Corn," *Soil Sci. Soc. Amer. Proc.,* 22:533, 1958.

United States. In Europe, a wide range of opinions exist among the countries concerning the degree of water solubility required.[32] In England, water solubility of phosphorus is highly regarded and most phosphorus fertilizers are essentially all water soluble. Denmark favors high water solubility, in the vicinity of 60%. Lower emphasis is placed on water solubility in Sweden, Norway, and Finland where farmers readily use nitric phosphates with low water solubility. Most agronomists in those countries agree that the water-insoluble phosphorus should be in the dicalcium phosphate form, not tricalcium phosphate or apatite.

3.6.8 Urea-Ammonium Phosphates

TVA has studied, developed, and introduced on a limited scale a new class of phosphorus compounds, urea-ammonium phosphates (UAPP), formulated from urea and merchant-grade, wet process phosphoric acid. This process involves ammoniation of the phosphoric acid to produce an anhydrous melt from the heat of the reaction. The melt is produced in a T-reactor and is granulated in a pug mill that provides the working action needed to induce crystallization of polyphosphates produced in high-temperatures. Urea solution or prilled urea is added to the pug mill for mixing.

The polyphosphate content of these materials is controlled between 20 and 40% of the total phosphorus to facilitate granulation. No dryer is required because most of the water in the process is removed by the heat of ammonia-phosphoric acid reaction. The physical characteristics and storage properties of these products are reported to be better than those of similar products made by conventional processes. Chemical and physical characteristics of a common analysis urea-ammonium phosphate are given in Table 3-16. Storage tests conducted by TVA have indicated that these urea-based materials are completely compatible with triple superphosphate. Normally, urea mixed with triple superphosphates leads to decomposition of both materials with loss of ammonia. In addition to the 28-28-0 analysis of urea-ammonium phosphate reported in Table 3-16, TVA has also produced a 35-17-0 and a 21-42-0, all solids.

Urea-ammonium phosphates have good potential in bulk blending. The 28-28-0 grade, for instance, matches well in particle size to that of diammonium phosphate, triple superphosphate, and most of the potassium materials used in bulk blending. Use of a high N:P ratio UAPP could simplify bulk blend segregation problems by elimination of one material in the mixture. Urea-ammonium phosphates can also be used in the production of suspension fertilizers.

Agronomically, field studies in several states in the United States have shown that UAPP behaves similarly to urea and ammonium phosphates. As a source of phosphorus, UAPP is equivalent to other water-soluble phosphorus

---

[32] A. V. Slack, "It's Time to Consider Nitric Phosphates. IV. Agronomic Considerations," *Farm Chemicals*, June 1967, 124.

Table 3-16. Chemical and Physical Properties of TVA Granular Urea-Ammonium Phosphate

Property	Specification	Range
Chemical analysis, %		
Total nitrogen	28.0 minimum	28–29
Urea nitrogen, % of total	80.0	
Ammonia nitrogen, % of total	20.0	
Available P_2O_5	28 minimum	28–29
Polyphosphate P, % of total P	—	20–40
Water-soluble P, % of total P	—	99–100
Moisture	0.4	—
Biuret, %	0.5	—
Bulk density, lb/ft³	48	—
pH of 10% solution	4.9	—
Critical humidity, % relative humidity	—	50–55

SOURCE: TVA Preliminary Use Manual, Urea-Ammonium Phosphate, May 1976.

sources. Urea-ammonium phosphates should not be placed in direct seed contact due to the presence of the urea and the associated salt problems that are discussed in Chapter 7.

3.6.9 Other Phosphorus Fertilizers

Several mixtures of the phosphatic fertilizers discussed earlier exist in the fertilizer market. A mixture of monoammonium phosphate and ammonium sulfate having the analysis 16-20-0 is used in large tonnages in the United States. This material has essentially the same agronomic characteristics as MAP, but some researchers have experienced problems in germination when this material was placed in direct seed contact.

A mixture of MAP and DAP was formerly used in large quantities in the United States but has largely disappeared with the advent of DAP (see Fig. 3-21). The analysis of that mixture is 16-48-0.

Recently, developmental work has progressed to the pilot plant stage in the production of both potassium orthophosphates and potassium polyphosphates in the United States, but production has not occurred on a large scale. Studies of potassium polyphosphate (KPP) reported by Armbruster et al.[33] indicated that KPP is as effective as ammonium orthophosphate, ammonium polyphosphate, and triple superphosphate as a phosphorus source for corn and grain sorghum. Water solubility of phosphorus in KPP formulations has varied from about 25 to as high as 79% of the total phosphorus.

[33] J. A. Armbruster, L. S. Murphy, L. J. Meyer, P.J. Gallagher, and D. A. Whitney, "Field and Growth-Chamber Evaluations of Potassium Polyphosphate," *Soil Sci. Soc. Amer. Proc.*, 39:144, 1975.

Most of the other phosphorus sources that have been examined or used to some degree over the years have been generally classified as thermal phosphatic fertilizers. These materials require large amounts of heat in their production and include basic slag, a by-product of steel production, primarily tetracalcium phosphate $(Ca_4P_2O_9)$. Basic slag is best suited for long-season crops with extensive root systems grown on acid soils. None of the phosphorus is water soluble but 80 to 85% is soluble in alkaline ammonium citrate. Most of the use of this material has occurred in Europe and the USSR.

Many attempts have been made to develop methods of making phosphorus in rock phosphate available without treatment with acid. Processes used in this regard have involved calcining (burning) the rock phosphate to remove fluorine or fusion with silica and alkali salts. These materials have found use in only a few areas.

Metaphosphates including calcium metaphosphate and potassium metaphosphate have been examined but are largely now off the fertilizer market in the United States. Calcium metaphosphate was used for a time and had the advantage of a high phosphorus concentration, about 27% P (62% P_2O_5). Production of this material $[Ca(PO_3)_2]$ has the disadvantage of requiring the production of elemental phosphorus. Also, the material will not react with ammonia so it cannot be used in mixed fertilizers where ammonia is the desirable nitrogen source due to its lower cost.

3.7 PHOSPHORUS RESPONSES IN CROPS

Plant responses to applied phosphorus have been under investigation by researchers for well over a century. Long-term experiments such as those at Rothamstead, England, the Morrow plots at the University of Illinois, the Jordan plots at Pennsylvania State University, and the Sanborn Field at the University of Missouri have demonstrated the effects of phosphorus on crop yield and quality. Evaluation and discussion of all such experiments is beyond the scope of this section and is not attempted. Most recent reports of phosphorus responses in various crops and the efficiency of various sources of phosphorus are discussed. Evaluation of the effectiveness of methods of phosphorus placement is relegated to Chapter 7.

3.7.1 Yield Responses

Yield responses to applied phosphorus are directly tied to soil characteristics and requirements of the crop. Soil analyses have served well as the primary means of determining crop needs. Plant tissue analyses and deficiency symptoms have served as other diagnostic techniques. (see Chapter 1).

Corn. Phosphorus responses in crops have often increased in intensity as cultivation is continued for an extended period. A classic example of the effects of cultivation on magnitude of phosphorus response in irrigated corn is indicated in Fig. 3–27. Note that phosphorus response did not occur until

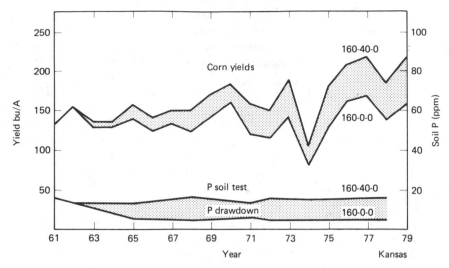

Fig. 3–27. Excellent phosphorus response in irrigated corn. Note that maximum yields with adequate phosphorus are continuing to increase while yields with nitrogen only have not exhibited such a trend. Phosphorus yield responses correlate closely with soil test phosphorus drawdown. About 4 years were required to produce a significant phosphorus response after initiation of the study. Soil test levels have been maintained by the annual application of 19 kg/ha of phosphorus (17 lb/A). (Source: R. E. Gwin and P. J. Gallagher, "Fertilization Studies with Irrigated Wheat, Corn and Grain Sorghum," Kansas Fert. Res. Report of Prog. 372, Dec., 1979, 75.)

about 4 years after irrigated corn production had been underway. The grain yield response parallels a decline in available soil phosphorus noted at the bottom of the figure.

Barber at Purdue University has studied corn responses to phosphorus on a soil in Indiana that tested low in phosphorus (approximately 6 ppm extractable phosphorus) initially. During the last 10 years the difference between phosphorus- and no-phosphorus-treated areas has averaged 1800 kg/ha (1600 lb/A) of grain. After 23 years of phosphorus applications at rates of 0, 24, and 48 kg/ha of phosphorus, soil test values were 6, 32, and 48 ppm extractable phosphorus, respectively, demonstrating the effects of phosphorus applications both on yield and soil phosphorus availability.

Recommendations for application of phosphorus in a given area are based on soil fertility research that correlates phosphorus responses to soil analyses and field experiences. Such research must be conducted at many locations to produce reliable recommendations. Attempts to stretch recommendations to areas not covered in the soil test correlation research can lead to disappointing results. Also, it is dangerous to attempt to translate recommendations in general to all areas of a given type of soil unless some idea of the available phosphorus exists. An example of such a translation problem would exist if one attempted to predict phosphorus responses on the soil noted in Fig. 3–27 without knowing the available phosphorus content of the soil. For the first

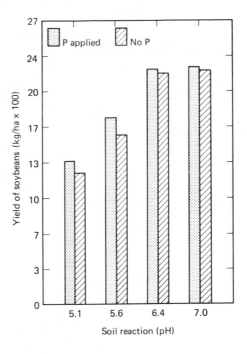

Fig. 3-28. Yield of soybeans in this Louisiana study increased with increasing soil pH but yield response to applied phosphorus was smaller at higher pH levels. This indicated greater availability of soil phosphorus to the plants as pH increased. (Source: W. J. Peevy et al., "The Influence of Soil Reaction, Residual Soil Phosphorus and Fertilizer Phosphorus on the Yield of Soybeans Grown on Olivier Silt Loam," Louisiana Agri. Exp. Sta. Bul. 669, 1972.)

several years after initiation of corn cultivation on such a soil, mineralization of organic phosphorus coupled with release of phosphorus from the native soil minerals would have made responses disappointing. Soil analyses would have predicted such a situation. The most scientific approach is to utilize soil types as a basis for correlating soil tests with crop responses (see Chapter 1).

Soybeans. Soybeans have been reported to respond better to previous phosphorus applications for other crops that have produced higher soil test levels for phosphorus. However, responses to direct application of phosphorus are common. Research conducted in Louisiana has demonstrated the relationships of phosphorus response in soybeans to phosphorus soil test levels and to soil pH. Peevy at al.[34] established an investigation on Olivier silt loam with pH values ranging from 5.1 to 7.0. Phosphorus soil test values ranged from 13 to 60 ppm, the result of continued phosphorus applications for 20 years in a previous experiment. Figure 3-28 shows how response to applied phosphorus declined in magnitude as the pH increased, an indication of the effects of liming on phosphorus availability discussed in Section 3.3.4. From that same experiment, Fig. 3-29 demonstrates the correlation between phosphorus soil test values and applied phosphorus at varying soil pH levels. Generally, greater phosphorus responses were recorded at low soil phosphorus levels and at low pHs.

[34] W. J. Peevy, B. E. Newman, J. E. Sedberry, Jr., and R. H. Brupbacher, "The Influence of Soil Reaction, Residual Soil Phosphorus and Fertilizer Phosphorus on the Yield of Soybeans Grown on Olivier Silt Loam," Louisiana Agri. Exp. Sta. Bul. 669, December 1972, 20 pp.

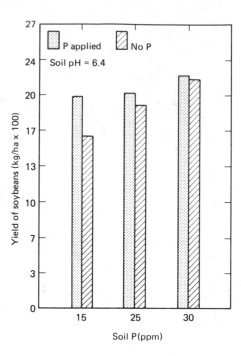

Fig. 3-29. Residual soil phosphorus affected the magnitude of soybean response to applied fertilizer phosphorus in this study. Higher soil test levels (residual phosphorus) result in lowered recommendations for fertilizer phosphorus. (Source: W. J. Peevy et al., "The Influence of Soil Reaction, Residual Soil Phosphorus and Fertilizer Phosphorus on the Yield of Soybeans Grown on Olivier Silt Loam," Louisiana Agri. Exp. Sta. Bul. 669, 1972.)

Similar types of soybean responses to applied phosphorus have been reported from other states in the United States. Phosphorus applications in a 24-year experiment in Indiana resulted in an average 22% yield increase (660 kg/ha) over the no-phosphorus control yield of 2960 kg/ha of beans in the last 10 years of the study.

Sorghum. Although sorghums are known to be efficient in foraging for nutrients in the soil, they respond to applied phosphorus when phosphorus soil test levels are low. An indication of phosphorus responses in irrigated grain sorghum is presented in Table 3-17. The phosphorus responses in grain sorghum to that of corn are similar and, for practical considerations, the two crops may be treated very much alike in terms of fertilizer phosphorus applications. Recommendations for grain sorghum are slightly less than those for corn. As an example, Kansas and Nebraska recommendations for grain sorghum average 20 to 25% less for sorghum than for corn.

Wheat. Wheat responses to applied phosphorus have also been well correlated to soil phosphorus levels. Some documentation of such responses has been summarized by Vitosh and Warncke at Michigan State University. Their relationships among soil test values for phosphorus, phosphorus application rates, and expected yield response in wheat are noted in Fig. 3-30. As the soil test levels increase, less phosphorus is needed to reach the set yield goal.

Wheat has also been shown to respond with time increasingly to phos-

Table 3-17. Irrigated Grain Sorghum Responses to Nitrogen and Phosphorus

| | Grain Yield (kg/ha) 1968–77 Means | | |
| | P (kg/ha) | | P yield Response |
N (kg/ha)	0	19	(kg/ha)
0	4766	4920	154
44	6264	6920	656
90	6941	7490	549
134	6812	7699	887
178	6656	7950	1294
224	6543	7953	1410

SOURCE: R. Gwin, Jr., and P. J. Gallagher, "Effects of Nitrogen, Phosphorus and Potassium on Yield of Irrigated Grain Sorghum," Kansas Fert. Res. Rept. Progress 372, 1979, p. 75.

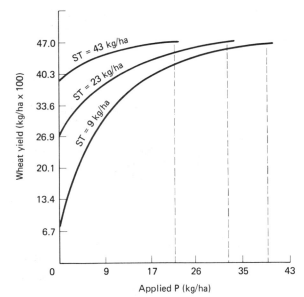

Fig. 3-30. Relationship between expected wheat yield and applied phosphorus fertilizer at three phosphorus soil test (ST) levels. Yield potential in this case was 4700 kg/ha (4192 lb/A). (Source: M. L. Vitosh and D. Warncke, "Fertilization of Wheat," Mich. Ext. Bul. E-1067, 1977.

phorus applications. Spratt and McCurdy[35] reported increasing levels of phosphorus response in hard red winter wheat in Saskatchewan as time progressed (Fig. 3-31). Excellent phosphorus responses in hard red winter wheat in Kansas have been reported (Fig. 3-32).

Rice. Phosphorus applications to rice have been studied in many areas of the world with significant contributions to the understanding of fertilization

[35] E. D. Spratt and E. V. McCurdy, "The Effects of a Variety of Long-Term Soil Fertility Treatments on the Phosphorus Status of a Clay Chernozem," *Canadian Journal of Soil Science*, 46:29, 1966.

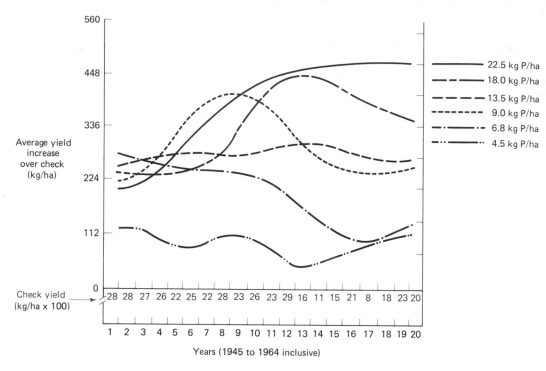

Fig. 3–31. Yield increases in wheat following summer fallow as affected by fertilization with monoammonium phosphate. Note that greatest effects were recorded at the highest rate of continual phosphorus application. (Source: E. D. Spratt and E. V. McCurdy, "The Effects of a Variety of Long-Term Soil Fertility Treatments on the Phosphorus Status of a Clay Chernozem," *Canadian J. Soil Sci.*, 46:29, 1966. Used with permission.)

needs coming from research in southeast Asia. Duangpatra and De Datta[36] estimated that 800,000 to 1,000,000 hectares of rice production land are phosphorus deficient in Indonesia, and they also note that phosphorus applications have caused definite grain yield increases in the states of Bihar, Orissa, Assam, and in central India, particularly on nonalluvial clay soils. Similar studies in Thailand demonstrated dramatic responses of rice to phosphorus when nitrogen and phosphorus were applied together. Recent studies in the Philippines by Duangpatra and De Datta indicated significant increases in grain yield and plant phosphorus concentrations from applications of various phosphorus fertilizers (Table 3–18).

Similar rice responses to applied phosphorus have occurred in the United States rice-producing areas of California, Texas, Louisiana, and Arkansas.

[36] P. Duangpatra and S. K. De Datta, "Urea-Ammonium Phosphate and Other Phosphorus Sources for Flooded Tropical Rice," *International Rice Commission Newsletter*, 18:(4), December 1969.

Fig. 3–32. Phosphorus responses in wheat are strongly related to the soil test levels of phosphorus. Soil test phosphorus at this location in central Kansas was very low. Cold conditions shortly after planting tended to accentuate plant growth response to applied phosphorus. The fertilizer phosphorus in this case was knifed into the soil preplant in combination with a nitrogen solution (UAN). (Courtesy L. S. Murphy.)

Mikkelsen and Lindt[37] among others have reported on rice responses to phosphorus.

Cotton. Yield responses of cotton to phosphorus fertilization is typically expressed as an increase in early fruit set. The increase in early fruit yield is usually greater than the total yield increase. Alabama studies reported by Cope[38] include significant cotton responses to applied phosphorus along with increased soil phosphorus test values from continued use of fertilizer phosphorus. Maples and Keogh[39] reported that three experiments on soils with medium phosphorus soil test values produced significant yield increases when phosphorus was applied. Yield increases were 220 kg/ha or less of seed cotton.

Sugar beets. Heavy applications of phosphorus on sugar beets have been a common practice in many beet-producing areas of the world due to recognition of the effects of phosphorus on sugar yield. Although an in-depth review

[37] D. S. Mikkelsen and J. H. Lindt, Jr., "Rice Fertilization," *Rice J.,* 69:74, 196.

[38] T. Cope, "Response of Cotton, Corn and Bermudagrass to Rates of N, P, and K." *Auburn Agr. Expt. Sta. Circ. 181,* 1970.

[39] R. Maples and J. L. Keogh, "Phosphorus Fertilization Experiments with Cotton on Delta Soils of Arkansas," Ark. Agri. Exp. Sta. Bul. 781, June 1973, p. 21.

Table 3-18. Effects of Sources and Rates of Phosphorus on the Grain Yield of Flooded Rice

| Treatment N-P-K (kg/ha) | Grain Yield | | Treatment Mean (kg/ha) | Treatment N-P-K (kg/ha) | Grain Yield | Treatment N-P-K (kg/ha) | P-Source |
| | P-Source | | | | P-Source | | |
	UAP[a] (29-30-0) (kg/ha)	Superphosphate (0-40-0) (kg/ha)			UAP (25-35-0) (kg/ha)		UAP (34-17-0) (kg/ha)
0-0-0	—	—	3726	0-0-0	3726	0-0-0	3726
116-0-50	—	—	5519	100-0-50	5270	102-0-0	5319
116-13-50	5901	5674	5788	100-15-50	5966	102-0-50	5088
116-25-50	6584	6269	6426	100-30-50	6062	102-7-50	5951
116-38-50	5873	6440	6156	100-45-50	6257	102-15-50	5658
116-50-50	6277	6570	6424	100-61-50	5805	102-22-50	5543
116-50-0	—	6189	6189	100-61-0	6405		
Source mean	6159	6228	—		6099		5717
Analysis of variance Treatment[b]							
Standard error	236	312			223		210
LSD (5%) kg/ha	745	961			687		662
CV (x) %	7	9			7		7

[a] Urea-ammonium phosphate (samples supplied by The Tennessee Valley Authority of the United States).

[b] Significant at 1% probability level.

SOURCE: P. Duangpatra and S. K. DeDatta, "Urea-Ammonium Phosphate and Other Phosphorus Sources for Flooded Tropical Rice," *International Rice Comm. Newsletter*, 18(4), 1969.

of this practice is not included in this section, recent data collected by the University of Nebraska demonstrate higher sugar yields from phosphorus use when the element was limiting. Results of that study reported in Table 3-19 indicate the effects of phosphorus on sugar yield.

Vegetables. Short growing periods and more restricted root systems of many vegetable plants result in higher requirements for phosphorus by these crops than other crops discussed earlier. For instance, consistent yield increases in potatoes have been noted even when the soil phosphorus test was high. Tomatoes, sweet corn, cabbage, edible beans, head lettuce, cucumber, eggplant, and sweet potato have responded significantly to increased phosphorus supplies. Nishimoto et al.[40] produced significant yield increases in several vegetable crops in Hawaii by increasing the concentration of phosphorus in the soil solution to given levels by the application of triple superphosphate. Lettuce, for instance, produced only 1% of maximum yield

[40] R. K. Nishimoto, R. L. Fox, and P. L. Parvin, "Response of Vegetable Crops to Phosphorus Concentrations in Soil Solution," *J. Amer. Soc. Hort. Sci.*, 102:705, 1977.

Table 3-19. Phosphorus Effects on Recoverable Sugar in Sugar Beets in Nebraska

Location	0	22	P (kg/ha) 45	67	112	224	Analysis of Variance	LSD $P_{.05}$
			Sucrose (kg/ha)					
Busch — 1974	6834	6606	7419	7627	8338	7415	Sig. at 5%	759
Sakarda — 1974	8259	8581	9527	7925	10128	9113	Sig. at 5%	1237
Miller — 1974	5978	6010	6432	5919	5313	6126	Sig. at 10%	650
Heimbouch — 1974	3044	3716	3218	3457	4112	4526	Sig. at 10%	952
Buehler — 1975	7924	8413	8766	8560	9246	8780	Sig. at 5%	531

SOURCE: G. Varvel, G. Peterson, and F. Anderson, "Phosphorus Nutrition of Sugar Beets," Soil Sci. Prog. Report 1976, Dept. of Agronomy, Univ. of Neb., pp. 17-1 to 17-5.

when grown on control plots with low available soil phosphorus. Frequently, quality as well as quantity of vegetable crops is improved by adequate phosphorus fertilization; and quality often determines their price.

Forages. Responses of legumes, particularly alfalfa, to applications of phosphorus when grown on low phosphorus soils has been documented repeatedly. The effects of phosphorus application on growth of alfalfa are indicated in Fig. 3-9. Whitney et al.[41] reported significant yield increases in sprinkler-irrigated alfalfa grown on a slightly acid sand in Kansas (Table 3-20).

Yield increases in alfalfa grown on land leveled for irrigation in Washington State ranged up to 10 metric tons/ha (4.5 tons/A) from applications of 150 kg/ha of phosphorus. Yield increases of 4.5 metric tons have been reported in Virginia from application of 100 kg/ha of phosphorus.

Cool-season grasses such as orchardgrass, bromegrass and tall fescue have also produced striking yield increases when deficient conditions were overcome by applications of phosphorus. Applications of nitrogen are much less effective when phosphorus is deficient on these forages. Yield increases of 9 metric tons/ha of bermudagrass in Louisiana have been reported when phosphorus was applied at 74 kg/ha phosphorus. Orchardgrass and bromegrass yield increases of 4 metric tons/ha are not uncommon when adequate phosphorus is supplied on phosphorus-deficient soils.

Responses of forages, like those of other crops, must be correlated to soil tests on specific soil types in order to be predictable. Lack of understanding of the soil phosphorus supplying capability leads to disappointing yields and poor results from the fertilization program. (see Chapter 1).

[41] D. A. Whitney et al., "Potassium Effects on Yield and Composition of Corn, Alfalfa, and Wheat," Kansas State University Report of Progress, Potash/Phosphate Institute, February 1978, 21 pp.

Table 3-20. Effect of Phosphorus Fertilization of Irrigated Alfalfa on Forage Yield and Protein Production

P	K	Hay Yield (mt/ha)		Protein Yield (kg/ha)	
(kg/ha)		Kanza[a]	Marathon[a]	Kanza	Marathon[a]
0	66	18.1	17.3	3404[a]	3290
17	66	20.9	19.7	4015	3751
35	66	20.9	21.2	3993	4154
52	66	21.6	20.6	4180	3993
	LSD .05	0.6		99	

[a] Two alfalfa varieties.

SOURCE: Unpublished data, Jim Ball and George Ten Eyck, Sandyland Exp. Field, Kansas State University.

Forest Crops. Although much of the attention to forest fertilization has been given to the use of nitrogen, significant yield increases and improved rate of growth from the use of phosphorus have been recorded in many areas of the world. Especially in the southern United States, foresters estimate that production could be dramatically increased by phosphorus fertilization of pine plantations growing on poorly drained soils in the Southeast; 1,600,000 ha are estimated to be potentially responsive to nitrogen and phosphorus. Graphic evidence of response of these species of phosphorus fertilization is presented in Florida Extension Circular authored by Laird.[42] Note the difference in tree growth in Fig. 3-33 as a result of phosphorus fertilization.

3.7.2 Crop Quality Effects

In addition to yield increases, phosphorus can also exert some influence on crop composition or quality. Definitions of quality of a crop vary depending on how the material is to be used. Higher concentrations of phosphorus in forage or grain represent a quality increase from the standpoint of animal or human nutrition. Whitney[43] reported large increases in phosphorus concentrations in oat grain in Iowa when phosphorus was added to a basic nitrogen-potassium treatment (Table 3-21). Dumenil[44] also indicated important increases in the phosphorus concentration in corn, oats, and soybeans as soil test levels for phosphorus increased (Table 3-21).

Grasses are frequently low in phosphorus content, particularly when little attention has been given to the phosphorus status of the soil. Phosphorus fer-

[42] C. R. Laird, "Fertilization of Slash Pines on Poorly-Drained Soils in Northwest Florida," Florida Ext. Circ. 378, October 1972.

[43] D. A. Whitney, "Making Oats Better," Proceedings 20th Annual Iowa Fertilizer and Agricultural Chemical Dealers Conference, Dept. of Agronomy, Iowa State University, Sec. 7-1, 1968.

[44] L. Dumenil, "What Affects P and K Soil Fertility Levels?" Proceedings 26th Annual Iowa Fertilizer and Agricultural Chemical Dealers Conference, Iowa Ext. Circ. 904, 1974.

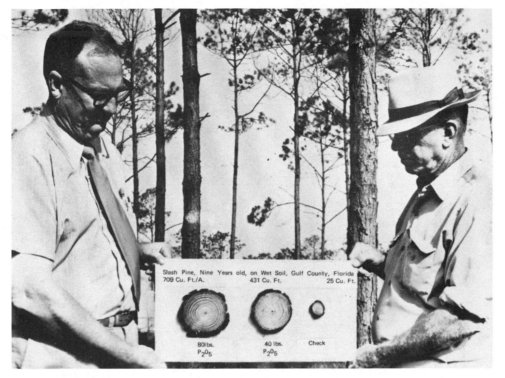

Fig. 3-33. Florida foresters examining effects of application of phosphorus fertilizer to slash pine trees on very poorly drained soils in Gulf County, Florida. All three trees are the same age, 9 years. (Source: Courtesy C. R. Laird, Fla. Coop. Ext. Serv.)

tilization can be an effective means of improving the phosphorus content of forages both for grazing and hay. Black et al.[45] and Reynolds at al.[46] reported that phosphorus fertilization of range grasses in south Texas not only resulted in improved animal reproductive performance but also increased the carrying capacity and beef production of the range (Table 3-22). Increasing the phosphorus concentration in the forage lowers the amount of supplemental phosphorus that must be added to the ration or to the animals' mineral supplement. Indications of the effects of phosphorus fertilization of grasses on yield and phosphorus concentration in a perennial cool-season grass (tall fescue) and a warm-season annual (sudangrass) are given in Table 3-23. In addition to increasing the yield and the phosphorus concentration in grasses, supplying additional phosphorus when soils are deficient can significantly increase the ability of the crop to produce protein. Note the effects of additional phosphorus on the protein yield of bromegrass in Fig. 3-34.

[45] W. H. Black, L. H. Tash, J. M. Jones, and R. J. Kleberg, "Comparison of Methods of Supplying Phosphorus to Range Cattle," USDA Tech. Bul. 856, 1949.

[46] E. B. Reynolds, J. M. Jones, J. H. Jones, J. F. Fudge, and R. J. Kleberg, "Methods of Supplying Phosphorus to Range Cattle in South Texas," Texas Agr. Exp. Sta. Bul. 773, 1953.

Table 3-21. Fertilizer Phosphorus Effects on Phosphorus Concentrations in Grain

Fertilizer N-P-K (kg/ha)	Oats[a] Yield (kg/ha)	Oats[a] Grain P (%)	P Soil Test (kg/ha)	P in Grain[b] Corn P (%)	P in Grain[b] Oats P (%)	P in Grain[b] Soybeans P (%)
60-0-30	3369	0.29	11	0.19	0.25	0.45
60-20-30	4301	0.42	22	0.24	0.32	0.54
			45	0.31	0.40	0.64
			90	0.35	0.44	0.72
			112	0.36	0.45	0.73

[a] D. A. Whitney, "Making Oats Better," Proceedings 20th Annual Iowa Fertilizer and Agricultural Chemical Dealers Conference, Dept. of Agronomy, Iowa State University, Sec. 7-1, 1968.

[b] L. Dumenil, "What Affects P and K Soil Fertility Levels?" Proceedings 26th Annual Iowa Fertilizer and Agricultural Chemicals Conference, Iowa Ext. Circular 904, 1974.

Table 3-22. Methods of Providing Supplemental Phosphorus for Cattle

Group	Weaned Calves (kg/ha)
Control, no supplemental P	104
Bone meal supplement	130
Disodium phosphate in drinking water	160
P-fertilized range	197

SOURCE: W. H. Black, L. H. Tash, J. M. Jones, and R. J. Kleberg, "Comparison of Methods of Supplying Phosphorus to Range Cattle," USDA Tech. Bul. 856, 1949.

Table 3-23. Effects of Phosphorus on the Yield and Composition of Grasses

Sudangrass[a] P Applied (kg/ha)	Sudangrass[a] Yield (kg/ha)	Sudangrass[a] P Concentration (%)	Tall Fescue[b] N (kg/ha)	Tall Fescue[b] P (kg/ha)	Tall Fescue[b] K (kg/ha)	Tall Fescue[b] Yield (kg/ha)	Forage Composition N (%)	Forage Composition P (%)	Forage Composition K (%)	Protein Yield (kg/ha)
0	1,546	0.104	100	0	25	7,571	1.06	0.10	1.42	502
33	4,256	0.116	100	22	25	8,154	1.05	0.22	1.15	535
65	6,160	0.124	100	43	25	9,587	1.01	0.23	1.36	605
130	7,146	0.136	150	0	25	6,406	1.41	0.10	1.36	564
260	10,214	0.167	150	22	25	8,579	1.12	0.21	1.21	600

[a] D. W. James, G. E. Leggett, and A. I. Dow, "Phosphorus Fertility Relationships of Central Washington Irrigated Soils," Wash. Agr. Exp. Sta. Bul. 688, 1967.

[b] G. L. Kilgore, "Effects of Nitrogen, Phosphorus, Potassium and Lime on the Yield of Tall Fescue," Kansas Fertilizer Research Report of Progress 224, 1974, pp. 84–85, 100.

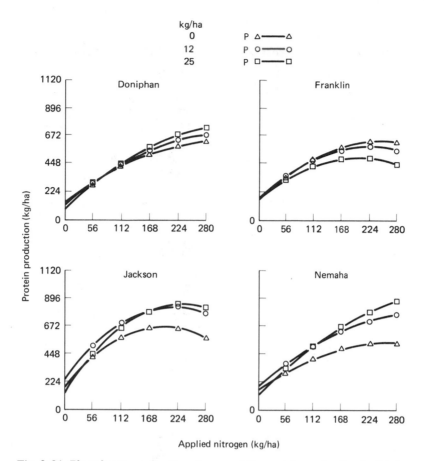

Fig. 3–34. Phosphorus applications increased the protein production in this brome-grass study in Kansas. Bromegrass is quite responsive to applied phosphorus when soil test levels of phosphorus are low. The Franklin County location in this study had higher soil test phosphorus values and phosphorus had no positive effect on protein production. Higher total protein yield from phosphorus application was due primarily to higher yield of forage with acceptable protein content, not higher protein concentrations. Nitrogen produced both higher protein concentration and higher protein yield. (Source: C. N. Gruver, "The Effect of Nitrogen and Phosphorus Fertilization on Yield and Quality of Bromegrass (*Bromus inermis* Leyss),'' M.S. Thesis, Department of Agronomy, Kansas State University, 1971.)

Alfalfa responds to phosphorus fertilization with improved forage quality as well as higher yields. Large increases in phosphorus concentration of alfalfa were noted in Kansas when low soil phosphorus was supplemented by fertilization. Percent protein and total protein yields were also increased by phosphorus fertilization. (Table 3–24).

Research in Virginia reported by Griffith[47] indicates that both phosphorus

---

[47] W. K. Griffith, "Fertilizing for Quality Gains Dollars," Folder D–277, Potash/Phosphate Institute, 1977.

Table 3-24 Phosphorus Fertilization Effects on Forage Yield, Protein Yield and Phosphorus Content of Irrigated Alfalfa — Kansas

Preplant[b] P (kg/ha)	Annual P (kg/ha)	Cut 1[a] Forage Yield (kg/ha)	Protein Yield (kg/ha)	P (%)	Cut 2 Forage Yield (kg/ha)	Protein Yield (kg/ha)	P (%)	Cut 3 Forage Yield (kg/ha)	Protein Yield (kg/ha)	P (%)	Cut 4 Forage Yield (kg/ha)	Protein Yield (kg/ha)	P (%)	Cut 5 Forage Yield (kg/ha)	Protein Yield (kg/ha)	P (%)
0	0	4973	1076	0.244	4390	923	0.242	3718	786	0.261	2240	446	0.220	2352	360	0.172
139	0	6160	1432	0.283	4883	1054	0.265	4234	977	0.310	2531	481	0.302	2486	427	0.248
139	17	6339	1467	0.327	5040	1125	0.314	4592	973	0.306	2643	520	0.315	2554	444	0.262
139	35	5980	1358	0.358	4816	1086	0.319	4614	999	0.317	2800	546	0.327	2688	481	0.281
139	52	6384	1486	0.361	5197	1132	0.302	4614	1043	0.314	2531	487	0.348	2509	453	0.292
0	17	5869	1366	0.288	4861	1028	0.270	4390	1008	0.305	2419	498	0.265	2374	431	0.213
0	35	6294	1479	0.329	4973	1134	0.297	4592	1034	0.309	2509	481	0.299	2554	444	0.243
0	52	6384	1509	0.361	4838	1058	0.295	4346	996	0.314	2486	487	0.338	2486	428	0.264
LSD.05		269	—	0.02	179	—	0.02	313	—	0.02	179	—	0.02	157	—	0.01

[a] All cuttings at 1/10 bloom.

[b] Preplant P applied in fall prior to seeding.

SOURCE: J. Ball, G. Ten Eyck, L. S. Murphy, D. Kissel, and G. Posler, "Management of Irrigated Alfalfa," Kansas Fert. Res. Report of Progress 372, 1979, pp. 167–168.

and potassium have significant effects on soybean quality as well as yield. Phosphorus applied at the rate of 194 kg/ha of phosphorus 3 years prior to the crop being considered increased yields 336 kg/ha and reduced the percent of shriveled grain from 20.8 to 12.5%. Dockage was reduced from $4.33/100 kg to $1.98/100 kg. Application of 372 kg/ha of potassium along with the phosphorus increased yields another 269 kg/ha, reduced shriveled and diseased grain to 1.3% and reduced the dockage to zero.

Phosphorus effects on sugar beet quality and sugar production are discussed in Table 3–19.

3.7.3 Relative Effectiveness of Phosphorus Fertilizers

Aside from the agronomic effects of phosphorus fertilizers placed in direct seed contact discussed in Section 3.6.6 and the question of water solubility of phosphorus sources, questions continue to arise over the relative efficiency of phosphorus compounds as fertilizers. Much of this discussion has centered around the relative efficiency of ammonium orthophosphates and ammonium polyphosphates. Some of the evidence relative to this question should be examined.

Polyphosphates' properties of sequestering micronutrient metals described in Section 3.6.6 have led to much discussion of the effect of this phenomenon on micronutrient absorption. Admittedly, this has little to do with phosphorus availability *per se* but, in the discussion of orthophosphate and polyphosphate efficiency, this is still frequently included. Examination of most of the data available, indicates slight if any difference in the absorption of micronutrients by plants when these elements, particularly zinc and iron, are supplied in polyphosphate fertilizers. Some work conducted at TVA under controlled conditions has suggested that iron applied in a polyphosphate is more available than when applied in other formulations. Field studies of such applications have been variable in their results. Numerous zinc investigations in the field in Nebraska and Kansas have failed to indicate any consistent benefit of zinc application in ammonium polyphosphate formulations other than the formulation benefit produced by higher concentrations of the metal in polyphosphate solutions versus orthophosphate solutions (see Table 3–14 and Chapter 6).

The primary questions, however, concerning the effectiveness of phosphorus sources have little to do with micronutrient questions. Reactions with soil components and solubility of the material govern the basic efficiency. An excellent example of the type of study that has received much attention around the world and has added much to the appreciation of the capabilities of phosphorus sources is the one presented in Fig. 3–35.

Several researchers have compared the agronomic effectiveness of ammonium polyphosphate (APP) and ammonium orthophosphate (AOP) fertilizers using several crops. Results of these comparisons have not been consistent.

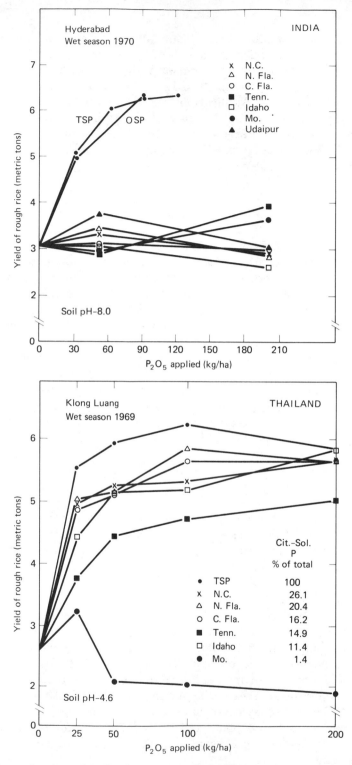

Fig. 3–35. Effects of various phosphorus sources on the yield of rough rice. (Source: O. P. Engelstad, J. G. Getsinger, and P. J. Stangel, "Tailoring of Fertilizers for Rice," TVA Bul. Y–52, p. 56, 1972.)

Using cabbage as a test plant, Hensel[48] indicated that the total yield and weight per head increased about 5% when APP was used as compared to ammonium orthophosphate (AOP). Work[49] conducted under flooded and nonflooded soil conditions indicated that superphosphate, an orthophosphate source, and liquid APP were more effective phosphorus sources than solid APP under flooded conditions for lowland rice. No differences were observed between superphosphate and liquid or solid APP under nonflooded conditions.

In some Nebraska studies,[50] inconsistent results between locations were observed when AOP was compared to APP for corn. Differences were attributed to the many types of soils used. Only about half of the locations studied responded to added phosphorus in terms of increased yields and/or tissue phosphorus content.

Studies of APP and AOP as phosphorus sources for barley in England showed that, on 60 of 98 soils studied, phosphorus uptake was greater from AOP treatments than from APP. This effect was considered to be due to soil temperature differences, with low temperatures being associated with low phosphorus uptake from APP treatments. Soil temperatures lower than 30° to 35°C (86° to 95°F) were suggested as possibly restrictive to APP hydrolysis and production of orthophosphate ions, the form in which phosphorus is absorbed by the plant. Sutton and co-workers[51] stated that the 30° to 35°C (86° to 95°F) temperature range is probably the optimum for polyphosphate hydrolysis.

An early experiment comparing sodium orthophosphate and polyphosphate as phosphorus sources for barley, wheat, and oats concluded that these sources were equal in their ability to supply phosphorus to plants. Kida[52] also conducted studies with rice, buckwheat, radishes, barley, and oats, using phosphorus sources of various solubilities. He determined that the water solubility of the phosphorus sources affected the availability of phosphorus more than did the form of phosphorus.

Kansas workers[53] studied MAP and APP in solution cultures under growth-chamber conditions using corn as the test crop. Monoammonium phosphate was slightly more effective than APP only under low zinc condi-

[48] D. R. Hensel, "Fertilizer Placement on Cabbage," *Soil Crop Sci. Soc. Florida Proc.*, 27:227, 1968.

[49] E. C. Cherian, G. M. Paulsen, and L. S. Murphy, "Nutrient Uptake by Lowland Rice Under Flooded and Non-Flooded Soil Conditions," *Agron. J.*, 60:544, 1968.

[50] E. J. Penas, "Comparative Evaluation of Ammonium Orthophosphate and Ammonium Polyphosphate as Carriers of Phosphorus and Zinc for Corn," M.S. Thesis, Dept. of Agronomy, University of Nebraska, 1967.

[51] C. D. Sutton, D. Gunary, and S. Larsen, "Pyrophosphate as a Source of Phosphorus for Plants," *Soil Sci.*, 101:199, 1966.

[52] Y. Kida, "Fertilizing Value of Pyrophosphates and Metaphosphates," *J. Sci. Agr. Soc.*, 245:61, 1923.

[53] R. B. Ganiron, D. C. Adriano, G. M. Paulsen, and L. S. Murphy, "Effect on Phosphorus Carriers and Zinc Sources on Phosphorus-Zinc Interaction in Corn," *Soil Sci. Soc. Amer. Proc.*, 33:306, 1969.

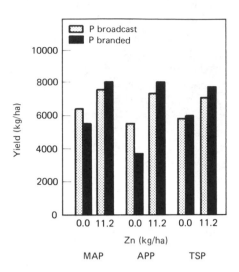

Fig. 3-36. Irrigated corn yields responded dramatically to zinc applications and applications of phosphorus in eastern Kansas. When zinc was deficient, greater yield depression was noted when ammonium polyphosphate was the phosphorus source. However, when adequate zinc was supplied, mono-ammonium phosphate, ammonium phosphate, and triple superphosphate were equal in efficiency as phosphorus sources. (Source: D. C. Adriano and L. S. Murphy, "Effects of Ammonium Polyphosphates on Yield and Chemical Composition of Irrigated Corn," *Agron. J.*, 62:561, 1970 by permission of the American Society of Agronomy.)

tions. Adriano and Murphy[54] noted that irrigated corn grown on zinc- and phosphorus-deficient acid fine sandy loam was injured more from zinc deficiency when ammonium polyphosphate was banded near the seed than when comparable amounts of phosphorus as monoammonium phosphate were placed in the same position (Fig. 3-36). However, when adequate amounts of zinc were supplied, the two forms of phosphorus produced comparable plant growth responses and grain yields. Webb (see footnote 10 of this chapter) found in several field studies involving soils of slightly acid to basic pH that the trend was toward superiority of APP over MAP in yield responses, although the differences were not statistically significant. Differences between liquid and solid APP were not significant in terms of leaf phosphorus concentrations or grain yield. More recently Schield et al. (see footnote 17 of this chapter) reported that field and growth chamber studies comparing polyphosphate and orthophosphate fertilizers on soils with pH ranging from 7.2 to 8.0 indicated nonsignificant yield differences between polyphosphate and orthophosphate fertilizers. Some significantly higher phosphate and zinc concentrations in sorghum and corn leaf tissues were recorded from the use of APP versus AOP; however, the differences were inconsistent.

Work in the USSR by Janischevski et al.[55] complicates the picture even further because their work indicated superiority for APP over ammonium orthophosphates in several Greyearth soils (Aridisols) particularly when the studies were run for several years. Janischevski's work was conducted under controlled conditions.

[54] D. C. Adriano and L. S. Murphy, "Effects of Ammonium Polyphosphates on Yield and Chemical Composition of Irrigated Corn," *Agron. J.*, 62:561, 1970.

[55] F. V. Janischevski, M. A. Prokoscheva, V. M. Zchestokova, and G. V. Poljakova, "Effectiveness and Utilization of Liquid and Solid Polyphosphates of Ammonia on Greyearth," *Agrochemistry* (USSR), No. 5, pp. 3-12, 1975.

In summary, both ammonium polyphosphates and orthophosphates seem to be excellent sources of phosphorus. Variations in comparable performance between these two classes of phosphorus fertilizers do exist but no clear-cut situation seems to warrant recommendation of one type of material over the other. More research is needed in this area to understand better the soil microbiological effects on availability of phosphorus sources and the reaction compounds produced when the various phosphorus sources are added to soils with and without the presence of other macronutrients and micronutrients. Until the time that more definitive information is available, recommendations considering essential equality of processed phosphorus sources for most crops seem to be consistent with good crop production.

SUMMARY

Phosphorus chemistry and biology in the soil and in plant nutrition is complicated and much remains to be discovered despite considerable effort in research over the last century. Phosphatic fertilizers available for supplementation of the native soil phosphorus include processed and unprocessed rock phosphate plus several forms of commercially processed, higher solubility compounds. Research has demonstrated the value of increasing availability of the phosphorus by converting the raw apatite minerals in rock phosphate to ammonium, calcium, and potassium phosphates. Processed phosphorus fertilizers have been demonstrated to be essentially equal in terms of their ability to supply phosphorus to plants although characteristics of the materials dictate that they be applied in different manners. Organic sources of phosphorus including animal residues and municipal waste can be highly effective and economical where available locally and when moved only short distances. Otherwise, the relatively low phosphorus contents of these phosphorus sources tends to increase their costs quickly.

PROBLEMS

1. Describe phosphorus deficiency symptoms in corn.

2. What is the primary form of phosphorus in normal superphosphate? Triple superphosphate?

3. Outline briefly the processes involved in reversion or fixation of phosphorus fertilizers in soils.

4. Demonstrate your knowledge of the relationship of soil pH to phosphorus availability to plants.

5. Which ammonium phosphate placed in direct seed contact with wheat poses the most hazard to germination in a soil with a pH of 7.8? Why?

6. Demonstrate your knowledge of the benefits of the phenomenon of sequestration that occurs in liquid fertilizers containing polyphosphate.

7. Compare the changes in the oxidation states of nitrogen and phosphorus in plant metabolism.

8. Calculate the phosphorus concentration in *pure* diammonium phosphate, mono-ammonium phosphate, and triammonium pyrophosphate.

9. Briefly compare the agronomic effectiveness of phosphorus in triple superphosphate, monoammonium phosphate, and ammonium polyphosphate.

10. Outline the locations of phosphorus deposits in the United States and briefly note the nature or geologic origin of the most important of these deposits.

REFERENCES

Alexander, M., *Introduction to Soil Microbiology,* 2nd ed., New York: John Wiley & Sons, 1977.

Baker, J. M., B. B. Tucker, and L. G. Morrill, "Effects of Sources of Phosphorus Under Varying Temperature and Moisture Regimes on the Emergence of Winter Wheat," *Soil Sci. Soc. Amer. Proc.,* 34 (1970) 694.

Barber, S. A., and R. K. Stivers, "Phosphorus Fertilization of Field Crops in Indiana — Research Prior to 1963," Purdue Res. Bul. 759, (1963) 28 pp.

Blair, G. J., C. P. Mamaril, and M. H. Miller, "Influence of Nitrogen Source on Phosphorus Uptake by Corn from Soils Differing in pH," *Agron. J.,* 63 (1971) 235.

Fried, M., and H. Broeshart, *The Soil-Plant System.,* New York: Academic Press, 1967.

Hemwall, J. B., "The Fixation of Phosphorus in Soils," *Advan. Agron.,* 9 (1957) 95.

Hendrix, J. E., "The Effect of pH on the Uptake and Accumulation of Phosphate and Sulfate Ions by Bean Plants," *Amer. J. Bot.,* 54 (1967) 560.

Hood, J. T., and L. E. Ensminger, "The Mechanism of Ammonium Phosphate Injury to Seeds," *Soil Sci. Soc. Amer. Proc.,* 29 (1965) 320.

Jones, U. S., *Fertilizers and Soil Fertility,* Reston, Va.: Reston Publishing Co., 1979.

Kamprath, E. J., "Residual Effect of Large Applications of Phosphorus on High Phosphorus Fixing Soils," *Agron. J.,* 59 (1967) 25.

Kang, B. T., and M. Yunusa, "Effect of Tillage Methods and Phosphorus Fertilization on Maize in the Humid Tropics," *Agron. J.,* 69 (1977) 291.

Riley, D., and S. A. Barber, "Effect of Ammonium and Nitrate Fertilization on Phosphorus Uptake as Related to Root-Induced pH Changes at the Root-Soil Interface," *Soil Sci. Soc. Amer. Proc.,* 35 (1971) 301.

Sauchelli, V., *Chemistry and Technology of Fertilizers,* New York: Reinhold, 1960.

Sdobnikova, O. V., and Y. M. Kaptzinel, "Effectiveness and Coefficient of Utilization of Phosphorus in Poly- and Orthophosphates," Proc. Inter. Sci. Symp. on Methods for Effect. Util. of Mineral Fert., May 21–26, 1973. Moscow, USSR.

Sinha, M. N., and R. K. Rai, "Efficiency of Fertilizer Utilization by Wheat Using ^{32}P as a Tracer," *Indian J. Agron.,* 21 (1976) 180.

Soon, Y. K., and M. H. Miller, "Changes in the Rhizosphere Due to NH_4^+ and NO_3^- Fertilization and Phosphorus Uptake by Corn Seedlings (*Zea mays* L.)," *Soil Sci. Soc. Amer. J.,* 41 (1977) 77.

Thien, S. J., and W. W. McFee, "Influence of Nitrogen on Phosphorus Absorption and Translocation in *Zea mays*," *Soil Sci. Soc. Amer. Proc.*, 34 (1970) 87.

Tisdale, S. L., and W. L. Nelson, "Soil Fertility and Fertilizers," 3rd ed., New York: Macmillan, 1975.

Van Wazer, J. R., "Phosphorus and Its Compounds," Vol. II., New York: Interscience, 1961.

Whiteaker, G., C. Gerloff, W. H. Gabelman, and D. Lindgren, "Intraspecific Differences in Growth of Beans at Stress Levels of Phosphorus," *J. Amer. Soc. Hort. Sci.*, 101 (1976) 472.

Wilbanks, J. A., M. C. Nason, and W. C. Scott, "Liquid Fertilizers from Wet-Process Phosphoric Acid and Superphosphoric Acid," *J. Agri. and Food Chem.*, 9 (1961), 174.

CHAPTER FOUR

Potassium Fertilizers

4.1 INTRODUCTION

Potassium, one of the "big three" macronutrients, is absorbed by higher plants in larger amounts than any other mineral element except nitrogen and, in some cases, calcium. It is the third figure in a fertilizer grade such as 10–20–10, which refers to percentage potassium oxide (K_2O). The term *potash* when used in connection with fertilizers usually refers to potassium oxide, written chemically as K_2O. Due to the custom of many years and state and federal laws, the potassium content of fertilizers in the United States is given in terms of K_2O, even though there is no K_2O as such in the material. There is some agitation to have the potassium content expressed as percentage potassium (K) rather than percentage potassium oxide (K_2O). It is likely that this will take place when the United States changes to the metric system.

The element potassium occurs in abundance over widespread areas of the earth's surface as a component of various rocks, minerals, and brines. Potassium is the seventh most abundant element in the crust of the earth. Potassium in nature never occurs in its elemental state (K) or as potassium oxide (K_2O) but is always combined with other elements. The primary minerals that are generally considered to be the original sources of potassium are the potash feldspars: orthoclase and microcline, both $KAlSi_3O_8$; muscovite, $KAl_3Si_3O_{10}(OH)_2$; and biotite $K_2(Mg, Fe)_2Al_2O_{10}(OH)_2$.

The preparation of potassium carbonate or *potash* by leaching and concentrating wood ashes was the subject of the first United States patent, issued to

Samuel Hopkins in 1790. It was signed by George Washington. The process used about 5 acres of timber to produce 1 ton of potash. The term *potash* is said to have derived from the manufacture of this product by the leaching of hardwood ashes in large iron pots.

4.2 MINING AND RECOVERY OF POTASSIUM SALTS

Over the eons of time, as rocks near the surface of the earth weathered, potassium was released to form soluble salts. These salts found their way into rivers and eventually into the seas and oceans. Certain bodies of water were closed off to become inland seas. Eventually the water evaporated and salt deposits remained. Practically all the potassium fertilizers produced today are derived from bedded deposits of water-soluble potassium minerals or brines of ancient seas. The Great Salt Lake in Utah and the Dead Sea between Jordan and Israel are examples of continuing salt accumulation because of no outlets.

Potassium salts were found in the Stassfurt deposits in 1839 and became the basis of much of the early German chemical industry. In 1914, at the start of World War I, the only deposits being worked and the only ones of any extent known were in Germany. As shipments from Germany were completely cut off, potassium was recovered in the United States from salty lake waters, from plant residues, from greensand silicates (glauconite), and from other types of deposits.

In the United States, Searles Lake in California, a mixed crystalline mass, was the first deposit worked. A plant was completed in 1914 but did not begin operation until 1916. In 1925 oil well cores in New Mexico in the Permian basin revealed potassium salts, and several deposits at Carlsbad, New Mexico were later developed (Fig. 4-1).

The world's largest known body of high-grade potassium ore was discovered in Saskatchewan, Canada. The deposits were 900 to 1500 m (3000 to 5000 ft) down. In addition to these North American sources, potassium is produced in various parts of the world including East and West Germany, France, Spain, Poland, USSR, Italy, Israel, Jordan, Chile (South America), and Zaire (Africa).

The compositions of the commercially important potassium minerals are listed in Table 4-1. These ores are believed to supply more than 95% of the total fertilizer produced annually. Sylvite (KCl) is the predominant source of fertilizer potassium in most producing areas. The ore that is mined is sylvinite, a mechanical mixture of sylvite (KCl) and halite (NaCl). Carnallite ($KCl \bullet MgCl \bullet 6H_2O$) is a major source of potassium ore in Israel. Kainite ($KCl \bullet MgSO_4 \bullet 3H_2O$) is the predominant material mined in Italy and is also mined to some extent in Germany. Langbeinite ($K_2SO_4 \bullet 2MgSO_4$) is an important potassium ore in the United States. The only deposit of nitre (KNO_3) of commercial significance is located in Chile, South America.

Underground deposits of solid potassium minerals are mined by two methods: solid ore and solution. The largest percentage of potassium ore is

Fig. 4–1. Aerial view of potash mine and plant near Carlsbad, New Mexico. After the crushed solid ore is brought to the surface, it is further pulverized and the KCl separated from NaCl by flotation or crystallization processes. (Courtesy Duval Corporation, Carlsbad, New Mexico.)

removed by the solid mining method. Extraction of potassium brines from concentrated deposits in lakes constitute another recovery process.

In the solid mining method, potassium ore is removed by opening a shaft to the strata containing the potassium salts. The crystallized rocklike salts are broken and pulverized by continuous mining machines, conveyed to the shaft pit, and hoisted to the surface in huge skips (Figs. 4–2 and 4–3). The material is usually a mixture of sylvite (KCl) and halite (NaCl) along with clay and small amounts of other impurities. The mining process creates many miles of underground tunnels using what is referred to as the *room-and-pillar*

Table 4–1. Commercially Important Potassium Minerals

Mineral	Composition	Approx. Plant Nutrient Content (%)	
		K_2O	K
Sylvite	KCl	63.2	52.5
Sylvinite (sylvite + halite)	KCl·NaCl mixture	Variable	Variable
Carnallite	KCl·MgCl·6H$_2$O	17.0	14.1
Kainite	KCl·MgSO$_4$·3H$_2$O	18.9	15.7
Langbeinite	K$_2$SO$_4$·2MgSO$_4$	22.6	18.8
Nitre	KNO$_3$	46.5	38.6
Polyhalite	K$_2$SO$_4$·MgSO$_4$·2CaSO$_4$·2H$_2$O	15.5	12.9

SOURCE: K. C. Kapusta, "Potassium Fertilizer Technology." In *The Role of Potassium in Agriculture,* American Society of Agronomy, Madison, Wis. (1968), pp. 23–52.

Fig. 4-2. Potassium ore being loaded by an underground loader into a shuttle car. (Courtesy Duval Corporation, Carlsbad, New Mexico.)

method. After the crushed solid ore is brought to the surface, it is further pulverized and the potassium chloride separated from sodium chloride by flotation or crystallization processes.

In the solution method, wells are drilled into the potash strata. A hot brine saturated with sodium chloride is pumped down to dissolve the minerals and the resultant solution is pumped to the surface. The brine is cooled, potassium chloride precipitates and is further refined. This method allows for deeper deposits to be mined and more complete recovery of the potassium deposits.

Recovery from brines is practiced in such areas as Searles Lake in California and the Great Salt Lake and Bonneville Salt Flats in Utah. The raw brine solution is concentrated by forced evaporation or solar evaporation in ponds. Preparation of the final potash products is then accomplished by standard principles. Potassium sulfate is a major product of the Great Salt Lake operation.

4.3 POTASSIUM REACTIONS IN SOILS

Most soils are comparatively high in *total* potassium, usually containing more of the element than any of the other nutrients. However, the quantity of potassium held in an easily *available* form at any one time is relatively small.

Fig. 4-3. Underground shuttle car dumping potassium ore into a crusher where it is then crushed and transported on a conveyer belt to the shaft pit. The ore is hoisted to the surface in huge skips. (Courtesy: Duval Corporation, Carlsbad, New Mexico.)

An equilibrium exists among the various forms of potassium in the soil as shown in Fig. 4-4. The various forms of potassium that occur in soils can be classified on the basis of availability to plants in three broad general groups: (1) relatively unavailable, (2) slowly available, and (3) readily available. Most potassium (90% to 98%) is in the relatively unavailable form; therefore, the other forms are more significant from the standpoint of plant nutrition and crop production.

4.3.1 Relatively Unavailable Forms

The greatest portion of all soil potassium in the soil is relatively unavailable. The unavailable forms are those occurring in the primary potassium-bearing minerals. Minerals containing most of this potassium are the feldspars and micas. Because these silicate minerals are resistant to weathering, they release only small quantities of potassium during a single crop season. However, their regular and prolonged contribution to the total available potassium in the soil is of considerable importance. Potassium is gradually released or converted to more available forms through action of solvents such as water, carbonic acid, and organic acids, on the minerals.

4.3.2 Slowly Available Forms

The slowly available potassium consists of potassium fixed within clay minerals such as illite, vermiculite, and chlorite. These clay minerals are made up of sheets of silica (SiO_2) and alumina (Al_2O_3). Soil potassium can be trapped and held between the silica and alumina layers of these minerals. Potassium held in this manner is not easily released and, consequently, it is very slowly available for nutrition of growing plants. Potassium in this form cannot be readily replaced by ordinary cation exchange processes and is referred to an *nonexchangeable* or *fixed* potassium.

4.3.3 Readily Available Forms

Readily available potassium constitutes only a small percentage of the total present in an average soil. It consists of two forms: (1) potassium ions in the soil solution and (2) exchangeable potassium adsorbed on the soil colloid surfaces. Exchangeable potassium is in equilibrium with the soil solution potassium, but there is only a very small portion in the soil solution—perhaps one to two % of the total potassium and 10% of the exchangeable potassium. Although most available potassium is in an exchangeable form, soil solution potassium is absorbed more readily by higher plants and also is more subject to leaching loss.

These two forms of readily available potassium are in dynamic equilibrium, which is important from a practical standpoint. Potassium absorption from the soil solution by plants results in a temporary disruption of the equilibrium. Some exchangeable potassium ions move immediately from the soil colloids into the soil solution, replacing those taken up by the plant. Equilibrium is again reestablished. On the other hand, when water-soluble potassium fertilizers are added, the reverse of this reaction may occur by direct adsorption of fertilizer potassium by soil colloids. Because of constant potassium removal from the soil system by higher plants and leaching, it is doubtful in the strictest sense of the word that a true or static equilibrium is ever obtained; perhaps dynamic equilibrium is a more appropriate term.

The soil solution and the exchangeable potassium are the major sources of potassium absorbed by plants. They are also the forms of potassium that are measured in soil analysis for determining the amount available for plant growth in a particular soil.

4.4 POTASSIUM FIXATION AND RELEASE

As previously indicated, only a very small percentage of the total potassium in a soil is in a form available to plants at any given time because the reactions shown in Fig. 4–4 are constantly taking place. With the application of potassium fertilizer, potassium first goes into the soil solution and soon after

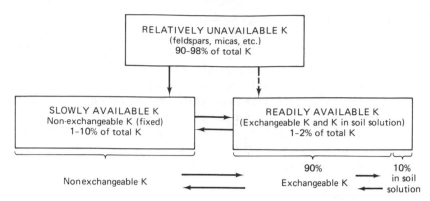

RELATIVELY UNAVAILABLE K
(feldspars, micas, etc.)
90–98% of total K

SLOWLY AVAILABLE K
Non-exchangeable K (fixed)
1–10% of total K

READILY AVAILABLE K
(Exchangeable K and K in soil solution)
1–2% of total K

Nonexchangeable K

90%
Exchangeable K

10%
in soil
solution

Fig. 4–4. The three forms of potassium are considered to be in equilibrium. Any change in the status of one form tends to be offset by an appropriate shift in the system to restore a new balance.

most of it goes into the exchangeable and some to the nonexchangeable forms. As crops remove the readily available potassium, the reactions are reversed and exchangeable potassium goes into solution. As a result, there is constant fixation and release of potassium in the soil.

The potassium cycle is illustrated in Fig. 4–5. During weathering, physical, chemical, and biological forces act on the parent materials and break them down into finer fractions, largely sand-, silt-, and clay-size particles. This breakdown results in the release of elements, including potassium and the formation (synthesis) of different clay minerals. Most of the total potassium in-

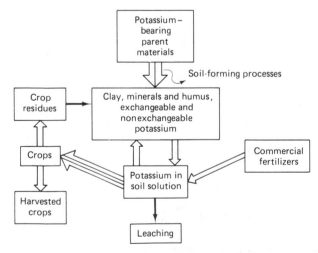

Fig. 4–5. Potassium cycle in soil. The schematic drawing shows the sources of soil potassium fixation and how the levels of available potassium are depleted and replenished. There is little or no leaching of potassium except in very sandy soils or organic soils. (Source: H. M. Miller, L. Murdock, and K. Wells, "Potassium in Kentucky Soils," University of Kentucky, Ext. Bul. AGR–11, 1973, 15 pp.)

herited from the parent material is in the nonexchangeable and exchangeable forms (Fig. 4–4).

The relative amounts of sand, silt and clay fractions found in a soil depend on the kind of parent material (sandstone, limestone, shale, or mica) from which the soil is derived. Potassium fixation and release is greatly influenced by the relative amounts of these fractions and the kinds of clay minerals present in the soil.

The sand and silt fractions of most soils are made up largely of quartz. Other minerals, mainly feldspars, in these fractions may contain potassium and other nutrients, but because the particle size is relatively large, the particles dissolve very slowly and the rate of potassium release is low. Also, because of the physical and mineralogical nature of silt and sand, their ability to fix potassium is very low.

4.4.1 Clay Minerals

Clay minerals (the dominant materials in the clay or colloidal fraction) in a soil are relatively active in fixing and releasing potassium. The different types of clay minerals vary in their capacity to fix and release potassium. There are several distinctly different clays that can be found in soils, but they may be classified for purposes of simplicity into three main types: kaolinite, illite and/or vermiculite (formed by weathering of mica minerals), and montmorillonite. Each clay mineral has its own characteristics with respect to potassium fixation and release. In addition, each clay mineral contains different amounts of native potassium, which is bonded between the clay layers.

Because of their crystal structure and the location and amount of negative charges within the crystals, illite and vermiculite clays are capable of adsorbing potassium from the soil solution and trapping it between layers of the clay particle (Fig. 4–6). The distance between the plates of illite and vermiculite is fixed and the clay is nonswelling; therefore, the potassium ions adsorbed there are physically trapped. The potassium cations are fixed or trapped in this way because of the relationship of their size to the hexagonal cavities in the silica sheets of two adjoining mica or vermiculite layers. This fixed (nonexchangeable) potassium is not immediately available to plants but is slowly released as the levels of exchangeable and soil solution potassium become lower.

Soils containing predominantly kaolinite clay have less exchangeable potassium to release than soils that have a higher percentage of the illite and vermiculite type of clay. Note in Fig. 4–6 that kaolinite does not have potassium trapped between the layers of the clay mineral. Montmorillonite clay can hold large amounts of exchangeable potassium but will fix only a small percentage of it. Montmorillonite has variable distances between the plates and swells on wetting, thus, exposing more surface for potassium ion adsorption. Therefore, most of the potassium held by montmorillonite clay is in an available form. In general, potassium fixation is not a serious problem in

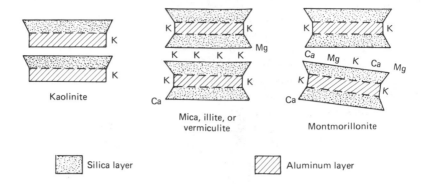

Kaolinite

Mica, illite, or
vermiculite

Montmorillonite

Silica layer   Aluminum layer

Fig. 4-6. The structure of the three main types of clay minerals. The potassium held by the kaolinite clay is adsorbed on the jagged edges of the plates and is exchangeable or available for plant use. The potassium held by the illite and vermiculite clays are adsorbed on the surfaces of and between the clay crystals or layers. The potassium held between the layers is largely nonexchangeable, whereas that held on the surface is exchangeable. The potassium held by the montmorillonite is also held on the surface and between the layers. Most of this potassium is exchangeable.

montmorillonite clays and may be regarded as one means of storing potassium in the soil.

Soil parent materials influence both the type of clay minerals found in soils and the soil's ability to supply potassium to plants. For example, soils derived from calcareous shales are high in exchangeable potassium. These soils generally contain some illite and vermiculite in their clay and fine silt fraction, as well as some potassium-bearing feldspars. Soils derived from limestone formations contain mixtures of clay minerals and are medium in their ability to release potassium. Soils derived from sandstone and acid shales are low in their ability to supply potassium due to a low percentage of clay in these soils and dominance of kaolinite among the clay minerals present.

4.4.2 Role of Cation Exchange Capacity (CEC) in Potassium Availability

Cation exchange capacity can be thought of as the nutrient holding capacity of a soil. Soil colloids with negative charges attract and hold cations. Because of the negative charges on the humus particle, humus can also adsorb cations in much the same way as the clay minerals hold them. The contribution of the clay mineral fraction to the cation exchange capacity is dependent on both the kinds and amounts of minerals in the soil. Although the clay minerals and humus account for most of the CEC, the finer fractions of silt can also have a limited number of exchange sites.

The CEC is expressed as millequivalents (meq) of hydrogen that can be held (adsorbed) by 100 grams of soil (oven-dry basis). This may vary from less than 5 meq/100 g in sandy soils to over 30 meq/100 g in soils high in clay and

Table 4–2. Example of Annual Potassium Recommended for a 10,000-kg/ha (160 bu/A) Yield Goal for Corn as Influenced by Soil Test and CEC

Soil Test Value (pp2m of K)	CEC (meq/100 g soil)		
	10	20	30
	K_2O as pp2m (lb/A)		
50	130	150	170
150	90	110	130
250	50	70	90

NOTES:
pp2m = parts per two million = lb/A.
kg/ha = pp2m × 1.121.

SOURCE: B. L. Schmidt et al., "Agronomy Guide," The Ohio State University, Coop. Ext. Serv. Bul. 472, 1978–79.

organic matter. A sandy soil might have 5% or more of the CEC saturated with potassium and a soil high in clay might have only 2% or less.

For example, a sandy loam with a CEC of 5 meq/100 g and 5% saturated with potassium would have 0.25 meq of exchangeable potassium (198 pp2m or 222 kg/ha or 198 lb/A).[1] A clay loam soil with a CEC of 25 and only 2% saturated with potassium would have 0.5 meq of exchangeable potassium (390 pp2m or 437 kg/ha or 390 lb/A) or twice as much potassium. One milliequivalent of potassium is equal to (780 pp2m or 874 kg/ha or 780 lb/A).

In general, the higher the CEC, the greater the amount of potassium that must be present in order to provide sufficient amounts of potassium for adequate plant nutrition. It appears that the sufficiency levels in soils should vary with CEC but not merely as a simple percentage of it.[2] The Ohio Soil Testing Laboratory increases the amount of potassium fertilizer recommended for a certain soil test as the CEC increases. Table 4–2 shows an example of the annual potassium recommendation for corn expressed as pp2m potash (lb/A).

Of the clay minerals, kaolinite has the lowest CEC (5 to 15 meq/100 g of clay). The CEC of illite is intermediate (10 to 45 meq/100), and vermiculite and montmorillonite clay minerals are relatively high (60 to 150 meq/100 g clay). The CEC of humus is about 140 meq/100 g. The values are for pure clay minerals or humus.

Cations on exchange sites are held rather loosely on the edges or surfaces of the clay mineral or humus particles and are constantly being replaced by

[1] pp2m = lb/A, lb/A × 1.121 = kg/ha. A soil is considered to have 2,000,000 lb of soil per acre 6 in. (15 cm). Soil test values are usually reported in ppm (parts per million) or in pp2m (parts per 2 million).

[2] E. O. McLean, "Exchangeable K Levels for Maximum Crop Yields on Soils of Different Cation Exchange Capacities," *Communications in Soil Science and Plant Analysis*, 7(9):823–38, 1976.

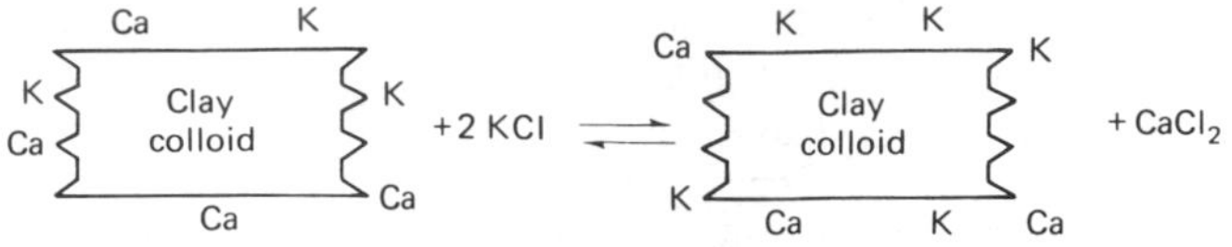

Fig. 4-7. Cation exchange reaction. Note that when potassium is added as fertilizer, by mass action, the potassium replaces calcium on the clay, allowing the calcium to combine with the chloride in the soil solution. The process is reversible. Other exchangeable cations such as H^+, Mg^{2+}, Na^+, and NH_4^+ can be replaced in a similar manner to Ca^{2+}.

other cations. They occupy exchange sites because they are balancing the negative charges of the clay minerals and humus fractions in the soil. For this reason the reactions are reversible.

The importance of cation exchange capacity (CEC) is that it prevents or reduces the leaching of fertilizer components such as potassium, ammonium, magnesium, calcium, and other cations. Cation exchange is a means by which the soil can store potassium and other cations that may be released later to plants. The interaction of potassium and other cations, such as calcium and magnesium, with soil colloids is referred to as *cation exchange*. This is shown in Fig. 4-7.

4.4.3 Potassium Leaching Losses

Because positively charged potassium ions (K^+) are held in an exchangeable and available form on negatively charged clay particles, potassium is not easily lost from the soil through leaching. Some leaching may take place on very sandy soils because they do not contain enough clay particles or humus to hold potassium. Organic soils (mucks and peats) do not hold potassium as strongly as other positively charged nutrients, such as calcium. Therefore, loss of potassium by leaching is one reason sandy and organic soils often test low in available potassium.

In general, most studies have indicated that very little leaching of potassium takes place on silty or clayey soils. In fact, a study conducted in Texas on a loamy sand indicated little potassium movement below 46 cm (18 in.) after 9 years of potassium fertilization (Fig. 4-8). This means that the supply of available potassium can be increased by potassium fertilization. The more clay or organic matter a soil contains, the greater its ability to store potassium applied as fertilizer for future use in crop nutrition.

4.5 POTASSIUM IN PLANT NUTRITION

Potassium is found in cell and plant fluids. It is only weakly bound and not thought to be a part of fixed organic compounds in the plant. Potassium is very readily absorbed by the plants. A major part of the absorbed potassium exists in the cell sap in soluble form. It is very mobile in the plant and moves readily from older tissues to the growing points of roots and shoots.

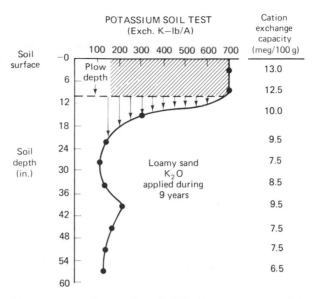

Fig. 4-8. After 9 years of potassium fertilization at recommended rates, little potassium movement has occurred below 46 cm (18 in.) on a loamy sand in Texas. [*NOTE:* lb/A = parts per 2 million (pp2m).] (Source: John Box, "Potassium and Crop Production," Texas Agric. Ext. Serv. Bul. MP-1007, 1972, 19 pp.)

Potassium is usually taken up earlier than nitrogen and phosphorus and uptake increases faster than dry matter production (Fig. 4-9). This means that the potassium accumulates early in the growing period and then is translocated to other plant parts. At maturity, the corn grain should contain not more than one-third of the total potassium that is in the aboveground parts of the plant.

4.5.1 Function in Plants

A large amount of biochemical research has been conducted relative to the supply of potassium in plant nutrition and the metabolic malfunctions that occur under conditions of potassium deficiency. Crops produce poorly when too little potassium is supplied for their optimum growth; yet it has not been found as a part of the actual structure of plants. Potassium performs many functions in plants. The most important is the role potassium has in the process of photosynthesis.

Photosynthesis is complex and many variables influence it. Potassium seems to relate to photosynthesis in a number of ways. For example, potassium supply and carbon dioxide (CO_2) uptake are related. Potassium is thought to regulate stomatal openings (tiny pores) in the leaf so that carbon dioxide can enter.[3]

[3] W. A. Jackson and R. J. Volk, "Role of Potassium in Photosynthesis and Respiration." In *The Role of Potassium in Agriculture,* American Society of Agronomy, Madison, Wis. (1968), pp. 109–145.

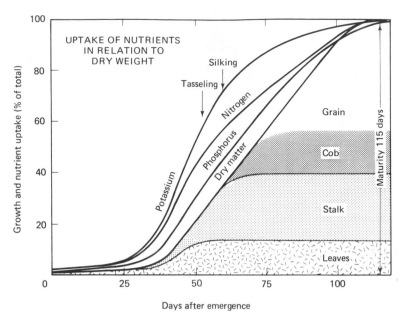

Fig. 4–9. The rate of uptake of potassium by corn is faster than uptake of nitrogen or phosphorus or dry matter accumulation. (Source: J. J. Hanway, "Growth and Nutrient Uptake by Corn," Iowa Extension Pamphlet No. 277, 1960, 4 pp.)

It has long been known that potassium is associated with plant strength and resistance to lodging (Figs. 4–10 and 4–11). Potassium is involved in maintaining structural integrity of cellular components and perhaps even the cellular membranes. A deficiency of potassium results in plants with less physical strength due to decreased structural integrity (Fig. 4–12). The decrease in structural integrity may be due to a progressive breakdown of parenchyma in the brace roots and stems of potassium-deficient plants.[4]

Potassium is necessary for normal lignin and cellulose development, which gives strength and stiffness to plants, enabling them to stand upright with reduced lodging. Brace root development is usually more extensive on corn receiving adequate potassium as compared to plants with potassium deficiency. This is usually expressed as an increase in total number of brace roots as well as greater permeation of roots in a larger soil volume (Fig. 4–13). Potassium encourages plant root development. As an example, plants well supplied with potassium have an abundance of roots to utilize soil moisture efficiently; whereas plants grown under low potassium levels have very few roots, thereby resulting in low water use efficiency. Increased root growth improves the drought resistance of most plants.

Crops receiving high rates of nitrogen, coupled with inadequate potassium, are also more prone to lodging. The incident of stalk rot in corn and grain sorghum has been shown to be interrelated with nitrogen and

[4] W. C. Liebhardt and J. T. Murdock, "Effect of Potassium on Morphology and Lodging of Corn," *Agron. J.*, 57:325–28, 1965.

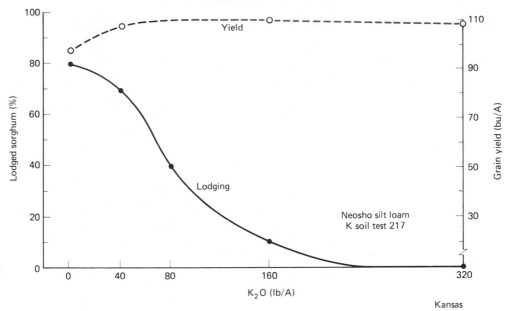

POTASH REDUCES SORGHUM LODGING

Fig. 4–10. In a Kansas experiment, sorghum lodging decreased very significantly as potash rates went up. The experiment was conducted on a Neosho silt loam with a soil test of 217 pp2m potassium. [*NOTE:* lb/A x 1.121 = kg/ha; bu/A x 62.776 = kg/ha; pp2m = parts per 2 million = lb/A.] (Courtesy Potash and Phosphate Institute.)

potassium rates. Stalk rot tends to increase with increasing rates of nitrogen and to decrease with increasing applications of potassium (Fig. 4–14). Root systems of potassium-deficient plants are weakened and disease susceptible. Nitrogen-potassium balance is important in certain soils. For example, the most profitable rate of nitrogen depended on higher rates of potassium to achieve the highest yield of corn in Illinois (Fig. 4–15).

Potassium has been given credit for several important roles in plant nutrition associated with the quality of a product. Potassium increases the starch content of grains, making plump kernels with high test weights, and also increases the sugar content of fruits, beets, and sugarcane. Potassium promotes protein development, which is very important in all plants, and improves the quality of forage plants. It increases the thickness of the epidermal layer of cells, adding to insect and disease resistance and to the shipping and keeping qualities of thin-skinned fruits and vegetables. Translated into practical terms, the effects of potassium upon plant growth show up in many ways of special interest to the producer and to the consumer.

More than 40 enzyme systems of plants, animals, and microorganisms require potassium for normal activity.[5] These enzymes are involved in

[5] H. J. Evans and C. G. Sorger, "Role of Mineral Elements with Emphasis on the Univalent Cations," *Ann. Rev. Plant Physiol.,* 17:47–76, 1966.

Fig. 4-11. Strong stalks and root systems are a major factor behind today's high corn yields. Research has shown low potash as one of the main causes of poor root development and stalk lodging. (Courtesy Potash and Phosphate Institute.)

N   NP   NPK

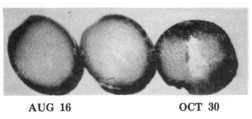

AUG 16   OCT 30

SEPT 20

Fig. 4-12. Photographs showing the relative effects of fertility (above) and maturity (below) on corn stalk breakdown. The top series of photos indicate that with added potassium, a solid healthy cornstalk developed. The bottom series of photos illustrate that without adequate potassium the stalk tissue steadily breaks down during the growing season. (Courtesy Potash and Phosphate Institute.)

Fig. 4–13. The corn plants receiving potassium fertilizer developed more brace roots and covered a larger soil volume than the corn plants with potassium deficiency. (Courtesy Potash and Phosphate Institute.)

Fig. 4–14. Examination of grain sorghum stalks revealed a strong relationship between potassium rate and the incidence of stalk rot. Stalk rot continued to decline through the highest potassium rate in the experiment. [*NOTE:* 320 lb/A = 358 kg/ha.] (Courtesy David Whitney, Kansas State University.)

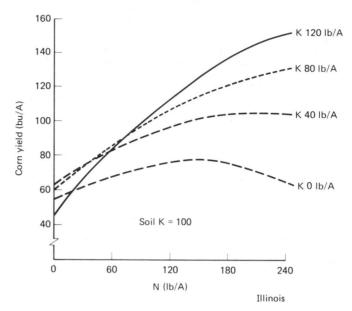

Fig. 4-15. When nitrogen was added to a potassium-deficient soil, the best nitrogen rates boosted yields to only 100 bu/A (6278 kg/ha) with 50 lb K₂O (40 lb K). But when 145 lb K₂O (120 lb K) was applied, 240 lb nitrogen (269 kg/ha) was profitable —getting over 150 bu/A (9416 kg/ha). More nitrogen calls for more potassium to get the most out of today's high-yield crops (Illinois). [*NOTE:* bu/A x 62.776 = kg/ha (corn); lb/A x 1.121 = kg/ha.] (Courtesy Potash and Phosphate Institute.)

photophosphorylation, glycolysis, oxidative phosphorylation, respiration, protein synthesis, and glycogen or starch synthesis. These are all vital plant processes, so it is no mystery that potassium is a major essential plant nutrient.

Another important function of potassium has been shown to be in be in the realm of water relations of plants, particularly in the leaves. Measurements of water loss per unit leaf area have revealed that potassium-deficient plants usually transpire more than plants with adequate potassium. Plants receiving adequate potassium tend to have a slower transpiration rate than potassium-deficient plants (Fig. 4-16). When exposed to hot dry winds the plants with adequate potassium apparently close their stomata much more quickly than potassium-deficient plants.

Water is vital to the movement of plant nutrients such as potassium from the soil to the root. Research has shown that when rainfall is short, additional potassium helps to maintain the yield of soybeans (Fig. 4-17). This type of response has also been shown to be true for corn (Table 4-3). Potassium increased the yield of corn by 2448 kg/ha (39 bu/A) when rainfall amounts were low and by 3014 kg/ha (48 bu/A) when rainfall amounts were high. There was only a 502 kg/ha (8 bu/A) increase when the rainfall was optimum.

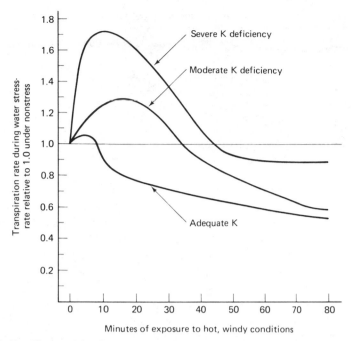

Fig. 4–16. Plants with adequate potassium lose less moisture because they have a slower transpiration rate. When exposed to hot, dry windy conditions, they apparently close their stomata much more quickly than potassium-deficient plants, as this Montana barley study indicates. (Courtesy Potash and Phosphate Institute.)

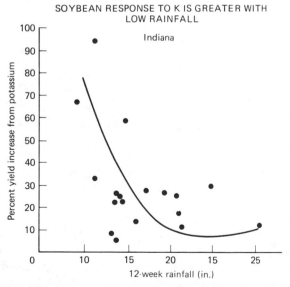

Fig. 4–17. Water is vital to potassium movement from the soil to the root. The data indicates that the less the rainfall after planting, the greater the percent yield increase from potassium fertilizer. When rainfall is short, additional potassium can help to maintain yields (Indiana). (Courtesy Potash and Phosphate Institute.)

Table 4–3. The Response of Corn to Potassium Fertilizer in Indiana

Rainfall (cm)	Yield of Corn (kg/ha)		
	− K Fertilizer	+ K Fertilizer	Increase
Low — 18 (7.1 in.)	5713 (91 bu/A)	8161 (130 bu/A)	2448 (39 bu/A)
Medium — 45 (17.7 in.)	9291 (148 bu/A)	9793 (156 bu/A)	502 (8 bu/A)
High — 65 (25.7 in.)	5775 (92 bu/A)	8789 (140 bu/A)	3014 (48 bu/A)

SOURCE: Potash and Phosphate Institute.

4.5.2 Deficiency Symptoms

Because potassium is mobile and moves to the younger leaves when the supply is short, the deficiency symptoms first appear in the older leaves. Nearly every species of plant has its own specific pattern, but common signs of potassium deficiency are:

1. The older leaves show a light green to a "scorched" effect on the margins and tips. In severe cases, these dead areas may fall out, leaving ragged edges. In corn, small grains, and grasses, "firing" starts at the tips of the leaves and proceeds down from the edges or margins, usually leaving the midribs green.
2. Root development is poor and stalks are weak. Plants tend to bend and break over and lodge (Fig. 4–18).
3. Plants grow slowly.
4. Seeds and fruits are small and shriveled.
5. Plant resistance to certain diseases is reduced.

Certain plants may exhibit other symptoms. For instance, clover and alfalfa develop white spots on the leaf margins as early symptoms. Later, if the deficiency continues, the margins will "scorch." Scorched leaf edges of corn or grain sorghum are typical of potassium deficiency. In addition, ears or heads do not fill properly and a poor yield of low-quality grain results.

Deficiency symptoms should not be used as a means of determining potassium fertilizer needs, because by the time the symptoms manifest themselves the crop yield has been reduced drastically and profitable production for the current season has been jeopardized. If a grower waits until deficiency symptoms are present, the soil potassium has been too low for some time. It is much better to use a method of diagnosis—such as soil tests and/or plant analyses—that detect potassium needs long before deficiency symptoms occur (see Chapter 1). Also, one should realize that some plant species may die or disappear from a field without ever showing any visible deficiency symptoms. For example, in a field of grasses and legumes, if potassium is deficient in the soil, the legume will be forced out. This may happen even though no ob-

Fig. 4–18. Strong stalks and root systems are a major factor behind today's high corn yields. Research has shown low potassium as one of the main causes of poor root development and stalk lodging. (Courtesy Potash and Phosphate Institute.)

vious deficiency symptoms ever appear in the legume plants. Grass has a greater ability to compete for potassium. Solid plantings of legumes such as alfalfa require high potassium levels to maintain a satisfactory stand.

4.6 UTILIZATION OF POTASSIUM BY PLANTS

Potassium is a monovalent cation that is absorbed in larger quantities by plant roots than any other cation. During periods of rapid growth, the soil must be able to supply large amounts of potassium. For example, corn and alfalfa may require about $3\frac{1}{2}$ kg/ha (3 lb/A) of potassium per day in peak periods of growth and corn about 9 kg/ha (8 lb/A). Because of the large quantities of potassium absorbed by the plant, many soils are rapidly depleted of available potassium by cropping so that the supply of potassium to the plant root becomes the principle limiting factor in maximizing yields.

Plants contain most of the same kinds of ions present in the soil solution; however, there is little relation between the relative quantities of absorbed ions in soil and in the plant root. Plants commonly contain 10 times as much potassium as calcium; yet the level of available calcium in the soil may be 10 times that of potassium.[6] Even though plant roots are highly selective in the absorption of nutrients, calcium has been shown to compete with potassium for entrance into the plant. Hence, a calcareous soil or a recently limed soil may require a higher level of potassium in the soil in order to maintain an adequate level of potassium in the plant.

4.6.1 Movement of Potassium to Plant Roots

Plant roots absorb nutrients as a result of root interception, mass flow of nutrients, and diffusion of nutrients to the root. The roots of plants do not contact more than about 3% of the soil volume; therefore, the amount of potassium absorbed as a result of root interception is small. The absorption of soil moisture by plants cause a mass movement of soil moisture and dissolved nutrients toward the roots. Mass flow is considered to account for only limited amounts of potassium absorbed by plants. The diffusion of ions through the water films around soil particles and roots is believed to be the mechanism responsible for the greatest amount of absorption of potassium.

The amount of water in the soil through which diffusion may take place is a prime factor in determining the rate of diffusion. If soil moisture levels are low and moisture films are thin, then diffusion is slow. Likewise, if the concentration gradient of potassium or other nutrient levels in the soil solution is low, then diffusion is low. The amount of water in the soil through which diffusion of potassium ions may take place is the prime factor in determining the rate of diffusion of potassium. Increased soil moisture content results in increased diffusion as well as increased potassium uptake by plants. Conversely, inadequate soil moisture may result in a deficiency of potassium in plants, even though adequate available potassium is present in the soil.

4.6.2 Soil Temperature and Potassium Uptake

The amount of potassium in soil solution has been shown to be related directly to the temperature. Consequently, soil temperature influences the uptake of potassium (Fig. 4–19). This influence has been shown to be rather significant; that is, a few degrees increase in soil temperature can result in a rather large increase in extractable potassium. In addition, the rate of release of nonexchangeable potassium to the exchangeable form has been found to be directly temperature dependent.[7] To these influences, the increased metabolic activity of plants at warmer temperatures must be added. The net result is an important positive influence of temperature conditions on potassium nutrition of plants.

[6] S. A. Barber, "Mechanism of Potassium Absorption by Plants." In *The Role of Potassium in Agriculture,* American Society of Agronomy, Madison, Wis. (1968), pp. 293–310.

[7] T. Haagsma and M. H. Miller, "The Release of Non-exchangeable Soil Potassium to Cation-exchange Resins as Influenced by Temperature, Moisture, and Exchanging Ion," *Soil Sci. Soc. Amer. Proc.,* 27:153–156, 1963.

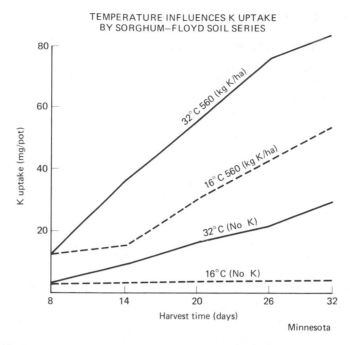

Fig. 4-19. Temperature influences potassium uptake by sorghum. Potassium rates of 0 and 560 kg/ha (500 lb/A) and two temperatures (16° and 32°C, 61° and 90°F) were used in this experiment. (Courtesy Potash and Phosphate Institute.)

4.6.3 Soil Aeration and Potassium Uptake

Several factors can cause poor soil aeration. Poor soil structure (either natural or reduced by improper tillage), naturally waterlogged conditions, and dense or impervious layers in the soil can all result in poor aeration. Research results have shown that the potassium content of plants is decreased as the oxygen content of the soil air is decreased.[8] The amount of oxygen in the soil atmosphere required to maintain normal potassium uptake apparently is not the same for all plants but appears to be in the range of 5 to 20%. Root development is restricted in compacted soils and in soils with a saturated or waterlogged condition. It is not known whether the decrease in the amount of potassium taken up into a plant is a result of decreased potassium uptake or a result of decreased transport from root to shoot. There is some indication that the transport aspect is involved, but is it reasonable to suspect that perhaps both potassium uptake and potassium transport are unfavorably affected by a limited supply of oxygen for plant roots. Stated in another way, artificial drainage may be necessary before adequate potassium nutrition is possible.

[8] G. W. Thomas and B. W. Hipp, "Soil Factors Affecting Potassium Availability." In *The Role of Potassium in Agriculture*, American Society of Agronomy, Madison, Wis. (1968), pp. 269-91.

4.6.4 Soil pH and Potassium Uptake

The effect of soil pH on fixation and the release of soil potassium is a controversial subject. Research results are often contradictory. The problem is important because of its relation to the practice of liming in a soil management program.

Calcium and potassium have the ability to replace each other in the cation exchange system. Soils with a high degree of base saturation lose less potassium to leaching than low-base saturated soils.

Liming is a common practice for increasing base saturation. The effect of liming on potassium behavior is explained on the basis of cation exchange phenomena. Liming has two effects upon exchange properties that tend to retard the possible loss of potassium by leaching. The addition of lime ($CaCO_3$) removes (Al^{3+}) by precipitation as aluminum hydroxide from competition with potassium (K^+), thereby opening up new sites on the exchange complex. This gives potassium and calcium an opportunity to occupy the newly vacated positions left by aluminum. This greatly increases the potassium held by the exchange sites and reduces the potassium in the soil solution. The reduction in solution-phase potassium reduces the leaching hazard. In such cases there is a net loss of potassium from the solution and it could be inferred that liming decreased potassium availability to the plant.

From a practical standpoint, however, the increase in plant vigor, depth, and volume of soil occupied by plant root systems on limed soils as compared with unlimed soils more than offsets the reduced potassium availability due to liming. Correcting pH of a very acid soil to a near neutral soil pH has been shown to increase the uptake of potassium in most cases (see Chapter 8).

4.6.5 Luxury Consumption of Potassium

The term *luxury consumption* is something of a misnomer. The term implies that plants tend to absorb more of an element than is required for optimum growth without a corresponding increase in yield or crop quality. This suggests that there has been an inefficient and uneconomical use of an element. However, today's higher yields along with corresponding higher concentrations of potassium in plants, and other nutrients as well, suggest that the luxury consumption is less meaningful than it was 1 or 2 decades ago. For example, a 1.0 to 1.5% potassium content in alfalfa was once thought to be adequate; however, a potassium content of 2 to 3% is now considered to be necessary in order to maintain high yields and good stands consistently.

4.7 POTASSIUM FERTILIZER MATERIALS

The sources of potassium fertilizer are presented in Table 4–4. The most common potassium fertilizer used on field crops is potassium chloride. In 1977 it accounted for 97% of the potassium fertilizer used in the United States. It is

Table 4-4. Principal Sources of Potassium Fertilizers

Material	Chemical Formula	Approximate Composition	
		Oxide Basis $N-P_2O_5-K_2O$	Elemental Basis $N-P-K$
Potassium chloride	KCl	0-0-60	0-0-50
Potassium sulfate	K_2SO_4	0-0-50	0-0-42 + 18% S
Potassium magnesium sulfate	$K_2SO_4 \cdot 2MgSO_4$	0-0-22	0-0-18 + 22% S + 11% Mg
Potassium nitrate	KNO_3	13-0-44	13-0-37
Potassium hydroxide	KOH	0-0-75	0-0-62

the least expensive source of potassium and is usually recommended except when a crop also needs sulfur or magnesium. Certain special crops require the use of the sulfate form of potassium to maintain crop quality. For example, tobacco should be fertilized with the sulfate form of potassium because it will not burn properly when the chloride form is used as the potassium source.

4.7.1 Potassium Chloride

Potassium chloride, or muriate of potash as it is commonly called, is a crystalline, water-soluble product containing about 50 to 52% potassium (K) (60–63% K_2O). It is lower in cost and more soluble than most of the other potassium carriers. It is more concentrated than other forms, and for most crops it is just as good a source of potassium as any on the market. As a clear liquid fertilizer, about 8% potassium is the maximum concentration that can be obtained with this material. In suspension fertilizers, where attapulgite clay is used to keep the fine crystals in suspension, grades containing up to 25% potassium (K) may be made with potassium chloride (KCl).

4.7.2 Potassium Sulfate

Potassium sulfate is a water-soluble white crystalline salt containing 42 to 44% potassium (50–53% K_2O), 18% sulfur, and less than 2.5% chloride. It is produced mainly to satisfy the trade requirement for a potash fertilizer containing a low percentage of chloride. Potassium sulfate is more expensive per unit of potassium than the chloride form. It is a preferred source for certain crops such as tobacco, and potatoes that are to be used for chipping purposes. The reasoning behind the preference of potassium sulfate for potatoes is that excessive amounts of chlorides result in a rubbery rather than a crisp potato chip. Potassium sulfate has lower solubility and is not used in clear liquid fertilizers. However, it can be used in suspensions.

4.7.3 Potassium-Magnesium Sulfate

This product is manufactured and marketed under various trade names. It contains 18% potassium (22% K_2O), 11% magnesium, and 22% sulfur. It is an excellent source of water-soluble potassium and magnesium and is of special importance in soils where magnesium is also deficient.

4.7.4 Potassium Nitrate

Potassium nitrate contains about 13% nitrogen and 37% potassium (44% K_2O). In the past most of the nitrate forms of potassium used in the United States came from Chile where it was obtained as a by-product in the refining process of nitrate of soda. It is now made by chemically combining potassium salts with nitric acid. Potassium nitrate (KNO_3) has lower solubility at low temperatures but is more soluble than potassium chloride (KCl) above 21°C (70°F). As a liquid fertilizer, about 5% potassium is the maximum that can be obtained in a clear solution with potassium nitrate.

4.7.5 Other Potassium Fertilizers

Potassium hydroxide (KOH) is a caustic, water-soluble product containing about 62% potassium (75% K_2O). Because of its high cost, potassium hydroxide has been limited chiefly to specialty liquid fertilizers. Use of potassium hydroxide as a source of potassium in liquid fertilizers allows the production of liquid fertilizers with analyses high in potassium.

Potassium phosphate is made by reacting phosphoric acid with caustic potash (KOH) or potassium carbonate (K_2CO_3) and may be prepared as potassium orthophosphate (K_3PO_4), potassium monohydrogen phosphate (K_2HPO_4), potassium dihydrogen phosphate (KH_2PO_4), or potassium metaphosphate (KPO_3). Potassium phosphates vary greatly in solubility and some are used in specialty grades of clear liquids, where price of materials is not a major factor. They are a source of both potassium and phosphorus.

Potassium carbonate (K_2CO_3) containing 56% potassium and potassium bicarbonate ($KHCO_3$) containing 39% potassium do not have much commercial use as fertilizers. High price and limited availability are the major factors. Near seacoasts, some organic gardeners may use seaweeds (kelp) as a source of potassium. Wood ashes are also rich in potassium (see Chapters 8 and 10).

4.8 APPLICATION OF POTASSIUM FERTILIZERS

Generally speaking, if the need for potassium is large, it is best to broadcast and incorporate most of the potassium before planting. Fertilizers containing potassium can also be applied in the fall or winter well ahead of planting. In most cases, a surface or topdressed application of potassium for some crops such as alfalfa or grasses gives excellent results.

In order to achieve maximum efficiency, potassium may be applied in localized bands at, or just prior to, time of planting. However, band placement of nitrogen and/or potassium fertilizers (KCl in particular) produces a high concentration of soluble salts in the deposition zone. Hence, if the material is placed with or too near the seed, severe plasmolysis of seedlings can occur with resulting loss of stand and yield. Also potassium applications directly above or directly below the point of seed placement are not recommended. Placement of fertilizer 5 cm (2 in.) to the side and 5 cm below the seed eliminates the possibility of seed damage, and seedling root development can occur normally without injury from high concentrations of fertilizer salts. Within a short period (2 weeks) and with favorable soil moisture, the salt concentration diffuses through a larger volume of soil and is diluted by soil moisture so that the hazards to tender plant tissue no longer exist.

Crops with a long growth period and/or perennial crops are likely to benefit from split or delayed application of potassium fertilizers. Small frequent applications are preferable where practical to large, infrequent applications on perennial crops. The same recommendation applies to crops growing in low cation exchange capacity soils, such as sands, and organic soils where leaching losses occur (see Chapter 7).

4.9 POTASSIUM CONVERSION FACTORS

Table 4.5 Potassium Conversion Factors

To Convert from Column A	To Column B	Multiply Column A by
Potassium (K)	Potassium (K_2O)	1.205
Potassium (K_2O)	Potassium (K)	0.830

SUMMARY

Extensive deposits of soluble potassium salts are found in many areas of the world. The potassium salts and brines found in these deposits are of a high degree of purity and lend themselves to mining operations for the production of potassium salts, usually termed *potash* salts by the trade.

The various forms of potassium that are found in the soil are classified on the basis of availability to plants and are designated as relatively unavailable, slowly available, and readily available. The unavailable forms are those occurring in the primary potassium-bearing minerals. The slowly available forms are the result of potassium ions interacting with certain clay minerals and becoming trapped or fixed. The readily available forms are made up of exchangeable and water-soluble potassium.

Potassium performs many functions in plants. It has an important role in the process of photosynthesis. Of all the plant nutrients, potassium is mentioned most often in relation to crop quality. Low starch content in potatoes, low oil content in soybeans, low protein content in forage plants, low sucrose content in sugarcane and sugar beets, weak stalks and lodging problems, and many more are some of the end results when crops are grown in a potassium-deficient environment.

Plant availability of soil potassium is greatly influenced by the type and amount of clay mineral present, the soil moisture content, soil temperature, soil aeration, and soil pH. Potassium is absorbed in larger quantities by plant roots than any other cation. The application of large amounts of potassium fertilizer may result in high uptake of potassium. This has been termed *luxury comsumption;* however, with today's higher yields along with higher concentrations of potassium in plants it does not appear to be a serious economic problem.

The principal fertilizer materials containing potassium are potassium chloride, potassium sulfate, potassium-magnesium sulfate, and potassium nitrate. The agronomic effectiveness of the principal potassium fertilizers is about equal. Obviously, potassium-magnesium sulfate would be more effective on a magnesium-deficient soil than would potassium chloride or potassium sulfate.

PROBLEMS

1. The various forms of potassium can be classified on the basis of availability to plants in three broad general groups. What are they?

2. Explain how potassium may become "fixed" in soils.

3. Potassium has been given credit for several important roles in plant nutrition. What are some of these roles?

4. Name the common sources of fertilizer potassium.

5. How many pounds of potassium chloride would be required for a recommendation of 80 pp2m (lb/A) of (a) potassium (K)? (b) potassium (K_2O)? (c) Make the same calculation for potassium sulfate.

NOTE: Problems 6 to 10 are given with the solutions to provide a better understanding of cation exchange capacity (CEC) of soils.

6. Calculate the equivalent (eq) and millequivalent (meq) weights of the following ions:

Ion	Ionic wt (g)	Equivalent wt (g)	Millequivalent (meq) wt (mg)
H^+	1.00	1.00/1 valence = 1.00 g	1.00 mg
Ca^{2+}	40.08	40.08/2 = 20.04 g	20.04 mg
Mg^{2+}	24.32	24.32/2 = 12.16 g	12.16 mg
K^+	39.10	39.10/1 = 39.10 g	39.10 mg
NH_4^+	18.02	18.02/1 = 18.02 g	18.02 mg
Na^+	22.99	22.99/1 = 22.99 g	22.99 mg

7. Assume that soil to a depth of 15 cm over an area of 1 ha weighs 1.764×10^6 kg/ha.
   (a) Calculate the amount of exchangeable potassium ions (in kilograms) that is equivalent to 1 meq of potassium ions per 100 g of soil.
   (b) Express that value also in terms of parts per million (ppm) potassium (K^+).

   Solution:

   (a) meq wt K = 39.10 ÷ 1 = 39.1 mg/meq

   1 meq K/100 g soil = 39.1 mg K/100 g soil
   $$= 391 \text{ mg/kg soil}$$

   $$\frac{391 \text{ mg K}}{1 \text{ kg soil}} = \frac{X \text{ mg K}}{1.764 \times 10^6 \text{ kg soil}}$$

   $$X = \frac{391 \text{ mg K} \times 1.764 \times 10^6 \text{ kg soil}}{1 \text{ kg soil}}$$

   $$X = 690 \times 10^6 \text{ mg K} = 690 \times 10^3 \text{ g K} = 690 \text{ kg K}$$

   $$\therefore \frac{1 \text{ meq K}}{100 \text{ g soil}} = \frac{690 \text{ kg K}}{1.764 \times 10^6 \text{ kg soil}} = \textbf{690 kg K/ha to 15 cm}$$

   (b) Calculating ppm K (concentration of exchangeable K^+):

   $$\frac{690 \text{ kg K/ha}}{1.764 \times 10^6 \text{ kg soil/ha}} = \frac{X \text{ kg K}}{1 \times 10^6 \text{ kg soil}}$$

   $$X = \frac{690 \text{ kg K} \times 1 \times 10^6 \text{ kg soil}}{1.764 \times 10^6 \text{ kg soil}}$$

   X = 391 ppm K^+ (concentration)

8. A soil has a cation exchange capacity (CEC) of 15 meq/100 g soil. If that soil were 5% saturated with exchangeable potassium ions, (a) how many milligrams of potassium ions would be present per 100 g of soil? (b) How many kilograms of exchangeable potassium ions would be present in 1.764×10^6 kg of soil? (c) How many ppm potassium ions are exchangeable?

   Solution:

   (a) 15 meq CEC/100 g soil x 5% saturation = 0.75 meq K/100 g soil

   0.75 meq K/100 g soil × 39.1 mg K/meq = **29.3 mg K/100 g soil**

   (b) 29.3 mg K/100 g soil = 293 mg K/1000 g soil = 293 mg K/kg soil

   $$\frac{293 \text{ mg K}}{\text{kg soil}} = \frac{X \text{ mg K}}{1.764 \times 10^6 \text{ kg soil}}$$

   $$X = \frac{293 \text{ mg K} \times 1.764 \times 10^6 \text{ kg soil}}{1 \text{ kg soil}}$$

   $$X = 517 \times 10^6 \text{ mg K} = 517 \times 10^3 \text{ K} = \textbf{517 kg K}/1.764 \times 10^6 \textbf{ kg soil}$$

   (c) 29.3 mg K/100 g soil = 293 mg K/1000 g soil = 293 mg K/1,000,000 mg soil = **293 ppm K**

9. A soil analysis is returned to you from a commercial laboratory. You are asked to interpret it and supply information on the exchangeable cations (bases) present. The analysis sheet shows the following values:

Calcium Ca^{2+} = 15 meq/100 g soil
Potassium K^+ = 0.5 meq/100 g soil
Magnesium Mg^{2+} = 1.5 meq/100 g soil
Sodium Na^+ = 0.2 meq/100 g soil
Ammonium NH_4^+ = 0.1 meq/100 g soil
Cation exchange capacity (CEC) = 20 meq/100 g soil

(a) Calculate the percent base saturation as well as the saturation with each cation.
(b) Calculate the milliequivalents of hydrogen ions by difference and then calculate percentage hydrogen ion saturation.
(c) Calculate the percentage saturation of the exchange capacity of the soils by Ca^{2+}, K^+, Mg^{2+}, Na^+, and NH_4^+.

Solution:

(a) Percent base saturation:
15 meq Ca + 0.5 meq K + 1.5 meq Mg + 0.2 meq Na + 0.1 meq NH_4 = 17.3 meq of these cations = 17.3 meq of bases

$$\frac{17.3 \text{ meq bases/100 g soil}}{20 \text{ meq CEC/100 g soil}} \times 100 = \textbf{86.5\% base saturation}$$

(b) meq of H (by difference):
20 meq CEC − 17.3 meq bases = 2.7 meq H/100 g soil

$$\% \text{ H saturation} = \frac{2.7 \text{ meq H/100 g soil}}{20 \text{ meq CEC/100 g soil}} \times 100$$

$$\% \text{ H saturation} = \textbf{13.5\%}$$

(c) Percent saturation by various ions:
15 meq Ca/100 g soil ÷ 20 meq CEC/100 g soil × 100 = 75.0%
0.5 meq K/100 g soil ÷ 20 meq CEC/100 g soil × 100 = 2.5%
1.5 meq Mg/100 g soil ÷ 20 meq CEC/100 g soil × 100 = 7.5%
0.2 meq Na/100 g soil ÷ 20 meq CEC/100 g soil × 100 = 1.0%
0.1 meq NH_4/100 g soil ÷ 20 meq CEC/100 g soil × 100 = 0.5%

Total   86.5%

10. Fertilizer applications add exchangeable cations to the soil. Calculate what percent of the exchange capacity of the following soils is represented by the various nutrient applications of all those ions adsorbed on the exchange complex. Refer to Problem 6 for milliequivalent weights. (Refer to Problem 7 for amount of soil involved.)

Soil Texture	CEC (meq/100 g)	Nutrient Applied (kg/ha)	Percent Saturation with	
(a) Silty clay loam	26.0	180 K^+	K^+	_____
(b) Silt loam	16.1	100 N	NH_4^+	_____
(c) Loamy sand	6.9	50 Mg^{2+}	Mg^{2+}	_____
(d) Clay	33.3	200 N	NH_4^+	_____
(e) Sand	2.8	800 Ca^{2+}	Ca^{2+}	_____

Solution:

[Parts (a) and (b) only]:

(a) $$\frac{180 \text{ kg K}}{1.764 \times 10^6 \text{ kg soil}} = \frac{180 \text{ mg K}}{1.764 \times 10^6 \text{ mg soil}} = \frac{180 \text{ mg K}}{1.764 \times 10^3 \text{ g soil}} = \frac{18 \text{ mg K}}{176.4 \text{ g soil}}$$

$$\frac{18 \text{ mg K}}{176.4 \text{ g soil}} = \frac{X}{100 \text{ g soil}}$$

$$X = \frac{100 \text{ g soil} \times 18 \text{ mg K}}{176.4 \text{ g soil}} = 10.2 \text{ mg K/100 mg soil}$$

10.2 mg K ÷ 39.1 mg K/meq K = 0.26 meq K/100 g soil

(0.26 meq K/100 g ÷ 26.0 meq CEC/100 g) × 100 = **10% K$^+$ saturation**

(b) $$\frac{10 \text{ mg NH}_4^+}{176.4 \text{ g soil}} = \frac{X}{100 \text{ g soil}} = X = 5.7 \text{ mg NH}_4^+/100 \text{ g soil}$$

5.7 mg NH$_4^+$ ÷ 18.02 mg/meq NH$_4^+$ = 0.31 meq NH$_4^+$/100 g soil

(0.31 meq NH$_4^+$ ÷ 16 meq CEC/100 g) × 100 = **1.9% NH$_4^+$ saturation**

REFERENCES

Barber, S. A., "Relation of Fertilizer Placement to Nutrient Uptake and Crop Yield: II. Effects of Row Potassium, Potassium Soil Level, and Precipitation," *Agron. J.,* 51 (1959) 97–99.

Barrett, W. B., C. F. Engle, and R. M. Smith, "Factors Influencing the Levels of Exchangeable Potassium in Gilpin and Cookport Soils," West Virginia Univ. Agr. Exp. Sta. Bul. 622T (1973).

Bolt, G. H., M. E. Summer, and A. Kamphorst, "A Study of the Equibria Between Three Categories of Potassium in an Illitic Soil," *Soil Sci. Soc. Amer. Proc.,* 27 (1963), 294–99.

Box, John, "Potassium and Crop Production," Tex. Agr. Ext. Bul. MP–1007 (1972).

Kilmer, V. J., S. E. Younts, and N. C. Brady, eds., "The Role of Potassium in Agriculture," American Society of Agronomy, Madison, Wis. (1968).

Miller, Harold, Lloyd Murdock, and Kenneth Wells, "Potassium in Kentucky Soils," Univ. of Kentucky Ext. Bul. AGR–11 (1973).

Skogley, E. O., "Potassium in Montana Soils," Montana Agri. Exp. Sta. Research Report No. 88, (1976).

Tisdale, S. L., and W. L. Nelson, *Soil Fertility and Fertilizers,* New York: Macmillan Publishing Co., 1975.

Walsh, L. M., "Soil and Applied Nutrients," Univ. of Wisconsin Fact Sheet No. A2521, (1973).

CHAPTER FIVE

Secondary Nutrients

5.1 INTRODUCTION

Calcium (Ca), magnesium (Mg), and sulfur (S) are designated *secondary elements* because plants require them in fairly substantial quantities for normal growth, but these elements are generally not added as fertilizers in large quantities. The term *secondary* may not be appropriate because there is an increasing tendency for calcium, magnesium, and sulfur to be grouped with nitrogen, phosphorus, and potassium as macronutrients. This move results in just two categories: macronutrients (N, P, K, Ca, Mg, and S) and micronutrients (Cu, Fe, Mn, Zn, Cl, B, and Mo). However, in order to be consistent with most literature sources, this textbook continues to refer to calcium, magnesium, and sulfur as secondary nutrients. Actual requirements and uptake of any or all of these secondary elements varies with the plant species, soil conditions, and environment. Deficiencies of magnesium and sulfur are becoming increasingly more common. Research indicates that concentration levels of these three secondary nutrients needed in plants for normal metabolic activity are generally higher than the levels reported 10 to 20 years ago.

5.2 CALCIUM

The largest concentration of any of the secondary nutrients in plants is generally calcium. Some plants contain higher concentrations of calcium than potassium. Calcium is important not only as a plant nutrient but also for its role in the correction of soil acidity (see Chapter 8). In acid soils, there may be an excess of soluble iron, aluminum, and/or manganese in combination with shortages of the basic cations calcium and magnesium.

Calcium is relatively abundant in soils and rarely limits crop production. It makes up about 3.6% of the earth's crust. It is present in such minerals as feldspar, amphibole, pyroxene, dolomite, calcite, apatite, and gypsum. The calcium content of soils ranges from 0.5% in coastal plain sands to over 5% in soils of arid regions or those developed from limestone, marl, or chalk.

5.2.1 Calcium Reactions in Soils

Calcium occurs as a part of relatively insoluble primary and secondary minerals. The calcium (Ca^{2+}) ion also exists in soil solution and is adsorbed on the surface of mineral and organic colloids. Plants absorb calcium from the soil solution; also calcium adsorbed on the surface of soil colloids is considered to be exchangeable and available to plants.

Most exchangeable calcium is available to plants, although the calcium in primary and secondary minerals is only slowly available. Because of this, analysis for total soil calcium is not closely related to calcium uptake and plant nutrition and growth.

Calcium is normally the predominant positively charged cation (Ca^{2+}) on soil particles because it is held more tightly than magnesium (Mg^{2+}) and potassium (K^+). Also, the soil parent material usually contains much more calcium than magnesium or potassium. Calcium does not readily leach through soils because it is held tightly on the surface of soil colloids.

The exchangeable calcium may range from 30 to 90% of the total exchange capacity of a soil. The total cation exchange capacity (CEC) of soils ranges from 4 to 60 meq/100 g (milliequivalents per 100 grams) of soil. One milliequivalent of calcium is equivalent to 400 pp2m (400 lb/A or 448 kg/ha).[1] Quantities of exchangeable calcium vary greatly between soils and within the soil profile. The poorly drained soils contain the most exchangeable calcium. Fine-textured soils have significantly more calcium than sandy soils. The surface horizons of most soils contain less exchangeable calcium than the subsoil horizons because of leaching.

Calcium is an essential element for plants; it also is very important in soils for other reasons. It improves soil structure by aggregating the colloidal clay and humus particles and making the soil more granular. This improves air and

---

[1] pp2m = parts per 2 million = lb/A; lb/A x 1.121 = kg/ha.

water penetration into soil, thus providing a better place for the growth of plant roots and soil organisms. Calcium corrects soil acidity and neutralizes the by-products of organic decomposition (see Chapter 8). As a general rule, soils with a favorable pH (6.0 or above) contain an adequate level of calcium for the growth of plants. However, a pH test does not necessarily indicate the amount of available calcium.

5.2.2 Calcium in Plant Nutrition

Calcium is a part of every plant cell. Much of the calcium that is found in plants occurs as calcium pectates in and along the walls of cells of the leaves and stems. Such concentrated deposits of calcium thicken and strengthen these parts of the plant. Plants and plant parts vary greatly in their content of calcium, depending on the nature of the plants and the condition under which they were grown. The flower and seed parts of a plant are usually low in calcium. A relatively large part of the calcium content of plants is located in the leaves. The aboveground portion of most mature grain crops and grasses contains from 0.25 to 0.5% calcium. It occurs in sizable quantities, up to 4%, in tobacco leaves. The aboveground parts of cotton, soybean, and alfalfa plants average 2.0% calcium.

Calcium is a relatively immobile element within a plant. It does not easily redistribute in a plant subject to stress from calcium shortages. It does not move from old leaves to new; hence, a continuous supply of calcium is essential. The seed contains too little calcium to supply the plant beyond seedling emergence.

Calcium in plants has been ascribed to function in a number of different roles. It promotes root and leaf development and is a constituent part of the cell walls where it occurs as a necessary insoluble calcium pectate. It improves general plant vigor and increases stiffness of straw. Calcium is necessary for cell elongation, protein synthesis, and normal cell division (mitosis). It influences the uptake of water and other plant nutrients. Calcium regulates the translocation of carbohydrates, regulates cell acidity and permeability, and serves as an antagonist for some other cations. It encourages grain and seed production. Calcium is important for the formation and functioning of root nodule bacteria (*Rhizobia*) in legumes.

5.2.3 Calcium Deficiency Symptoms

Calcium deficiency in field crops is seldom encountered. Symptoms of calcium deficiency are difficult to determine until the deficiency has become critical. Deficiency symptoms show up first in the young leaves and growing points, because calcium is an immobile element in the plant. The young leaves and roots often become crooked, wrinkled, short, or bunched together. This is sometimes associated with a weak stem structure. The lack of growth may cause nitrogen concentration in the young leaves and give a dark green color. In extreme cases, the leaves of the terminal bud curl, turn light green, and eventually die, the leaf tips and margins being affected first.

In corn, deficiency is indicated by inability of the leaves to unfold or emerge. Although legumes have a high calcium requirement, distinct calcium deficiency symptoms are rarely seen under field conditions.

5.2.4 Calcium Fertilizer Materials

In the past, mixed fertilizers contained appreciable amounts of calcium. The major sources of this calcium were fertilizers such as normal superphosphates (refer to Chapter 3). In addition, limestone, dolomite, and gypsum were added as a "filler" to produce a certain nutrient anlysis. In the manufacture of mixed fertilizer, both fluid and solid, the use of phosphate material has shifted from normal to triple superphosphate and to ammoniated phosphates and phosphoric acid. The shift to higher analysis in mixed fertilizer manufacture has decreased the need for lime as a filler and, therefore, decreased the amount of calcium applied with mixed fertilizers. The presence of calcium in clear liquid fertilizers containing ammonium phosphate causes the precipitation of phosphorus and is, therefore, avoided by manufacturers of this type of product.

The increased use of ammonium fertilizers and high-analysis, low-calcium, mixed fertilizers, along with increased yields, means that more frequent limestone applications are needed to maintain adequate soil pH levels. A tremendous amount of data indicate that finely ground limestone is an effective material for correcting or preventing calcium deficiencies as well as for reducing soil acidity and its detrimental effects on plant growth.

In arid regions, gypsum (calcium sulfate) is often used as a source of calcium. The calcium in gypsum is fairly soluble and may be recommended to supply calcium as a plant nutrient. In the case of sodic soils, gypsum is recommended as a soil amendment to supply calcium to replace exchangeable sodium (see Chapter 9). As a soil amendment, gypsum improves soil structure by promoting flocculation and granulation and thus speeds up water penetration into "puddled" or dispersed soils. Agricultural gypsum is a good source of both calcium and sulfur (Fig. 5–1) and has the added advantage that it has little effect on soil pH.

Gypsum is an important source of calcium in the production of peanuts. Peanuts are frequently grown in very sandy acid, low-calcium soils with a low exchange capacity. The peanut plant has a high requirement for calcium at the time the nuts are setting and developing from the pegs (*gynophore*). Gypsum also helps to control the pod rot fungus (*Pythium*) in peanuts. Garden crops that have a high calcium requirement such as tomatoes, peas, and beans can be furnished calcium as gypsum without making the soil too basic for those plants that do best in slightly acid soils.

Fertilizers and soil amendments represent the most important supplemental sources of calcium used in crop production. Organic complex sources of calcium are also available but are not widely used. The organic complex sources of calcium have been used for peanuts and soybeans that are grown on

Fig. 5-1. Agricultural gypsum ($CaSO_4 \cdot 2H_2O$) is a relatively low-cost source of both calcium and sulfur. Agricultural gypsum fines are available for truck loading at the National Gypsum Company's Sun City quarry located near Medicine Lodge, Kansas, as well as many other locations. (Courtesy D. A. Whitney, Kansas State University.)

acid soils to supply a mobile form of this nutrient where the use of lime has been omitted or is considered undesirable. The major sources of calcium are shown in Table 5-1.

5.2.5 Calcium in Animals

Calcium is a component of every animal cell and a very important constituent of bones and teeth. Calcium and phosphorus are frequently referred to as bone minerals for livestock. About 99% of the calcium and 80% of the phosphorus of the body are present in bones and teeth. The calcium and phosphorus occur in approximately a 2:1 ratio in bones and teeth. Calcium deficiencies in a diet can cause rickets in children, brittle bones among older people, and defective eggshells in birds. In many countries, milk is a major source of calcium for human diets.

Adding limestone to soils to correct soil acidity and to supplement available calcium in the soil can have substantial indirect effects on human and animal nutrition. Foods and feeds high in calcium reduce the opportunities for deficiencies.

Table 5-1. Calcium Contents of Lime, Fertilizer, and Soil Amendment Materials

Carrier	Chemical Formula	Calcium (Ca) (%)
Lime Material		
Blast furnace slag	$CaSiO_3$	29.3
Calcitic limestone	$CaCO_3$	31.7
Dolomitic limestone	$CaCO_3 + MgCO_3$	21.5
Hydrated lime	$Ca(OH)_2$	46.1
Marl	$CaCO_3$	24.0
Precipitated lime	CaO	60.3
Fertilizers		
Calcium cyanamide	$CaCN_2$	38.5
Calcium nitrate	$Ca(NO_3)_2$	19.4
Phosphate rock	$3Ca_3(PO_4)_2 \bullet CaF_2$	33.1
Superphosphate, normal	$Ca(H_2PO_4)_2 + CaSO_4 \bullet 2H_2O$	20.4
Superphosphate, triple	$Ca(H_2PO_4)_2$	13.6
Soil Amendment		
Gypsum	$CaSO_4 \bullet 2H_2O$	22.5

5.3 MAGNESIUM

The amount of magnesium found in soils varies but is usually abundant in most soils. It comprises about 1.9% of the earth's crust and occurs in the minerals amphibole, biotite, chlorite, dolomite, montmorillonite, olivine, pyroxene, serpentine, and vermiculite. Soils developed from coarse-grained rocks low in these minerals tend to be low in magnesium, whereas fine-textured soils developed from rocks high in magnesium minerals contain greater amounts.

5.3.1 Magnesium Reactions in Soils

Magnesium is a positively charged cation (Mg^{2+}) and as such is held on the surface of clay and organic matter particles. This exchangeable magnesium is available to plants, and it does not leach readily from the soil. The magnesium of the primary and secondary minerals is relatively insoluble and, therefore, relatively unavailable to plants. This illustrates why total soil magnesium is not closely related to plant growth.

Magnesium deficiencies are found on acid sandy soils and organic soils containing free calcium carbonate or marl. In sandy soils, magnesium deficiency is accentuated by high levels of available potassium. High concentrations of potassium in the soil solution, caused by recent fertilization, tend to interfere with the uptake of magnesium by plants. Magnesium deficiencies can develop on soils that have been limed for many years with materials low in magnesium, such as calcitic limestone or marl.

Fig. 5-2. Magnesium sulfate (MgSO₄) increased the yield on corn grown on an acid sandy soil on the Research Farm located at the University of Nigeria, Nsuka, Nigeria. The plot on the right received 112 kg/ha (100 lb/A) of MgSO₄. A soil test for magnesium indicated only 8 ppm of exchangeable magnesium (Mg). (Courtesy Kenneth Payne, Michigan State University.)

Magnesium deficiency in plants is often found on soils that are very high in available potassium. Although the antagonistic effect of potassium on the absorption of magnesium is well-known,[2] there is also a thermodynamic interaction between these two cations at various soil moisture levels due to their difference in charge. This interaction has been demonstrated for wheat grown on several soils at varying moisture levels.[3] As the soils dry from a nearly water-saturated condition, there is a change in the proportion of the two cations in the soil solution that contributes to very low magnesium uptake under nearly water-saturated conditions. The very high level of available potassium in the soil interferes with the uptake of magnesium by plants. On sandy and loamy soils, the application of magnesium fertilizers is often effective in increasing crop yield and the concentration of magnesium in the crop (Fig. 5-2). However, on fine-textured soils, especially those with substantial reserves of potassium, the application of a magnesium fertilizer may not result in higher yields or higher magnesium concentrations in the plant.

There is a relationship between soil pH and magnesium uptake by plants.

[2] G. A. Taylor and G. B. Smith, "Nutritional Studies with Sweet Corn in Blair County, Pennsylvania," *Amer. Soc. Hort. Science Proc.*, 74:454–59, 1959.

[3] D. L. Karlen, "Influence of Soil and Climatic Factors on Forage Quality Indices of Grass Tetany in Ruminants," Ph.D. Dissertation, Kansas State University, 1978.

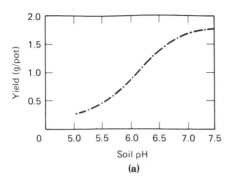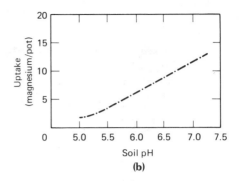

Fig. 5-3. An acid soil (Muskingum series) from Columbiana County, Ohio, was limed to the various soil pH's and alfalfa was then grown in greenhouse pots. (a) This graph shows the yield of alfalfa with increasing soil pH; (b) this graph shows the uptake of magnesium by alfalfa with increasing soil pH. (Source: J. B. Jones and F. Haghiri, "Magnesium Deficiency on Columbiana County Soils," Ohio Agri. Exp. Sta. Research Cir. 116, 1963, 18 pp.)

Liming with calcitic limestone has been shown to increase both the yield and uptake of magnesium in alfalfa (Fig. 5-3). Magnesium uptake can be increased with magnesium fertilization on acid soils using magnesium sulfate ($MgSO_4$), magnesium oxide (MgO), potassium magnesium sulfate ($K_2SO_4 \bullet 2MgSO_4$), and dolomite ($CaCO_3 \bullet MgCO_3$). However, at low pH's large additions may be necessary to provide adequate available soil magnesium for normal plant nutrition.

For the potato grower the correction of magnesium deficiency can be a difficult problem. Many growers are not willing to lime for fear of providing conditions for the development of potato scab (*Streptomyces scabies*). The very high rates of potassium fertilization frequently necessary for potato production can lead to magnesium deficiency on low pH soils. Sizable additions of $MgSO_4$ and other sources including MgO, $K_2SO_4 \bullet 2MgSO_4$, and dolomite are used on acid soils in potato-producing areas. Crop rotation should be employed to control potato scab because low soil pH and heavy applications of potassium does not always ensure best yields and potatoes free from scab. Potatoes are responsive to higher soil pH's and less nutritional difficulties are likely to arise when a more moderate potassium fertilizer plan is followed. Magnesium deficiency on potatoes can be controlled by liming with dolomitic limestone and by use of other magnesium sources such as $MgSO_4$, MgO, and $K_2SO_4 \bullet 2MgSO_4$.

5.3.2 Magnesium in Plant Nutrition

It has been said that the most important chemical reaction in nature is photosynthesis through which plants convert carbon dioxide from the air, water from the soil, and energy from the sun into simple sugars. The green pigment in plants, called *chlorophyll,* converts light energy into chemical energy and is essential for photosynthesis.

Chlorophyll contains about 2.7% magnesium as an essential constituent. About 10% of the magnesium in plants is found in the chloroplasts. Therefore, the most important function of magnesium in plants is the formation of chlorophyll. No other plant nutrient can substitute for magnesium. Iron is necessary for the formation of chlorophyll, but it is not a constituent of the molecule.

It is of interest to note how closely plants and animals are associated through the comparison of chlorophyll of plants and the hemoglobin of the red corpuscles of the blood. These compounds are chemically similar. Magnesium is the central element in chlorophyll and iron holds the same key position in hemoglobin. Chlorophyll-a is given the formula $C_{55}H_{72}O_5N_4Mg$. The formula of hematin is $C_{34}H_{33}O_5N_4Fe$ and resembles that of chlorophyll when iron (Fe) is substituted for magnesium (Mg).

In addition to its role as a constituent of chlorophyll, magnesium aids in the formation of many plant compounds, such as sugars, proteins, oils, and fats. It regulates the uptake of other plant nutrients, especially phosphorus, and is involved in the translocation and metabolism of carbohydrates. It acts as a carrier for phosphorus, particularly into the seeds. Magnesium is also a specific activator of a number of enzymes including certain of the transphosphorylases, dehydrogenases, and carboxylases.

The accumulation of magnesium from the soil is strongly affected by the species of plant. Leguminous plants, such as clovers, beans, and peas, usually contain more magnesium than grasses, tomatoes, corn, and other nonleguminous plants regardless of the level of available magnesium in the soil where they grow.

5.3.3 Magnesium Deficiency Symptoms

Because of magnesium's vital role in the chlorophyll of plants, a deficiency of this element results in chlorosis (yellowing) of the plant leaves. The first symptom is a loss of a healthy, green color between veins. As the deficiency becomes more severe, the leaves may turn reddish purple. Because magnesium readily moves from lower to upper parts of the plant, the deficiency first appears on the lower leaves. In corn, the lower leaves yellow between the veins (Fig. 5–4). Chlorosis starts on the leaf margins and progresses inward interveinally. Although deficiency symptoms show most frequently on small plants, they usually develop too late for any remedial treatment of the current crop. If such characteristics occur where plant tissue and soil tests are low, it is relatively certain that the symptoms truly reflect magnesium deficiencies.

The greatest likelihood of finding a magnesium deficiency is with a combination of the following conditions: (a) a highly leached sandy soil, low in magnesium; (b) a crop with high magnesium requirements; (c) high rates of

Fig. 5-4. Magnesium deficiency in corn. Older leaves show interveinal chlorosis. Symptoms usually appear early but may later disappear. Severe deficiency causes stunting. (Courtesy R. H. Follett.)

nitrogen and potassium fertilizers; and/or (d) liming with calcitic limestone (very low magnesium content).

5.3.4 Magnesium Fertilizer Materials

The most economical way of correcting magnesium deficiency on an acid soil is to apply dolomitic limestone. Finely ground dolomitic limestone is an effective material for correcting or preventing magnesium deficiency as well as reducing soil acidity and supplying calcium. However, if the pH is already high (above 6.0) or if the crop being grown requires an acid soil (such as potatoes), then other carriers of magnesium such as Epsom salt ($MgSO_4$) or potassium magnesium sulfate ($K_2SO_4 \cdot 2MgSO_4$) must be used. Epsom salt is the preferred source on soils that are very high in available potassium. Epsom salt and potassium magnesium sulfate contain water-soluble magnesium. Such materials may be preferable where a quick crop response is required to correct a magnesium deficiency.

Dolomite and magnesium oxide (MgO) are much less soluble than other materials. These materials should be used sometime before a crop is planted to allow sufficient time for them to react with the soil. The most common magnesium carriers are shown in Table 5-2.

5.3.5 Magnesium in Animals — "Grass Tetany"

Magnesium is a part of every animal cell. Natural toxicities from this element are not recognized, although on occasion deficiencies occur. With scientific feeding programs, deficiencies are not likely.

Table 5–2. Magnesium Carriers

Carrier	Chemical Formula	Magnesium (Mg) (%)
Dolomite	$MgCO_3 + CaCO_3$	8–20
Epsom salt	$MgSO_4 \cdot 7H_2O$	9.6
Kieserite	$MgSO_4 \cdot H_2O$	18.3
Magnesium ammonium phosphate	$MgNH_4PO_4 \cdot H_2O$	14.0
Magnesium oxide (magnesia)	MgO	55.0
Potassium magnesium sulfate	$K_2SO_4 \cdot 2MgSO_4$	11.2

The most common cases of magnesium deficiencies in animals are in lactating cows. "Grass tetany" is the common name for acute magnesium deficiency. Cattle are more susceptible than sheep or goats, and females are the most susceptible, especially when lactating or pregnant.[4] Grass tetany generally strikes animals grazing grass or small-grain pasture during the winter if lactating cows are being fed on low-magnesium grass hay. Cattle grazing pastures containing monocultured high-yielding grass or small-grain forages have a greater incidence of grass tetany than those grazing pastures containing some legumes or forbs, which are normally higher in magnesium.

The disease is often referred to as "hypomagnesemic tetany" and in New Zealand as "grass staggers." It occurs worldwide in the temperate zones and particularly on spring pastures under intensive management. Not all the factors causing grass tetany are understood. The grasses eaten by the animals are usually low in magnesium and high in potassium. Frequently the concentration of the ammonium form of nitrogen and certain organic acids is also high. Ammonium nitrogen when ingested by the animal would tend to deplete carbohydrate reserves in the rumen, generate free ammonia, and raise the rumen pH, further reducing the availability of magnesium to the animal and cause grass tetany. Additions of excessive nitrogen and potassium fertilizers have also increased the incidence of grass tetany.

The disease is seasonal. Severe outbreaks occur during some years but may be uncommon in cattle grazing the same pasture in other years. Some of the problems in "wheat pasture poisoning" are undoubtedly due to the low availability of magnesium in the wheat. There may be certain factors associated with wheat tissue interfering with magnesium uptake. An unusually high proportion of ammonium or amide forms of nitrogen might be involved in this interference in the utilization of magnesium by ruminants. Climatic conditions favoring the formation and persistence of ammonium nitrogen, rather than nitrate nitrogen, in the soil and its subsequent uptake might be closely involved in grass tetany. Magnesium soil applications have

[4] D. L. Grunes, "Grass Tetany of Cattle and Sheep." In *AntiQuality Components of Forages*, Crop Sci. Soc. of Amer., Madison, Wis. (1973), pp. 133–140.

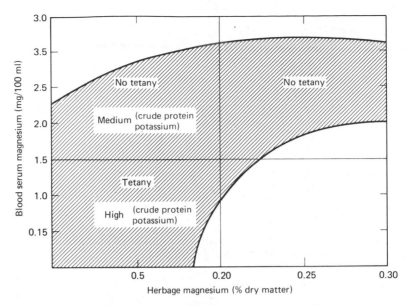

Fig. 5-5. Low magnesium (Mg) in the blood serum of cattle is related to magnesium content of the herbage (forage). This results in a disease called *grass tetany* (hypomagnesemia). High nitrogen (NH_4^+) and high potassium forages may cause grass tetany. (Source: Duval Sales Corporation, Houston, Texas.)

been quite ineffective, however, in increasing magnesium content in wheat.[5]

It is generally agreed that the disorder is due to a deficiency of blood serum magnesium. Figure 5–5 illustrates the relationship between blood serum magnesium and herbage magnesium. Affected animals become nervous, develop leg muscle spasms and convulsions, and fall. If the disorder is detected in time and the animals are injected with magnesium, they recover and return to normal within a few minutes (Fig. 5–6). Where the problem is not detected, the animals usually die during or after a convulsion.

To prevent grass tetany, cattle should be provided with a mineral supplement high in magnesium. Good pasture fertilization also helps to prevent grass tetany. Fertility programs must be balanced with additions of magnesium along with nitrogen, phosphorus, and potassium fertilizer. Dusting pastures with magnesium oxide (MgO) is useful to increase the direct intake of magnesium by ingestion of vegetation with MgO adhering to it. Rates of 15 to 30 lb/A (17–34 kg/ha) of MgO can be used, with lower rates suited to cases where cattle are moved to new pastures every 2 or 3 days. The use of dolomitic limestone also tends to increase the magnesium levels in pasture grasses and legumes (Fig. 5–7). The use of a mineral supplement containing magnesium plus an appetizer like molasses or a daily feeding of a high magnesium legume hay prevents grass tetany. Magnesium supplements

[5] C. L. Harms, "Magnesium Effects on Wheat Forage Composition Involved in the Tetany Syndrome," Ph.D. Dissertation, Kansas State University, 1976, 103 pp.

Fig. 5–6. Injecting calcium-magnesium gluconate into the bloodstream of a cow down with grass tetany. (Courtesy D. L. Grunes, USDA-SEA-AR, Ithaca, N.Y.)

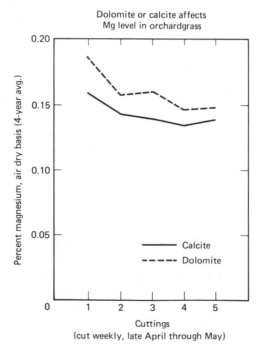

Dolomite or calcite affects
Mg level in orchardgrass

Fig. 5–7. Dolomite increased early-season magnesium levels in orchardgrass more than calcite (West Virginia). (Courtesy Potash and Phosphate Institute.)

should be started several weeks before the usual tetany period to get the animals accustomed to them.

5.4 SULFUR

Sulfur (S), like calcium (Ca) and magnesium (Mg), is essential for all plants and animals and is classified as a secondary element because it is required in smaller amounts than the major elements but in much greater quantities than the micronutrients. However, many crops use as much sulfur as phosphorus.

In recent years, deficiencies of sulfur in crop plants have been reported in soils from many parts of the world. It has been predicted that greater areas will become increasingly sulfur-deficient in the future because of the expanding use of sulfur-free fertilizers, increasing crop yields that make greater demands on soil nutrients, more intensive cropping (double cropping, irrigation) the implementation of air-pollution-control schemes, and the decreasing use of sulfur-containing fungicides and insecticides.

5.4.1 Sulfur Reactions in Soils

Most soil sulfur is in the organic matter and therefore is concentrated in the surface soil. Sulfur transformations are similar to those of nitrogen. Sulfur in organic combinations is not available to plants. As with nitrogen, the organic sulfur is converted into available sulfate sulfur by many species of soil bacteria when environmental conditions favor the decomposition of organic matter. In humid-temperate regions, from 1 to 3% of total soil sulfur is mineralized to available sulfur each year.

As shown in Fig. 5–8, organic sulfur (proteins) and reduced sulfide (S^{2-}) sulfur are oxidized to form available sulfate (SO_4^{2-}) sulfur in warm well-aerated soils. This process of sulfofication is very similar to the ammonification and nitrification processes that change organic nitrogen into available ammonium (NH_4^+) and nitrate (NO_3^-) nitrogen. The bacterial process requires aerobic autotrophic organisms (mostly genus *Thiobacillus*) that obtain their energy from the oxidation of sulfur.

Sulfate (SO_4^{2-}) is immobilized when assimilated by bacteria during the decomposition of crop residues rich in carbon. Also, it can be changed into unavailable sulfide sulfur by several genera of bacteria under reducing conditions caused by poor drainage and waterlogged (anaerobic) conditions. However, the immobilized or reduced sulfur is usually only temporarily unavailable. As the soils become warmer and as the aeration improves, this unavailable sulfur is again oxidized to form the available sulfate (SO_4^{2-}) form of sulfur.

Sulfur is removed from the sulfur cycle by harvesting crops or by leaching. Crop removal varies from about 11 kg/ha (10 lb/A) of sulfur for grain crops to over 22 kg/ha (20 lb/A) for legumes. There is considerable variation in sulfate levels in the upper portions of root zones depending on the amount of precipitation. Because the sulfate ion has a negative charge, it is not readily

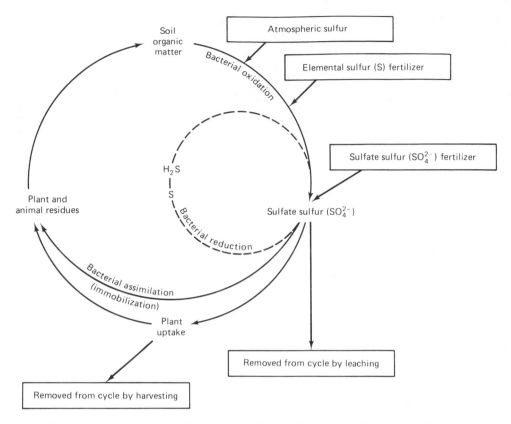

Fig. 5–8. The sulfur cycle. (Source: L. M. Walsh, "Soil and Applied Sulfur," University of Wisconsin Extension Series A2525, 1975, 4 pp.)

adsorbed by soil colloids and is subject to leaching, especially on sandy soils. The sulfate ion is not so easily leached as the nitrate (NO_3^-) ion. However, it is much more subject to leaching loss than phosphorus, potassium, calcium, or magnesium.

Soil incorporation of plant residue low in sulfur can cause sulfur deficiencies in crops growing on these soils during decomposition of the residue.[6] When carbonaceous energy sources (straw, stalks, or other plant residue) are added to the soil, the soil microorganisms multiply rapidly and consume much of the available sulfur supply in the production of new cells. This process is similar to the immobilization of available nitrogen when an energy source is added to the soil (see Chapter 10).

Because the topsoil usually contains most of the sulfate sulfur in the soil, practices that reduce erosion are means of reducing sulfur deficiency. Because much of the sulfur is held in the organic fraction of the soil, maintenance of the organic matter level is effective in preventing sulfur deficiency.

[6] B. A. Stewart, L. K. Porter, and F. G. Viets, "Effects of Sulfur Content of Straws on Rates of Decomposition and Plant Growth," *Soil Sci. Soc. Amer. Proc.*, 30:355–358, 1966.

5.4.2 Sulfur From Rainwater and Irrigation Water

There are several potential sources of sulfur in the atmosphere. The burning of coal and (to a lesser extent) oil and gas result in release of sulfur dioxide (SO_2) to the atmosphere. Another source of atmospheric sulfur is the anaerobic decomposition of organic matter in swamps that releases hydrogen sulfide (H_2S). This atmospheric sulfur is removed from the air and deposited on land in rainwater.

The amount of sulfur added to the soil varies by location. For example, the amount of sulfur reported in rainfall for a highly industrialized area near Gary, Indiana, was measured as 142 kg/ha of sulfur (127 lb/A) annually.[7] By comparison, in Wisconsin, rural and urban stations received an average of 16.5 and 42.1 kg/ha of sulfur per year (14.7 and 37.6 lb/A), respectively.[8] Crops that are located near sulfur-releasing industries may receive all of their sulfur requirement from precipitation (rain and snow). On the other hand, crops grown in areas that are a long distance from industrial areas may only receive a small fraction of their total sulfur requirement from precipitation (see Chapter 8).

The amount of sulfur contributed to a crop from irrigation water may often equal or exceed that crop's sulfur requirements. Only 3 of the 47 wells analyzed in the San Luis Valley of Colorado (Table 5-3) tested less than 11 parts per million (ppm) of sulfate sulfur (SO_4-S). None of the wells tested less than 11 ppm (10 lb/A-ft of sulfate sulfur) for Scott County, Kansas. Two of the rivers in Michigan tested less than 11 ppm sulfate sulfur (10 lb/A-ft of sulfate sulfur) in the water (Table 5-4). The majority of the samples shown in Tables 5-3 and 5-4 contributed more than most crops require during the growing season. For example, a high-yielding corn crop requires about 22 to 34 kg/ha (20–30 lb/A) of sulfur. If the corn receives 0.12 or 0.25 ha-m (1–2 A-ft) of a water containing 30 ppm (27 lb/A-ft of sulfate sulfur) of water, then it has received an ample supply of sulfur.

The data in Tables 5-3 and 5-4 illustrate that irrigation water from many wells and streams can be a good source of sulfur. Sulfur recommendations are quite often suggested for irrigated crops without considering the sulfur levels in the irrigation water. Thus, if there is a need to know the sulfur potential of water from a specific source, chemical analysis is essential.

5.4.3 Sulfur in Plant Nutrition

Sulfur is a part of every plant cell and is absorbed from soil primarily as the sulfate (SO_4^-) ion. Sulfur is readily translocated within a plant and is closely associated with nitrogen (N) because both are involved in the synthesis of essential sulfur amino acids. The two sulfur-containing amino acids (cys-

---

[7] B. R. Bertramson, M. Fried, and S. L. Tisdale, "Sulfur Studies of Indiana Soils and Crops," *Soil Sci.*, 70:27–41, 1950.

[8] R. G. Hoeft, D. R. Keeney, and L. M. Walsh, "Nitrogen and Sulfur in Precipitation and Sulfur Dioxide in the Atmosphere in Wisconsin," *J. Environ. Quality*, 1(2):203–08, 1972.

Table 5-3. Sulfate-Sulfur Contributions to Soils from Irrigation Wells Located in the San Luis Valley of Colorado and in Scott County, Kansas

Range of Sulfate–Sulfur (SO$_4$-S)		Location and Number of Samples	
ppm	lb/A-ft	San Luis Valley, Colorado	Scott County, Kansas
<11	<10	3	0
11–22	10–20	5	0
23–33	21–30	8	3
34–55	31–50	13	21
56–110	51–100	7	14
>110	>100	11	7
	Total samples	47	45

NOTES:

lb/A-ft = ppm x 0.907.

kg/ha/30.5 cm (1 ft) = 3.05 x ppm.

Hectare-meter = acre-feet x 0.12335.

SOURCE: R. H. Follett, Irrigation Water Analysis Summary, Colorado State University (1970). Data for Scott County, Kansas was summarized from Kansas Geological Survey, Chemical Quality Series 2, Chemical Quality of Irrigation Water in West-Central Kansas (1975).

Table 5-4. Average Dissolved Sulfate-Sulfur Levels in Michigan Rivers

River	County	Sulfate-Sulfur (SO$_4$-S)	
		ppm	lb/A-ft
Sturgeon	Houghton	5.1	4.6
Escanaba	Delta	16.8	15.2
Pine	Charlevoix	22.0	20.0
Elk	Antrim	10.0	9.1
Cheboygan	Cheboygan	11.2	10.2
Thunder Bay	Alpena	12.7	11.5
Flint	Saginaw	63.0	57.1
Cass	Saginaw	81.0	73.4
Rouge	Wayne	30.0	27.2
Raisin	Monroe	81.0	73.5
St. Joseph	Berrien	43.0	39.0
Grand	Ottawa	85.0	77.1
Muskegon	Muskegon	32.5	29.5

NOTES:

lb/A-ft = ppm × 0.907.

kg/ha/30.5 cm (1 ft) = 3.05 × ppm.

Hectare-meter = acre-feet × 0.12335.

SOURCE: L. S. Robertson, M. L. Vitosh, and D. D. Warncke, "Essential Secondary Elements: Sulfur," Michigan State University Ext. Bul. E-997 (1976).

teine and methionine) combine with other amino acids to form plant proteins. Sulfur is necessary for the formation of vitamins and the synthesis of some hormones and glutathione. Although sulfur is not a constituent of chlorophyll, it apparently influences the synthesis of that pigment. Sulfur is also present in proteins as disulfide bonds that maintain the particular structural formation of enzymes.

Sulfur is a part of several other organic compounds of plant origin. It is present in glycosides that give characteristic odors and flavors to mustard, onion, and garlic. Sulfur deficiency decreases the formation of nodules on legumes and reduces the activity of nodule nitrate reductase.[9] Therefore, nitrogen fixation is severely restricted in the absence of sufficient amounts of this element.

5.4.4 Sulfur Deficiency Symptoms

Plants that are sulfur-deficient generally have a light green color that resembles nitrogen deficiency. As the deficiency becomes more severe, chlorosis may occur on the entire plant. Plants are also stunted. Sulfur-deficient plants may exhibit delayed maturity. The crops most likely to show a sulfur deficiency are those grown in sandy, low-organic-matter soils.

Legumes, especially alfalfa, which has a high sulfur requirement, normally is the first crop to respond to sulfur fertilization (Fig. 5–9). Sulfur-deficient alfalfa leaves are long and slender. Affected plants do not branch normally, resulting in a thin stand. Symptoms may not appear on alfalfa until the stand is 2 or 3 years old. If high-analysis phosphates or other sulfur-free fertilizers are applied, the symptoms are more likely to occur earlier.

Corn and grasses may exhibit a striping of the upper leaves. Deficiency symptoms are noted most frequently on young plants. As the growing season advances, symptoms may disappear completely because of bacterial transformation of sulfur from organic matter. Sulfur deficiencies can also appear on fruit trees (Fig. 5–10).

5.4.5 Sulfur Fertilizer Materials

A large number of well-known materials contain sulfur (Table 5–5). Depending on the source, the sulfur content of some materials such as gypsum may vary from the values shown. Low-analysis phosphate carrier such as 0–20–0 contain as much as 12% sulfur as a result of the manufacturing process. Many high-analysis phosphate fertilizers contain very little sulfur.

5.4.6 Sulfur in Animals

Sulfur is a part of every animal cell and all fluids. Animals require sulfur in the form of sulfur-containing amino acids. In ruminant animals such

[9] F. J. Wooding, G. M. Paulsen, and L. S. Murphy, "Response of Nodulated and Non-nodulated Soybean Seedlings to Sulfur Nutrition," *Agron. J.*, 62:277–80, 1970.

Fig. 5–9. Alfalfa prior to second harvest in Wisconsin. Zero sulfur on the left and 28 kg/ha (25 lb/A) on the right. Blanket application of potassium at 235 kg/ha (210 lb/A). (Courtesy The Sulphur Institute.)

Fig. 5–10. Cluster of sulfur-deficient pear tree leaves on the right compared to normal leaves on the left (Washington State). (Courtesy The Sulphur Institute.)

Table 5-5. Sulfur Carriers

Carrier	Chemical Formula	Sulfur (S) (%)
Ammonium phosphate sulfate	Varies	10–14
Ammonium polysulfide	Varies	36–45
Ammonium sulfate	$(NH_4)_2SO_4$	23.7
Ammonium thiosulfate	$(NH_4)_2S_2O_3$	26 (60% aqueous)
Copper sulfate	$CuSO_4$	12.8
Epsom salt	$MgSO_4 \cdot 7H_2O$	14.0
Iron sulfate (copperas)	$FeSO_4$	11.5
Gypsum	$CaSO_4 \cdot 2H_2O$	16.8
Manganese sulfate	$MnSO_4$	14.5
Potassium magnesium sulfate	$K_2SO_4 \cdot 2MgSO_4$	22.0
Potassium sulfate	K_2SO_4	17–18
Sulfur, elemental	S	30–99.6
Sulfur dioxide	SO_2	50.0
Sulfuric acid	$(100\% \ H_2SO_4)$	32.7
Superphosphate, normal	$Ca(H_2PO_4)_2 + CaSO_4 \cdot 2H_2O$	11.9
Superphosphate, triple	$Ca(H_2PO_4)_2 + CaSO_4 \cdot 2H_2O$	1.4
Urea ammonium sulfate	Varies	4–13
Zinc sulfate	$ZnSO_4 \cdot H_2O$	18.0

as cattle, sheep, and goats, the microorganisms in the rumen synthesize the required amino acids from sulfur in inorganic forms. Therefore, the sulfur requirement of a cow or sheep can be based upon the *total* amount of all forms of sulfur taken in by the animal. However, for monogastric animals, including man, the sulfur requirement is expressed in terms of the sulfur amino acids, methionine and cysteine.

The swarming population of microorganisms, located in the rumen of cattle and sheep, requires sufficient sulfur in inorganic forms such as elemental sulfur and sulfate to form the sulfur-containing amino acids cysteine and methionine that are needed to build microbial protein. In addition, sulfur is involved in other ways in the normal growth of rumen microorganisms. The sulfur that is first used by microorganisms in the rumen of cattle and sheep eventually becomes available to the host ruminants for their nutritional requirements. Upon the natural death of rumen microorganisms, some of the released sulfur is used in the production of meat, milk, and wool.

Cattle and sheep require adequate sulfur in their diets because this mineral is a vital constituent of the amino acids cystine and methionine, which are essential for the production of proteins in meat, milk, and wool. Sulfur is also needed for the formation of certain vitamins such as thiamine (B_1) and biotin, as well as for the functioning of many enzymes. It is also found throughout the body in several compounds containing sulfate: e.g., (a) chondroitin, an important component of cartilage, bone, tendons, and walls of blood vessels; (b) taurine and its derivative taurocholic acid, which is found in bile acid; and (c) heparin, a blood anticoagulant. In addition, sulfate sulfur has an important role in the detoxification of many substances produced in the body.

Volatile fatty acids produced in the rumen are utilized by cattle and sheep as a source of energy and their increased production in the rumen is a measure of an increase in the efficiency of feed utilization by the animal. One of the major sources of volatile fatty acids is from the fermentation of cellulose. Production of volatile fatty acids increases when supplements of sulfur are added to rations low in this mineral.

One result of sulfur deficiency in crops is a tendency for nonprotein nitrogen (amides, ammonium, etc.) and nitrate to accumulate at levels sufficiently high to seriously lower feeding quality.

Ample levels of dietary sulfur have the added advantage of increasing nitrogen-use efficiency by cattle and sheep. Nitrogen (protein) is probably one of the most expensive components of ruminant diets. When cattle and sheep are fed rations with a wide ratio of nitrogen to sulfur, the animals are unable to utilize the nitrogen fully and as a result they waste significant portions of this costly nitrogen nutrient. A dietary nitrogen to sulfur ratio of about 10:1 or slightly less normally results in the best utilization of nitrogen or protein. Recent reports show up to 25% greater nitrogen utilization by ruminants when the nitrogen to sulfur ratio is narrowed to 10:1.

5.5 CONVERSION FACTORS FOR CALCIUM, MAGNESIUM, AND SULFUR

Table 5–6. Conversion Factors for Calcium, Magnesium, and Sulfur Equivalents[a]

Column A	Column B	To Convert from A to B Multiply by	To Convert from B to A Multiply by
Calcium (Ca)	Calcium oxide (CaO)	1.40	0.71
Calcium (Ca)	Calcium carbonate ($CaCO_3$)	2.50	0.40
Calcium (Ca)	Gypsum ($CaSO_4 \cdot 2H_2O$)	4.30	0.23
Magnesium (Mg)	Magnesium oxide (MgO)	1.66	0.60
Magnesium (Mg)	Magnesium carbonate ($MgCO_3$)	3.47	0.29
Sulfur (S)	Sulfur dioxide (SO_2)	2.00	0.50
Sulfur (S)	Sulfuric acid (H_2SO_4)	3.06	0.33
Sulfur (S)	Gypsum ($CaSO_4 \cdot 2H_2O$)	5.37	0.19

[a] Calculated, using the following atomic weights: Ca, 40.08; Mg, 24.30; O–16.00; S, 32.06.

SUMMARY

Calcium, magnesium, and sulfur are commonly referred to as secondary elements. This does not mean they play a secondary role in plant nutrition. They are just as important in plant growth as the primary elements, but they are used by plants usually in smaller amounts and generally are not as limiting in the soil. However, a deficiency of any of the secondary elements can limit plant growth to the same extent as a deficiency of any of the primary elements.

In the past, we were not too concerned about secondary element fertilization because considerable amounts of these elements were present in the fertilizers commonly used. For example, superphosphate was a primary source of phosphorus for formulating mixed fertilizers. Superphosphate contained considerable quantities of gypsum (calcium sulfate). Therefore, in addition to its phosphorus content, it contained an average of 20% calcium and 12% sulfur. Because of greater economy, in recent years there has been a trend toward the use of higher-analysis fertilizers. In formulating the higher-analysis fertilizers, it is necessary to use a more concentrated source of phosphorus than superphosphate, such as concentrated superphosphate or diammonium phosphate. Concentrated superphosphate contains less calcium and sulfur than normal superphosphate and ammonium phosphates contains only traces of these elements.

The secondary elements have many important functions in plants. For example, calcium is necessary for protein synthesis and cell elongation. Magnesium is an integral part of the chlorophyll molecule. Sulfur is necessary for the formation of vitamins and the synthesis of some hormones. Therefore, with many crops the availability of these secondary nutrients is of major importance.

QUESTIONS AND PROBLEMS

1. Calcium, magnesium, and sulfur have some very important functions in plants. List five functions or roles in plant nutrition for each secondary element.

2. What are some common sulfur fertilizers?

3. Explain the causes of "grass tetany."

4. What is the most economical way of correcting magnesium deficiencies on acid soils?

5. Explain how sulfur transformations in soils are similar to those of nitrogen.

6. An animal ingests 18 kg of a certain forage in a ration. The forage at the time of consumption contained 65% moisture. Sulfur analysis of the forage (dry matter basis) indicated a concentration of 0.15% sulfur. How many milligrams of sulfur were ingested by the animal?

   Solution:

   18 kg forage × 35% dry matter = 6.3 kg dry matter
   6.3 kg × 0.15% S = 0.00945 kg S = **9450 mg S**

7. A water sample from an irrigation well in eastern Colorado contains 20 ppm sulfur as sulfate (SO_4^{2-}). A farmer is using this water on a sandy soil, low in organic matter.

During the season he will apply about 38 cm (15 in.) of water for irrigated corn using a sprinkler system. (a) Calculate the amount of sulfur that will be added per hectare during the season. (b) If a lab that analyzed his soil recommended 25 kg/ha of sulfur (22.3 lb/A), how should he modify his planned fertilizer program in light of the water analysis?

Solution:

(a) Water applied:

38 cm (or 0.38 m) $\times$ 10,000 m²/ha = 3800 m³
m³ of water = 100 cm $\times$ 100 cm $\times$ 100 cm = 1 $\times$ 10⁶ cm³
1 $\times$ 10⁶ cm³ water = 1 $\times$ 10⁶ g water
3800 m³ water $\times$ 1 $\times$ 10⁶ g water/m³ = 3.8 $\times$ 10⁹ g/ha
3.8 $\times$ 10⁹ g water/ha = 3.8 $\times$ 10⁶ kg water/ha
3.8 $\times$ 10⁶ kg water/ha $\times$ 20 ppm S (20 mg S/kg water) =
76 $\times$ 10⁶ mg/ha S = 76 $\times$ 10³ g/ha S = **76 kg/ha S**

(b) 25 kg/ha S is recommended and there will be 76 kg/ha applied in irrigation water. Therefore, *no fertilizer S application is necessary.*

NOTE: In very sulfur-deficient conditions, additional sulfur preplant or as a starter might be necessary. Prewatering would have added some sulfur that would be available to the corn.

8. A crop of alfalfa (dryland) produced 6 metric tons of hay per hectare over an entire season. If that yield was calculated at 12.5% water and if the forage on a dry matter basis contained 0.20% sulfur, how much sulfur would have been removed per hectare by the alfalfa?

Solution:

6 mt = 6000 kg ha/hay

6000 kg/ha of hay $\times$ 87.5% dry matter = 5250 kg/ha of dry matter

5250 kg/ha of dry matter $\times$ 0.20% S = **10.5 kg/ha of S removed in the forage**

9. How much magnesium (Mg^{2+}) was removed by the alfalfa crop in Problem 8, if the alfafa contained 0.38% Mg^{2+}? Assume that the water content of the hay was the same.

10. Ammonium thiosulfate, a liquid nitrogen-sulfur fertilizer has an analysis of 12% nitrogen and 26% sulfur. If 40 kg/ha of sulfur were recommended for a crop of irrigated alfalfa and if the producer planned to use ammonium thiosulfate as the sulfur source in a liquid fertilizer program, how much would be needed totally for a 65-ha field (160 A)? Give your anwer in kilograms of ammonium thiosulfate.

Solution:

40 kg/ha of S $\times$ 65 ha = 2600 kg S

2600 kg S ÷ 26% S in ammonium thiosulfate = **10,000 kg ammonium thiosulfate**

11. A farmer limed an area with 3000 kg/ha of a liming agent with an effective calcium carbonate (ECC) content of 80%. He later learned that the ECC was made up of 10% magnesium carbonate ($MgCO_3$) and 78% calcium carbonate ($CaCO_3$) for a total

calcium carbonate equivalent rating of 90%. $MgCO_3$ is 1.19 times as effective as $CaCO_3$ due to its lower equivalent weight. If he needed to add 100 kg/ha of Mg^{2+} because of a low soil exchangeable Mg^{2+} level, would the lime (dolomitic lime) be adequate as a Mg^{2+} source?

Solution:

3000 kg lime $\times$ 80% ECC = 2400 kg/ha of ECC
10% of ECC is $MgCO_3$; 2400 kg ECC $\times$ 10% = 240 kg/ha of $MgCO_3$
$$\% \text{ Mg in } MgCO_3 = \frac{\text{atomic weight of Mg (24)}}{\text{molecular weight of } MgCO_3 \text{ (84)}} \times 100 = 28.6\% \text{ Mg}$$
240 kg $MgCO_3$ $\times$ 28.6% Mg = 68.6 kg/ha of Mg
100 kg/ha of Mg needed
100 − 68.6 = **31.4 kg/ha of Mg still needed**

REFERENCES

Adams, Fred, "Field Experiments with Magnesium in Alabama — Cotton, Corn, Soybeans, Peanuts," Auburn University Bul. 472 (1975).

Allaway, W. H., "The Effect of Soils and Fertilizers on Human and Animal Nutrition," Agriculture Information Bulletin No. 378 (1975).

Bauman, W. E., and E. L. Whitehead, "Agricultural Use of Gypsum," Oklahoma State University Extension Fact No. 2214 (1970).

Elkins, C. B., R. L. Haaland, C. S. Hoveland, and W. A. Griffey, "Grass Tetany Potential of Tall Fescue as Affected by Soil O_2," *Agron. J.*, 70 (1978) 309–11.

Grunes, D. L., and H. F. Mayland, "Controlling Grass Tetany," USDA Leaflet No. 561 (1975).

Jones, U. S., "Calcium, Magnesium and Finely Ground Limestone," Clemson University, Agronomy and Soils Research Series No. 89 (1970).

Jones, U. S., "Sulfur—Essential for Protein," Clemson University, Agronomy and Soils Research Series No. 55 (1971).

Miller, H. F., and G. D. Corder, "Secondary and Micronutrient Element Needs for Field Crops in Kentucky," University of Kentucky Extension Cir. 613 (1967).

Overdahl, C. J., and R. P. Schoper, "Magnesium for Minnesota Soils," University of Minnesota Extension Folder 440 (1978).

Reid, R. L., G. A. Jung, I. J. Roemig, and R. E. Kocher, "Mineral Utilization by Lambs and Guinea Pigs Fed Mg-Fertilized Grass and Legume Hays," *Agron. J.*, 70 (1978), 9–14.

Robertson, L. S., D. R. Christenson, and D. D. Warncke, "Essential Secondary Elements: Magnesium," Michigan State University Extension Bulletin E-994 (1976).

Robertson, L. S., D. R. Christenson, and D. D. Warncke, "Essential Secondary Elements: Calcium," Michigan State University Extension Bulletin E-996 (1976).

Robertson, L. S., M. L. Vitosh, and D. D. Warncke, "Essential Secondary Elements: Sulfur," Michigan State University Extension Bulletin E-997 (1976).

Terman, G. L., "Atmospheric Sulphur — The Agronomic Aspects," Technical Bulletin Number 23. The Sulphur Institute, Washington, D.C. (1978).

U.S. Plant, Soil and Nutrition Laboratory Staff, Ithaca, N.Y., "The Effect of Soils and Fertilizers on the Nutritional Quality of Plants," Agricultural Information Bulletin No. 299 (1965).

Voss, R. G., "Sulfur — An Essential Secondary Nutrient," Iowa State University Pm-801 (1977).

Walsh, L. M., "Soil and Applied Calcium," University of Wisconsin Extension Series A2523 (1973).

Walsh, L. M., "Soil and Applied Magnesium," University of Wisconsin Extension Series A2524 (1973).

Wilcox, G. E., and J. E. Hoff, "Grass Tetany: An Hypothesis Concerning its Relationship with Ammonium Nutrition of Spring Grasses," *J. of Dairy Science,* 57 (9) (1974) 1085–89.

CHAPTER SIX

Micronutrients

The micronutrients have erroneously been called *minor* elements, suggesting that their role in plant nutrition is of some minor nature compared to the *major* nutrients or macronutrients. That is hardly the case. Micronutrients are required in smaller quantities for normal plant nutrition, but their role is equally important and deficiencies of micronutrients lead to severe depression in plant nutrition, growth, and yield.

In this chapter, the soil chemistry, plant nutrition, deficiency symptoms, fertilizer sources, methods of application, and crop responses of micronutrients are discussed. Elements classified as micronutrients include zinc (Zn), iron (Fe), copper (Cu), manganese (Mn), boron (B), molybdenum (Mo), and chlorine (Cl).

6.1 ZINC

Geochemically, zinc exists primarily in sulfide and silicate minerals. Some carbonates of zinc occur but the principal zinc mineral ore is sphalerite, zinc sulfide (ZnS). Zinc does substitute for magnesium to some extent in silicate minerals. The most common form of zinc silicate is hemimorphite $[Zn_4(OH)_2Si_2O_7 \bullet H_2O]$. The average abundance of zinc in the earth's crust is 70 to 80 ppm compared to 50,000 to 60,000 for iron, 1000 ppm manganese, 50 to 60 ppm copper, 10 ppm boron, and less than 2 ppm molybdenum.

In all of these forms, zinc exists as a divalent cation (Zn^{2+}). Zinc is not likely to undergo reduction in nature due to its electropositive nature. Natural occurrences of metallic zinc (native zinc) are rare if they indeed do occur. In comparison, copper is much more easily reduced and native occurrence of metallic copper is relatively common.

6.1.1 Soil Chemistry

Ion Species and pH

As a general statement, the solubility of soil zinc compounds can be expected to decline as the pH increases. This phenomenon also relates to the availability of other metals in the soil including iron, manganese, and copper. Lindsay and Norvell[1] studied the solubilities of various zinc minerals in soil including zinc hydroxide [$Zn(OH_2)$] and zinc carbonate ($ZnCO_3$), and they expressed the relationship of zinc solubility to soil hydrogen ion concentration in the following reaction:

$$Zn^{2+} + soil \longrightarrow Zn\text{–soil complex} + 2\ H^+$$

They noted that a zinc-soil complex that is apparently responsible for fixation of zinc in soils is about 100,000 times less soluble than either zinc hydroxide or zinc carbonate. Although this zinc-soil complex was originally thought to be a type of zinc silicate, that idea has now been discounted and the exact nature of the complex remains undetermined.

Lindsay[2] pointed out that several soluble zinc complex ion species are represented by the zinc in the soil solution that is in equilibrium with the soil zinc. Those ions include Zn^{2+}, $Zn(OH)^+$, $Zn(OH)_2$, $Zn(OH)_3^-$, and some $Zn(OH)_4^{2-}$ although the latter (zincate) is rare. Below pH 7.7, the dominant soluble species is Zn^{2+}; above that pH, $Zn(OH)_2$ is dominant. Solubility of these species is pH-dependent as noted. At pH 5, the concentration of Zn^{2+} in soil solution is approximately 6.5 ppm but this drops to 7 parts per trillion at pH 8.0. The activity of Zn^{2+} decreases about a hundredfold for each unit increase in pH. It is easy to see, then, why zinc deficiencies frequently are associated with high-pH soils. Lindsay points out that at low pH values some exchangeable Zn^{2+} may be present in soils, but at high pH levels soil solution zinc is so low that little Zn^{2+} is held by the exchange complex.

Viets, Boawn, and Crawford[3] examined the effects of soil types on zinc absorption by plants and noted the strong effect of pH on availability. As soil pH declined, zinc uptake by grain sorghum was increased (Fig. 6–1). Using

[1] W. L. Lindsay and W. A. Norvell, "Equilibrium Relationships of Zn^{2+}, Fe^{3+}, Ca^{2+} and H^+ with EDTA and DTPA in Soils," *Soil Sci. Soc. Amer. Proc.*, 33:62, 1969.

[2] W. L. Lindsay, "Inorganic Phase Equilibria of Micronutrients in Soils." In *Micronutrients in Agriculture*, J. J. Mortvedt et al. (Ed.), Soil Science Society of America, Madison, Wis., 1972.

[3] F. G. Viets, Jr., L. C. Boawn, and C. L. Crawford, "The Effect of Nitrogen and Types of Nitrogen Carrier on Plant Uptake of Indigenous and Applied Zinc," *Soil Sci. Soc. Amer. Proc.*, 21:197, 1957.

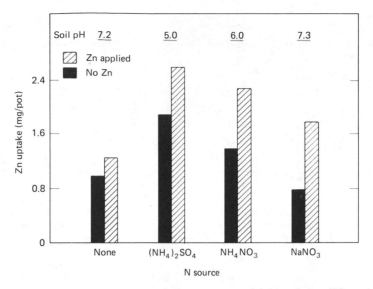

Fig. 6–1. Soil pH can have a strong effect on the availability of zinc. When zinc is incorporated into various nitrogen fertilizers, the resulting pH in the vicinity of the nitrogen-zinc application may increase or decrease zinc uptake. Nitrogen alone had little effect on zinc uptake when supplied as sodium nitrate but ammonium nitrate and ammonium sulfate simultaneously decreased soil pH and increased zinc uptake. Zinc applied with the acid-forming ammonium forms of nitrogen was more effective than that applied with sodium nitrate. See Chapter 2 for effects of nitrogen sources on soil pH. (Source: F. G. Viets, Jr., L. C. Boawn, and C. L. Crawford, "The Effect of Nitrogen and Types of Nitrogen Carrier on Plant Uptake of Indigenous and Applied Zinc," *Soil Sci. Soc. Amer. Proc.*, 21:197, 1957 by permission of the American Society of Agronomy.)

different nitrogen sources that imparted different pH's to the soil (see Chapter 2) also affected zinc uptake. The relationship of zinc availability to soil pH should not be too heavily emphasized, however. Soil tests confirmed by greenhouse and field investigations have definitely shown that deficiencies of zinc can exist even on slightly acid soils. The dramatic zinc response of irrigated corn in Fig. 6–2 occurred on a soil of pH 6.7. Soil tests had indicated severe zinc deficiency on the area. The point is that zinc deficiencies can and do exist at many different levels of soil pH but are *more common* when the pH is high and when the organic matter content of a high pH soil is low.

Adsorption by Clays. Soils can adsorb zinc on their exchange complex provided there is some soluble source of zinc to be adsorbed. Several researchers[4] have reported that montmorillonite clay can adsorb zinc beyond the values of the cation exchange capacity (CEC), particularly at near-neutral or alkaline pH values. Some have suggested that such high adsorption is due

[4] F. T. Bingham, A. L. Page, and J. R. Sims, "Retention of Cu and Zn by H-Montmorillonite," *Soil Sci. Soc. Amer. Proc.*, 28:351, 1964.

Fig. 6–2. Zinc responses can be spectacular. Corn is quite responsive to applied zinc when the element is deficient in the soil, particularly when growing conditions are cool and wet. (Courtesy Dr. Roscoe Ellis, Jr., Kansas State University.)

to the formation of precipitates such as hydroxides within the lattice of the clay. Treatments with strong acids have demonstrated, however, that this type of adsorption is probably not related to substitution of zinc into the octahedral layer (hydroxyl-containing layer) of the clay as has been suggested.

Organic Matter Reactions. Soil organic matter forms some very stable products with zinc. Carboxyl (COOH) and phenolic groups are important in the bonding of zinc with organic matter. Studies at Purdue University[5] indicated that, through the involvement of carboxyl and phenolic groups, humic and fulvic acid were involved in the zinc-binding process by organic matter. Although research has shown that adsorption of zinc by organic matter may involve carboxyl and phenolic groups in the process, considerable research will be required to unravel the facts surrounding metal adsorption and complexing by organic matter. If some type of chelating effect or sequestering effect is exerted on zinc and other metals by organic matter, such a complexing could protect the zinc from further reactions with inorganic soil components and could significantly increase plant availability of indigenous or added zinc. It is significant to note that zinc deficiencies are frequently associated with low organic matter soils. Addition of organic matter to such soils has materially improved the availability of zinc. Applications of manure can improve zinc availability. In addition to adding organic matter, such applications also add significant amounts of zinc.

Oxidation-Reduction State. The reactions of zinc in the soil, although complex, do not involve oxidation or reduction of the zinc itself. The valence of the zinc ion is $+2$, but the effective valence of the ion can be affected by association with varying numbers of hydroxyl ions. The effective valence can

[5] F. L. Himes and S. A. Barber, "Chelating Ability of Soil Organic Matter," *Soil Sci. Soc. Amer. Proc.*, 21:368, 1957.

range from $+2$ (Zn^{2+}) to -2 [$Zn(OH)_4^{2-}$], the zincate ion. Contrast the oxidation state of zinc to that of iron and manganese in Sections 6.2.1 and 6.4.1.

Phosphorus-Zinc Interactions. High amounts of phosphorus are known to induce zinc deficiency in plants, and some researchers have suggested that the phenomenon is due to the formation of insoluble zinc phosphates in the soil. Lindsay[2] examined the theory of this possible reaction and noted that formation of a compound such as $Zn_3(PO_4)_2 \cdot 4H_2O$ could hardly be considered as a deterrent to zinc uptake by plants due to the relatively high solubility of this compound. Lindsay even suggested that this compound would make a good zinc fertilizer, a fact that was later substantiated by TVA researchers using very finely ground material that was well mixed with the soil.

Zinc deficiencies are common in citrus crops grown in Florida on high phosphorus soils (see Chapter 3) and in Tennessee on high phosphorus soils cropped with corn. Added fertilizer phosphorus can also induce zinc deficiency.[6,7] The nature of the relationship between phosphorus and zinc is not fully understood, but it is known that this is basically a physiological problem rather than a soil chemistry phenomenon. Even then, zinc deficiency is not always induceable on high pH soils. Researchers in Utah and Washington were unable to produce phosphorus-induced zinc deficiency in crops as divergent as peaches and beans by very heavy applications of phosphorus.

Apparently, problems in the translocation of zinc from the roots to the shoots of affected plants that occur during phosphorus-induced zinc deficiency may be due to the formation of insoluble zinc phosphates at the surface of the roots or may involve the formation of iron phosphates in the same areas. Note the heavy accumulation of radioactive ^{65}Zn isotope in the roots of pea beans grown in a high phosphorus solution culture in Fig. 6–3 compared to the plants grown in a lower phosphorus system in that same figure. The presence of ^{65}Zn is shown by the dark areas of the plant, indicating exposure of X-ray film by the gamma radiation.

The large number of studies conducted on the phosphorus-zinc interaction in plants suggest that the problem is at least partially due to physiological relationships. For instance, some very dramatic changes in zinc concentrations in the tops of plants have been recorded when only very small applications of phosphorus have been made. Stukenholtz et al.[8] recorded significant changes in zinc concentration in corn with the addition of only 5 ppm phosphorus as monocalcium phosphate but noted only small zinc concentration decreases as the phosphorus application was increased to 80 ppm. This and results of similar studies have tended to dismiss the contention that a critical phosphorus-zinc ratio exists in plant tissue. On the other hand, potato

[6] W. Judy, G. Lessman, T. Rozycka, L. Robertson, and B. Ellis, "Field and Laboratory Studies with Zinc Fertilization of Pea Beans," Mich. Agr. Exp. Sta. Quart. Bul., 46:386, 1964.

[7] D. C. Adriano and L. S. Murphy, "Effects of Ammonium Polyphosphates on Yield and Chemical Composition of Irrigated Corn," *Agron. J.*, 62:561, 1970.

[8] D. D. Stukenholtz, R. J. Olsen, G. Gogan, and R. A. Olson, "On the Mechanism of Phosphorus-Zinc Interaction in Corn Nutrition," *Soil Sci. Soc. Amer. Proc.*, 30:759, 1966.

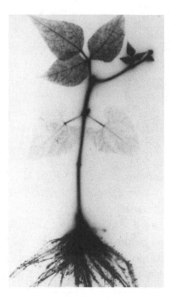

Fig. 6–3. The amount of phosphorus supplied to plants can influence the translocation of zinc from the roots to the tops. Radioactive zinc in these plants was used to expose film giving a graphic indication of the distribution of zinc in the plants. Note that the intensity of the exposure was greater in the tops of plants receiving a low amount of phosphorus (left) versus the greater intensity of exposure from the roots of the high phosphorus plant on the right. Note also the high concentration of zinc in the new growth areas of both plants. (Courtesy Dr. D. C. Adriano, Savannah River Ecology Lab, University of Georgia.)

studies involving a phosphorus-induced growth problem produced data indicating that the concentration of zinc in either leaf or stem tissue was not related.[9] This metabolic malfunction correlated better with P:Zn concentration ratios in these same tissues.

Apparently, significant differences among species of crop plants occur in the nutritional relationship of phosphorus to zinc. Phosphorus-zinc imbalances are a significant production factor that can be readily overcome by the addition of the required amounts of zinc. Some indication of the visual effects of phosphorus on Zn nutrition of corn is indicated in Fig. 6–4. Treatments that tend to improve phosphorus absorption such as banding of fertilizer phosphorus (see Chapter 3) may produce an even more severe zinc deficiency than broadcasting the same amount of phosphorus. Note in Fig. 6–5 that banded phosphorus fertilizer induced a more severe stunting due to deficient zinc than did a comparable phosphorus application that was broadcast. Plants in the background that received adequate zinc were quite normal. On the other hand, the likelihood of phosphorus effects on zinc nutrition are

[9] L. C. Boawn and G. E. Leggett, "Phosphorus and Zinc Concentrations in Russett Burbank Potato Tissue in Relation to Development of Zinc Deficiency Symptoms," *Soil Sci. Soc. Amer. Proc.*, 28:229, 1964.

Fig. 6-4. Applications of large amounts of phosphorus can induce even more severe zinc deficiency in many plants. In these cases zinc was already deficient by soil analysis. Application of a large amount of phosphorus to correct a phosphorus deficiency without adding the necessary zinc produced plants that were more chlorotic and stunted (left). Plants which received both phosphorus and zinc grew normally (right). (Courtesy Dr. Roscoe Ellis, Jr., Kansas State University.)

Fig. 6-5. Zinc deficiency effects can vary dramatically on the same site depending on climatic conditions. The photograph on the left indicates the severity of zinc deficiency in corn, accentuated by high applications of phosphorus, in a year with a cold wet condition in the early growing season. Plants in the background received zinc. The same site one year later (right) shows almost no zinc deficiency in corn because the growing conditions following planting were much warmer and drier. (Courtesy L. S. Murphy.)

increased when soil conditions are cool and wet and the plant is young. Such conditions tend to depress zinc uptake by plants whether or not phosphorus is involved.

Zinc Distribution in the Soil Profile. Although the total zinc concentration in soil is usually quite uniform, most of the available zinc is associated with the horizon having the highest organic matter content, the surface soil.

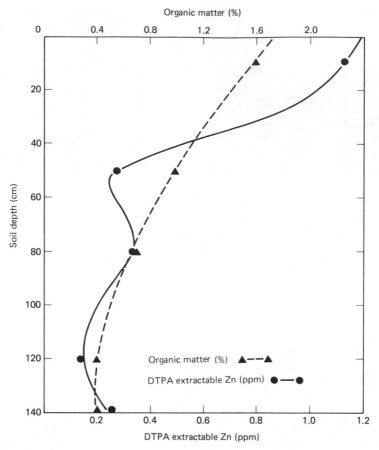

Fig. 6-6. Zinc distribution in the soil frequently follows this type of pattern. Zinc concentration and organic matter content are closely allied in this soil profile. The lower portions of the soil profile have been exploited for available zinc by native plant growth and the zinc concentrated in the surface organic layer. (Courtesy Dr. Roscoe Ellis, Jr., Kansas State University.)

Apparently, the absorption of zinc by plants over time has led to this accumulation. When plants die, the zinc is returned to the organic matter layer. Root uptake of zinc from the lower parts of the profile may be responsible for a sharp decline in available zinc as depth increases (Fig. 6-6). Below the root zone the zinc concentration again increases.

The fact that available zinc tends to accumulate in the surface soil suggests that leaching of zinc in soils with percolating water is slight. A study at the University of California[10] demonstrated that surface-applied zinc in a soluble inorganic form was only slightly leached even in a permeable soil. On the

[10] A. L. Brown, B. A. Krantz, and P. E. Martin, "Plant Uptake and Fate of Soil-Applied Zinc." *Soil Sci. Soc. Amer. Proc.*, 26:167, 1963.

other hand, the application of a chelated form of zinc permitted more movement of the element. That point is discussed further in Section 6.1.4.

6.1.2 Zinc Nutrition of Plants

Zinc Uptake. Zinc uptake by plants has been determined to be controlled primarily by diffusion in the soil solution. Barber[11] used a zinc isotope to demonstrate that zinc is depleted in the immediate vicinity of plant roots and established that diffusion gradients do exist around plant roots. Elgawhary et al.[12] estimated that more than 95% of the zinc required by corn is supplied by diffusion. Available zinc, however, is not represented only by the zinc in soil solution. Studies have indicated that highly significant amounts of readily solubilized zinc are present in the soil and this supply of *labile* zinc readily replenishes that absorbed by plants from the soil solution. The amount of labile zinc varies with soil type and explains why some soils could have a greater ability to supply zinc than others. Two factors control zinc uptake, a concentration factor which represents the concentration of zinc in the soil solution and a capacity factor which represents the soil's ability to replenish zinc in the soil solution.

Frequently, sandy soils are low in available zinc because of a low amount of labile zinc in native soil minerals. A lack of organic matter, typical of well-drained sandy soils, further reduces the amount of labile zinc present. Follett and Lindsay[13] reported a high correlation between organic matter and chelate-DTPA extractable zinc. On the other hand, peat and muck soils are also occasionally deficient in zinc. This condition may occur through depletion of zinc in the surface layer, by plant removal, and by physical separation of the mineral zinc portion from the root zone by some condition such as a high watertable. In addition, peat and muck soils may also have a high pH that further contributes to lower concentrations of labile zinc.

Zinc uptake is strongly affected by soil temperature. Soil zinc levels that are adequate in one season may be deficient the following year due to a lower soil temperature regime and subsequent reduced root development during the cold period. Zinc deficiency symptoms are usually more pronounced during the cool, early parts of the growing season. Later, these deficiency symptoms may diminish in intensity or disappear entirely.

Bauer and Lindsay[14] determined that higher soil incubation temperatures greatly increased the available zinc as measured by corn uptake. However, this available zinc increase could not be measured by chemical extraction pro-

[11] S. A. Barber, J. M. Walker, and E. H. Vasey, "Mechanisms for the Movement of Plant Nutrients from the Soil and Fertilizer to the Plant Root," *J. Agr. Food Chem.*, 11:204, 1963.

[12] S. M. Elgawhary, W. L. Lindsay, and W. D. Kemper, "Effect of Complexing Agent and Acids on the Diffusion of Zinc to a Simulated Root," *Soil Sci. Soc. Amer. Proc.*, 34:211, 1970.

[13] R. H. Follett and W. L. Lindsay, "Changes in DTPA-Extractable Zinc, Iron, Manganese and Copper in Soils Following Fertilization," *Soil Sci. Soc. Amer. Proc.*, 35:600, 1971.

[14] A. Bauer and W. L. Lindsay, "The Effect of Soil Temperature on the Availability of Indigenous Soil Zinc," *Soil Sci. Soc. Amer. Proc.*, 29:413, 1965.

cedures that tend to extract many times more zinc than is removed by the crop. Similar results were obtained by Gallagher et al.[15] Incubation of an acid soil at a temperature of 30°C released more zinc than did an incubation temperature of 5°C. Plant growth temperatures exerted a greater influence on corn zinc uptake than did soil incubation temperature, however. A similar study on a calcareous soil, though, did not show a consistent effect of higher soil incubation temperature on release of zinc. Growth temperature effects on zinc uptake have been noted frequently in the literature.

Several possible explanations for the effect of temperature on zinc uptake are: (1) plant root systems are less well developed during cold temperatures, subsequently the zone of soil explored by the roots is diminished and less soil solution is exploited; (2) because diffusion processes are also slowed by cold temperatures and movement of zinc into the depleted zone from other areas of the soil solution is reduced; and (3) low temperatures also restrict the activities of soil organisms; for this reason less zinc is released from organic matter to replenish that removed by plants from the soil solution.

The effects of a combination of soil temperatures and continued cropping on the intensity of zinc deficiency symptoms in corn in the field can be seen in Fig. 6-5. Cold, wet soil conditions produced much more severe zinc deficiencies than did warm, dry conditions on the same area the following year. Admittedly, continued cropping into the second year also resulted in slightly increased available zinc concentration in the surface soil.

Plant uptake of zinc is strongly affected by plant species and varieties. Recognized differences in the susceptibility of different species to zinc deficiency are noted (Table 6-1). Corn, sorghum, and beans (*Phaseolus vulgaris*) generally are highly susceptible to zinc deficiencies. Interestingly, within these three species, significant differences also exist among varieties in terms of zinc deficiency susceptibility. One of the most frequently referenced varietal differences in zinc susceptibility is that of Saginaw and Sanilac varieties of navy beans (pea beans). Ellis[16] noted that Sanilac was more susceptible to zinc deficiency. Later work in Michigan also indicated that Saginaw is both more zinc efficient and more tolerant of excess amounts of zinc. These same varieties also demonstrated differential absorption of iron and phosphorus, Sanilac containing much higher concentrations of both of these elements compared to Saginaw. Similar differences in the magnitude of phosphorus effects on zinc uptake have been identified in soybeans, whereas corn research has also demonstrated genetically controlled zinc efficiency.[17]

Absorption Mechanisms. Zinc absorption (uptake) is accomplished by passive and active absorption. Passive *ab*sorption results from electrostatic

[15] P. J. Gallagher and L. S. Murphy, "Effects of Temperature and Soil pH on Effectiveness of Four Zinc Fertilizers," *Comm. Soil Sci. Plant Anal.*, 9:115, 1978.

[16] B. G. Ellis, "Response and Susceptibility." In *Zinc Deficiency — A Symposium, Crops and Soils*, 18(1):10, 1965.

[17] A. H. Halim, C. E. Wasson, and R. Ellis, Jr., "Zinc Deficiency Symptoms and Zinc and Phosphorus Interactions in Several Strains of Corn (*Zea Mays L.*)," *Agron. J.*, 60:267, 1968.

Table 6–1. Relative Responses of Selected Crops to Zinc

High Response

Beans	Onions
Corn, field	Sorghum
Corn, sweet	

Medium Response

Barley	Sudangrass
Potatoes	Sugar beets
Soybeans	Table beets
	Tomatoes

Low Response

Alfalfa	Oats
Asparagus	Peas
Carrots	Peppermint
Clover	Rye
Grass	Spearmint
	Wheat

SOURCE: L. S. Robertson and R. E. Lucas, "Zinc," Mich. Ext. Bul. E-1012, August 1976.

*ad*sorption of zinc ions on cell walls and other surfaces in the apparent free spaces of roots. This attachment of zinc to the cell surfaces is nonspecific, is not metabolically linked, and in fact represents an adsorption process very much like the attraction of cations to clay surfaces. This adsorbed (exchangeable) zinc may represent as much as 90% of the total zinc held by plant roots. Active *absorption,* on the other hand, is highly selective and is metabolically linked (see Chapter 2). Factors that lower the metabolic activity (such as metabolism of sugars) of the roots also tend to depress the uptake of actively absorbed nutrients. Temperature and oxygen supply are factors that can affect active absorption. Effects of these two factors and others on plant uptake of phosphorus were discussed in Chapter 3.

Many of the early, short-term experiments designed to study zinc absorption were conducted for such a short period of time that only the adsorption of zinc on root cell surfaces was examined. More recent studies[18] have reported higher concentrations of zinc in xylem exudates than were present in the nutrient solution containing the zinc. This suggests active absorption. Other researchers have reported that zinc uptake was reduced by low temperature and certain *antimetabolites.* Based on the results of the more recent investigations and the strong relationships of zinc deficiency in the field of soil temperatures, most workers agree that zinc absorption is largely an active process.

[18] J. E. Ambler, J. C. Brown, and H. G. Gauch, "Effect of Zinc on Translocation of Iron in Soybean Plants," *Plant Physiol.,* 46:320, 1970.

Location of Zinc in the Plant. Zinc has an intermediate mobility in the plant compared to some of the other elements, but its deficiency symptoms in most species are most strongly expressed on new tissue. Roots may have a higher concentration of zinc than stem or leaf tissues (see Fig. 6–3). In fact roots may show luxury accumulation of zinc under conditions of high zinc availability even though translocation occurs readily from the roots to the leaves. Plant species and varieties vary in their ability to absorb zinc from a nutrient source and also vary in their ability to translocate zinc from the roots to the shoots. Luxury accumulation by the roots can partially alleviate zinc deficiencies that may occur later in the growing season if the roots expand into an area of low available zinc.

Concentrations of zinc in the plant leaves are widely variable depending on species, variety, and growing conditions. Usually, levels of zinc in plants range from around 10 to 100 ppm. Zinc concentrations are normally highest in the very young plants and decrease with age due both to dilution effects and eventually to translocation to the seed. Decreases in zinc concentrations in plants late in the growing season could represent a reduction in the ability of the soil to supply the element. Variations that occur in the zinc concentration due to the variables mentioned makes selection of a critical level difficult. For corn as an example, concentrations in leaf tissue early in the growing season (about the eight-leaf stage) of less than 15 ppm should be viewed as probably deficient. Extremely deficient plants may only show 7 to 9 ppm *total* zinc. Interpretation of tissue analyses for zinc is dependent on many factors including age of the plant, location of the sample on the plant, species, and moisture supply. Considerable experience is necessary to interpret results of tissue analyses for micronutrients in general.

Metabolic Roles of Zinc. Enzymatic roles of zinc have been established by numerous studies. *Metalloenzymes* that involve zinc as a part of their total system and activity include those listed in Table 6–2. Zinc is recognized as an essential component of a number of dehydrogenases, proteinases, and peptidases. In these roles, zinc then can exert an influence on electron transfer reactions including those in the *Krebs* cycle and subsequently on energy production in the plant. Other general types of reactions affected by zinc's enzymatic roles include protein synthesis and degradation. Several investigations have shown that one of the earliest and possibly most important events in the processes associated with zinc deficiency is a decrease in the levels of RNA (ribonucleic acid) and the ribosome content of cells. Ribosomes are the location of much of the cell RNA and are involved in protein synthesis.

Table 6–2. Some Zinc-Containing Enzymes in Plants

Carbonic anhydrase	Malic dehydrogenase
Alcohol dehydrogenase	Aldolase
Glutamic dehydrogenase	Lactic dehydrogenase
Carboxypeptidase	DNA polymerase I

Associated with the role of zinc in protein synthesis is its apparent role in the activity of tryptophan synthetase. Tryptophan is one of the amino acids and is involved in the eventual synthesis of the growth control compound indoleacetic acid. Observations of zinc-deficient plants suggest that they are deficient in growth control regulators.

6.1.3 Deficiency Symptoms

Corn. Zinc deficiency in corn is one of the most definite and recognizable of the micronutrient deficiency symptoms. Symptoms usually appear within the first 2 weeks after emergence on severely deficient areas and are characterized by development of a broad band of chlorotic tissue on one or both sides of the center midrib of the leaf (Fig. 6–7). The chlorotic area is most pronounced toward the base of the leaf. Young leaves are the most severely affected. More severe deficiency conditions produce interveinal chlorosis on both sides of the center midrib.

In addition to chlorosis, lack of sufficient growth regulators in the plant *may* be responsible for stunting and shortening of the internodes. Similar conditions are produced with deficiencies of other micronutrient metals including iron, manganese, and copper but, in combination with the chlorosis just described, this characteristic may be helpful diagnostically for zinc. Plants showing this effect tend to produce an "umbrella" effect with most of the leaves emerging from short internodes on the stalk.

Abnormal development of the leaf edges also is associated with zinc deficiency. Crinkled leaf edges accompanied by an accumulation of anthocyanin (reddish-purple) pigments at the leaf edge is further evidence of severe zinc deficiency (compare with phosphorus deficiency symptoms on corn in Chapter 3).

All symptoms are more pronounced early in the growing season, particularly on cold, wet soils. As the plant develops, further root extension may allow exploration of soil with more available zinc with subsequent diminished

Fig. 6–7. Zinc deficiency in corn is characterized by severe chlorosis of the upper leaves that develop early in the growing season. Leaves show a broad band of chlorotic tissue on either side (or both sides) of the midrib. Plants are stunted due to severe shortening of the internodes. Severely affected plants may show chlorosis on all leaves. (Courtesy Dr. Gary Hergert, University of Nebraska, North Platte Exp. Sta.)

intensity of the symptoms. Root development in severely affected plants, however, is stunted and uptake of other elements is also reduced.

Both severe and mild zinc deficiency in field corn and sweet corn tend to reduce yields even with decreasing deficiency symptoms by mid or late season. Delayed maturity and reduced yield of marketable ears of sweet corn usually result.

Grain Sorghum. Zinc deficiency symptoms in grain sorghum are similar to those in corn except that the incidence of deficiency in this crop is observed less frequently than in corn. Grain sorghum apparently has a greater ability to extract zinc from the soil than does corn. Nevertheless, severely deficient conditions tend to produce a similar chlorosis with development of the characteristic white areas on one or both sides of the center midrib. Stunting, shortening of the internodes, and development of the anthocyanin pigmentation are all associated with the deficiency in sorghum but are less frequently expressed.

Sorghum maturity can be dramatically affected by zinc deficiency (Fig. 6-8). Although yields may not be severely affected under conditions of mild deficiency, late-season maturity can produce high sorghum grain moisture content at harvest, particularly if the crop is subject to an early season frost.

Fig. 6-8. Zinc deficiency can exert significant effects on plant maturity. The grain sorghum plant on the left did not show visible zinc deficiency symptoms but developed seed heads several days later than the plant on the right that received adequate zinc. Even if no yield differences occurred, higher drying costs could be associated with the zinc-deficient sorghum. (Courtesy Dr. David Whitney, Kansas State University.)

This cannot be used as an infallible diagnostic technique however, because other elements, especially phosphorus, exert a similar effect on maturity (see Chapter 3).

Small Grains. Small grains tend to show some of the same deficiency symptoms as corn and grain sorghum. Rice is the most commonly affected small grain but zinc deficiencies have also been noted on wheat. Rice deficiency symptoms appear in the early stages of plant development and are expressed as a sudden blighting of the oldest leaves of the seedlings. Lesions produced on the leaf surface are surrounded by chlorotic areas that may become white. These lesions and chlorotic areas are parallel to the leaf veins and extend from the leaf sheath up the midribs of the oldest true leaves. Eventually, the base of the leaf blades and midveins become chlorotic. Even when these symptoms are expressed on the outer leaves, younger leaves may remain green. Cool, cloudy weather for extended periods may contribute to zinc deficiency symptom development.

Zinc deficiency symptoms in rice are often not developed until the permanent floodwater is applied. Initially, the lower leaves of affected plants become limp and float on the water surface. These leaves rapidly turn yellow-orange (bronzing) and develop small brown flecks. In severely affected areas, seedlings begin to disappear below the water surface particularly in deep-water areas (thus the term *deepwater disease*). Stands eventually thin and open-water areas appear.

Zinc deficiencies of wheat are expressed as chlorosis on older leaves with interveinal chlorosis developing toward the leaf edges. Under severe deficiency conditions, plants are stunted with shortened internodes. Tillering is severely reduced under such conditions. In general, deficiency symptoms in small grains other than rice are less well pronounced than those for corn or sorghum and may easily be confused with other nutrient deficiencies. Symptoms may be verified by plant analysis.

Soybeans, Navy (Pea) Beans, and Pinto Beans. Zinc deficiencies in soybeans have not been widely observed but are becoming more common as soybean production is extended into calcareous soils in many parts of the world, particularly the western corn belt in the United States. Severe deficiencies in this crop are expressed shortly after emergence with the gradual development of chlorosis of the younger leaves. Chlorosis under extremely deficient conditions may extend to all leaves on the plant. Intensity of the chlorosis also tends to become more pronounced under severely deficient conditions as the season progresses.

Affected leaves tend to show total chlorosis without strong expression of green veins such as is the case with manganese deficiencies. It is frequently difficult to distinguish between zinc and iron deficiencies in soybeans because both conditions may exist simultaneously and tend to be developed under the same type of soil conditions. Iron chlorosis tends to be more severe early in the growing season. Plant analysis is the only sure way to determine an existing zinc deficiency where iron deficiency is also suspected. Plant analysis for iron may not show a low level of this element even when chlorosis is due to improper plant use of iron (see Section 6.2).

Stunting also accompanies zinc deficiency in soybeans. This in itself is hardly a reliable diagnostic tool but should be expected in conjunction with the chlorosis described earlier. Yields are dramatically reduced under severe deficiency conditions. Soybean seeds may also contain much lower amounts of zinc under severe deficiency conditions.

Navy beans and pinto beans have similar zinc deficiency symptoms. Beans deficient in zinc emerge light green in color. When the deficiency is severe, the areas between the veins of the leaves becomes pale green and then yellow near the tips and outer edges. The leaves in the early stages are deformed, small, and crinkled. In later stages of development, these leaves become necrotic and appear to have been killed by either herbicide or sunscald. Zinc-deficient bean plants are extremely stunted (Fig. 6-9) and yield essentially nothing under severe conditions. Under slightly milder conditions, terminal blossoms set pods that later abort. Maturity of beans in general is much delayed.

Cotton. Incidence of zinc deficiency in cotton has been largely relegated to high pH soils and areas where land leveling has taken place for irrigation. Cotton tends to show deficiency symptoms somewhat similar to other dicotyledonous plants (dicots) such as beans. Deficiency symptoms appear early in the growing season and are typified by a general stunting of the plants.

Fig. 6-9. Zinc deficiency in pea beans (navy beans) can have severe effects on plant growth and yield. This picture from the "thumb" area of Michigan demonstrates the effects of zinc deficiency and also shows the effects of cultivar on susceptibility to zinc deficiency. Sanilac (right) is much more susceptible than Saginaw. Differences between cultivars in susceptibility to other nutrient deficiencies also exist in other crops. (Courtesy Dr. B. D. Knezek, Michigan State University.)

Stalk elongation tends to be retarded and the plant assumes a bushy appearance due to shortened internodes. Shortly after appearance of the first true leaves, a general interveinal chlorosis develops. Older leaves become thickened and brittle and tend to be cupped upward. Fruiting processes begin in zinc-deficient cotton plants, but flowers and squares are usually aborted. Less severe deficiencies may allow set of additional squares later in the season that may eventually mature into bolls. Late-season growth may eventually produce a plant with leaves clustered toward the top and with a bare stem below. Overall effects of zinc deficiency on cotton production is to delay maturity greatly and to reduce yields significantly because of the loss of the lower bolls.

Potatoes. Zinc deficiency on potatoes results in rosetting or *little leaf* formation. Leaves become thick and brittle and the leaf midrib becomes S-shaped that gives a fern-leaf appearance. Interveinal areas gradually change from green to yellow. Leaflets are dwarfed and curl upward. With severe deficiency, the entire leaflet may become yellow and necrotic tissue may develop around the margins and leaf tips. Generally, foliage and yield of tubers are greatly reduced under zinc-deficient conditions.

Citrus, Deciduous Fruits, and Nuts. Zinc deficiency in citrus is confined to new growth areas. Under severe deficiency conditions, leaf size is drastically reduced and twig length is shortened. New growth has a bushy, upright appearance. Less severe deficiencies have proportionally less effect on leaf size.

Leaf chlorosis is a distinctive zinc deficiency symptom. Deficient leaves develop a creamy-yellow interveinal chlorosis. The chlorosis on terminal growth becomes more severe with each new flush of growth. Symptoms may gradually disappear under mild cases of deficiency but as zinc deficiency persists, chlorosis may become persistent. Severe deficiency may cause new leaves to become almost totally devoid of chlorophyll. Terminal growth areas show dieback of the smaller twigs and eventually larger twigs also die.

Severe deficiency produces a drastic decrease in fruit production and fruit quality. Fruits are prematurely yellow, small, and deformed and the pulp is dry and woody. Less severe deficiencies may decrease yield of fruit but have relatively little effect on fruit quality.

Zinc deficiencies in deciduous fruit and nut trees generally affect the growing points. Typically the terminal is dwarfed, nodes are short, and a rosette effect is produced in the arrangement of the leaves. Leaves are generally small and narrow and develop interveinal chlorosis. The small-leaf effect is known as *little leaf* and is frequently noted in pears and pecans. In addition to pears and pecans, apples, peaches, cherries, English walnuts, and almonds show these characteristics. Tung-oil tree leaves show a characteristic wavy margin effect and a lopsided development, one side of the leaf being larger than the other.

Although deciduous fruit and nut trees are rarely killed by zinc deficiency, severe dieback of new growth can occur and production decreases.

6.1.4 Fertilizer Sources

The common sources of zinc fertilizers available for application are listed in Table 6-3. In addition to the manufactured compounds commonly used, animal manures and sewage sludge contain appreciable quantities of zinc. All sources available should be considered in attempting to correct soil zinc deficiencies.

Manufactured fertilizers can be classified into two broad categories, inorganic and organic. Primary inorganic sources of zinc include zinc sulfate ($ZnSO_4$), the most commonly used zinc fertilizer; zinc oxide (ZnO), and zinc-ammonia complex (Zn-NH_3). Other compounds are listed but these are of minor importance relative to these three. It is not uncommon to find both zinc sulfate and zinc oxide present in the same formulation of a zinc fertilizer.

Inorganic Zinc Sources. Much continued discussion surrounds the relative efficiency of the inorganic sources of zinc. As early as 1935, Barnette et al.[19] reported that corn on sandy soils in Florida responded well to zinc sulfate but noted that zinc oxide was not so efficient in correcting the problem. Boawn, Viets, and Crawford[20] studied the efficiency of soil treatments of various types of zinc carriers using grain sorghum as a test plant. Their investigation indicated that zinc sulfate and zinc oxide were about equal in correcting zinc deficiency in sorghum on a fine sandy loam, pH 7.2. Mortvedt and Giordano[21] found that zinc oxide was a good source in liquid fertilizers at a soil pH of 7.3 and superior to the same source applied in a granular form due to a better distribution of zinc in the soil.

Salako et al.[22] also examined the availability of several organic and inorganic sources of zinc and reported higher yields of soybeans from zinc sulfate applications than from applications of zinc oxide to a soil with a pH of 7.1. Results from a number of other studies indicate that zinc oxide and zinc sulfate are comparable when finely divided zinc oxide is used, such as the form that would be mixed in fluid fertilizers. Higher pH soils may tend to somewhat depress the availability of zinc from ZnO. Granular zinc oxide or zinc oxide coated on the surface of solid fertilizers may be less available due to a lower specific surface area. Of these two compounds, zinc sulfate is the more commonly utilized inorganic zinc source because of its higher water solubility.

[19] R. N. Barnette and J. D. Wagner, "A Response of Chlorotic Corn Plants to the Application of Zinc Sulfate to the Soil," *Soil Sci.*, 39:145, 1935.

[20] L. C. Boawn, F. Viets, Jr., and C. L. Crawford, "Plant Utilization of Zinc from Various Types of Zinc Compounds and Fertilizer Materials," *Soil Sci.*, 83:219, 1957.

[21] J. J. Mortvedt and P. M. Giordano, "Crop Response to Zinc Oxide Applied in Liquid and Granular Fertilizers," *J. Agr. Food Chem.*, 15:118, 1967.

[22] E. A. Salako, L. S. Murphy, P. J. Gallagher, and R. Ellis, Jr., "Research Shows Soybeans Need Zinc," *Fertilizer Solutions*, 19(6):96, 1975.

Table 6-3. Some Manufactured Sources of Zinc

Source	Formula	Zinc (%)
Zinc sulfate monohydrate	$ZnSO_4 \cdot H_2O$	36
Zinc oxide	ZnO	78–80
Zinc carbonate	$ZnCO_3$	52
Zinc ammonia complex	$Zn\text{-}NH_3$ [a]	10
Zinc chelate	$Na_2ZnEDTA$	9–14
Zinc ligninsulfonate	—	5–12
Zinc polyflavonoid	—	7–10

[a] Zinc and ammonia form a Werner-type complex $Zn(NH_3)_4^{2+}$.

On the other hand, addition of zinc oxides to fluids, both clear liquids and suspensions, has been attractive due to the generally lower price of zinc oxide per unit of applied zinc.

Zinc-ammonia complex is one of the newest additions to the ranks of zinc fertilizers. This material usually contains about 10% zinc and 10% nitrogen. It is essentially manufactured from zinc sulfate and aqua ammonia. Much of the zinc in this material is present as a Werner complex, comprised of one atom of zinc and four molecules of ammonia with a net positive charge of +2. In any event, this liquid source of zinc has proved to be effective, producing about the same type of yield response as zinc sulfate.

Recently, another inorganic zinc source, zinc nitrate, has proved effective in foliar applications for several tree crops, particularly pecans. Inclusion of urea-ammonium nitrate in the solution formulations significantly increased zinc absorption.[23]

Organic Zinc Sources. The most common organic sources of zinc available for use as fertilizers include zinc ethylenediaminetetraacetic acid (ZnEDTA), zinc ligninsulfonate, and zinc polyflavonoids. The latter two are wood by-products in paper production and are called *natural organic complexes.* Numerous investigations of the relative effectiveness of inorganic versus organic (*chelated*) sources of zinc have been carried out with a large number of crops. Generally, zinc chelates are somewhat more available than the inorganic sources, with ligninsulfonates and polyflavonoids intermediate. The degree of increased efficiency is still open to question. Some results suggest an efficiency ratio as wide as 10:1; that is, 1 kg of zinc supplied by a chelate such as ZnEDTA is as effective as 10 kg of zinc supplied as zinc sulfate. Other studies, show only about a 2 or 3:1 efficiency ratio favoring ZnEDTA. Still other studies have shown essentially no difference between inorganic and organic zinc sources.

Greater effectiveness of chelated sources of zinc at high soil pH levels is due to the high stability of some chelates in alkaline soils. In the case of the

[23] J. B. Storey, M. Smith, and P. Westfall, "Zinc Nitrate Opens New Frontiers of Rosette Control," *Pecan Quarterly,* 8(1):9, 1974.

ligninsulfonates and polyflavonoids, unless the zinc is truly chelated by the organic molecule, zinc ions may be quickly lost to the soil solution through exchanges for iron or calcium.

Attempts to arrive at an efficiency factor between organic and inorganic zinc sources are often misleading because soil pH, soil texture, and methods of application affect the comparison. Chelates may be more effective when used on very high pH, calcareous soil. On the other hand, on very acid soils, chelates may exchange with iron so efficiency of chelated zinc is no different than of inorganic zinc. Band placement favors the availability of chelated zinc sources because of generally lower rates of application and necessity of placing the nutrients near the seed. Broadcast applications may favor inorganic sources because of more widespread distribution of a larger amount of material. The high solubility of chelated zinc applied to a very coarse-textured soil may result in lower effectiveness due to leaching of the soluble zinc source.

In conclusion, the efficiency ratio between organic and inorganic zinc sources for plants is hardly constant. Although some 10:1 ratios may exist favoring the chelates, more workers suggest that the efficiency ratio is about 2 or 3:1. Boawn[24] demonstrated this fact quite clearly in his studies with corn and red beans.

Zinc in Macronutrient Fertilizers. Inclusion of zinc in macronutrient fertilizers has become a common practice in recent years. Such inclusions eliminate the expense of a separate application for the micronutrients and reduce the problems of segregation of micronutrients in bulk blends. Granulation of zinc into some macronutrient fertilizers such as solid ammonium polyphosphate (APP), triple superphosphate (TSP), and urea has produced good efficiency of the applied zinc.[25] However, incorporation of zinc sulfate or zinc oxide into diammonium phosphate (DAP), urea-ammonium phosphate (UAP), and nitric phosphates (NP) produced lower availability. In the same study, Mortvedt and Giordano determined that ZnEDTA was not detrimentally affected when granulated into macronutrient fertilizers.

Field studies of the effectiveness of zinc granulated in macronutrient fertilizers, particularly APP and monoammonium phosphate (MAP), have given mixed results. Michigan studies with navy beans conducted by Judy et al.[6] indicated that zinc oxide was more effective when incorporated into granular APP than when applied in MAP. APP also was more effective than MAP as a phosphorus source even without applied zinc. On the other hand, studies in Nebraska by Penas[26] employed both zinc sulfate and zinc oxide granulated

[24] L. C. Boawn, "Comparison of Zinc Sulfate and Zinc EDTA as Zinc Fertilizer Sources," *Soil Sci. Soc. Amer. Proc.,* 37:111, 1973.

[25] J. J. Mortvedt and P. M. Giordano, "Availability to Corn of Zinc Applied with Various Macronutrient Fertilizers," *Soil Sci.,* 108:180, 1969.

[26] E. J. Penas, "Comparative Evaluation of Ammonium Orthophosphate and Ammonium Polyphosphate as Carriers of Phosphorus and Zinc for Corn," M.S. Thesis, Dept. of Agronomy, University of Nebraska, 1967.

Table 6–4. Reduced Effectiveness Can Result from Coating Zinc Sources on Diammonium Phosphate

Treatment (ppm)		Carrier[a]		Method of Zn Application	Grain Yield (kg/ha)	Leaf Concentration 8-Leaf Stage	
P	Zn	P	Zn			P (%)	Zn (ppm)
26	0	AOP DAP	—	—	2,634	0.26	22.6
26	0.5	AOP DAP	EDTA	Coated	4,453	0.16	35.1
26	1.0	AOP DAP	EDTA	Coated	5,018	0.16	36.3
26	2.0	AOP DAP	ZnSO₄	Coated	4,265	0.19	20.7
26	4.0	AOP DAP	ZnSO₄	Coated	4,140	0.18	17.7
26	2.0	AOP DAP	ZnO	Coated	2,132	0.19	19.4
26	4.0	AOP DAP	ZnO	Coated	3,450	0.23	22.1
26	2.0	AOP DAP	Zn-NH₃	Coated	3,450	0.24	18.0
26	4.0	AOP DAP	Zn-NH₃	Coated	4,579	0.18	21.8
26	4.0	AOP DAP	ZnSO₄	Separate	10,411	0.35	—

[a] L. S. Murphy.

NOTE: Note the difference in yield and phosphorus concentration of irrigated corn where P and Zn were applied separately. Soil pH was 7.4. Geary County, Kansas. 1971.

into both ammonium orthosphosphates (AOP) and APP. Both fertilizers were comparable as phosphorus sources and as carriers of zinc. Instability of the zinc-polyphosphate complex apparently tends to equalize the availability of APP-zinc and AOP-zinc mixtures in the soil (see Chapter 3).

Relatively more problems have been encountered in coating of zinc on the surface of phosphatic fertilizers than when the zinc is granulated with the macronutrient source. Problems in zinc coating of macronutrient materials are not due to physical distribution problems of zinc within the mixture but rather are due to formation of reaction products on the particles' surface. No problems in the literature have been noted with the coating of zinc on nitrogen sources such as ammonium nitrate and ammonium sulfate. Zinc oxide coated on urea has not performed well in some cases, usually on high pH soils.

X-ray diffraction analyses of the phosphorus-zinc reaction products have indicated the formation of compounds such as $Zn_3(NH_4)_2(P_2O_7)_2 \cdot 2H_2O$ and $Zn_3(PO_4)_2 \cdot 4H_2O$. Relatively small effective surface area of large particles causes slow dissolution of these zinc compounds, and dissolution is hampered even further by high soil pH. Allen and Terman[27] reported relatively poor performance of these zinc compounds under greenhouse conditions and noted the detrimental effects of large particle size. Data in Table 6–4 demonstrate the effect of coating diammonium phosphate (DAP) with four zinc sources and using the coated material to correct zinc deficiency in corn on a high pH (7.4) soil. Yields were extremely low where the coated phosphorus material was

[27] S. E. Allen and G. L. Terman, "Response of Maize and Sudangrass to Zinc in Granular Micronutrients," *Intl. Soc. Soil Sci.*, Trans. Comm. II and IV, p. 255, Aberdeen, Scotland, 1966.

used. At the eight-leaf stage of sampling, both phosphorus and zinc concentrations in leaf tissue were low although the phosphorus concentrations were relatively lower than the zinc concentrations. Note, however, the differences in yield that occurred when these treatments were compared to an adjacent study (3 m away) where the same amount of phosphorus as DAP and one of the same zinc sources, $ZnSO_4$, were applied separately. Obviously, coating under those conditions was not effective. However, on an acid soil (pH 6.8), no problems were encountered in either phosphorus or zinc uptake the same year. Greenhouse studies conducted later with these same coated materials and others that had been recently formulated indicated that zinc availability from coatings on phosphorus fertilizers is not a function of time of coating.

Miscellaneous Sources. Zinc can also be present in significant quantities in some waste materials. Zinc concentrations in animal manures or sewage residues are extremely variable, and analyses should be performed to determine if an adequate or *excessive* amount of zinc will be applied in a planned treatment. Applications of animal wastes, particularly hog and cattle manures, generally provide significant quantities of zinc. Further beneficial effects on zinc availability may be obtained from the organic matter in manure and sewage sludge (see Chapter 10).

6.1.5 Methods of Application

Four basic methods of zinc application are: (1) broadcast and incorporated prior to planting, (2) banded with or near the seed at planting, (3) seed coating, and (4) foliar application. *Broadcast applications* of zinc are effective for most row crops but the relative efficiency of broadcast versus the other methods of application depends on several factors including soil conditions, crop, climatic factors (soil moisture and temperature), and rate of zinc application. Because zinc moves only slightly in the soil, broadcast application must be incorporated into the soil to obtain maximum efficiency.

Broadcast applications of zinc have been effective in Kansas (Table 6-5), Texas, Nebraska, Arkansas, Louisiana, and California investigations to name a few. Studies have generally shown that small amounts of zinc should be placed close to the seed and not broadcast because there is less chance of the plant roots coming into contact with the low rate of broadcast material early in the growing season.

Banded applications of zinc are quite effective and in fact are favored as a means of application particularly in the northern corn belt states in the United States. Michigan recommendations for navy beans and other vegetables call for banded applications. There is little doubt that zinc placed near the seed is more effective under some conditions such as cold, wet soils at the time of planting. On the other hand, the differences between banded and broadcast applications for row crops further south are only slight with moderate to high rates of zinc (4–8 kg/ha Zn). With rates of 1 kg/ha or less of zinc, banded applications are the most effective.

Table 6-5. Effect of Rates and Methods of Application of Zinc and Phosphorus on the Yield and Zinc Content of Irrigated Corn — Pottawatomie County, Kansas

		Method of P-Zn Application			
		Broadcast Preplant		Banded at Seeding	
P	Zn	Yield	Zn Content	Yield	Zn Content
(kg/ha)		(kg/ha)	Leaves (ppm)	(kg/ha)	Leaves (ppm)
0	0	8,222	20		
0	22	7,633	45	6,843	38
39	0	7,815	12	7,476	8
39	5.5	10,688	18	9,778	17
39	11	11,340	24	11,384	24
39	22	8,938	28	10,951	30
LSD$_{(.05)}$		2,722	8	2,722	8

SOURCE: R. Ellis, Jr., "Effect of Rates and Methods of Application of Zinc and Phosphorus on the Yield and Zinc Content of Irrigated Corn," *Kansas Fertilizer Handbook* — 1967, p. 74.

Seed coating with zinc compounds is a relatively new process that presently is relegated almost solely to rice. Studies in Arkansas[28] and California have shown significant zinc responses from zinc coated on the rice seed. The close proximity of zinc to the rice seed is mandatory on zinc-deficient soils. More soluble zinc sources even when placed in the same soil zone with the seed tend to diffuse out of the seed zone when rice is seeded into flooded fields by aerial application. Some preliminary seed-coating work has been carried out with corn, beans (*Phaseolus vulgaris*), and soybeans without definitive results.

Foliar applications of zinc are quite effective and have been commonly used on vegetables, citrus, and deciduous fruits and nuts. Frequently, soil applications of zinc for tree crops produce a slow recovery from the deficiency because of the fact that it is difficult for the zinc to come into contact with the plant roots. Usually, foliar applications are considered emergency treatments to be applied when zinc deficiency is detected in a growing crop. Effectiveness of foliar applications in increasing the zinc concentrations of pecan leaves is shown in Table 6-6.

Foliar applications should be applied as soon as deficiencies are detected. Delay in application causes increasing reductions in yields, particularly in annual crops. Because rates of application are much lower with foliar treatments, repeated sprays may be necessary, which increases the costs of foliar applications. In conclusion, foliar applications of zinc are no substitute

[28] L. Thompson and N. R. Kasireddy, "Zinc Fertilization of Rice by Seed Coating," *Rice. J.*, July–August, 1975, p. 28.

Table 6–6. Effects of Foliar Applications of Zinc on the Zinc Concentration in Pecan Leaves [Note the effect of the inclusion of nitrogen solution in the spray on zinc content of the leaves (Texas)]

Zn Source	Composition of Solution (grams of Zn/100 l)	UAN[a]	Leaf Zn (ppm)
Control	—	No	28
$ZnSO_4$	89	No	454
$ZnSO_4$	89	Yes	734
$Zn(NO_3)_2$	45	No	275
$Zn(NO_3)_2$	45	Yes	976
$Zn(NO_3)_2$	90	No	976
$Zn(NO_3)_2$	90	Yes	1062

[a] UAN—Urea-ammonium nitrate solution included at 0.5% by weight. Zinc solutions applied in five applications: bud break, plus four additional weekly applications.

SOURCE: J. B. Storey, "Zinc Nitrate Opens New Frontiers of Rosette Control," *Pecan Quarterly*, 8(1):9, 1974. Used with permission.

for a good soil fertility program that includes applying zinc *when and where deficient,* as indicated by a soil test. Foliar treatments should be viewed strictly as supplemental.

Injection of zinc into the trunks of zinc-deficient pecan trees has been demonstrated to be effective in Georgia. Worley et al.[29] reported that replicated trials with zinc solution injection (involving application rates of 1 g of zinc per 2.5 cm of trunk circumference) increased leaf zinc from 38 to 230 ppm. Injections made in April were still effective in August. Soil zinc applications prior to the injections (1 kg Zn/tree) were ineffective until the following year.

Rates of Application. It is impossible to cover all recommended rates of zinc application for all crops here but information in Table 6–7 indicates broadcast, banded, and foliar rates of application for some crops. Actual determination of needs should be based on soil and/or plant analysis, correlated with the results of field research. Indiscriminate use of any nutrient can be both costly and occasionally detrimental; "shotgun" applications are never recommended.

Residual Effects. Zinc applications of sufficient magnitude over a number of years eventually produces an accumulation of zinc in the soil. Availability of residual zinc is good and continued soil testing can take the amount of

[29] R. E. Worley, R. H. Littrell, and S. G. Polles, "Pressure Trunk Injection Promising for Pecan and Other Trees," *Hortscience*, 11(6):590, 1976.

Table 6–7. Some Recommended Rates of Zinc Application for Various Crops

Crop	Zinc (kg/ha)	Source	Method of Application	Comments
Corn	5–11 1–2	$ZnSO_4$, ZnO, Zn-NH$_3$ Zn chelate	Broadcast or banded Banded	Use lower rates for higher soil test values
Sorghum	4–9 1–2	$ZnSO_4$, ZnO, Zn-NH$_3$ Zn chelate	Broadcast or banded Banded	Use lower rates for higher soil test values
Soybean	2–4 1–2	$ZnSO_4$, ZnO, Zn-NH$_3$ Zn chelate	Broadcast or banded Banded	Use lower rates for higher soil test values
Rice	8–11 1	$ZnSO_4$, ZnO, Zn-NH$_3$ Zn chelate	Broadcast preplant Banded	Use lower rates for higher soil test values
Dry beans	4–5 0.5–4	$ZnSO_4$, ZnO, Zn-NH$_3$ Zn chelate	Broadcast or banded Banded	Use lower rates for higher soil test values
Citrus	0.6 kg Zn/ 100 l H$_2$O	$ZnSO_4$	Foliar	Wet foliage, repeat until symptoms disappear
Pecans	30–60 g Zn/100 l H$_2$O; 1000 l/ha of trees	Zn(NO$_3$)$_2$	Foliar	Five applications start- ing at bud break, then repeated weekly
Snap beans, onions, lima beans, potatoes	0.7–1.3 0.3–0.7 0.2	Zn chelate Zn chelate Zn chelate	Broadcast Banded Foliar	Repeat foliar until symptoms disappear or leaf analysis con- firms adequate Zn

residual zinc into account in future recommendations. Short-term residual effects following very small rates of soil application may be small due to reactions of the applied zinc with the soil and subsequent poor extraction by plant roots. Applications in the range of 1 kg/ha or less per year of zinc exert minimal residual effects. On the other hand, applications in the range of 2 to 20 kg of zinc show detectable effects for varying lengths of time. Length of residual zinc effects are related to rate of application, soil type, and crop.

Boawn[30] showed that residual effects from rates of application as low as 1.5 ppm of zinc were still evident by soil extraction 7 years after the initial application. After the third year, extractable zinc for each application rate apparently approached a static level greater than that of the original soil. This is good evidence for the wisdom of repeated soil testing.

6.1.6 Crop Responses

Corn. Irrigated corn often responds to applications of zinc. This type of response is frequently noted in the Plains States and far west in the United States. Such dramatic responses *do not* occur every year even on the same site

[30] L. C. Boawn, "Sequel to Residual Availability of Fertilizer Zinc," *Soil Sci. Soc. Amer. Proc.*, 40:467, 1976.

Table 6–8. Grain Sorghum Response to Applied Zinc — Sedgwick County, Kansas

Applied Zn (kg/ha)	Grain Yield (kg/ha)	Leaf Zn (ppm)
0	4077	18
0.6	4767	20
1.1	5018	20
2.2	5268	24
4.5	5331	26
9.0	4704	25

NOTE: All Zn was applied pre-plant, broadcast and incorporated by disking. Leaf samples collected at 8-leaf stage, youngest fully emerged leaf.

(see Fig. 6–5). Climatic differences and continued cropping can turn a spectacular difference one year into a very mediocre one the next.

Grain Sorghum. Sorghum responses to zinc applications are less spectacular than those of corn, because this crop apparently is better able to exploit the soil's available zinc, but responses can be significant. (Table 6–8). Important effects of zinc applications on sorghum maturity are noted in Fig. 6–8.

Soybeans. Although the zinc requirement of soybeans has been recognized for some time, few studies have been conducted in the field. The magnitude of zinc response in soybeans under very deficient conditions is shown in Table 6–9.

Rice. Rice responses to zinc both in the United States and in Southeast Asia are frequent and significant. Data from the Arkansas Rice Experiment Station in Table 6–10 indicate both the magnitude and value of the response. Like other crops, this response was typical of a given location in a given year. Other conditions may have produced crop responses of different magnitudes.

Table 6–9. Soybean Responses to Zinc under Irrigated Conditions (Kansas)

Zinc (kg/ha)	Yield (kg/ha)	Yield Increase (kg/ha)	Value of Increase per Hectare	Cost of Zn per Hectare	Return from Zn Application	Return per Dollar Invested in Zn
0	2016	—				
2.2	3091	1075	$255.30	$ 7.90	$247.40	$31.32
4.5	3360	1344	319.12	15.81	303.32	19.18
9	2755	739	175.52	31.62	143.90	4.55

NOTE: Zinc cost calculated at $3.52/kg. Soybeans valued at $236.87/metric ton or $0.236/kg.

Table 6–10. Rice Responses and Returns to Zinc Fertilization (Arkansas)

Zn Treatment	Yield (kg/ha)	Yield Increase (kg/ha)	Value of Increase	Cost of Zn	Return to Zn ($/ha)	Return/ Dollars Invested
Control	3584	—	—	—	—	—
ZnEDTA 0.84 Zn kg/ha	5633	2050	$400.78	$27.22	$373.06	$14.45
ZnEDTA 1.68 Zn kg/ha	6675	3091	539.58	55.44	484.14	9.73
ZnSO₄ 5.6 Zn kg/ha	6059	2475	483.86	16.50	467.36	29.32
ZnSO₄ 11.2 Zn kg/ha	6563	2979	582.39	33.00	549.39	17.64

NOTE: ZnEDTA, $33/kg Zn; ZnSO₄, $3.30/kg Zn; Rice, $195.50/metric ton (1000 kg).

SOURCE: B. R. Wells, Lyell Thompson, G. A. Place, and P. A. Shockley, "Effect of Zinc on Chlorosis and Yield of Rice Grown on Alkaline Soil," Arkansas Agr. Exp. Sta. Report 208, 1973.

Pecans. Zinc deficiencies in pecans and citrus lead to serious reductions in yield. Data in Table 6–11 from the Georgia Experiment Station demonstrate zinc effects on pecan nut yields.

6.1.7 Conversion Factors

To work effectively with zinc fertilizers, they must be considered on the basis of their zinc content rather than on the basis of the percentage of some zinc compound present. For instance, if a by-product material such as a flue dust containing 20% ZnO is to be compared to commercial grade ZnSO₄ that contains 36% zinc, then the percentage of zinc in the material must be known to make a conversion to the common method of expression, percentage

Table 6–11. Five-Year Average Yield Response of Stuart Pecan Trees to Zinc Sulfate on Limed Soil (Georgia)

ZnSO₄ (kg/tree)	Zn (kg/tree)	Total Cost of Lime, N, and Zn (per tree/yr)	5-yr Average Yield (kg/tree/yr)	5-yr Average Gross Returns (per tree/yr)
0	0	$2.57	17	$12.21
1.4	0.5	2.83	42	29.57
2.7	1.0	3.08	51	36.00
4.5 (every 5 yr)	1.6	2.75	46	32.79
4.5 (repeated when rosette developed)	1.6	2.75	54	38.25

NOTE: Average price of pecans during study: $0.71/kg.

SOURCE: O. L. Brooks. "Yield and Growth Response of Stuart Pecan Trees to Zinc Sulfate and Nitrogen," Georgia Agr. Exp. Sta. Cir. N.S. 40, 1964.

of zinc (Zn). The following example shows how this conversion can be carried out:

$$\text{Zn concentration in ZnO} = \frac{\text{atomic wt. Zn}}{\text{molecular wt. ZnO}} = \frac{65.37}{81.37} = 80.34\% \text{ Zn}$$

Original material is 20% ZnO; therefore the zinc content of the flue dust is 20% × 80.34% Zn = 16.07% Zn. If the study required 10 kg/ha (8.9 lb/A) of zinc, the amount of flue dust required to provide that amount is

$$\frac{10 \text{ kg ha/Zn}}{16.07\% \text{ Zn}} = 62.2 \text{ kg/ha (55.6 lb/A)}$$

The amount of commercial $ZnSO_4$ to be applied per hectare (acre) for the same rate of Zn would be

$$\frac{10 \text{ kg ha/Zn}}{36\% \text{ Zn}} = 27.8 \text{ kg/ha (24.8 lb/A)}$$

By law, zinc fertilizers sold in interstate or intrastate commerce must report the percentage of zinc they contain. Similar conversions can be made between any two materials by reducing them to the common denominator, percentage of zinc (Zn).

6.2 IRON

Iron is the most abundant element on earth and the fourth most abundant element in the crustal rocks. The element can exist in the metallic form (Fe) or in two oxidized forms, ferrous (Fe^{2+}) and ferric (Fe^{3+}). Iron oxides, silicates, sulfides, and carbonates exist in the earth's crust. The occurrence of these minerals is determined by environmental conditions. Ferric oxide is the most abundant form of iron in the surface environment. More reduced forms of iron, both Fe and Fe^{2+}, are rapidly oxidized to the ferric form with the formation of thermodynamically stable ferric oxide (Fe_2O_3). In sedimentary rocks, a common occurrence of iron is in the form of hydrous silicates such as glauconite. The principal form of iron in oxidized parts of soils is ferric oxide. In soil horizons containing high organic matter, reduced forms of iron (Fe^{2+}) may exist. Wetness and other oxygen stress conditions are also conducive to the presence of ferrous iron.

Problems with iron nutrition of plants have been reported in many areas of the world with many crops. Estimates have suggested that as much as one-third of the world's land surface is calcareous and potentially associated with iron nutrition deficiency of crops. The following sections address the soil chemistry, functions in plants, deficiency symptoms of iron and correction techniques for iron deficiencies.

6.2.1 Soil Chemistry

Probably the most important aspect of iron chemistry in soils is not the amount of the metal that is present (10,000 to 100,000 ppm of iron) but the solubility and availability to plants of that which is present. The solubility of iron in soils is largely controlled by the hydrous oxides of ferric iron. This form of iron, Fe^{3+}, is the common product of aerobic soil reactions. In general, the solubility of such compounds as ferric hydroxide [$Fe(OH)_3$] is very low and is only slightly affected by the addition of soluble iron compounds to the soil.

The solubility of ferric iron in soil solution varies with pH and declines rapidly with increasing pH, thus resulting in iron nutritional problems in high pH soils. Minimum values for the solubility of ferric iron are reached over the broad pH range of 6.5 to 8.2. As the soil pH increases from a value of about 4, each unit increase in pH results in a thousandfold decline in activity of the ferric ion in soil solution.

Iron in soil solution can exist in a number of forms, depending on the pH of the system. Above pH 8, ferric iron is associated with hydroxyl ions to form the $Fe(OH)_4^-$ species. At lower pH values, other species that may exist include $Fe(OH)_2^+$, $FeOH^{2+}$, Fe^{3+}, and $Fe_2(OH)_2^{4+}$.

In addition to different equilibrium products of ferric iron with hydroxyl ions, an equilibrium also exists between ferric and ferrous iron. Ferrous iron minerals and compounds are relatively unstable in soils in comparison to the compounds formed by ferric iron. The ferrous silicates, carbonates, and hydroxide minerals are too soluble to persist in well-aerated soil for any length of time and subsequently do not represent good forms of supplemental soil iron for plants due to the rapid oxidation of the available ferrous iron to relatively unavailable ferric forms.

The equilibrium between ferrous (Fe^{2+}) and ferric ions (Fe^{3+}) also involves relatively insoluble ferric hydroxide and is affected both by soil solution pH and the partial pressure (degree of aeration) of oxygen in the soil atmosphere. That equilibrium can be represented by the following equation:

$$Fe(OH)_3 + 2\ H^+ \rightleftharpoons Fe^{2+} + \tfrac{1}{4}O_2 + 2\tfrac{1}{2}H_2O$$
$$\text{Gas}$$

Whenever the amount of oxygen in the soil atmosphere declines, the solubility of iron in the soil *tends* to increase because conditions favor the reduction of ferric iron to the ferrous form that is slightly more soluble. Practically speaking, however, the changes in solubility are so slight that this has no appreciable effect on iron availability. Soil pH has a relatively greater effect on solubility of ferrous compounds. As the pH decreases one unit, the solubility of ferrous iron or, more correctly, the activity of ferrous iron in the soil solution increases a hundredfold. At a pH of 7.8 and a partial pressure of oxygen of 0.2 atmosphere, the activities of Fe^{2+} and Fe^{3+} in the soil solution are equal. The concentration under those conditions is 1×10^{-21} Molar. By comparison, the solubility of phosphorus (see Chapter 3) that is considered to be relatively low might be as high as 0.1 ppm. Expressed as the $H_2PO_4^-$ ion, that concentration would be 2.1 ppm or about 3.3×10^{-3} Molar, which is roughly 3 billion billion times more concentrated than iron in the soil solution.

Under some very reduced conditions, such as waterlogged soils, it is possible for the availability of iron to be greatly increased by reduction of significant amounts of ferric iron to the ferrous form. Ponnamperuma and co-workers[31] reported that in highly reduced soils, concentrations of iron in soil solution both as ferrous and ferric may reach levels as high as 1×10^{-3} Molar.

The reactions of iron with phosphorus in soil were discussed in Chapter 3. Iron can greatly reduce the availability of phosphorus due to the insoluble nature of iron phosphates of various types. Despite the effect on phosphorus availability, this reaction has essentially no effect on availability of iron because total iron in soil is usually much greater than total phosphorus. This should not be taken to suggest that there is no iron-phosphorus interaction in plant nutrition. However, evidence suggests that the interaction is physiological in nature and not due to a depressed solubility or availability of iron in soil.

Iron soil chemistry is significantly modified by the presence of natural or synthetic chelates. The ability of soil organic matter to form stable combinations with various metal ions including iron has been well established. Stevenson and Ardakani[32] discussed this subject in detail. Organic matter reactions with metal ions can prevent the normal conversion of these metal ions to insoluble complexes such as iron hydroxides. Therefore, reactions with organic matter or incorporation of a metal such as iron into a synthesized chelate prior to soil application may dramatically improve availability of the metal to plants. Compounds that may react with metals in soil organic matter include organic acids, polyphenols, amino acids, proteins, polysaccharides, humic acids, and fulvic acids. The humic acid fraction of organic matter tends to form insoluble combinations with metals such as iron, whereas metals found in soluble organic complexes are mainly associated with compounds such as the relatively simple organic acids. Citric acid is a good example. Fulvic acid complexes with metals also tend to have a relatively high water solubility.

Metals such as iron apparently are chelated by organic compounds in the soil through reactions with carboxylic, phenolic, alcoholic, amino, and aldehyde functional groups. Generally speaking, *chelates* are organic compounds that contain a metal ion bound with varying degrees of strength but in such a fashion as to be protected from reactions with other inorganic soil components. The term *chelate* is derived from the Greek word *chele* meaning *claw*. The metal ion is considered to be contained within a ring structure or

[31] F. N. Ponnamperuma, E. M. Tianco, and T. A. Loy, "Redox Equilibria in Flooded Soils. I. The Iron Hydroxide Systems," *Soil Sci.*, 103:374, 1967.

[32] F. J. Stevenson and M. S. Ardakani, "Organic Matter Reactions Involving Micronutrients in Soils." In *Micronutrients in Agriculture*, J. J. Mortvedt (Ed.), Soil Science Society of America, Madison, Wis., 1972, p. 79.

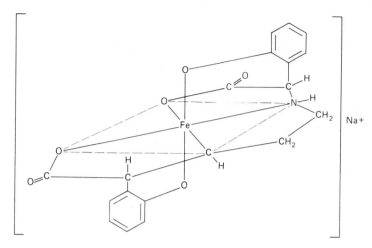

Fig. 6-10. Structure of an iron chelate, FeEDDHA. Note the position of iron in the center of the structure.

claw of the organic molecule (Fig. 6-10). In such a form the metal has a relatively high water solubility and is available to plants but its activity or reactivity with inorganic components has been significantly diminished.

In order for a *ligand* (the organic portion of a chelate) to modify significantly the availability or mobility of a metal such as iron in the soil, the chelate formed from the ligand and the metal must have sufficient stability to exist in equilibrium with the concentrations of the ligand and other metals in the soil solution. The stability of iron chelates in soils has always been of great interest in plant nutrition because of the need to supply iron to many crops subject to iron deficiency. Several manufactured chelating agents or ligands have a relatively high specificity for iron (Fe^{3+}) and generally form chelates of high stability.

The stability with a given metal can be modified by soil pH and the presence of other metals such as calcium (Ca^{2+}). For instance, ethylenediaminetetraacetic acid (EDTA) chelates iron (Fe^{3+}) almost exclusively below pH 6.3 and would be a good iron source for plants below that soil pH. However, as the pH rises, this ligand tends to react with calcium, displacing the iron and subsequently becomes a very poor source of iron at soil pH levels where iron deficiencies normally exist. On the other hand, the chelating agent EDDHA [ethylenediamine di(o-hydroxy-phenylacetic acid)] is totally selective for iron over a pH range of 4 to 9 and subsequently is a much better chelated source of iron for soil application under calcareous conditions. Another chelating agent, DTPA (diethylenetriaminepentaacetic acid), selectively chelates with iron up to a pH between 7 and 7.5. Above that pH, the calcium chelate becomes dominant and FeDTPA would represent a potentially less effective means of adding iron to the soil.

6.2.2 Iron Nutrition of Plants

Metabolic Roles of Iron

Iron has long been recognized as essential for the maintenance of chlorophyll in plants. The primary deficiency symptom of this element is the disappearance or lack of chlorophyll in leaves. As early as 1844, observations of plant nutritional deficiencies linked iron with chlorophyll. Recent research has established a rather specific role of iron in the RNA metabolism of chloroplasts, the chlorophyll-containing bodies of the leaves. Iron-deficient chloroplasts contain less than half as much RNA and chloroplast ribosomes as are contained in the same structures from iron-sufficient cells. Iron is not a part of the chlorophyll molecule, but through its role in the chloroplast it does influence chlorophyll synthesis.

Iron is an essential constituent of hemoglobin and of cytochromes and other components of respiratory enzyme systems (cytochrome oxidase, catalase, peroxidase). Although hemoglobin is normally considered of prime importance in animals rather than plants, this compound is present in the nodules of legumes and is significant in the process of symbiotic nitrogen fixation.

The roles of iron in respiratory reactions in plants are similar to those found in the animal system. Various heme-iron compounds, the cytochromes, are important electron carriers in the respiration system (metabolism of sugars). The cytochromes act as catalysts in the major pathway of biological oxidations. Cytochromes are iron-porphyrin proteins of several different types, most of which are attached to mitochondria of the cell. One member of the cytochromes, cytochrome oxidase, is the terminal catalyst in the sense that it affects the direct union of oxygen with electrons initially derived from metabolic substrates such as sugars. Cytochrome oxidase and the other members of the cytochrome group undergo repetitive oxidation and reduction of the iron contained in their structure as an integral feature of their function.

Iron also functions in plants in certain proteins known as *ferredoxins*. These non-heme (to distinguish them from proteins such as the cytochromes) iron-sulfur proteins have small molecular weights and are involved in key metabolic processes such as nitrogen (N_2) fixation, photosynthesis, and electron transfer in chloroplasts of green plants, and they are directly linked to the formation of reduced coenzymes that are subsequently involved in energy-requiring reactions elsewhere in the plant. One such reaction is the reduction of nitrate-nitrogen discussed in detail in Chapter 2.

Interactions with Other Elements. Phosphorus is frequently involved with iron nutrition of plants. Most researchers believe that this interaction, like that of zinc and phosphorus, is not related to some precipitation process in the soil that limits the availability of iron. Instead, the uptake or utilization of iron may be depressed by the presence of large amounts of phosphorus in the growth medium. As an example, Watanabe et al.[33] reported stunted, iron-deficient corn plants when the phosphorus level in nutrient solution was increased from 0.2 to 0.6 millimoles (mM) and iron was held constant at 4 ppm. When more iron was supplied, the phosphorus effect was overcome.

Plants showing an iron chlorosis resulting from the use of large amounts of phosphorus in the growth medium may show relatively normal iron concentrations in the tissue but quite high concentrations of phosphorus. Other investigations have indicated lower iron concentrations when excessive phosphorus was supplied in excessive quantities. Still other experiments have produced data showing very high concentrations of iron and phosphorus in or on the surface of roots relative to the concentrations of both elements in the leaves. These latter data were interpreted as demonstrating an accumulation or precipitation of iron phosphates on the root surfaces. That theory has been supported by investigations that verified an iron phosphate coating on root surfaces through the use of scanning electron microscope techniques and X-ray fluorescence.

Some investigators have suggested that the phosphorus effect on iron absorption/utilization may be due in part to competition between phosphorus and organic ligands within the plant that function to maintain the iron in a soluble, mobile form. Iron associated with phosphorus is evidently less mobile. Electrophoretic techniques have indicated that iron translocates in plants chiefly as iron citrate.[34]

Metabolic functions of iron in plants also are affected by the supply of zinc. Watanabe et al.[33] reported that corn grown in nutrient solutions experienced a depression in growth as zinc was increased from 0.75 to 2.25 micromoles (μM) when iron was held constant at 2 ppm. Iron deficiency symptoms existed at all levels of phosphorus in that experiment and disappeared only when the iron concentration was doubled to 4 ppm. However, concentration of iron in the plants and iron uptake were not depressed by zinc. On the contrary, Rosell and Ulrich[35] reported that increasing the zinc supply to sugar beets from essentially 0 to 12 ppm in the nutrient solution resulted in a reduction in iron concentration in beet leaves from 900 to 90 ppm. Brown and Jones[36] studied nutrient interactions in grain sorghum and noted that as zinc was supplied to correct a soil deficiency, iron concentrations in the plant tissue tended to decline significantly. Iron applications to the soil, however, had no significant effect on zinc concentrations in the plants.

There is no doubt that nutritional interaction between iron and zinc does exist in plants, but the explanation for this interaction is not completely

[33] F. S. Watanabe, W. L. Lindsay, and S. R. Olsen, "Nutrient Balance Involving Phosphorus, Iron and Zinc," *Soil Sci. Soc. Amer. Proc.,* 25:562, 1965.

[34] L. O. Tiffin, "Translocation of Iron Citrate and Phosphorus in Xylem Exudate of Soybean," *Plant Physiol.,* 45:280, 1970.

[35] R. A. Rosell and A. Ulrich, "Critical Zinc Concentrations and Leaf Minerals of Sugar Beet Plants," *Soil Sci.,* 97:152, 1964.

[36] J. C. Brown and W. E. Jones, "Fitting Plants Nutritionally to Soils. III. Sorghum," *Agron. J.,* 69:410, 1977.

understood. However, a mere reduction in iron concentrations in a plant by zinc applications may not necessarily be detrimental, particularly if zinc applications overcome a deficiency and induce more plant growth. Iron concentrations may thereby be reduced by a dilution effect. Plant concentrations of iron also are not as well correlated to deficiency of iron as concentrations of zinc are to zinc deficiency symptoms. Plants that show severe iron deficiency symptoms may contain much higher concentrations of iron than plants with no deficiency symptoms.

Iron-manganese interactions have been reported. Chlorotic plants growing on acid soils with high manganese levels have been made greener by foliar applications of iron. High concentrations of manganese in the growth medium apparently interfere with iron absorption, translocation, and utilization. Some have suggested that the similarity of the metabolic roles of iron and manganese may be related to their interaction but others discount this possibility.

Applications of chelated forms of iron to acid muck (organic) soils (Histosols) may produce an even more severe manganese deficiency when the soil was already manganese-deficient. Under acid conditions, chelates such as EDTA form iron complexes that are quite stable. These soluble iron complexes increase iron uptake by the plants with the resulting exclusion of manganese. Both MnEDTA and FeEDTA produce more severe manganese deficiencies under such conditions because soil iron quickly replaced the manganese in the MnEDTA, again increasing iron uptake and depressing absorption of manganese.

Copper-iron interactions have been noted in citrus and other plant species. Continued applications of copper beyond that needed for normal plant nutrition can lead to iron chlorosis. Long-term use of copper-containing spray mixtures for fungus disease control in citrus has been identified as a cause for iron chlorosis on acid soils. Apparently this is related to proper absorption and utilization of iron. Phosphorus and copper may act together in limiting iron availability under conditions of very high accumulation of both phosphorus and copper.

Iron-molybdenum interactions have been observed in several plant species but are not well understood. Although the effects of molybdenum on iron are somewhat variable, most observations noted that a high supply of molybdenum in the growth medium produced iron chlorosis. Higher concentrations of iron in growth medium enhanced growth under these conditions in both tomato and red clover. Berry and Reisenauer[37] reported that molybdenum-deficient tomato plants showed the lowest uptake of iron. Iron uptake was enhanced at marginally adequate levels of molybdenum but at a still higher level iron uptake was depressed.

Mechanisms of Iron Uptake. Iron uptake mechanisms are similar to those reported for other cations. Absorption of iron increases with the iron concen-

[37] J. A. Berry and H. M. Reisenauer, "The Influence of Molybdenum on Iron Nutrition of Tomato," *Plant Soil*, 27:303, 1967.

tration in the growth medium. Plant metabolic activity affects iron absorption in the same manner as that described for zinc. Low temperature in the root zone depresses iron absorption and is considered to be one of the factors in development of more severe deficiency symptoms on susceptible crops when soils are cold and wet. Metabolic inhibitors of various types such as dinitrophenol (DNP) have been shown to disrupt or depress iron uptake.[38]

Absorption of iron by different plant species and different varieties within a species varies with the ability of the plant to reduce iron from the ferric to the ferrous form. This reduction according to the work of Chaney et al.[39] appears to be necessary before iron can be absorbed. Reduction appears to take place at the outer surface of the root cell membrane and results from a supply of electrons from inside the cell. This reductive capacity is increased by iron stress.

Besides supplying the reductive capability for converting ferric to ferrous iron, plants may further modify the environment in the immediate vicinity of the root both by excreting chemically active substances and by absorbing water and ions. Root exudates include such materials as organic acids, amino acids, and bicarbonate and hydrogen ions in addition to polysaccharides. These compounds or ions may increase the availability of iron and other metals in the immediate vicinity of the roots by providing a chelating effect that improves iron solubility and availability.

Apparently, iron is separated from a chelating ligand prior to the absorption step. Studies have failed to show the presence of the ligand in decapitated plants when iron was supplied as a certain chelate despite the fact that iron concentrations in plants increased. Therefore, chelates aid iron absorption primarily through protection of the iron from other inorganic soil reactions and possibly improving mobility in soil solution.

Iron Translocation in Plants. Plants deprived of an iron source soon show chlorosis, which develops in the new growth areas. Older tissues remain green, demonstrating that this element is essentially immobile in the plant tissue. Therefore, a continuous supply of iron to the new growth areas via the xylem is necessary for normal plant development. Root uptake is the normal means of iron entry into the translocation system but this process can be supplemented by iron applications to the leaves. Leaf applications may lead to iron translocation downward to the roots *via* the phloem.

The ability of organic acids to complex or chelate iron in the soil solution is also used as a means of maintaining iron in a soluble form within the plant's translocation system. Present information suggests that citrate (citric acid) is the major compound in keeping iron mobile in plants. Several studies have indicated that iron in xylem exudate occurs mainly in negatively charged forms.[34] Ferric citrate has been identified as this negatively charged, anionic

[38] D. Branton and L. Jacobson, "Iron Transport in Pea Plants," *Plant Physiol.*, 37:539, 1962.
[39] R. L. Chaney, J. C. Brown, and L. O. Tiffin, "Obligatory Reduction of Ferric Chelates in Iron Uptake by Soybeans," *Plant Physiol.* 50:208, 1972.

Table 6-12. Susceptibility of Crops to Iron Chlorosis[a]

High Susceptibility	Moderate Susceptibility	Low Susceptibility
Blueberries	Alfalfa	Apples
Citrus	Barley	Millet
Field beans	Corn	Potatoes
Flax	Cotton	Sugar beets
Forage sorghum	Flax	
Grain sorghum	Grass (most)	
Grapes	Oats	
Ornamentals	Rice	
Peaches	Vegetables	
Peanuts	Wheat	
Soybeans		
Sudangrass		
Tomatoes		
Walnuts		

[a] Susceptibility varies with genotype in many instances.

form of iron in several plant species. Ferritin, an iron-protein complex, has been identified in plants and may serve as a form of stored iron in cotyledons of seeds.

Translocation of iron within and from leaves treated with an iron spray is not uniform. If a surfactant is not used in such sprays, a spotty greening effect may result. Local utilization of the applied iron results in a regeneration of chlorophyll in the affected area but there is little movement to other areas of the leaf.

Iron-Genotype Interactions. Differential responses of plant genotypes of micronutrients have been reviewed in detail by Brown et al.[40] They suggest that plant breeding offers the possibility of controlling a problem that is extremely difficult with soil and foliar applications of iron. Genetic control has been widely used in deciduous fruits and citrus to control iron chlorosis by grafting iron-efficient root stocks to desirable scions. Plant breeders and physiologists have noted differences in iron nutrition efficiency in tomatoes, corn, cotton, grain sorghum, and soybeans (Table 6-12).

Brown and Jones[36,41,42] demonstrated significant genetic differences in species as well as in cultivars of a given species, which makes it possible to

[40] J. C. Brown, J. E. Ambler, R. L. Chaney, and C. D. Foy, "Differential Responses of Plant Genotypes to Micronutrients." In *Micronutrients in Agriculture,* J. J. Mortvedt et al. (Ed.), Soil Science Society of America, Madison, Wis., 1972, p. 389.

[41] J. C. Brown and W. E. Jones, "Fitting Plants Nutritionally to Soils. I. Soybeans," *Agron. J.,* 69:399, 1977.

[42] J. C. Brown and W. E. Jones, "Fitting Plants Nutritionally to Soils. II. Cotton," *Agron. J.,* 69:405, 1977.

select genotypes capable of withstanding some degree of low nutrient availability. Because iron deficiency is usually due not to the amount of iron present in the soil but rather to the form of the iron present or to the plant's ability to absorb iron, genetic control of deficiencies has seemed particularly well fitted to iron nutrition. Brown and Jones noted distinct differences in iron chlorosis tolerance among cultivars of soybean[41] and sorghum. Reduction of ferric iron to ferrous iron at the root surface has been demonstrated to be genetically linked. Reducing compounds are released by the roots of some plants, especially iron-efficient soybeans under iron stress. Similar differences in cultivars also were shown in regard to zinc.

In summary, soil scientists, physiologists, and plant breeders note the challenge and potential of adapting plant genotypes to nutritional availability through genetics. Better coordination of plant physiology, biochemistry, genetics, and soil chemistry is necessary for success of such efforts.

6.2.3 Deficiency Symptoms

Corn and Grain Sorghum. Grain sorghum or the sorghums in general are more subject to iron chlorosis than is corn. Iron deficiency, or chlorosis as it may be more properly called, is particularly likely on highly calcareous soils. This point should be taken into account in trying to diagnose iron chlorosis.

Iron chlorosis is typified by severe interveinal chlorosis that develops on the younger tissue first but can affect the entire plant (Fig. 6–11). Veins of affected leaves may remain green, at least greener than the surrounding tissue. On severely affected areas, the loss of chlorophyll may be complete and the leaves are almost white. Regrowth of sorghums tends to show the chlorosis even more severely than the original crop. Yields are severely reduced. Heads are few and small.

Fig. 6–11. Grain sorghum is among the most sensitive crops to iron deficiency. Plants are severely stunted and show interveinal chlorosis. Eventually, the leaves may turn almost white. Symptoms are first expressed on the younger leaves but all leaves may later be affected. (Courtesy L. S. Murphy.)

Symptoms of iron deficiency on corn are very much like those of sorghum and no particular distinguishing characteristics other than those on sorghum can be listed. Yield reductions can be significant.

In general, chlorosis is more likely to occur on poorly drained areas in a field or on areas low in organic matter and simultaneously high in calcium carbonate. Areas cut in leveling for irrigation or eroded areas in the western states of the United States are particularly subject to iron chlorosis development. Climatic conditions can also affect the severity of the chlorosis. Cold, wet conditions are more conducive to chlorosis. Weather has enough effect on iron chlorosis to cause plants in some areas not normally chlorotic to become chlorotic when the conditions previously listed are met.

Small Grains. Iron chlorosis in small grains is quite similar to those in sorghum and corn. Leaf blades develop yellow stripes between green veins and the upper leaves continue to turn more yellow over the entire leaf blade. In very severe cases, the leaves may become almost white and growth is stunted. This problem has been noted fairly frequently with oats and rice and more recently with wheat. Problem areas are usually calcareous; however, deficiencies also may occur on noncalcareous soils. Rice grown on high pH, nonsaline, high sodium soils under flooded conditions has occasionally responded to applications of ferrous sulfate. Yield reductions can and do occur under severely chlorotic conditions.

Soybeans and Dry Beans. Iron deficiency in soybeans is more common on cold wet soils low in organic matter. Areas of iron deficiency in Iowa, Minnesota, eastern South Dakota, Nebraska, and Kansas are usually associated with high pH and frequently with poor drainage. Leveling for irrigation exposes calcareous, low organic matter soils in the western states, and iron deficiencies are common in soybeans on such soils. Zinc deficiencies also occur under similar conditions simultaneously with iron deficiency in the western states.

Iron chlorosis in soybeans initially appears as yellowing between the veins of the younger leaves. As the intensity of the chlorosis increases, the entire leaf and stalk becomes yellow and eventually almost white (Fig. 6-12). Manganese deficiencies in beans could be confused with iron chlorosis early in the development of both, but iron chlorosis eventually leads to loss of chlorophyll in the veins. Iron deficiencies may be somewhat separated from zinc problems in the fact that iron chlorosis appears very early in the growth of the plant, whereas zinc deficiency symptoms may develop a little more slowly. Plants showing chlorosis may not necessarily have low concentrations of iron in the leaves. Foliar application of a solution of ferrous sulfate results in rapid greening of leaves and can serve as an additional diagnostic tool where both iron and zinc chlorosis appears.

Dry beans frequently develop iron deficiency symptoms on calcareous soils in the western states. The chlorotic pattern is essentially identical to that of soybeans. Stunting is severe in both soybeans and dry beans when

Fig. 6-12. Iron deficiency or chlorosis in soybeans is characterized by strong interveinal chlorosis of the upper leaves. Chlorosis develops early in the growing season. Severe cases may affect the entire plant. Affected plants are stunted and yield poorly. (Courtesy L. S. Murphy.)

chlorosis is severe. Yields are drastically reduced in both crops following symptoms of iron chlorosis.

Cotton. Iron deficiency is characterized by a pale green to bright yellow chlorosis between the veins of the younger leaves. Leaves may show a netted appearance from the interveinal chlorosis. New growth and final yields are reduced. The same types of soil conditions that caused iron deficiency symptoms in soybeans, corn, and sorghum would be expected to produce iron chlorosis in cotton.

Potatoes. Chlorosis of the younger leaves is characteristic of iron deficiency in potatoes. Chlorosis develops between the veins but leaf tips and margins may remain darker green than the area at the base of the leaf. Affected leaves may tend to cup upward. Eventually, affected tissue may become almost white.

Citrus, Deciduous Fruits, and Nuts. Iron chlorosis is one of the oldest problems associated with citrus production. Unlike iron chlorosis on the other crops, iron deficiencies in these three crops occur on acid sandy soils in addition to calcareous soils. Problems on acid sandy soils in Florida were recognized as being related to very large amounts of copper in the soil. Regardless of how iron shortages in the plant developed, deficiency symptoms or chlorosis patterns are very similar to those of other dicots. Leaves showing the problem are the newer growth. Interveinal chlorosis results in green veins on a yellow background. Eventually, the leaf becomes almost white. Twigs may be small and the amount of new twig growth is reduced.

Grapes frequently show iron chlorosis but the symptoms are really little different than already described. Iron deficiency in grapes is most severe on the youngest leaves and develops early in the growing season. Interveinal chlorosis eventually results in a total loss of chlorophyll.

Peaches and various nut species show iron chlorosis in the new growth areas with typical interveinal chlorosis. Iron chlorosis in pecans, for instance, can be recognized from zinc deficiency symptoms because of a lack of a rosetting effect typical for zinc.

6.2.4 Fertilizer Sources

Supplemental iron for crops can be supplied from: (1) animal manures or organic matter of various types and (2) manufactured iron compounds. The latter category can be further divided into organic and inorganic classes. Animal manures vary widely in their iron content, that value frequently being a function of the amount of soil that is mixed with the manure. Recommendations in many of the western states in the United States recognize the benefit not only of the iron *per se* contained in the manure but also the value of the organic matter that is applied simultaneously. Animal manure applications of 25 to 50 metric tons/ha (10–20 tons/A) for several years provide not only more iron but perhaps more importantly provide organic acids and other compounds that may be able to chelate iron and improve the availability of the iron already present in the soil. Any decaying plant or animal residue provides the same benefit. Where animal manures are available in large quantities close by, their application may provide one of the most effective and economical methods of controlling iron chlorosis (see Chapter 10).

Inorganic Sources. Soil applications of inorganic iron sources of all types have proved to be generally ineffective due to reactions of iron compounds with soil components. Some of the compounds that have been examined as inorganic sources of iron are listed in Table 6-13. These compounds include both ferrous and ferric iron and have a relatively wide range of solubilities in water. Ferrous sulfate has been the most commonly used inorganic iron source both for soil and foliar application. As a class, the inorganic iron compounds are much lower in price per unit of iron than are the organic materials.

Organic Iron Sources. Organic iron compounds available on the fertilizer market include chelates synthesized from hydrocarbons and compounds with chelating or sequestering capabilities produced from wood processing. Among the chelates produced from hydrocarbons, particularly ethylene, are EDTA, EDDHA, DTPA, and HEDTA.

These materials have been used both for soil and foliar applications. Compared to inorganic sources, the prices of chelates are many times higher but effectiveness is also improved. Comparisons of effectiveness among the various chelates must take into account the soil conditions on which the chelate is to be used. For instance, FeEDTA is effective only under acid soil conditions and is a very poor source of iron under calcareous conditions. FeEDDHA, on the other hand, is a good iron source over a wide range of pH conditions and in the presence of large amounts of calcium. This iron chelate is stable from a pH of about 4 through 9. FeDTPA is somewhat intermediate in effectiveness and

Table 6–13. Partial List of Manufactured Iron Fertilizers

Source	Formula	Iron (% Fe)
Inorganic sources		
Ferrous sulfate	$FeSO_4 \cdot 7H_2O$	20
Ferric sulfate	$Fe_2(SO_4)_3 \cdot 4H_2O$	20
Ferrous carbonate	$FeCO_3 \cdot H_2O$	42
Ferrous ammonium sulfate	$(NH_4)_2SO_4 \cdot FeSO_4 \cdot 6H_2O$	14
Iron frits	Combination	40
Organic sources — hydrocarbon chelates		
FeDTPA	(see text for full names of compounds)	10
FeEDTA	"	9–12
FeEDDHA	"	6
FeHEDTA	"	5–9
Organic sources — paper by-product chelates and sequestrants		
Ligninsulfonate		5–11
Methoxyphenylpropane complex		5
Polyflavonoid		6–10

stability between EDTA and EDDHA. Holmes and Brown[43] reported that only DTPA and EDDHA were effective in correcting iron chlorosis in soybeans grown on *calcareous* soils.

The iron chelates or sequesterants derived from the paper industry are primarily iron ligninsulfonates, iron polyflavonoids, and iron methoxyphenylpropane (MPP). These compounds differ greatly depending on the type of process used in extracting them from wood or bark and the processes used in further refinement. Some exert a true chelating effect, whereas others exert an effect somewhat similar to that of polyphosphates (see Chapter 3). Due to the variations in source and processing, no specific formula can be assigned to these classes.

In terms of effectiveness, the ligninsulfonates, polyflavonoids, and methoxyphenylpropanes are somewhat intermediate in effectiveness between the inorganic materials and the synthetic chelates. That generalization must be taken with several "grains of salt" too because, under some conditions, performance may equal, or surpass, the synthetic chelates. These natural organic complexes are intermediate between the cheap inorganic sources of iron and the very expensive hydrocarbon chelates. Efficiency ratios between the inorganic iron sources, the wood by-product materials, and synthetic chelates are extremely difficult if not impossible to establish. For a generalization, however, it is probably much wider than the 2 to 3:1 ratio for zinc

[43] R. S. Holmes and J. C. Brown, "Chelates as Correctives for Chlorosis," *Soil Sci.*, 80:167, 1955.

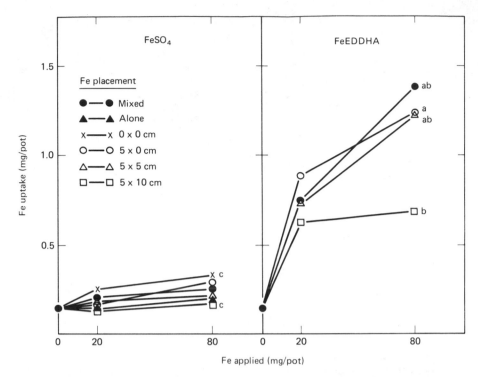

Fig. 6-13. Iron uptake by grain sorghum plants was improved at higher rates of Fe application in this greenhouse study when either ferrous sulfate or iron chelate (FeEDDHA) were applied in an ammonium polyphosphate suspension fertilizer, 12-40-0. Placement close to the seed was more effective than that placed 5 cm (2 in.) to the side and 10 cm (4 in.) below the seed. (Source: J. J. Mortvedt and P. M. Giordano, "Response of Grain Sorghum and Corn to Fe." TVA Bul. Y-100, 1975, p. 16.)

chelates versus inorganic zinc sources due to the greater tendency of iron to react with soil components.

Iron in Macronutrient Fertilizers. Researchers in the National Fertilizer Development Center of TVA have pioneered in studies of the effectiveness of various micronutrients applied in macronutrient fertilizers. Mortvedt and Giordano[44] reported that ferric and ferrous sulfate may be effective when soil-applied with fluid APP fertilizers (Fig. 6-13). Soil applications of iron sulfates with several granular fertilizers were not effective, however. Iron EDDHA chelate applied in the same manner was also effective, but the relatively high costs of the chelates tend to diminish the importance of the fact.

Some investigations of the effects of ammonium thiosulfate (ATS) (12% N, 26% S) on availability of iron applied as ferrous sulfate in a fluid APP in-

---

[44] J.J. Mortvedt and P. M. Giordano, "Response of Grain Sorghum to Iron Sources Applied Alone or with Fertilizers," *Agron. J.,* 63:758, 1971.

dicated that this reducing agent may have some value in maintaining iron availability for an extended period. Forage yields and iron uptake by grain sorghum from a band application of $FeSO_4$-APP suspension fertilizer in a greenhouse study on a calcareous soil were doubled over those of APP alone. Inclusion of ammonium thiosulfate with this mixture resulted in another 15 to 25% increase in crop response. Extraction of soil around the fertilizer bands after harvest showed that inclusion of ammonium thiosulfate resulted in increased iron movement from the fertilizer band. Mortvedt and Giordano[45] noted, however, that placement of such a mixture containing ATS must be at least 3 cm from the seed to prevent toxicity to the seedlings.

In summary, some improvement of iron availability may occur when iron sulfates are mixed with suspension APP fertilizers but concentrations are too low when added in true solutions. It may be possible to apply iron sulfate successfully with APP suspensions but variability in the field responses to such applications make this practice questionable.

Some success has been reported in correcting iron chlorosis by soil acidulation using by-product sulfuric acid. The amounts of acid needed are dependent on soil composition and acid strength. This procedure is difficult due to problems in handling the very corrosive acid. Close proximity to available sources of sulfuric acid make this technique slightly more practical. Soil acidification over small areas can also be carried out by the use of elemental sulfur. A comparable technique is described in Chapter 9 for reclamation of saline and sodic soils.

6.2.5 Methods of Application

Iron applications have historically been made to soil or to foliage. Recently, additional methods of application have been developed including injection of soluble iron compounds into irrigation water and coating of certain iron compounds on seeds.

Soil applications of iron have been studied for many years. Applications of ferrous sulfate to soils in citrus groves was recommended over 60 years ago in an attempt to overcome iron chlorosis. In general, however, soil applications of iron in either the inorganic or organic form have been quite disappointing. The few studies reporting successful soil applications of such compounds as ferrous sulfate, ferric sulfate, or chelates such as FeEDDHA have involved very high rates of application, frequently 100 kg/ha of more Fe in the case of the inorganic forms. Chelate applications have been somewhat more effective when soil applied than have inorganic iron sources, but the amount of iron in the chelated form required for correction of at least part of iron chlorosis has usually been too expensive to be practical. Any soil applications should be

[45] J. J. Mortvedt and P. M. Giordano, "Grain Sorghum Response to Iron in a Ferrous Sulfate-Ammonium Thiosulfate-Ammonium Polyphosphate Suspension," *Soil Sci. Soc. Amer. Proc.*, 37:951, 1973.

made prior to seeding. Applications in combination with acid-forming nitrogen fertilizers may enhance iron efficiency.

Foliar applications of iron have been generally more effective in correcting visible effects of iron chlorosis than have soil applications. Obviously, foliar applications are not subject to soil chemical reactions but have disadvantages because repeated foliar applications are frequently necessary. The success of foliar applications of iron for relatively low-value crops such as corn, soybeans, and grain sorghum may frequently be more apparent than real. Although foliar applications may cause the chlorosis to diminish in intensity or even disappear with repeated applications, indelible damage to the crop already may have occurred and yield responses may be insufficient to pay for the cost of the treatment. Sorghum frequently responds poorly in yield to iron applications even though chlorosis has been diminished. Foliar applications have involved both inorganic and organic sources of iron. Recommendations for several crops include a 3 to 4% $FeSO_4$ solution with sufficient rates of application to wet the foliage. Some recommendations have specified from 280 to 370 l of a 4% $FeSO_4$ solution per hectare (30–40 gal/A).

Some very notable responses to foliar application of iron for soybeans have been reported recently from Minnesota by Randall.[46] Applications of only 0.17 kg/ha of iron (0.15 lb/A Fe) as FeEDDHA within 3 to 7 days after symptoms appeared produced dramatic increases in yield (Table 6-14). In the case of the studies reported, applications were generally made at the second trifoliate stage. Waiting until beans are at the fourth or fifth trifoliate stage produced essentially no yield response.

Some recommendations for correction of iron chlorosis are summarized in Table 6-15. Foliar applications are most frequently recommended. Foliar applications of iron or other nutrients should wet the entire leaf surface for maximum effectiveness. Iron translocation from one spot to another in leaves is not particularly good, and incomplete coverage of the leaf surface results in a spotted appearance. Some indication of the localized effects of iron treatments on sorghum leaves are given in Fig. 6-14 where a solution of ferrous sulfate was painted on the leaf surface. Such localized effects can be minimized with relatively low rates of water application by including a surfactant in the spray. Use of surfactants in foliar fertilization is common and many such materials are available in the market.

Recently, some success has been recorded in correction of iron chlorosis in grain sorghum by coating seed with an iron lignin sulfonate prepared by a fermentation process to remove sugars. Although information on this subject has not been published at this time, positive results have been obtained from investigations carried out by Texas A&M University and by New Mexico State University. Coatings of the sorghum seed have involved up to 12% by weight of the seed of a material containing 10% iron (Fe). Apparently, the

[46] G. W. Randall, "Correcting Iron Chlorosis in Soybeans," Minn. Agr. Ext. Soils Fact Sheet 27. Undated.

Table 6–14. Yield Responses to Foliar Applications of Iron on Soybeans with Iron Chlorosis Symptoms[a]

| Minnesota County | pH | Soluble Salts (mmhos) | Yield (kg/ha) | | | | Statistical Significance[e] | BLSD[f] (.05) (kg/ha) |
| | | | No Iron | Iron Once | | Iron Twice[d] | | |
				Early[b]	Late[c]			
				1973				
Waseca	7.9	0.6	1825	2822	1996	2728	99	591
Waseca	8.0	0.6	2123	2318	1996	2352	90	
				1974				
Waseca	8.0	0.6	202	585	309	766	95	363
				1975				
Blue Earth	7.7	0.6	2332	2876		2822	99	215
Brown	8.1	0.8	2285	2554		2654	95	262
LeSueur	7.6	0.8	2029	2970		3226	99	430
Nicollet	8.0	2.3	2312	2802		2876	90	
Renville	8.0	0.8	2426	2715		2641	ns	
Renville	8.1	0.6	1035	3152		3508	99	551
Renville	7.8	2.8	1378	2177		2224	99	296
Watonwan	7.9	0.7	2419	3387		3488	99	302
Watonwan	7.8	0.8	504	2345		2634	99	625

[a] Early applications of iron gave higher yield response. Application rates were 0.11 kg/ha of Fe in 1973–74 and 0.17 kg/ha of Fe in 1975.

[b] One foliar application generally at the second trifoliate stage.

[c] One foliar application generally at the fourth trifoliate stage.

[d] Two foliar applications generally at the second trifoliate stage and again 7–14 days later.

[e] Significant at the 99, 95, and 90% levels, respectively. ns = not significant at the 90 percent level.

[f] BLSD means Baysean least significant difference.

SOURCE: G. W. Randall, "Correcting Iron Chlorosis in Soybeans," Minn. Agr. Ext. Soils Fact Sheet 27, undated.

Fig. 6-14. Foliar applications of iron for grain sorghum can partially correct iron deficiency. It is necessary to add surfactants to the iron solution to be sure that the entire leaf is covered. This leaf shows that a solution of ferrous sulfate applied to the surface with a felt-tipped pen affected only the area covered. (Courtesy Dr. W. W. Harris, Ft. Hays, Kansas State University.)

Table 6-15. Some Recommended Rates of Iron for Selected Crops[a]

Crop	Source of Iron	Rate	Method of Application	Remarks
Vegetables	Fe chelates	0.5–1.1 kg/ha Fe	Foliar	Wet foliage, repeat as needed
Citrus	Fe chelates	12–24g Fe/tree	Broadcast	—
Grain sorghum and corn	$FeSO_4 \cdot 7H_2O$	0.79–1.25 kg Fe/100 l H_2O	Foliar	Three sprays of 280 l/ha
Field (dry) beans	$FeSO_4 \cdot 7H_2O$	595 g Fe/100 l H_2O 190–280 l/ha	Foliar	2-week intervals until symptoms disappear
Deciduous fruits	Fe polyflavonoid	60–100 g Fe/100 l H_2O	Foliar	Wet foliage, repeat as needed
Soybeans	FeEDDHA	0.17 kg/ha Fe	Foliar	Spray band over row at second trifoliate; 280 l/ha
Cotton	$FeSO_4 \cdot 7H_2O$	1.16 kg Fe/100 l H_2O	Foliar	Wet plants, repeat as needed

[a] Listings of particular iron fertilizers are not intended to be exclusive.

small amounts of iron thus applied in the immediate vicinity of the seed are effective in correcting the iron deficiency.

Iron application through sprinkler irrigation systems is not common but has been carried out successfully. This subject is discussed in some detail in Chapter 7. More research is needed in this area before such a practice would be widely recommended. Much of the applied iron would be ineffective in such an application due to soil contact. Iron applications in drip-irrigation systems, however, may pose a more effective means of application than through sprinkler systems. Drip irrigation systems for citrus in Israel have resulted in significant reductions in water use (one-third of the amount needed with a sprinkler system) and effective application of as little as 0.01 kg of iron per tree as FeEDDHA. A large number of active roots in the moist soil zone containing the added iron have allowed the increased efficiency as compared to the amount of iron recommended per tree (Table 6–15).

6.3 COPPER

The occurrence of copper (Cu) in the earth's crust is mainly in the form of sulfide minerals, although other less stable forms including silicates, carbonates, and sulfates exist. Copper has two valence states in its naturally occurring compounds, Cu^+ and Cu^{2+}, but occurs occasionally in the native, metallic state. Copper concentrations in soils range from 10 to 200 ppm, present primarily in an adsorbed form and as organic matter complexes. As long as soil conditions are conducive to oxidation, copper is present primarily as the cupric (Cu^{2+}) form and in that form is of primary importance in plant nutrition.

6.3.1 Soil Chemistry

Like the chemistry of iron and zinc, solubility of copper compounds in soil tends to decrease sharply as the soil pH increases. Lindsay[2] discussed inorganic phase equilibria of copper in detail and pointed out that the solubility of such copper minerals as azurite [$Cu_3(OH)_2(CO_3)$], malachite [$Cu_2(OH)_2(CO_3)$], and tenorite [CuO] as well as a copper-soil complexes declined at about the same rate as the soil pH increased.

As the soil pH increases, copper tends to become more complexed with hydroxyl-forming anions at the highest pH levels. Ion species that may exist in the soil solution at various pH values include Cu^{2+} at pH's below 7.3; above that pH, $CuOH^+$ is most common. From pH 7.6 and upward, some $Cu(OH)_3^-$ anions may exist but in smaller quantities than $CuOH^+$. Several researchers have suggested that the $CuOH^+$ ion may be important in copper adsorption reactions on clay minerals and organic matter.

Copper reactions with organic matter form very stable complexes. Research has suggested that carboxyl and phenolic groups attached to heterocyclic compounds in soil organic matter are important in copper bonding. Studies of organic soils in Ireland have indicated that as much as 60% of the copper exchange capacity of peat was due to the presence of phenolic hydroxy groups and most of the remainder is considered to be due to the presence of carboxylic groups. Others have suggested that humic and fulvic acid fractions of organic matter are important in adsorption of both copper and zinc in the soil.

The importance of natural chelating ligands in the soil relative to the availability of iron and zinc is similar to the relationships that pertain to copper availability. However, chelates of copper are normally more stable than chelates of zinc. Norvell[47] indicated that in a calcareous soil the stability of various chelates with copper decreased in order: CuDTPA > CuHEDTA > CuEDTA = CuEDDHA. The ratio of chelated copper to cupric ions in the soil solution is probably always greater than 10:1 and may be 1000:1 in alkaline soils.

Studies of copper adsorption in clay-organic matter systems have indicated that copper is adsorbed by the organic matter; thus suggesting that the bonding of copper to organic matter is stronger than bonding to clays. Because organic soils tend to be more copper deficient than mineral soils, Cu adsorbed on clay probably is more available to plants than that bonded to organic matter.

[47] W. A. Norvell, "Equilibria of Metal Chelates in Soil Solution." In *Micronutrients in Agriculture*, J. J. Mortvedt et al. (Ed.), Soil Science Society of America, Madison, Wis., 1972, p. 115.

Soil solution pH can significantly affect the amount of copper adsorbed by soils. As the pH increases, the amount of copper adsorbed by both clays and organic matter tends to decline. This fact supports the importance of phenolic hydroxylic groups as copper adsorption sites. As the pH of the soil solution increases, the hydroxyl concentration increases and dissociation of the hydroxyl groups is less likely. Subsequently, fewer negatively charged sites will exist and the dominant $CuOH^+$ ion (at pH levels above 7.3) will not be as readily attached. Much remains to be learned about the chemistry of organic matter-copper reactions; however, it is known that most of the soil's copper is associated with the organic layer and that very high organic matter contents in soils may give rise to copper deficiency.

Interactions with Other Elements. Copper availability in soils can be affected by a copper-phosphorus interaction similar to those already pointed out for zinc and iron. Prolonged use of excess phosphorus results in copper deficiencies in plants. Field observations of copper deficiency in citrus due to high concentrations of available phosphorus in the soil have been confirmed by soil analyses. High concentrations of phosphorus in growth media have been reported to reduce copper concentrations significantly in plant tissue.[48] Several different crops have been reported to be similarly affected. In all cases, however, application of adequate amounts of copper overcame this induced deficiency. The mechanism of the interaction is not well understood but is thought to be similar to the same problem with iron and zinc, that is, physiological or nutritional rather than some type of precipitation in the soil.

In Section 6.2, an interaction between copper and iron is discussed. High concentrations of available copper in the soil solution resulting from long-term applications of copper as a fungicide have been shown to produce an induced iron chlorosis (copper toxicity). Spencer[49] reported both on the copper-iron interaction and the copper-phosphorus interaction in citrus. Applications of iron decreased the copper-induced problem but never completely overcame it.

Copper-molybdenum interactions occur in plants as well as in animals. Giordano[50] reported that copper interferred with the normal role of molybdenum in nitrate reduction in tomatoes. Other studies have suggested that toxicities produced by excesses of either element can be overcome by application of the other. Again, this type of problem is nutritional in nature rather than due to soil reactions between the two elements.

Copper-zinc interactions also exist. Excessive rates of zinc applications have been reported to induce copper deficiency in wheat on coarse-textured

[48] F. T. Bingham and M. M. Garber, "Solubility and Availability of Micronutrients in Relation to Phosphorus Fertilization," *Soil Sci. Soc. Amer. Proc.*, 24:209, 1960.

[49] W. F. Spencer, "Effect of Copper on Yield and Uptake of Phosphorus and Iron by Citrus Seedlings Grown at Various Phosphorus Levels," *Soil Sci.*, 102:296, 1966.

[50] P. M. Giordano, H. V. Koontz, and E. J. Rubins, "C^{14} Distribution in Photosynthate of Tomato as Influenced by Substrate Copper and Molybdenum Level and Nitrogen Sources," *Plant and Soil*, 24:437, 1966.

soils. Copper and zinc concentrations in plant tissue tend to be inversely related.

6.3.2 Copper Nutrition of Plants

Metabolic Roles. Copper functions in plants are largely associated with enzymes. Some of the enzymes containing copper as a part of their system are listed in Table 6–16. The oxidase-catalyzing reactions that reduce both atoms of molecular oxygen to water, specifically cytochrome oxidase, ascorbic acid oxidase, and laccase, all contain copper. Plastocyanin is a protein component of the photosynthetic electron transfer chain and is present in chloroplasts. It contains one atom of copper per molecule. The catalytic unit of cytochrome C oxidase is thought to contain two distinct forms of copper in addition to cytochrome A. Evidence suggests that copper undergoes a valence change during electron transfer to molecular oxygen.

Copper Absorption. Copper absorption by plants has been characterized as an active process that is subject to low temperature and metabolic inhibitors. Higher concentrations of copper in the growth medium increase absorption up to a maximum concentration of about 0.1 mM. Bowen[51] has reported that zinc and copper compete for the same absorption sites on roots, a partial explanation for the observed copper-zinc interaction.

Copper Translocation. Copper deficiency symptoms have largely been associated with new growth areas of the plant and suggest that the element, once deposited, is relatively immobile. Electrophoretic studies of copper translocation reported by several researchers have indicated the presence of a copper anion in the xylem exudate of plants. Tiffin[52] has suggested that the compounds forming anions with copper may be amino acids.

Copper Concentrations in Plants. Copper occurs in all tissues of plants but is concentrated in leaves and new growth areas. Concentrations in leaves and petioles of various species range from about 3 to as high as 50 ppm. Robertson and Lucas[53] have suggested sufficiency ranges for copper for several plants (Table 6–17). Note that the lower levels of sufficiency are in the 6 to 11 ppm range.

Copper Toxicity. High concentrations of available copper in the soil can lead to depressed uptake of iron, zinc, and molybdenum. In addition, ex-

---

[51] J. E. Bowen, "Absorption of Copper, Zinc and Manganese by Sugar Cane Tissue," *Plant Physiol.,* 44:255, 1969.

[52] L. O. Tiffin, "Translocation of Micronutrients in Plants." In *Micronutrients in Agriculture,* J. J. Mortvedt (Ed.), Soil Science Society of America, Madison, Wis., 1972, p. 199.

[53] L. S. Robertson and R. E. Lucas, "Copper — An Essential Micronutrient," Mich. Coop. Ext. Cir. 776–3. Undated.

Table 6–16. Some Copper-Containing Enzymes (Proteins) in Plants

Ascorbic acid oxidase	Diamine oxidase
Penol oxidase	Cytochrome oxidase
Laccase	Plastocyanin

Table 6–17. Copper Sufficiency Ranges in Selected Crops

Crop	Location of Plant Sample	Copper (ppm)
Corn	Ear leaf at first silk	6–20
Soybean	Upper fully developed leaf prior to first flower	10–30
Navy bean	Upper fully developed leaf prior to first flower	10–30
Alfalfa	Top 15 cm sampled prior to first flower	11–30
Wheat	Upper leaves sampled prior to initial bloom	6–50
Sugar beet	Center fully developed leaf — midseason	11–40
Vegetables	Top fully developed leaf — midseason	8–20
Potato	Petioles of most recently matured leaf — midseason	7–30

SOURCE: L. S. Robertson and R. E. Lucas, "Copper — An Essential Micronutrient," Mich. Coop. Ext. Ser. Circular 776-3, undated.

cessive concentrations of copper in the plant may also interfere with normal utilization of phosphorus. Copper toxicities are generally not a major problem in crop production but are difficult to eliminate. Excessive application of copper as a fungicide spray for citrus has led to plant nutritional imbalances on acid, sandy soils in Florida.

6.3.3 Deficiency Symptoms

Crop plants vary widely in their susceptibility to copper deficiency and conversely in their response to applied copper (Table 6–18). Deficiencies are most commonly associated with soils high in organic matter such as peat or muck soils. In addition to the organic soils, copper deficiency has been reported on acid sands in Florida.

Corn. Copper deficiency symptoms appear first on the youngest leaves and tend to develop most frequently on young plants, although the symptoms may persist throughout the growing season. There is a general chlorosis of the affected leaves (Fig. 6–15). Chlorotic leaves tend to show some striping or interveinal chlorosis that is more pronounced toward the lower end of the leaves. Leaf tips may show severe chlorosis and necrosis and may bend backward along the midrib. Older leaves affected may show necrosis or dieback along leaf tips and margins in a pattern somewhat comparable to potassium deficiency symptoms. There is little chance of confusing copper for

Table 6–18. Sensitivity of Selected Crops to Copper Deficiency

	High sensitivity crops	
Alfalfa	Spinach	Wheat
Lettuce	Sudangrass	Citrus
Oats	Table beet	Pangolagrass
Onions		
	Medium sensitivity crops	
Barley	Clovers	Sugar beet
Broccoli	Cucumber	Sweet corn
Cabbage	Corn	Tomato
Carrot	Radish	Turnip
Cauliflower	Tung-oil	Sorghum
Celery		
	Low sensitivity crops	
Asparagus	Mint	Rye
Beans	Potatoes	Soybeans
Peas		

potassium, however, since potassium symptoms appear on the older leaves first. Severely affected plants are stunted and will not mature.

Small Grains. Copper deficiency symptoms in small grains are frequently noted on organic soils in the northern United States and in the eastern provinces of Canada (Fig. 6–16). Deficiency symptoms are characterized by general yellowing of the plant, leaf-tip dieback, and twisting of leaf tips. On severely copper-deficient peat soils, wheat ceases growth and dies after reaching the tillering stage. Wheat deficiency symptoms may include a pale green color of new leaves, a lack of leaf turgor, and an eventual yellowing and rolling of the leaves. Deficiency symptoms in oats include rolling at the tips of terminal or new leaves, chlorosis, and appearance of yellow-gray spots on the leaves. Some severely deficient plants produce a grainless head. On less defi-

Fig. 6–15. Copper deficiencies in corn are usually associated with organic soils. Deficient plants are stunted and chlorotic and older leaves may die back. (Courtesy The Fertilizer Institute.)

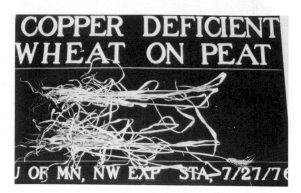

Fig. 6-16. Copper deficiency symptoms on small grains include general chlorosis, leaf-tip dieback and twisting of the leaves. Plants may die after reaching the tillering stage. (Courtesy Dr. G. W. Wallingford, Potash and Phosphate Institute.)

cient soils, small grains show no obvious deficiency symptoms but may respond to copper applications (Table 6–19).

Alfalfa. Deficiency symptoms occur on the youngest tissue with leaves turning a faded green with a grayish cast. Overall growth is stunted due to shortened internodes. Plants appear bushy and drought-stricken. Upper leaves may show necrotic areas around the edges. Terminal leaf petioles may show epinastic (downward) curvature. Leaflets tend to fold backward along the petioles and eventually wither and die.

Citrus, Deciduous Fruits, and Nuts. Copper deficiency symptoms in citrus were described (although unrecognized as such) prior to the Civil War. Unlike the deficiencies of other metals (iron, zinc, manganese), copper deficiency does not immediately result in chlorosis. Early growth symptoms appear on new tissue, particularly on trees that initially grow vigorously. Young trees deficient in copper may develop a few vigorous shoots with abnormally large leaves. Leaves may be dark green but yellow blotches appear on the green shoots beside or below leaf nodes due to accumulation of carbohydrates.

Table 6-19. Effect of Copper on Yields of Wheat and Barley Grown on Northwestern Minnesota Peat Soils

| County | Crop | Grain Yield | | Forage Yield | |
| | | No Copper | Copper | No Copper | Copper |
		(kg/ha)		*(kg/ha)*	
Clearwater	Wheat	1277	2419	5869	8198
Clearwater	Barley	1559	1828	—	—
East Polk	Wheat	403	2554	—	—
East Polk	Barley	1236	1451	—	—
Roseau	Wheat	0	1277	—	—
Roseau	Wheat	—	—	2139	6205

SOURCE: G. W. Wallingford and C. A. Simpkins, "Copper for Organic Soils," Minn. Ext. Folder 347, 1977.

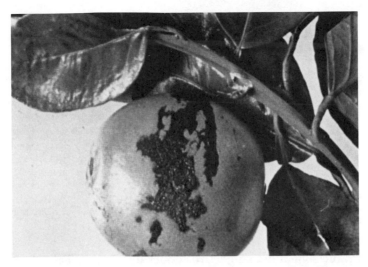

Fig. 6-17. Copper deficiency in citrus produces twisted leaves and weak twigs that grow from multiple buds. Such twigs frequently die back before developing fully. Fruit show the characteristic spotting and splitting due to copper deficiency. (Courtesy Dr. Herman Reitz, Florida Citrus Exp. Sta., Lake Alfred, Fla.)

These blotches eventually enlarge until the stem or shoot is girdled and they may be accompanied by swellings in the same areas that contain a water-soluble, brown gum.

As a result of girdling of the stem, leaves may become yellow-veined and drop off leaving a bare yellow twig that eventually dies back to the point where the shoot developed (Fig. 6-17). Several buds may develop at the point where the original bud or shoot died. This condition has been termed *multiple bud* and is associated with small, narrow, and elongated leaves. Persistence of the deficiency for several years eventually leads to dieback of large branches and the tree actually becomes smaller.

Copper deficiency also affects the appearance and quality of fruit. Gum accumulations in the outer layers of the peel on mildly deficient oranges produce a reddish brown stain. Under severe deficiency conditions, fruit discoloration may appear very early and fruits are misshaped and usually are dropped early in the season. Development of the gum beneath the peel causes the peel to become rigid and to split.

Copper deficiencies in tung-oil trees produces a cupping of the terminal leaves due to abnormal development of the leaf margins. Interveinal chlorosis may occur and become necrotic. These dead areas are eventually lost and the leaf has a ragged appearance. Many such affected leaves are shed prematurely. Dieback of new shoots may also occur, particularly in English walnuts.

Vegetables. Occurrences of copper deficiency in vegetables seem to be unusually frequent, because vegetables are frequently grown on the high

Fig. 6-18. Copper deficiencies in many vegetable crops are associated with organic soils due to complexing of copper by organic matter. Carrots did not grow in the absence of supplemental copper. The response in the foreground was due to a single application of copper sulfate as a foliar spray. These carrots were grown on an organic soil in Ireland. (Courtesy R. E. Lucas, Michigan State University.)

organic soils most subject to copper deficiency. Table beets, cabbage, carrots, celery, onions, peas, and tomatoes have been described as copper-deficient. Young beet leaves appear blue green; older leaves become chlorotic between the veins with the chlorosis beginning at the leaf tip and eventually spreading over the entire leaf surface. Tomatoes show a stunted shoot growth and poor root development. The upper leaves are unable to retain turgor and wilt permanently. Leaves may be curled or develop a dark bluish-green coloration and an eventual chlorosis.

Onions develop leaf chlorosis and thin bulb scales, and bulbs are soft and spongy due to a lack of turgor. Bulb color tends to be pale yellow rather than normal brown. Lettuce shows a similar lack of turgor in the heads. Leaves become chlorotic and bleached. The chlorosis appears first on the stem and the outer rim of the leaves. Leaves may become cupped. Carrots develop a stunted top growth (Fig. 6-18) and although chlorosis of the tops is not strongly expressed, roots may be poorly pigmented and poorly developed.

6.3.4 Fertilizer Sources

Like iron and zinc fertilizers, copper fertilizers are classified into inorganic and organic sources. These materials are listed in Table 6-20.

Inorganic Sources. Although many inorganic copper compounds are listed in Table 6-20, most of the studies involving copper fertilization with an inorganic source have employed copper sulfate. This material, $CuSO_4 \cdot 5H_2O$, is relatively cheap, has a good water solubility, and is widely available. Minnesota, Wisconsin, and Michigan recommendations equate copper oxides (CuO and Cu_2O) with copper sulfate as a means of overcoming deficiencies in organic soils in those states. Copper sulfate has a much higher water solubility than either cupric oxide or cuprous oxide but the oxides are both soluble in acidic conditions and thus are acceptable as copper sources on acid soils. Copper oxides and copper sulfate are available in crystalline form suitable for mix-

Table 6–20. Some Copper Fertilizers

Sources	Formula or Abbreviation	Copper (% Cu)
Inorganic sources		
Copper sulfate monohydrate	$CuSO_4 \cdot H_2O$	35
Copper sulfate pentahydrate	$CuSO_4 \cdot 5H_2O$	25
Cupric oxide	CuO	75
Cuprous oxide	Cu_2O	89
Cupric ammonium phosphate	$Cu(NH_4)PO_4 \cdot H_2O$	32
Basic copper sulfates	$CuSO_4 \cdot 3Cu(OH)_2$ (general formula)	13–53
Copper frits	—	40–50
Cupric chloride	$CuCl_2$	17
Organic sources		
Copper ethylenediaminetetra-acetic acid	$Na_2CuEDTA$	13
Copper hydroxyethylethyl-enediaminetriacetic acid	NaCuHEDTA	9
Cu polyflavonoid	CuPF	6
Cu methoxyphenylpropane	CuMPP	5–6

ing with other dry fertilizers. Low water solubility of the oxides does not lend these materials to easy incorporation into clear liquid fertilizers. Other inorganic copper compounds are relatively rare as fertilizers.

Organic Sources. Several organic forms of copper are available as fertilizers including synthetic chelates and complexed or sequestered compounds derived from wood by-products. CuEDTA has been available for some time and although most of the investigations carried out in the United States and elsewhere over the years have involved the use of inorganic sources of copper, CuEDTA is highly effective. Copper polyflavonoids and ligninsulfonates as well as copper methoxyphenylpropane are available and effective.

Copper chelates are somewhat more effective than the inorganic sources but few definite efficiency ratios have been set. Wisconsin recommendations have called for a 6:1 efficiency ratio of chelates versus inorganics. Costs of chelates are much higher than inorganic copper and are subsequently applied at lower rates (Table 6–21). These materials mix well with liquid fertilizers and are frequently used in those formulations.

6.3.5 Methods of Copper Application

Soil Applications. Copper, like zinc, is most commonly applied to the soil. Soil applications have a good residual effect and so heavy rates of copper application such as 50 to 280 kg/ha of $CuSO_4 \cdot 5H_2O$ have been used to correct deficiencies for several years.

Either broadcast or banded applications of copper are recommended

Table 6–21. Some Recommended Rates of Copper for Selected Crops

Crop	Source of Copper	Rate of Copper (kg/ha)	Method of Application	Comments
Small grains	$CuSO_4 \cdot 5H_2O$ or CuO	1–7	Banded	Higher rates on organic soils
		4–14	Broadcast	" "
	Cu chelates	0.5–2	Banded	" "
Corn	$CuSO_4 \cdot 5H_2O$ or CuO	4–14	Broadcast	" "
		1–3	Banded	" "
	Cu chelates	0.2–0.5	Banded	" "
Vegetables	$CuSO_4 \cdot 5H_2O$ or CuO	2–14	Broadcast[a]	" "
		1–4	Banded	" "
	Cu chelates	0.8–2	Broadcast	" "
		0.2–0.8	Banded	" "
Soybeans	$CuSO_4 \cdot 5H_2O$	3–6	Broadcast	" "
		2	Banded	
Citrus	$CuSO_4 \cdot 5H_2O$	7–25	Broadcast	Repeat in 5 years
Citrus	$CuSO_4 \cdot 5H_2O$	90 g Cu/100 l H_2O	Foliar	Annual applications
Small grains	Cu chelates	75 g Cu/185 l H_2O	Foliar	Annual applications

[a] Some recommendations call for 25 to 70 kg/ha copper in initial applications on organic soils. 45 kg/ha copper is considered to be a maximum allowable treatment on mineral soils to avoid copper toxicity.

(Table 6–21) but banded applications are *usually* intended as a 1-year treatment to be repeated the following season. Wallingford and Simpkins[54] have recommended 7 kg/ha of copper as a band application for small grains and have doubled that recommendation when broadcast treatments are to be used. They recommended, however, that not more than 44 kg/ha of copper should be applied in a single application in order to avoid toxic accumulations of copper in the soil. Heavy soil applications of copper on either mineral or organic soils should be followed by tissue and/or soil testing to determine if further applications are warranted.

Foliar Applications. Early researchers noted that Bordeaux sprays containing copper sulfate stimulated plant growth in addition to controlling the disease for which they were applied. Usually, foliar applications have involved use of $CuSO_4 \cdot 5H_2O$ but chelates and other organic sources also are adapted to foliar treatments.

The whole concept of foliar fertilization with copper should be critically examined as to where it fits into the scheme of plant nutrition. Because copper deficiencies can affect plant tissues very early in their development, cop-

[54] G. W. Wallingford and C. A. Simpkins, "Copper for Organic Soils," Minnesota Ext. Folder 347, 1977.

per should be applied prior to or during planting; both of these are soil applications. Foliar applications, then, are primarily intended to salvage a crop that is found to be copper-deficient. Adequate soil testing and/or plant analysis could prevent the original development of copper deficiency by alerting the grower to the need for copper in the planning stages of crop production.

Foliar applications are obviously much smaller in magnitude than soil treatments and must be repeated until the symptoms are eliminated. The frequency of application of foliar treatments depends on the crop and may be only annual in the case of some tree crops.

6.4 MANGANESE

Manganese is similar to iron in both its chemistry and geology. Manganese concentrations in the earth's crust, igneous and sedimentary rocks, and soil are second only to those of iron. Soil concentration of total manganese ranges from about 20 to 6000 ppm. Manganese has three common valences, Mn^{2+}, Mn^{3+}, and Mn^{4+}. Of these valences, the Mn^{2+} ion is the most common, particularly in reference to plant nutrition. In soils, neither total nor exchangeable manganese shows much correlation with bedrock composition, indicating a relatively high mobility for this element.

6.4.1 Manganese Soil Chemistry

Soil chemistry of manganese is quite complex because this element can be oxidized and reduced in the soil depending on conditions. Bacteria also have a role in the soil chemistry of manganese, further complicating an already complex inorganic chemistry. Oxides of manganese, for instance, may be nonstoichiometric with mixed valence states. Further, oxides of manganese may be coprecipitated with iron oxides giving even further complexity to its chemistry.

One of the most stable oxides of manganese is MnO_2. As the soil solution becomes more acid, the solubility of this compound is increased and the Mn^{4+} form may be reduced to Mn^{2+}. In the soil solution, complex ions of manganese may exist and may include $MnHCO_3^+$ and $MnOH^+$ in addition to Mn^{2+}, which is normally dominant. The activity of the Mn^{2+} ion or, stated another way, the availability of the Mn^{2+} ion increases 100X for each unit decrease in soil pH from a pH of 9 to 4. The relative concentrations of the other two complex ions increase but are less than the concentrations of Mn^{2+} even at pH 9.

Solubility of MnO_2 in soil solution is only slightly affected by concentrations of O_2 and CO_2 in the soil atmosphere/solution. Other compounds [$MnCO_3$, $Mn(OH)_3$, $Mn(OH)_2$, and $MnSiO_3$)] are all much more soluble than MnO_2 but their solubility is depressed with increasing soil pH.

The solubility of manganese in soils may be higher than might be calculated from some of these compounds due to the formation of coprecipitates with other metals, particularly iron. Deficiencies of manganese are not particularly common in calcareous soils despite the effects of pH on solubility of the various oxides and hydroxides.

Manganese adsorption by clay minerals in soils is relatively unimportant and represents only a transition phase between manganese ions in the soil solution and in precipitated forms of the elements such as MnO_2. Under very acid conditions, more exchangeable manganese may be present. Evidence for the adsorption of manganese by organic matter has accumulated and work has shown that manganese may form chelates with certain organic acids. As a general rule, manganese can be expected to form much less stable chelates in the soil with synthetic ligands than does zinc or copper. Below pH 6.3, none of the chelating agents such as DTPA, EDTA, or HEDTA produces a ratio of chelated to nonchelated Mn^{2+} greater than 1. In neutral to calcareous soils, DTPA, EDTA, and HEDTA form the most stable chelates with manganese. On the other hand, investigations by Geering et al.[55] have suggested that natural organic complexes in the soil maintain a chelated-manganese to Mn^{2+} ratio of 10 or 100 in acid, neutral, and calcareous soils. Above pH 7, only MnDTPA appears more available than in the naturally occurring chelating agents. Geering and co-workers estimated that from 84 to 99% of the manganese in displaced soil solution was present in organic complexes.

Oxidation-Reduction Reactions. The three oxidation states of manganese in the soil ($+2$, $+3$, and $+4$) are considered by some soil chemists to exist in equilibrium. Conversion from one oxidation state to another is dependent on soil pH, soil aeration, and the activity of soil bacteria. The reduction of Mn^{4+} to Mn^{2+} is directly related to soil pH over the range of about 3 to 8. More acid conditions tend to favor the formation of Mn^{2+} according to the following reaction:

$$MnO_2 + 2\ H^+ \rightleftharpoons Mn^{2+} + \tfrac{1}{2}\ O_2 + H_2O$$

Very acid soil conditions can lead to the presence of large, sometimes toxic, quantities of Mn^{2+} in soil solution. Some have considered that biological oxidation of manganese to the Mn^{4+} form is responsible for reduced availability to plants but others feel that the biological activity may be responsible for the formation of undetermined organic complexes of low availability.

Interactions with Other Elements. Iron-manganese interactions are discussed in Section 6.2. High concentrations of available manganese in the soil can lead to a depression in iron absorption and/or utilization. Toxic effects

[55] H. R. Geering, J. F. Hodgson, and C. Sdano, "Micronutrient Cation Complexes in Soil Solution. IV. The Chemical State of Manganese in Soil Solution," *Soil Sci. Soc. Amer. Proc.*, 33:81, 1969.

of excessive manganese produced by low soil pH may act through this relationship.

6.4.2 Manganese Nutrition of Plants

Metabolic Roles. Manganese is known to be a component of two enzyme systems, arginase and phosphotransferase. It is also known to substitute for magnesium in many of the ATP-dependent enzymes of glycolysis (sugar metabolism). In addition, manganese appears to participate in the oxygen-evolving system of photosynthesis. The manganese that functions in electron transport in chloroplasts does not appear to have the characteristics of specific manganese ions and has been suggested to be coordinated to a protein in the chloroplasts. The importance of manganese in the light reaction of photosynthesis, the Hill reaction, has been clearly identified but not totally defined. Manganese apparently has a role in photosystem II of photosynthesis, 3 to 4 atoms of manganese being present per photosynthetic II unit and is involved in the transfer of electrons from water to the photosynthetic II protein fraction. In summary, more information is needed to outline specifically the role of manganese in all enzyme systems identified but there is a strong connection between this element and electron transfer reactions. Manganese also has a role in the production of chlorophyll, although not as a component of the chlorophyll molecule.

Manganese Absorption. Manganese uptake or absorption by plants is governed by both passive and active mechanisms. Initial absorption is characterized as being related to concentration of manganese in the growth medium. This initial, rapid absorption is followed by a sustained, long-term absorption that is independent of concentration. The sustained absorption of manganese has all the characteristics of being metabolically controlled. It is severely inhibited by low temperature and by elimination of oxygen from the system. Also, very low pH has produced strong inhibition of manganese absorption in some plant species. Reported increases in manganese concentration in stem exudates of some species up to five times the concentration of manganese in the growth medium strongly suggest metabolic involvement. Data suggest that manganese uptake is similar to that of other cations and is metabolically linked.

Manganese Translocation in the Plant. Although citrate has been implicated strongly in the translocation of iron in the plant, this anion does not seem to be associated with manganese in the xylem of plants. Electrophoretic studies of plant stem exudate indicated that iron migrated in essentially the same fashion as Mn^{2+}. Phloem translocation of manganese is apparently possible because foliar applications of manganese compounds are capable of diminishing the severity of manganese deficiencies. A greater tendency for deficiency symptoms to develop on younger leaves of affected plants suggests

Table 6–22. Manganese Sufficiency Ranges in Selected Crops

Crop	Plant Part and Season Sampled	Sufficiency Range (ppm)
Corn	Ear leaf just before silking	20–150
Soybean	Upper mature leaf just before flowering	20–100
Alfalfa	Top growth — 15 cm, to flowering	30–100
Wheat	Upper leaves prior to first bloom	15–200
Sugar beet	Center mature leaf — midseason	20–150
Navy bean	Upper mature leaf just before flowering	20–150
Vegetables	Top fully developed leaf — midseason	30–200
Potato	Petioles from newly matured leaf at midseason	30–200

SOURCE: L. S. Robertson and R. E. Lucas, "Essential Micronutrients: Manganese," Mich. Ext. Bul. E–1031, 1976.

that once manganese has been delivered to a given area, translocation to new growth areas is slow or unlikely.

Genotype Interactions. Cultivar or varietal interactions of available soil manganese and plant requirements can be significant. Several investigators working with small grains have found that some cultivars are highly susceptible to manganese deficiencies, whereas other lines are completely tolerant of existing conditions. Oats are particularly susceptible to manganese deficiencies, and differences in absorption characteristics among cultivars persist under variable soil and environmental conditions. Because these differences do persist over a range of conditions, internal factors such as the reductive capacity of the roots may control manganese uptake. Iron, you will recall, is also subject to differences in the plants' abilities to reduce iron at the root surface or in its immediate vicinity.

Concentrations of Manganese in Plants. Robertson and Lucas[56] have summarized sufficiency ranges for several common crops (Table 6–22) and concentrations range from 20 ppm total manganese to 200 ppm. Manganese concentrations in plant tissue are a function of many factors including age of the plant, soil levels of manganese, and concentrations of other elements in the root environment, and soil temperature. Manganese usually is concentrated in leaves and stems. Seeds usually contain only small amounts of the element.

6.4.3 Manganese Deficiency Symptoms

Most of the deficient areas reported in the United States are found in humid areas, usually on soils with a high organic matter content. Soils that have been drained and brought under cultivation are frequently associated

[56] L. S. Robertson and R. E. Lucas, "Essential Micronutrients: Manganese," Mich. Ext. Bul. E–1031, 1976.

Table 6–23. Sensitivity of Selected Crops
to Manganese Deficiency

High Sensitivity

Bean	Pea	Soybean
Citrus	Peach	Spinach
Lettuce	Pecan	Sudangrass
Oat	Potato	Table beet
Onion	Radish	Wheat

Medium Sensitivity

Alfalfa	Celery	Sorghum
Barley	Clover	Spearmint
Broccoli	Cucumber	Sugar beet
Cabbage	Corn	Tomato
Carrot	Grass	Turnip
Cauliflower	Peppermint	

Low Sensitivity

Asparagus	Blueberry	Rye
Cotton		

with manganese deficiency. In fact, the deficiency of manganese has frequently been termed *reclamation disease.* High pH levels associated with high organic matter seem to be even more favorable to development of manganese deficiency symptoms. Liming acid, organic soils sometimes increases severity of manganese deficiencies. Manganese deficiencies have been reported in Wisconsin, Michigan, Minnesota, Florida, North and South Carolina, Ohio, Illinois, and New York as well as in several other states. As might be expected, crops show a wide range of susceptibility to manganese deficiency. A partial listing of manganese-sensitive crops is given in Table 6–23.

Corn and Grain Sorghum. Manganese deficiency symptoms in corn (Fig. 6–19) and sorghum are not so well defined as deficiency symptoms of zinc and iron in the same crops. Interveinal chlorosis accompanied by a general stunting is characteristic of both crops; however, more reports of deficiencies occur in corn than in grain sorghum. Chlorosis tends to be expressed first on the younger leaves indicating a general lack of mobility of manganese. Symptoms are somewhat similar to those of iron except that iron deficiencies seldom occur on high organic matter soils. Visual deficiency symptoms should be confirmed by tissue analyses.

Small Grains. Oats show the most severe deficiency symptoms of the small grains. Because many of the manganese-deficient areas in the United States are in the states surrounding the Great Lakes where oats are more commonly grown than the other small grains, it is partially coincidental that the deficiency has been noted so frequently with that crop. Nevertheless, oat

Fig. 6-19. Manganese deficiency in corn is not very common but occurs on both organic and mineral soils particularly in the northern part of the United States. Deficiency results in an interveinal chlorosis, an overall stunting, and internodes may be shortened. Higher soil pH levels usually result in lower available manganese. (Courtesy Dr. L. M. Walsh, University of Wisconsin.)

manganese deficiency symptoms are well defined. Mulder and Gerretsen[57] described the symptoms as marginal gray-brown necrotic spots and streaks appearing on the basal portion of leaves third from the top. The streaks eventually elongate and coalesce. As the necrotic spots extend across the leaves, the leaves may drop at that point. The ends of these affected leaves may remain green for an extended period. On older affected leaves, the necrotic spots are oval and gray-brown. *Gray-speck* disease and *halo blight* are terms that are also used to describe the symptoms.

Soybeans and Dry Beans. Deficiency symptoms of manganese in soybeans are an interveinal chlorosis that increases in intensity as the severity becomes more pronounced. As the deficiency becomes more severe, leaves may become pale green and eventually quite yellow (Fig. 6-20). Brown, necrotic spots develop as the deficiency becomes even more pronounced. Veins remain darker in the case of manganese deficiency as compared to iron deficiency. If plant roots explore soil with more available manganese later in the growing season, leaves that develop after the original symptoms have been expressed may be greener than those developed under deficient conditions. Depending then on the age of the plant, the chlorotic leaves may or may not be on the tops of the plants. Dry beans show essentially the same type of pattern as soybeans.

Cotton. Manganese deficiency in cotton is characterized by the development of a yellow-gray or reddish-gray color between green veins of the younger leaves. This deficiency has not been common in cotton but liming of long-cultivated, acid soils in the southeastern part of the United States has created some manganese deficiencies. Possibly a more common problem with manganese in cotton has been manganese toxicity. Cotton grown on very acid soils with large amounts of available manganese develops the condition

[57] E. G. Mulder and R. C. Gerretsen, "Soil Manganese in Relation to Plant Growth," *Adv. Agron.,* 4:221, 1957.

Fig. 6–20. Manganese deficiency in beans is characterized by increasingly severe interveinal chlorosis while veins may remain green. Eventually, the leaf may become totally chlorotic. Younger leaves are affected first. (Courtesy Dr. L. M. Walsh, University of Wisconsin.)

known as *crinkle leaf.* Leaves are deformed, cupped downward, and show a definite mottling. Later, necrotic areas develop along and between the veins. Leaves thus affected are brittle, thickened, and ragged.

Citrus, Deciduous Fruits, and Nuts. Manganese deficiency in citrus develops on the young leaves originally but, with persistence of the problem, older leaves may also show the typical interveinal chlorosis. The chlorosis tends to develop as light green areas along the leaf margins. Under severely deficient conditions, the chlorosis develops rapidly as leaves begin to enlarge. The irregular patches of chlorosis at the leaf margins contrast to dark green areas along the veins. Unaffected areas along the veins are wider than the veins themselves. As the leaves mature and the deficiency persists, the areas between the veins enlarge and become very light in color with whitish spots, especially in oranges. Older leaves on lemons may not be strongly affected. Symptoms may be most pronounced on the shaded portions of trees, lower branches on the north and northeast (northern hemisphere) and shaded leaves. Unlike copper, manganese does not seem to have an effect on fruit other than reduced production and a slightly lighter color.

Manganese deficiency symptoms in deciduous fruits such as peaches differ from symptoms on other broadleaf plants and do not affect the new growth areas. They may appear on any or all mature leaves as a mottling of tissue between the veins, but leaf tissue immediately adjoining the veins remains green, somewhat similar to those described for citrus. Chlorosis may be more complete near the leaf margin. Symptoms appear soon after the leaves are fully expanded and persist throughout the life of the leaf.

Most symptoms on leaves of nut trees are similar to those of deciduous fruits. In pecans, however, a shortening of the central vein in the leaf causes the leaf ends to be rounded. The leaves also are slightly crinkled, slightly cupped, and smaller than normal.

Grapes frequently show manganese deficiency, often at the same time that zinc deficiencies have developed. Unlike zinc deficiency, the distinct interveinal chlorosis with dark green bands along the veins tends to develop on basal leaves instead of on the terminal leaves of main and lateral shoots. Leaf size is not reduced nor is leaf morphology abnormal unless the conditions are very severe.

Vegetables. Manganese deficiency symptoms in most vegetable crops are somewhat similar. Tomatoes show a chlorosis developing in areas farthest from the veins in young leaves. As the condition becomes more severe, the leaves may become almost completely yellow with eventual loss of chlorophyll even along the veins. Necrosis may occur, beginning as small brown areas in the chlorotic zones farthest from the veins. Plants are also stunted and fail to flower properly, and yield is consequently depressed. Deficiency symptoms in celery, cabbage, and spinach are similar to those already described for tomatoes. Strong interveinal chlorosis of leaves at the growing tip is characteristic in spinach.

Beets. Sugar beets and, for that matter, table beets show essentially the same type of interveinal chlorosis described for the other crops. Veins and adjacent tissue remain green in sugar beets for an extended period of time despite chlorosis of the surrounding tissue. New leaf growth is most severely affected. Table beets develop a dark red to purple discoloration between the veins due to the large amounts of red pigments present. The discolored areas between the veins become necrotic in table beets. Sugar beets' leaves may become entirely chlorotic in very severe conditions and this causes some problems in distinguishing between manganese and nitrogen deficiencies. Tissue tests for nitrate nitrogen can distinguish between these two deficiencies.

6.4.4 Fertilizer Sources

Like zinc, iron and copper discussed in this chapter, manganese fertilizers include inorganic and organic materials. A partial listing of available sources of fertilizer manganese is given in Table 6–24.

Inorganic Fertilizers. Manganese sulfate, the most commonly used inorganic or organic source of manganese, has a high water solubility under both alkaline and acidic conditions. Manganese oxide (MnO) has a lower water solubility but performs well under acid conditions.

Organic Fertilizers. Organic sources of manganese have not been widely used due to the successful performance of the lower-cost inorganic sources. The same forms of chelated manganese exist that are discussed previously for

Table 6–24. Selected Manganese Fertilizers

Sources	Formula or Abbreviation	Manganese (% Mn)
Inorganic sources		
Manganese sulfate	$MnSO_4 \cdot 4H_2O$	24
Manganous oxide	MnO	68–70
Manganese carbonate	$MnCO_3$	31
Manganese chloride	$MnCl_2$	17
Manganese frits	—	10–25
Organic sources		
Manganese chelate	MnEDTA	12
Manganese polyflavonoid	MnPF	8
Manganese methoxyphenyl-propane	MnPP	5

zinc, iron, and copper. MnEDTA has been the most commonly used organic source, but recently ligninsulfonates and polyflavonoids have also entered the market. The efficiency of an available, soluble source of chelated manganese for a particular crop depends on soil conditions. Application of MnEDTA to soils deficient in manganese but high in iron results in rapid replacement of manganese by iron with subsequently high iron uptake and intensified manganese deficiency. Knezek and Greinert[58] noted this effect on manganese-deficient dry beans grown on a manganese-deficient muck soil.

Relative effectiveness of inorganic and organic sources of manganese is difficult to assess. Higher costs characterize the organic materials with MnEDTA usually higher than the wood-derived materials.

6.4.5 Methods of Application

Soil Applications. Conventional broadcast and banded applications of manganese sulfate are by far the most common application techniques recommended and used. Because manganese reacts readily with soil components, heavy rates of application intended to last for several years, such as are employed in zinc and copper applications, are not feasible with manganese. Recommended rates of application are lower for banded applications at planting compared to broadcast treatments (Table 6–25). Due to the high rates of fixation, some states do not recommend use of broadcast applications. Application in bands with acid-producing macronutrient fertilizers is sometimes more effective due to the localized acidifying effect of the macronutrient carrier.

Soil acidulation has been attempted as a means of supplying additional manganese, but costs have been prohibitive.

[58] B. D. Knezek and H. Greinert, "Influence of Soil Iron and Manganese EDTA Interactions upon the Iron and Manganese Nutrition of Bean Plants," *Agron. J.,* 63:617, 1971.

Table 6-25. Some Manganese Recommendations for Selected Crops

Crop	Source of Manganese	Rate of Application (kg/ha Mn)	Method of Application	Comments
Soybeans	$MnSO_4•4H_2O$	17–67	Broadcast	Annual application
	$MnSO_4•4H_2O$	6–22	Banded	Repeat as needed during season
	$MnSO_4•4H_2O$	1–2	Foliar	Repeat as needed during season
	MnEDTA	0.2–0.6	Foliar	Repeat as needed during season
Sugar beets	$MnSO_4•4H_2O$	22–90	Broadcast	Annual application
Onions	$MnSO_4•4H_2O$ or MnO	50–80	Broadcast	Annual application, check via soil test and plant analysis
	$MnSO_4•4H_2O$ or MnO	11–22	Banded	Annual application
Citrus, nuts	$MnSO_4•4H_2O$ or MnO	0.2–0.5 kg Mn/ 100 l H_2O	Foliar	Repeat as needed
Vegetables Snapbeans Spinach Cauliflower Celery Lettuce	$MnSO_4•4H_2O$ or MnO	10–12	Banded	Annual application
Corn, oats	$MnSO_4•4H_2O$ or MnO	17–67	Broadcast	Annual, but check soil test
		6–22	Banded	Annual application
Potatoes	$MnSO_4•4H_2O$	11–17	Banded	Annual application

Foliar Applications. Foliar applications of manganese may be recommended where regular spraying programs are in effect, where banded applications are not practical on high manganese-fixing soils, and where deficiency symptoms appear on an established crop. Foliar treatments should be considered as supplemental to soil applications. Inorganic ($MnSO_4$) and organic (MnEDTA) sources are both quite effective for foliar applications. Lucas[59] has recommended the foliar method of application as superior on soils that readily fix manganese. Randall and Schulte[60] compared manganese sources, rates of foliar application, and times of application of manganese for soybeans in Wisconsin and noted about a 3:1 ratio of efficiency between MnEDTA and $MnSO_4$. Manganese sulfate at 0.56 kg/ha of manganese was approximately equal to 0.17 kg/ha of manganese as MnEDTA. Multiple sprayings (two or three) were superior to a single application. In the final analysis, the feasibility of foliar applications of manganese is related to soil conditions, the regularity of a spray program, and time of discovery of the deficiency.

[59] R. E. Lucas, "Micronutrients for Vegetables and Field Crops," Mich. Ext. Bul. E–486, 1967.
[60] G. W. Randall and E. E. Schulte, "Manganese Fertilization of Soybeans in Wisconsin." Proc. Wisc. Fert. and Ag. Lime Conf., Madison, Wis., 1971, p. 4.

6.5 BORON

Boron and chlorine are the only nonmetals among the micronutrients. Boron always occurs in combinations with oxygen, usually coordinated with three oxygen atoms and occasionally with four. Although boron is found in some complex insoluble silicate minerals such as tourmaline, its primary minerals in terms of abundance are sodium and calcium borates. The common borates such as borax ($Na_2B_4O_7 \cdot 10H_2O$) and kernite ($Na_2B_4O_7 \cdot 4H_2O$) are quite soluble and tend to occur principally in evaporite deposits. Average concentrations of boron in the earth's crust are about 10 ppm. Concentrations in igneous rocks are in the same range but those values increase to 20 to 100 ppm for sedimentary rocks. Soil concentrations of total boron range from 5 to 80 ppm.

6.5.1 Soil Chemistry of Boron

Relatively little is known of the mineral forms of boron in soils. Some soils are known to contain appreciable quantities of tourmaline but nutrients in this borosilicate mineral containing sodium, lithium, aluminum, iron, and magnesium are essentially unavailable to plants without extensive weathering. Boron in the soil solution may exist as boric acid, H_3BO_3, which is only slightly dissociated up to a pH of 9.2. Above that pH, $H_2BO_3^-$ would become prominent. In all of these forms, boron remains in the B^{3+} oxidation state. At low concentrations, less than $0.1\ M\ H_3BO_3$, dissociation to $H_2BO_3^-$ is followed by hydration to form $B(OH)_4^-$; but dissociation is pH-dependent.

Boron is adsorbed more strongly by soil components than is chloride (Cl^-) or nitrate (NO_3^-). Apparently, boron adsorption by clay minerals occurs in a manner that is similar to adsorption of heavy metals such as zinc and iron. The major adsorption sites for boron in soils are thought to be iron and aluminum oxides and hydroxides that coat clay minerals, the clay minerals themselves (particularly the micaceous clay minerals), and magnesium hydroxy complexes on the surfaces of some primary silicate minerals. Adsorption by the hydroxy complexes is pH-dependent and is highest at pH 8 to 9 for iron complexes and highest at pH 7 for aluminum compounds. The availability of boron to plants is affected by soil pH not only in the adsorbed forms but also as it exists in association with organic matter.

Most of the soil's available boron is associated with soil organic matter.[61] Concentration of boron in soil organic matter has occurred through plant uptake of the element as it weathers from the primary forms such as tourmaline and is deposited in the soil's upper layers as plants die and decay. The types of reactions that occur between boron and organic matter are not well understood but probably revolve around the presence of hydroxyl groups on organic molecules such as phenols. The importance of hydroxyl groups in boron adsorption by organic soil components is strengthened by the known

---

[61] K. C. Berger, "Boron in Soils and Crops," *Advan. Agron.*, 1:321, 1949.

reaction of borates with phenols in plant nutrition and with iron and aluminum hydroxy compounds.

Factors Affecting Boron Availability. Of the micronutrients discussed to this point, boron has the greatest mobility in soils. Because boron is adsorbed by colloidal inorganic and organic fractions in the soil, sandy, acid soils may lose large quantities of applied boron in percolating waters. Higher quantities of clay and organic matter reduce leaching losses of boron.

Overliming acid soils to pH levels greater than 7.5 can result in lowered availability of boron. Excess liming may result in increased boron fixation or adsorption on clay minerals and through increased fixation by organic matter. Some researchers have suggested the formation of more insoluble calcium borates at high soil pH levels.

6.5.2 Boron Nutrition of Plants

Metabolic Roles. The biochemical role or roles of boron in plants is not well understood. Boron is not found as a portion of enzyme systems. One popular hypothesis is that boron functions by facilitating transport of sugars across cell membranes. This theory has been based on the ability of boric acid to complex with sugars and phenols that contain *cis*-diol groups in *in vitro* experiments. Phenylboric acid forms complexes with sugars on a one-to-one molecular basis. These complexes may be better able to move through the cell membrane due to their higher polarity. Some researchers have suggested that boron can complex with 6-phosphogluconate and subsequently modify the activity of the enzyme 6-phosphogluconate dehydrogenase, reducing the formation of compounds in the pentose shunt of sugar metabolism and leading to the production of phenolic compounds that have been observed to accumulate in boron-deficient plants.

One of the earliest morphological manifestations of boron deficiency in mung beans and tomatoes appears to be a slowdown in root extension, followed by a degeneration of meristematic tissue possibly due to a repressive effect of boron deficiency on cell division. This observation has prompted speculation of a relationship of boron to auxin synthesis or utilization. No specific metabolic role of boron in auxin metabolism has been identified; however, arylboric acids (derivatives of phenyl boric acid) have shown very definite effects on promotion of root growth.

In the absence of sufficient boron, polyphenolic compounds have been observed to accumulate in plants. These accumulations are related to the development of a brown color in plant tissues and lead to the speculation that boron is involved in the synthesis of cell wall components. Another physiological effect of boron deficiency in *in vitro* studies is increased uptake of certain labeled precursors of RNA into root tips (uridine and orthophosphate). This effect of boron has been likened to the observed effects of some plant hormones and has led to the speculation that boron metabolism may be in some way linked to synthesis and action of hormones.

Other observed effects of boron, although not completely documented, include influence on cell development and elongation of cells through control of polysaccharide formation. Boron has also been postulated to control the formation of starch, another polysaccharide, possibly preventing the excessive conversion of sugars into starch.

Boron effects on transpiration have also been recorded. Boron-deficient plants transpire less water, possibly due to higher concentrations of sugars and other hydrophyllic compounds in the plant cells. Reduced water absorption because of deficient boron was also considered to have a possible relationship with reduced transpiration. Although studies of the specific effects of boron in plant nutrition have been underway for over 60 years, more information is needed to determine the exact causes of the *observed* morphological effects of boron deficiency.

Absorption. Boron uptake by plants shows the same two-phase relationship described for other micronutrients. Early absorption is characterized as rapid and is followed by a steady linear phase. Boron uptake is temperature-dependent and is inhibited in different phases of the absorption by metabolic poisons such as cyanide and dinitrophenol. Boron uptake has also been inhibited competitively by hydroxyls, suggesting absorption in an anionic form. The effects of the antimetabolic compounds on uptake suggests that the process is connected to both the electron transport and to the oxidative phosphorylation systems depending on the concentration of boron in the growth medium.

Boron may be absorbed in one or more of its ionic forms, but because H_3BO_3 is more likely to exist at pH values below 9.2, uptake of the undissociated acid is also possible.

Translocation in the Plant. The form of boron translocated in the xylem from the roots to the upper portions of the plant is questioned. In simple aqueous solutions, boron probably occurs as the H_3BO_3 molecule, but the importance or role of sugar-borate complexes in sugar transport has been discussed frequently. There is little question of the existence of such complexes but relatively little is known about the stability of them, particularly at the relatively low pH values of xylem sap in plants.

6.5.3 Deficiency Symptoms

The relative sensitivity of various plant species to boron deficiency in the soil varies significantly. Table 6–26 lists some of the species that show greatest sensitivity to low boron. Cultivar differences in boron requirements also have been reported. Grape cultivars vary in their severity of boron deficiency symptoms. Root stocks determine boron sensitivity because sensitive cultivars grafted to nonsusceptible rootstocks and grown on a mildly boron-deficient soil do not develop deficiency symptoms. Table beets show extreme susceptibility to boron deficiency in some cultivars, where no deficiency was

Table 6–26. Relative Sensitivity of Selected Crops to Boron Deficiency

---

High Sensitivity

Alfalfa	Peanut
Cauliflower	Sugar beet
Celery	Table beet
	Turnip

Medium Sensitivity

Apple	Cotton
Broccoli	Lettuce
Cabbage	Parsnip
Carrot	Radish
Clovers	Spinach
	Tomato

Low Sensitivity

Asparagus	Pea
Barley	Peppermint
Bean	Potato
Blueberry	Rye
Cucumber	Sorghum
Corn	Spearmint
Grasses	Soybean
Oat	Sudangrass
Onion	Sweet corn
	Wheat

SOURCE: L. S. Robertson, R. E. Lucas, and D. R. Christenson, "Essential Micronutrients: Boron," Mich. Coop. Ext. Bul. E–1037, 1976.

noted with others on the same soil. Root rot is a common problem in susceptible cultivars of beets including both table beets and sugar beets.

Corn. Boron deficiency symptoms on corn in the field are not common although they may occasionally be observed. Boron deficiency results in a general stunting of the young plants due to shortening of the internodes but that, in itself, is not definitive. Leaves of young plants fail to emerge and unfurl properly and death of the growing point may occur. New leaves may show some white, irregularly shaped spots between the veins that become more numerous as severity of the deficiency increases. Eventually, these white, interveinal chlorotic spots assume the appearance of interveinal stripes. The affected tissues are waxy and are slightly raised above the surrounding tissue.

In mature corn, boron deficiency may be expressed as barren ears and subsequently decreased yields even though overt deficiency symptoms have not been observed during the growing season. Berger et al.[62] reported aborted

---

[62] K. C. Berger, T. Heikkinen, and E. Zube, "Boron Deficiency, a Cause of Blank Stalks and Barren Ears in Corn," *Soil Sci. Soc. Amer. Proc.*, 21:629, 1957.

flowering and abnormal ear development at boron concentrations in the growth medium sufficiently high to avoid production of leaf deficiency symptoms. Abnormal ears on boron-deficient plants may be shortened and bent with poorly developed, irregular rows of grain. Poor grain development may be only on one side of the ear.

Small Grains. Boron deficiency symptoms in small grains are rare and have not been adequately described. Levels of boron required for alfalfa or vegetables such as beets or celery can cause severe *toxicity* in small grains. Carry-over from banded applications of boron for a preceding row crop can have detrimental effects on small grains due to high concentrations of boron in a relatively small zone of soil. Boron toxicity on small grains is characterized by leaf-tip yellowing, interveinal chlorosis, and progressive scorching of the margins of the leaves. Toxicities are most likely to occur on acid, sandy soils. Toxicity symptoms of boron on small grain crops are notably similar to deficiency symptoms of copper and manganese.

Alfalfa and Red, Ladino, and Crimson Clovers. These forage legumes frequently show boron deficiencies, meaning that they are medium to high in sensitive to low levels of boron. Mild boron deficiency in alfalfa is expressed mainly as reduced flowering and seed set. Seed production is severely affected, but forage yields may be only slightly reduced. Under more severe conditions, boron deficiency produces a shortening of the internodes in the upper portions of the plants. Alfalfa is especially likely to show this effect. With shortening of the internodes, the upper leaves become rosetted, turn yellow over most of the leaf surface, and can be mistakenly identified as potassium-deficient or drought damaged. Actually, potassium deficiency does not appear on the same position of the plant, the chlorosis of potassium deficiency being expressed first on the older leaves at the base of the plant.

The frequent association of boron deficiency with moisture stress in alfalfa is a positional availability relationship. Because most of the soil's available boron is associated with the organic matter layer in the surface soil, poor moisture supplies cause the plant roots to absorb boron from the lower depths of the soil where less available boron is present. A rain will induce more boron absorption and subsequently new growth from axillary buds, which partially hide the chlorosis.

Deficiency symptoms in clovers are somewhat similar to those in alfalfa. A reddish to bronze discoloration occurs at the tips and margins of new leaves in red clover. Ladino leaves tend to be more yellowed like those of alfalfa. Youngest leaves are affected first and eventually these may become necrotic. Moisture relationships also hold for these crops as well as for alfalfa.

Peanuts. Production and quality of peanuts also may be seriously affected by boron deficiencies. Deficiencies in peanuts are expressed much more strongly in the nuts than in aerial portions of plants in the field. Boron deficiency symptoms in the nuts are dark depressions in the center of the nut.

Fig. 6-21. Effects of boron deficiency on cotton are pronounced. Note the effect of foliar applications of boron (left) in this Arkansas cotton crop. (Courtesy Dr. Woody Miley, Arkansas Coop. Ext. Serv.)

This condition is referred to as *hollow heart* and its presence greatly reduces quality and market value. Depending on severity of the deficiency, the condition can vary from a slight discoloration to a large dark cavity in the nut.

Cotton. Early symptoms are characterized by bands of discolored tissue around the petioles and the terminal bud often dies. Shortened internodes of secondary growth from axillary buds gives the plant a bushy appearance. Young leaves are frequently chlorotic but older leaves may remain green until frost. Older leaves are characteristically thick and leathery.

Flower buds of boron deficient plants become chlorotic. Squares frequently rupture and dry out. Squares and young bolls may be shed excessively resulting in very low yields (Fig. 6-21). Surviving bolls are frequently deformed and only partially open. Lint within such bolls may be dark and wet. Excellent descriptions of boron deficiency symptoms in cotton have been published by Miley and Woodall.[63]

Potatoes. Boron deficiency in potatoes may first become apparent in either the foliage or in the tuber, depending on the cultivar. Foliage deficiency symptoms appear at the growing point on the stem as shortened internodes and death of the growing bud. Growth of lateral buds gives the plant a bushy appearance. Deficiency of boron appears in tubers as stem-end browning, and severe deficiency causes strong russeting of the tubers. Tissue immediately below the skin may be dark and discolored.

Citrus, Deciduous Fruits, and Nuts. Boron deficiency in citrus may produce a variety of foliar symptoms including a yellowing or bronzing of leaves, accompanied by leaf thickening, brittleness, and downward curling. Upper

---

[63] W. N. Miley and W. E. Woodall, "Boron for Cotton in Arkansas," Ark. Agr. Ext. Ser. Leaflet 349, 1963.

leaf veins may be enlarged, corky, and ruptured. Research results have indicated a connection between boron deficiency, injured conducting tissue, and accumulations of carbohydrates in the leaves. Interruption of carbohydrate translocation begins in small twigs that may split the bark. Carbohydrate accumulations may then occur both in leaves and in fruit. As a further consequence of interference in carbohydrate translocation and metabolism, gum accumulations occur in twigs and fruit.

Gum accumulations in fruit is one of the most consistent and reliable deficiency symptoms. Citrus fruit may be deformed (lopsided) with large gum pockets in the white layer of the peel. Many dropped fruits result and ones not dropped may be hard and dry.

Apples, pears, and prunes show distinct boron deficiency symptoms. Apples develop internal cork in the fruit, particularly in relation to skin lesions. Fruits may show severe internal browning under very deficient conditions, specifically next to the core and fruit drop is common. New growth may die back and the bark develops corky patches and splits.

Pears experience drying and withering of blossoms (*blossom blast*) and dieback of the growing tips. Prunes experience a poor set of fruit in addition to development of elongaged, narrow leaves of reduced size. Growing tip dieback is common. Boron deficiencies seem to be more pronounced with application of nitrogen fertilizers.

Walnuts exhibit long, leafless shoots in the tops of the trees. Shoots become flattened and twisted at the tips and may be easily winter-killed.

Grapes are quite sensitive to boron deficiency and experience dieback of shoot tips, chlorosis of terminal leaves, and generally poor fruit set. Fruit development may produce many small seedless grapes accompanied by a few large berries. Grape clusters appear to abort or dry up when they are in bloom.

Vegetables. Beets of all types are sensitive to boron deficiency and exhibit both foliage and root deficiency symptoms. Leaves may be elongated and narrow showing a deeper red color. Leaves may die early followed by death of the growing point. A rosetted appearance of the foliage may develop due to the development of multiple crowns. Root symptoms appear as internal black spots or as black spots on the surface of the root. Eventually roots split and the black areas may completely encircle the root. Quality deteriorates to zero. Turnips and radishes exhibit similar symptoms.

Celery deficiency symptoms are characterized by cracked stems. Actual symptoms begin with mottling of bud leaves followed by development of brittleness and brown stripes in the stem. The stem or stalk eventually becomes cracked, curls outward, and becomes dark brown.

Cauliflower is another crop sensitive to boron deficiencies. Brittle leaves rolled downward or young leaves consisting almost entirely of enlarged, corky midribs are typical of increasing severity of boron deficiency. Deficient plants are stunted and the developing curd shows a brown discoloration. Stalks may be hollow.

Similar leaf brittleness, curling, and stem splitting are also characteristic

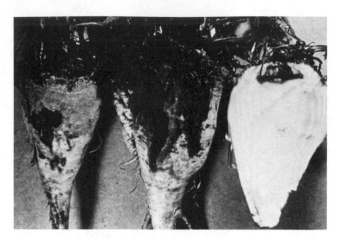

Fig. 6-22. Boron deficiency in sugar beets results in root or heart rot. The blackened areas in the cut beet on the right and the open lesions in the two beets on the left are typical of boron deficiency. (Courtesy The Fertilizer Institute.)

of deficiencies in broccoli. Carrots develop leaf bronzing and necrosis. Roots are small and badly cracked.

Sugar Beets. Deficiencies develop first in the young center leaves of the crown. Leaves are small and chlorotic and have short, brittle petioles. Prior to development of chlorosis, leaves are often darker green than normal. Petioles eventually crack, younger leaves die, and the growing point of the crown dies. Root rot is preceded by an internal browning in the vicinity of the crown. Root rot eventually hollows the beet, greatly reducing root and sugar yields (Fig. 6-22).

Positive diagnosis of boron deficiencies in many crops may require plant analyses. Some suggested boron sufficiency levels in crops are given in Table 6-27. Analyses of the leaves of fruit trees may be particularly helpful as a diagnostic tool.

6.5.4 Fertilizer

A list of boron fertilizers is given in Table 6-28. Unlike the other micronutrients discussed, boron fertilizers are all inorganic. The most common fertilizers are sodium borates, and differences between sources are due either to particle size or to the amount of water of hydration present in the compound. *Solubor,* a brand-name product, is a sodium borate with an intermediate amount of water of hydration and a small particle size, making it a practical source of boron for inclusion in liquid fertilizers.

Some work has been conducted in the southeastern United States with colemanite, a calcium borate mineral whose lower water solubility of that

Table 6-27. Suggested Boron Sufficiency Levels

Crop	Sample	Boron (ppm B)
Corn	Ear leaf at first silk	4–25
Soybeans	Top fully-developed leaves prior to bloom	21–55
Alfalfa	Top 15 cm prior to initial bloom	31–80
Sugar beets	Center fully developed leaf — midseason	26–80
Vegetables	Top fully developed leaf — midseason	30–60
Potatoes	Petioles, recently developed leaf — midseason	15–40
Navy beans	Top fully developed leaf — midseason	15–50
Prunes	Mid-terminal leaves, mid-August	30–50
Pears	Mid-terminal leaves, mid-August	25–50
Cotton	Top fully developed leaf at bloom	21–800

Table 6-28. Some Boron Fertilizers

Material	Formula	Boron (%)
Borax	$Na_2B_4O_7 \bullet 10H_2O$	11
Borate 48[a]	$Na_2B_4O_7 \bullet 5H_2O$	15
Borate 68[a]	$Na_2B_4O_7$	21
Solubor[a]	$Na_2B_4O_7$	20
Boric acid	H_3BO_3	17
Colemanite	$Ca_2B_6O_{11} \bullet 5H_2O$	10
Boron oxide	B_2O_3	31

[a] Brand name materials: 48 and 68 refer to B_2O_3 equivalent content, respectively.

material and subsequently reduced leaching of applied boron may make it useful on coarse, sandy soils. Otherwise, effectiveness of the few sources of boron fertilizers are quite comparable.

6.5.5 Methods of Application

Suitable boron materials are available for mixing with bulk blends, granulations, and fluid fertilizers. Boron materials of compatible particle size can be mixed effectively with bulk blends of solid materials. Similarly, boron can be effectively added to granular fertilizers without apparent undesirable side reactions. It is relatively easy to incorporate boron into fluid fertilizer formulations using some soluble source of the element such as *Solubor*.

Soil Application. Soil applications of boron are probably the most common method of supplying needed amounts of this element. Some recommended application rates for various crops are listed in Table 6–29. Broadcast applications are the most common methods of application. Banded applica-

Table 6–29. Boron Fertilizer Recommendations for Selected Crops[a]

Crop	Rate of Boron (kg/ha of B)	Method of Application	Comments
Cotton	0.2–0.6	Banded	*Do not* place in direct seed contact.
	0.6–1.1	Broadcast	Mix with macronutrient fertilizers.
	0.6	Foliar	Several sprays may be required. Begin application when plants are approximately 15 cm tall.
Alfalfa and clovers	1–4	Broadcast	Apply preplant or after first cutting; sandy soils may require annual applications. Less permeable soils may require B only once in 3 yr.
	0.6–1.1	Foliar	Apply when deficiencies have appeared. Not as effective as broadcast applications prior to symptoms.
Red beets	2.2–4.5	Broadcast	Preplant with macronutrients.
Sugar beets	1.1–2.8	Broadcast	Preplant with macronutrients.
Peanuts	0.3–0.6	Broadcast	Preplant with macronutrients; higher rates may cause yield reduction.
Soybeans	0.6–1.1	Broadcast	Preplant with macronutrients; higher rates may cause yield reduction.
Celery	0.6–1.1	Foliar	Multiple sprays totaling recommended rate are desirable.
	1.1–3.3	Broadcast	Preferable method of application.
	1.1	Side-dressed	Apply as soon as possible after observation of deficiency symptoms; less desirable than broadcast preplant.
Citrus	30–60 g/tree	Broadcast	Do not exceed recommended rates.
	6 g B/100 l	Foliar	Repeated sprays annually until symptoms disappear; check foliage analyses.
Apples	1.7–3.4	Broadcast	Apply every 3 years.
	48 g/100 l	Foliar	Annually at petal fall and 1 week later.

[a]Based on sodium borate ($Na_2B_4O_7$); several forms are available and all are effective.

tion rates are very critical due to the possibility of toxic effects if rates of application are too high. Placement directly with the seed is generally not recommended due to potential toxicity. Generally, soil applications are more effective if applied prior to seeding rather than after plant emergence.

Berger et al.[62] reported some success in sidedressing corn when the plants were about 30 cm (1 ft) high.

Rates of application are variable depending on the crop but in general are less than 3 kg/ha of boron (B). Recommendations vary somewhat with crop requirement but applications sufficient for high-requirement crops such as alfalfa and beets also suffice for crops such as corn as long as the treatments are broadcast. Crops having lower boron requirements such as cotton may be sufficiently supplied with as little as 0.6 kg/ha of B.

Foliar Treatments. Foliar applications of boron are most common on perennial crops or on those that routinely receive multiple sprayings. Deficiencies that develop during the growing season in annual crops may be treated successfully with foliar treatments. Most foliar applications involve the use of a soluble, finely divided source of sodium borate. Times of application of boron sprays vary with the crop; treatments should begin as soon as deficiencies are noticed. Repeat applications are usually required and are more effective than a single application of a greater amount of material.

Residual Effects. Boron leaches quite readily from acid, sandy soils and soluble borate applications to such soils may have minimal residual effects. On soils with higher clay and organic matter contents, residual effects may persist for several years, depending on the magnitude of the original application. Soil and tissue analyses are effective in determining the residual effects of boron.

Crop Responses. The magnitude of crop response to boron applications is highly variable including differences among plant species, varieties, and soil characteristics. Yield effects should also include quality effects such as those discussed concerning peanuts and sugar beets. Data presented in Table 6-30 represent attainable yield increases when soil test values for boron are low.

Table 6-30. Boron Responses in Selected Crops on Low Boron Soils

Treatments	Corn[a] (kg/ha)	Alfalfa[a] First Year (mt/ha)	Alfalfa[a] Second Year (mt/ha)	Soybeans[a] (kg/ha)	Wheat after[a] Soybeans (kg/ha)	Seed[b] Cotton (kg/ha)
Lime, P, K	4202	7.5	9.0	1996	3199	1902
Lime, P, K, B	4560	8.4	10.3	2278	3226	2488

[a]10-yr average, Kansas. L. Meyer and E. Beason, "Results of a Long-Term Rotation-Fertility Study," Kansas Fert. Research Report of Progress 194, 1972, p. 91.

[b]J. D. Lancaster, B. C. Murphy, B. C. Hunt, B. L. Arnold, R. E. Coats, R. C. Albritton, and Louie Walton, "Boron Now Recommended for Cotton," Miss. Agr. Exp. Sta. Bul. 635, 1962.

6.5.6 Conversion Factors

Expressions of boron in a fertilizer material should always be on an elemental basis but occasionally B_2O_3 is used to express the boron content of a material. Conversion from percent boron to percent B_2O_3 or vice versa can be made with the following factors:

$$0.3107 \times \% \ B_2O_3 = \% \ B$$
$$2.738 \times \% \ B = \% \ borax \ (Na_2B_4O_7 \cdot 10H_2O)$$
$$\% \ B \times 3.218 = \% \ B_2O_3$$

6.6 MOLYBDENUM

Molybdenum (Mo) is the least abundant of all the micronutrients either in the earth's crust or in soils. Concentrations of molybdenum in the crust are about 1 ppm, around 2 ppm in igneous rocks, and 0.2 to 3 ppm in sedimentary rocks. Soil concentrations usually range between 0.2 and 10 ppm. Some volcanic soils may be much higher in total molybdenum (300 ppm). Molybdenum can exist in several different oxidation states but those of most importance geologically and physiologically are +4 and +6. The most common mineral in igneous rocks is molybdenite (MoS_2). However, in minerals close to the earth's surface, including those in soils, molybdenum is an anion, molybdate. The most common form is the molybdate (MoO_4^{2-}) anion, which is mobile compared to the other heavy metals.

Molybdenum availability is generally high in alkaline soils, soils high in organic matter, and young soils derived from volcanic materials. Acid, highly leached soils such as those derived under forested conditions and acid sandstone-derived soils generally are low in available molybdenum.

6.6.1 Soil Chemistry

Molybdenum usually is present in soils chiefly as adsorbed molybdate. Ionic species in the soil may include $HMoO_4$ under very acid pH regimes as well as the more common MoO_4^{2-}. Solubility of calcium molybdate, for instance, increases 100X for each unit increase in pH. A soil-molybdenum complex of unknown composition also shows essentially the same increase in solubility with increasing pH. The relationship of available molybdenum to soil pH is exactly the *opposite* of the other micronutrient metals; i.e., the higher the pH the more readily available the molybdenum.

Of the forms of molybdenum present in the soil, Hodgson[64] considered adsorbed molybdenum to be the most important. Several researchers have noted

---

[64] J. F. Hodgson, "Chemistry of the Micronutrient Elements in Soils," *Advances in Agron.*, 15:119, 1963.

that hydrous iron oxides adsorb molybdenum much more strongly than aluminum oxides or various secondary silicate minerals. This adsorption by iron oxides is pH-related, increasing with decreasing pH in the range 7.75 to 4.45. Increasing hydroxyl concentrations at higher pH values in soil solution could suppress adsorption sites for another anion on the surface of hydrated iron oxides. In contrast to phosphorus adsorption in calcareous soils, molybdates do not form specific compounds with calcium and so the availability of molybdenum continues to increase regardless of the amount of calcium present in the soil.

Available molybdenum concentrations have been correlated with high levels of organic matter in some poorly drained soils. However, Hodgson[64] suggested that accumulation of molybdenum in soils along with organic matter is not necessarily an indication of a molybdenum-organic matter complex.

More information is needed to test the theories of the factors and compounds controlling molybdenum availability in soils. Despite theories of iron molybdate control of molybdenum availability, no direct evidence has been reported.

6.6.2 Molybdenum Nutrition of Plants

Metabolic Roles. The most widely discussed role of molybdenum in plant metabolism is in nitrate reductase activity, the reduction of nitrate nitrogen to nitrite or to an even more reduced state. This role is discussed in detail in Chapter 2 in relation to plant metabolism of nitrogen. Electron transport is a specific role assigned to molybdenum in that activity. In addition to the important function in nitrate reductase, nitrogenase, the enzyme responsible for the reduction of nitrogen from nitrogen gas (N_2) to an oxidation state equivalent to ammonium or amino nitrogen, has also been determined to require molybdenum. In the latter case, molybdenum apparently serves the same role in electron transport as in nitrate reductase. Interestingly, the proteins of these two enzymes are quite comparable in their physical-chemical characteristics. Of special significance is the fact that legume *Rhizobia* bacteria require molybdenum for N_2 fixation.

Because some plants require molybdenum even when all the nitrogen is supplied in a reduced form (urea-nitrogen, ammonium, or nitrate), some other process may also require molybdenum. Other enzymes, including xanthine oxidase and aldehyde oxidase require this element in animals.

Some plants have indicated an abnormal metabolism of phosphorus and ascorbic acid when grown in molybdenum-deficient conditions. This could be related to observations of a requirement for molybdenum by some hydrogenase enzymes in some lower plants. Molybdate has been noted to affect the chemical hydrolysis of orthophosphate and pyrophosphate esters and also influences the proportions of inorganic and organic phosphorus in plants. Addition of molybdenum to deficient plants results in the conversion of inorganic phosphorus, already absorbed, into organic forms. Such an effect could be in-

Table 6–31. Sufficiency Ranges for Molybdenum for Selected Crops

Crop	Sample and Time	Molybdenum (ppm Mo)
Corn	Ear leaf prior to silking	0.1–2.0
Soybeans	Fully matured leaf prior to flowering	1.0–5.0
Alfalfa	Top 15 cm prior to flowering	1.0–5.0
Wheat	Upper leaves prior to first bloom	0.03–5.0
Sugar beets	Center mature leaf — midseason	0.15–5.0
Vegetables	Top fully developed leaf — midseason	0.5–5.0
Potatoes	Petiole from newly matured leaf — midseason	0.5–4.0

direct through stimulation of cell growth and consumption of phosphorus ions already taken up.

Uptake and Translocation. Due to the probable dominance of the MoO_4^{2-} ion in soil solution, molybdenum uptake is probably in this form. Direct evidence for active absorption of molybdate is generally lacking despite the fact that its mechanism of absorption should be generally comparable to that of other anions such as sulfate, nitrate, and phosphate. Some short-term (6-hour) studies of molybdenum uptake by tomatoes have indicated depression of uptake by increased pH in solution culture. Sulfate has reportedly inhibited uptake and translocation of molybdate, whereas phosphate has enhanced translocation in plants. Sulfate competition with molybdate for absorption sites on the root surface in the first step of absorption is possibly due to the similarity of ion size and charge of sulfate and molybdate.

Plant Concentrations. Relatively little information is available on the concentrations of molybdenum in plants that are required for certain levels of plant growth, but values are generally quite low. Due to its requirement in nitrate reduction systems and in nitrogen fixation by legumes, the element is found in both the roots and leaves of plants (Table 6–31). Jones[65] reported that deficiencies are usually related to concentrations of 0.10 ppm Mo or less in dry matter. On the other hand he also notes that concentrations of several hundred parts per million also are found in normal plants. Such high concentrations in forages (> 15 ppm) are potentially *toxic to animals* (see Section 6.6.5).

6.6.3 Deficiency Symptoms

Basic work in the establishment of molybdenum requirements for higher plants was conducted following discovery of molybdenum-responsive

[65] J. B. Jones, Jr., "Plant Tissue Analyses for Micronutrients." In *Micronutrients in Agriculture*, J. J. Mortvedt et al. (Ed.), Soil Science Society of America, Madison, Wis., 1972, p. 319.

Table 6–32. Sensitivity of Selected Crops
to Molybdenum Deficiency

High Sensitivity	Medium Sensitivity	Low Sensitivity
Broccoli	Alfalfa	Asparagus
Cauliflower	Bean	Barley
Lettuce	Cabbage	Celery
Onion	Carrot	Corn
Spinach	Citrus	Grasses
Table beet	Clovers	Mint
Peanut	Peas	Oat
	Radish	Potato
	Soybean	Rye
	Sugar beet	Sorghum
	Tomato	Sudangrass
	Turnip	Sweet corn
		Wheat

areas in Australia. Problem areas in Australia were and are associated with the development of an animal grazing system on soils that were generally coarse textured and low in organic matter. Many of these soils have been derived from lateritic materials, high in iron oxides, the compounds capable of adsorbing molybdate. Much of this area is located in Western Australia. No general area of molybdenum deficiency exists in the United States. Molybdenum deficiencies have been enountered on acid, coarse-textured soils along the Atlantic Coast, the areas surrounding the Great Lakes, and in some areas in the Pacific Coast states. Liming practices have undoubtedly masked Mo deficiencies in many acid soils. Sensitivity to Mo deficiency varies among crop species (see Chapter 8) (Table 6–32).

Soybeans. Field deficiency symptoms in soybeans are not adequately described in the United States. Molybdenum responses in soybeans and other legumes have been recorded despite a lack of visible deficiency symptoms. Growth of soybeans in nutrient culture devoid of molybdenum has produced beans with deformed leaves, twisted on their stems and showing mild chlorosis. Some necrosis may also develop along the center midrib.

Alfalfa and Clovers. Molybdenum deficiency in most legumes is typical of that of nitrogen deficiency. Poor supplies of molybdenum lead to poor nitrogen (N_2) fixation and plants show a general chlorosis and stunting. Lower leaves in alfalfa and the clovers may become necrotic and drop off.

Vegetables. One of the first incidences of molybdenum deficiency in the field in the United States was reported in California with tomatoes grown in soils developed on serpentine, a magnesium-iron silicate mineral. Deficiencies on affected tomatoes were similar to those of nitrogen due to the role of molybdenum in nitrate reduction.

Fig. 6–23. Molybdenum deficiency in cauliflower produces rolled, twisted leaves commonly known as "whiptail." Molybdenum-deficient soils are usually acid or derived from iron oxide materials. (Courtesy Dr. C. H. Dearborn, USDA, SEA-AR, Palmer, Alaska.)

Molybdenum deficiency in cauliflower has been recognized in the field and leads to the development of a condition that has been described as "whiptail." Other members of the *Brassica* group such as broccoli also develop these symptoms, which are characterized by a cupping of the leaf margins of the younger leaves. The cupped leaves present a rolled appearance and may also have some interveinal chlorosis. In later stages of the deficiency, the leaves tend to twist around the central mid-rib causing the rolled appearance to be even more pronounced (Fig. 6–23).

Citrus. Molybdenum deficiency in citrus has been recognized in Florida. Symptoms develop on the underside of *older* leaves, consisting of reddish brown spots that become impregnated with a gum. Tissue surrounding the spots becomes yellow. Mild cases of deficiency result in a single spot per leaf, and severe deficiencies produce more spots per leaf. These leaves are usually dropped prematurely. Apparently, fruit appearance is unaffected by molybdenum deficiency.

6.6.4 Fertilizer Sources and Methods of Application

Fertilizer sources of molybdenum (Table 6–33) are quite limited and include sodium and ammonium molybdate as the primary materials. No organic sources are available on the fertilizer market. Both sodium and ammonium molybdates are quite soluble. Less soluble molybdenum sources, including molybdenum trioxide and molybdenum sulfide, have been applied in acid fertilizers such as superphosphate with some degree of success. The very low solubility of molybdenum sulfide even in macronutrient fertilizers makes it a relatively poor choice.

Methods of application include soil treatments, foliar applications, and seed coating. Only very low rates of molybdenum are required, so application

Table 6–33. Molybdenum Fertilizers

Source	Formula	Molybdenum (% Mo)
Ammonium molybdate	$(NH_4)_6Mo_7O_{24} \cdot 2H_2O$	54
Molybdenum trioxide	MoO_3	66
Molybdenum frit	—	30
Sodium molybdate	$Na_2MoO_4 \cdot 2H_2O$	39
Molybdenum sulfide	MoS_2	60

as a combination with macronutrients is most desirable to obtain uniform applications. Molybdenum can be granulated into various macronutrient materials or it can be sprayed on the surface of solid materials, thus improving the distribution of the small amount applied. Molybdenum has also been successfully added to sulfuric acid in the production of superphosphate (see Chapter 3). Soluble molybdenum sources can be effectively applied in liquid fertilizers. Rates of soil-applied molybdenum range from only a few grams to as much as 2 kg/ha of Mo, but most rates are about 50 to 100 g/ha.

Foliar applications are rarer but can be effective in correcting a deficiency in an established crop. As in the deficiency of any element, application soon after deficiencies are discovered improves efficiency of foliar treatments. Studies in Georgia by Boswell and Anderson[66] produced evidence of greater efficiency of early season foliar molybdenum applications of 560 g/ha for soybeans versus applications delayed until bloom or early pod development stages.

Seed treatments have been a common means of supplying molybdenum to legumes. Seed coating of soybeans, alfalfa, field peas, and other legumes has been effective at rates of about 50 to 100 g/ha of Mo. Some recommendations are indicated in Table 6–34.

Crop responses to molybdenum applications vary widely with sensitivity to the deficiency (see Table 6–32), soil type and soil pH.

Liming can frequently increase the availability of molybdenum and produce greater yield responses than the molybdenum application itself. Investigations in Louisiana indicated about 400 kg/ha yield increases in soybeans from molybdenum application, but liming produced a yield increase of over 900 kg/ha. Similarly, lime-molybdenum studies in Georgia produced 3100 kg/ha yield increases in alfalfa from molybdenum applied at the lowest lime treatment. Liming soil to the recommended level plus the addition of molybdenum produced a smaller response of 1900 kg/ha. Other reported soybean yield increases in Georgia from seed applications of molybdenum at a rate of about 90 g/ha of Mo were over 1000 kg/ha. Comparable foliar treatments produced slightly less, a yield increase of 940 kg/ha.

[66] F. C. Boswell and O. E. Anderson, "Effect of Time of Molybdenum Application on Soybean Yield and on Nitrogen, Oil and Molybdenum Contents," *Agron. J.*, 61:58, 1969.

Table 6–34. Some Recommendations for Molybdenum Application for Various Crops on Low Molybdenum Soils[a]

Crop	Rate of Molybdenum (g/ha of Mo)	Method of Application	Suggestions
Soybeans	200	Broadcast	Apply in a macronutrient fertilizer
	200	Seed coating	Use commercial preparation
	200	Foliar	Use wetting agent
Sugar beets	50–800	Broadcast	Apply in a macronutrient fertilizer
Field peas	50	Seed coating	Use commercial preparation
Cauliflower	20–30	Foliar	Use wetting agent
Citrus	1 g/tree	Foliar	Use wetting agent; repeat as indicated by soil test

[a]All sources were sodium molybdate ($Na_2MoO_4 \cdot 2H_2O$).

6.6.5 Toxicity to Livestock

Molybdenum at excessive concentrations in forage plants leads to toxicities in animals. This fact has been recognized for over 100 years. Affected animals develop diarrhea and lose weight, show a drop in milk production, and may eventually show bone deformities. *Molybdenosis* is the name given to this condition that apparently results from an antagonism of molybdenum on copper metabolism. The critical level of molybdenum concentration in forage grasses apparently is about 15 ppm Mo. Molybdenosis can be overcome by providing animals with supplemental copper or by copper application to the soil. The copper involvement in this relationship apparently occurs through the formation of a cupric thiomolybdate of low availability, at least in ruminants.[67]

6.7 CHLORINE

Chlorine has been the most recent addition to the list of essential micronutrients. Chlorine, or more correctly, chloride is the most widely distributed of the halogen elements in nature. The chloride ion (Cl^-), a component of soluble salts, is of primary interest in soil and plant chemistry. As an anion, this element is not readily adsorbed by soil components and is subsequently not fixed. Leaching of chloride is frequently used as a tracer for movement of other soluble anions such as nitrate. The oxidation state of chloride ions is not changed by soil bacteria.

[67] N. F. Suttle, "Trace Element Interaction in Animals," In *Trace Elements in Soil-Plant-Animal Systems*, D. J. D. Nicholas and A. R. Egan (Ed.), Academic Press, 1975, p. 271.

6.7.1 Plant Nutrition and Deficiency Symptoms

The essentiality of chlorine was discovered in 1954 by Broyer and co-workers,[68] who studied the effects of this element on tomatoes. Later this same group of scientists produced evidence of chlorine essentiality in lettuce, barley, alfalfa, field beans, sugar beets, and corn, to name a few of the species studied. All of these investigations were conducted in solution cultures.

Specific metabolic roles for chloride have been noted in noncyclic photophosphorylation and in the riboflavin phosphate pathway of cyclic photophosphorylation reactions of photosynthesis. These reactions are responsible for the capture and storage of light energy in the form of high-energy phosphate bonds (see Chapter 3).

Chloride deficiency symptoms are difficult to describe, for so few have actually been observed. Plants suffering from a severe deficiency of chloride have demonstrated chlorosis and necrosis of leaf areas. Leaf tips wilt followed by development of a bronze coloration that is followed by necrosis. Root development in solution culture has been notably restricted (Johnson et al. [69]). Uhlrich and Ohki[70] noted that chloride was required for the growth of sugar beets in artificial culture and described deficiency symptoms as an interveinal chlorosis on the blades of the younger leaves. They further described the deficiency symptoms as similar to those of manganese in the early stages of development that advanced into smooth, flat depressions in the interveinal areas. Veins in such areas appear raised.

The amount of chloride found in plants is many times higher than that of most of the other elements. Chloride concentrations may range from 2500 to 50,000 ppm (0.25 to 5.0%) chloride ion. Some estimates of chlorine deficiency suggest that 250 ppm is a critical level for tomatoes. Much more work is needed before definitive statements concerning critical chloride levels in plants can be made.

Chloride is normally present in the soil in sizable quantities, which are a function of the amount of soluble salts present, particularly sodium chloride. Estimates range from zero up to 1100 kg/ha of chloride. If this element is needed in supplemental amounts, several soluble sources are available. The most common is potassium chloride, which contains about 47% chlorine.

SUMMARY

Micronutrients known to be essential for plant growth include zinc, iron, copper, manganese, boron, molybdenum, and chlorine. Chlorine is the most recent

[68] T. C. Broyer, A. B. Carlton, C. M. Johnson, and P. R. Stout, "Chlorine—A Micronutrient Element for Higher Plants," *Plant Physiol.*, 29:526, 1954.

[69] C. M. Johnson, P. R. Stout, T. C. Broyer, and A. B. Carlton, "Comparative Chlorine Requirements of Different Plant Species," *Plant Soil*, 8:337, 1957.

[70] A. Uhlrich and K. Ohki, "Chlorine, Bromine and Sodium as Nutrients for Sugar Beet Plants," *Plant Physiol.*, 31:171, 1956.

addition to this group and the element about which least is known. No relative ranking in terms of importance can be assigned to any list of essential elements including the micronutrients. Most of these elements have been studied sufficiently so that their roles in plants are fairly well identified. Besides chlorine, the two other elements about which least is known (particularly in reference to soil chemistry and metabolic role in plants) are boron and molybdenum.

All micronutrients are metals except boron and chlorine. Availability of most micronutrients tends to decline with increasing soil pH. Molybdenum availability increases with increasing pH, whereas that of chloride is unaffected by soil pH.

Micronutrient fertilizers may be characterized generally as being more expensive per kilogram of nutrient than the macronutrient materials. This is not surprising because some of the metals are relatively rare and mining and processing is expensive. Limited requirements and limited consumption contribute to higher production costs per unit of nutrient.

Application rates of micronutrients are usually low enough so that both soil and foliar application are feasible. However, detection of micronutrient deficiency by soil analyses prior to seeding is desirable. Correction of deficiencies prior to their expression in plant morphological symptoms leads to better yields. The small amount of micronutrients required to overcome a deficiency also allows the possibility of seed coatings of some of these elements. Molybdenum and more recently iron apparently are effective in this method of application.

Returns for investment in micronutrients may be spectacular under deficient conditions, even greater than the returns produced by applications of macronutrients. However, responses to these elements are more subject to climatic conditions. Cold, wet conditions early in the growing season frequently are associated with micronutrient deficiencies.

PROBLEMS

1. Demonstrate your familiarity with the types of soil conditions that might be conducive to the presence of a zinc deficiency. Assume that the crop planned for production is corn. If soil analysis has demonstrated a deficiency of zinc, briefly outline a preferred means of correcting that problem including rates of nutrient to be applied, methods of application, and form(s) of zinc to be used.

2. Outline the types of soil conditions and climatic conditions that are conducive to the development of iron deficiency in grain sorghum.

3. Boron applications can be troublesome in some respects. Briefly tell why or how problems can develop in the process of fertilization to overcome a boron deficiency. Describe how you would avoid those problems.

4. Copper deficiencies are frequently associated with a soil component that is usually assumed to be conducive to availability of other micronutrient metals such as iron,

zinc, and manganese. What is that soil component, and how does it affect copper availability?

5. Manganese can exert a strong influence on the availability of another micronutrient, one that is very similar in terms of its chemistry. What is that element and how does manganese affect its availability? Also, outline your impression of the feasibility of applying MnEDTA to correct a manganese deficiency on a muck soil.

6. Molybdenum is required in small quantities by crop plants. Outline a practical program for correction of molybdenum deficiency in soybeans on a soil of pH 5.2.

7. A soil fertility recommendation calls for application of 8 kg of zinc (Zn), 5 kg of iron (Fe), and 5 kg of manganese (Mn)for a crop of grain sorghum in addition to 150 kg of nitrogen (N), 20 kg of phosphorus (P_2O_5), and 80 kg of potassium (K_2O) per hectare. You are called on to formulate a mixture of urea (45% N), DAP (18–46–0), and KCl (0–0–60) plus the sulfate forms of the micronutrients to supply these amounts of nutrients. How much of the various materials would be required for 100 ha of corn? All the materials will be applied preplant as a bulk blend. Assume that you have no segregation problems. What do you think of the recommendation, knowing nothing more than what you have been told? Explain any comments you make. Use the tables in this chapter for the composition of the metal sulfates.

CHAPTER SEVEN

Applying Fertilizers and Soil Amendments

More developments in techniques for fertilizer and soil amendment application have occurred in the last 10 to 15 years than in all the preceding history. Application techniques for gaseous, dry, and liquid fertilizers have all involved the development of new equipment, much of which has been associated with preplant application of nutrients. Equipment size now dwarfs anything conceived in the early 1960s. Flotation equipment designed for reducing soil compaction and operating on wet soils has revolutionized both the liquid and solid fertilizer and soil amendment markets. Ammonia application equipment has expanded in size and has rapidly been incorporated into tillage operations. Fertilization during irrigation has expanded rapidly with irrigation development in the Great Plains, Pacific Northwest, Southwest, and Southeast. Introduction of new tillage systems has dictated a reevaluation of fertilization and liming methods in many areas. All these subjects and more are discussed in the following sections.

7.1 APPLYING AMMONIA AND OTHER COMPRESSED GASES

Anhydrous ammonia is the basis for the modern nitrogen fertilizer industry (see Chapter 2). Direct application of this compressed gas as a nitrogen fertilizer has developed rapidly in the last 20 years and it now represents the most common nitrogen material in use in the United States for direct application (about 38% of total N used in 1979). One of the first uses of ammonia for direct application was injection into irrigation water; this is discussed in detail in Section 7.2. Other methods of application have far outstripped water application, however. Methods of applying ammonia directly in the soil are discussed here.

7.1.1 Early Soil Applications of Ammonia

Following the first use of ammonia in irrigation water in 1934, soil injection of ammonia was examined in California by Shell Development Company. Leavitt[1] was instrumental in the development of the first equipment and adapted a furrowing tool for the application of ammonia by welding tubes on the back of each shank of the furrower and trailing a hose a short distance behind the shank. Initial trials with this equipment indicated good soil retention of ammonia and provided the basis for development of the more sophisticated equipment in use today.

Application of ammonia in either soil or water required the development of some metering device that would allow controlled release of the compressed gas. Kortland[2] patented an early flowmeter involving the use of orifices to control the rate of flow, but the equipment was large and cumbersome and not readily adaptable to mobile application. Further research in this area led to the development of the more common needle valve types of metering devices widely used today. These meters are small, relatively cheap, and simple in construction, and they have a wide range of workable application rates. They operate on the basis of time control of flow so applicator speed and swath width must be taken into account to determine the rate of ammonia applied. After passing through the needle valve assembly the liquid ammonia partially vaporizes with a mixture of liquid and vapor being delivered to the soil. Flow to the points of the applicator must be divided in a manifold with sufficient pressure within the system to assure that all delivery lines receive the same amount of ammonia. Large openings in the ends of the delivery lines coupled with differences in hose lengths can produce variable rates of delivery to application points.

In some parts of the country where terrain is uneven, ammonia application involves the use of positive displacement pumps that act at a rate dependent on ground speed. Steep slopes in the field change the speed of the applicator making the use of constant flow, constant ground-speed-dependent equipment impossible. These pumps are essentially of the same design as pumps for liquid fertilizers but do include a heat exchanger. Pump delivery is split by a manifold prior to delivery to the various shanks. Before considering further the direct application of ammonia to the soil, some attention should be given to equipment needed for safe and efficient handling and transporting ammonia.

[1] F. H. Leavitt, "Process for Soil Fertilization," U.S. Patent 2 285 932, June 9, 1942, Shell Development Company.

[2] F. Kortland, "Flowmeter," U.S. Patent 2 038 511, April 21, 1936, Shell Development Company.

7.1.2 Ammonia Tanks

Because ammonia must be handled as a compressed liquified gas at normal temperatures, it was necessary to develop pressure vessels or tanks to transport the ammonia from the point of manufacture to the point of use. The first pressure vessels used for application of ammonia in irrigation water were carbon steel bottles or cylinders similar to those used for handling welding gases. These cylinders were too heavy to be handled easily, had no safety valves, and contained too little ammonia to be of practical use on large-scale applications. Limitations of these small cylinders led to the development of larger and larger pressure vessels capable of handling amounts of ammonia at least adequate for small fields.

The first tanks for the handling of bulk ammonia were considered large in the mid-1940s but held a maximum of only 800 kg of ammonia. These bulk tanks were still used primarily for application of ammonia in irrigation water. Later smaller vessels capable of being mounted on tractors for direct application were produced and held from 130 to 230 kg of ammonia. Development of trailer applicators quickly followed and involved the use of 1300 to 1700 kg capacity tanks. Today, applicator tanks and nurse tanks for transportation between dealer facilities and the field have increased to about 1000- to 2000-kg capacities taking into account the filling of tanks to about 85% of capacity as a safety precaution (Fig. 7–1).

All containers for the nonrefrigerated storage and transportation of ammonia must be constructed and tested under strict codes. These carbon steel tanks are designed for minimum pressures of 17.6 kg/cm^2 (250 lb/in.^{2}) and range up to 28 kg/cm^2 (400 lb/in.^{2}) for noninsulated tank cars. All pressure vessels are equipped with relief valves to protect the tanks from excess pressure and are set to discharge at a pressure less than the design maximum

Fig. 7–1. Anhydrous ammonia nurse tank used to transport ammonia from the dealer's storage to the field and now commonly used as a trailing tank for large field applicators. (Courtesy L. S. Murphy.)

of the tank. In addition, each tank is equipped with an 85% fixed-level gauge, pressure- and liquid-level gauges, and excess flow valves. The 85% fixed-liquid-level gauge is designed to determine when the tank has reached its 85% fluid fill capacity in the event of a failure of the liquid-level gauge. It is important not to fill high-pressure ammonia vessels with more than 85% of their capacity to allow for expansion of the liquid ammonia with rising temperatures. If a tank was filled to capacity while the ammonia was cool, higher daytime temperatures could cause expansion sufficient to produce rupture of the steel walls. The liquid-level gauge is a sight gauge attached to a float arm inside the vessel. The purpose of this gauge is to give the operator an indication of when the tank has reached capacity in filling or to indicate when the tank is empty.

Excess flow valves are installed in lines leading from the tank and are intended to close in the case of hose rupture. These valves are spring-loaded and set to trip at rates of flow higher than a predetermined level. They, like the pressure relief valve and the 85% fill gauges are important safety features in handling ammonia.

High-pressure vessels for ammonia storage, transportation, or application are comparable in their design. Basically, the only differences that exist in these tanks is their capacity and tank-operating pressures. As indicated earlier, applicator and nurse tanks range from 1000- to 2000-kg capacities (85% fill). Tank cars for ammonia range from about 19 metric tons to 60 metric tons and may be insulated or noninsulated (Fig. 7–2). Trucks have about 11- to 17-metric ton capacities. Barges with high-pressure tanks are constructed of several smaller tanks but have capacities in the range of 900 to 1600 metric tons.

Some of the equipment used for the transportation of anhydrous ammonia is also used in the off season for movement of liquified petroleum gases (LPG). When this type of interchange in use is planned for a barge, car, or truck, the

Fig. 7–2. Rail cars for ammonia transportation are of two main types. The standard, insulated car (left) has a capacity of about 11,000 gal (41,000 *l*) but is not filled to capacity allowing space for expansion of the ammonia on warming. The other type of car is the uninsulated, high-capacity car (right). Its nominal capacity is around 30,000 gal (113,000 *l*) and it too is not filled to capacity because of expansion of the ammonia on heating. (Courtesy L. S. Murphy.)

Applying Fertilizers and Soil Amendments

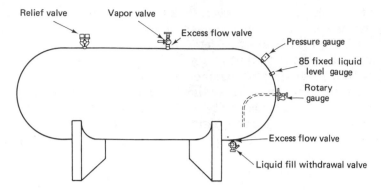

Fig. 7–3. Small storage tank for anhydrous ammonia showing valves and fittings. Location of valves and fittings varies depending upon the manufacturer.

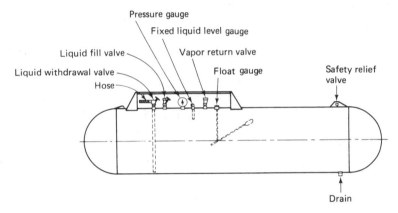

Fig. 7–4. Valves and fittings in a nurse tank are similar to those in a small storage tank, a truck, or a tank car.

metals used in the fittings must be compatible with ammonia. Brass (alloy of copper, tin, and zinc) fittings that are used for LPG must be replaced with stainless steel when the tanks are used for ammonia. Ammonia cannot come into contact with copper- and zinc-containing alloys without causing severe corrosion and eventual failure. Dealer ammonia storage and handling equipment as well as application equipment is essentially never used for LPG.

Some idea of the fittings associated with a small ammonia storage vessel at a dealer's place of business or nurse and applicator tanks is indicated in Figs. 7–3 and 7–4. Storage tanks often include temperature gauges that are not standard on nurse or applicator tanks.

Two other classes of ammonia tanks, medium and low pressure, are used at manufacturing and storage facilities and in barge transportation of ammonia. These two classes are differentiated by the operating vapor pressures common to them. Both classes involve refrigeration systems that keep the ammonia at a low temperature, lowering the vapor pressure and lowering the requirements for structural thickness of the tank walls. Medium-pressure vessels are usually used for storage at points of manufacture. These large spherical vessels range in size from about 400 to 2700 metric tons. Ammonia is used as its own refrigerant and consequently a compressor system is included in the design. Maximum pressure in these storage systems is about 3.5 kg/cm² (50 lb/in.²).

Low-pressure ammonia storage or cryogenic (cold) storage of ammonia involves refrigeration of ammonia to its boiling point, −33°C (−28°F). At this temperature, ammonia is a liquid at atmospheric pressure. This type of storage is usually used at manufacturing facilities and barge or pipeline terminals. Tanks of this type are very large, ranging from about 9000- to 30,000-metric ton capacities (Fig. 7–5). Pressure relief valves in this type of storage equipment are usually equipped with oxidation units to convert the ammonia into elemental nitrogen in case pressures climb above the design capacity of the tank. Obviously, tanks of this type are heavily insulated to reduce the requirements for the refrigeration system. Low-pressure storage has also been used in transporting ammonia via river and sea-going barges (Fig. 7–6).

Fig. 7–5. A modern anhydrous ammonia production facility showing cryogenic (cold) storage tanks. These large vessels store ammonia at its boiling point and have their own refrigeration system. Capacity of such vessels ranges from 9000 to about 30,000 metric tons (10,000 to 33,000 short tons) of ammonia in the liquid form. (Courtesy Farmland Industries.)

Applying Fertilizers and Soil Amendments

Fig. 7–6. A refrigerated ammonia barge. Anhydrous ammonia is kept in the liquid form at low pressure by refrigeration. The barge has its own refrigeration system. Ammonia is transported in this manner in refrigerated oceangoing ships as well. (Courtesy Brett and Garcia.)

7.1.3 Transferring Ammonia

For direct field application, ammonia must be transferred from the dealer's bulk storage tanks into nurse tanks for transportation to the field and then into applicator tanks. Three basic systems are involved in ammonia transfer. The first of these systems involves use of a vapor compressor to create differences in vapor pressure between the two vessels involved in the transfer. This type of system is normally used in dealer storage systems and is commonly used in unloading tank cars and trucks; however, smaller systems are available for use in the field. Vapor is withdrawn from the tank to be filled and is pumped into the tank being unloaded, creating a higher pressure in the unloading tank. Consequently, liquid ammonia flows into the storage tank by the force of the higher vapor pressure in the unloading tank. The same system is used to fill nurse tanks from the bulk storage. Capacities of these systems is sufficient to unload a 19-ton tank car in about 90 minutes. Nurse tanks can be filled by the same system in about 15 minutes. A general diagram of this type of transfer system is indicated in Fig. 7–7.

The second type of ammonia transfer system involves a rotary positive displacement pump that withdraws liquid ammonia from the tank being unloaded and delivers it to the vessel being filled. The liquid lines are complemented by vapor lines that transfer the vapor from the tank being filled to the tank being unloaded. This type of installation is cheaper than the compressor system but involves more maintenance. Complete removal of vapor

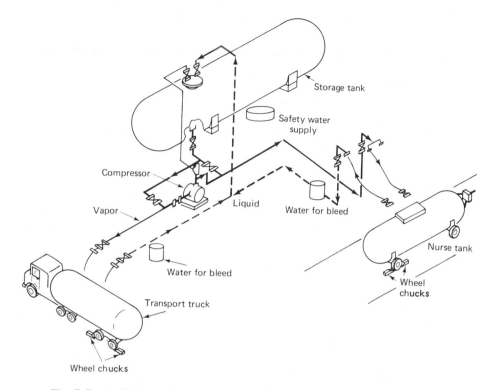

Fig. 7–7. A diagram of an ammonia storage facility showing systems for transferring ammonia from truck to storage tank or from storage tank to a nurse tank. Dashed lines are for vapor, solid lines, for liquid. The system is powered by a compressor. Note that water for safety is located at several locations. (Source: M. H. McVickar et al., *Agricultural Anhydrous Ammonia, Technology, and Use,* 1966 by permission of the American Society of Agronomy.)

from trucks and tank cars is impossible with this type of system; consequently, liquid transfer pumps are much less common than compressors.

The third and most common type of transfer system used in the field is one involving release of ammonia vapor to the atmosphere. This system does involve some cost to the person buying the ammonia because some ammonia that was weighed out in the nurse tank at the dealer's plant is lost in the transfer operation. Basically, the system involves the creation of a slight vapor pressure differential, a lower pressure in the tank being filled, by opening the vapor bleed valve. Liquid ammonia then will flow from the nurse tank to the applicator tank. Operators should stay upwind from the vapor bleed valve for safety reasons. As long as the pressure differential is not greater than 0.7 kg/cm² (10 lb/in.²), losses of ammonia in the transfer operation will not be more than about 2.5%. When costs are computed, however, ammonia loss by this system will eventually pay for a small gas-operated compressor system. Normally these small compressors are supplied on the nurse tanks by the dealer.

7.1.4 Safety

Direct application of ammonia on the farm or handling of ammonia at any facility must involve strict observance of safety precautions. Ammonia is not poisonous but it can have drastic effects on tissue that comes into contact with the liquid or high concentrations of the vapor. Ammonia is extremely irritating to eyes, lungs, and mucous membranes. Because ammonia is very soluble in water, it exerts a desiccating effect on tissues. Rapid evaporation of the liquid ammonia can cause freezing of tissues as well as desiccation. Lungs and eyes are particularly susceptible to ammonia damage and protection from ammonia concerns these two areas of the body first. Skin damage can also occur from liquid or vapor contact but the effects are less traumatic. Fortunately, the irritating odor and burning sensation from low concentrations of ammonia in the air will alert an individual to danger in advance of injurious concentrations.

Basic safety equipment for ammonia handling in the field includes rubber gloves, ammonia-type goggles or face shield, and water. Gloves and goggles should be worn when anyone is working with any aspect of ammonia transfer. An adequate supply of clean water is essential on all equipment both in the field and at the storage facility. All nurse tanks must be equipped with at least 20 l (5 gal) of clean water for use in case of an accident. First aid for ammonia burns always involves immediate rinsing with water.

In addition to goggles and gloves, all stationary storage facilities should also have the following ammonia safety equipment:

Approved ammonia gas mask with refill canister.

Rubber boots.

Rubber slicker or rubber suit.

First aid kit (ammonia type).

Safety shower or open water tank of at least 200 l (50-gal) capacity.

Additional equipment is dictated for storage terminals:

Self-contained breathing apparatus.

One-piece rubber or neoprene suit.

Transport trucks are required to have essentially the same equipment as a stationary field storage facility, merely substituting a 20-l water container for the safety shower.

First aid for ammonia contact to the skin or eyes should include immediate flushing of the affected areas with large amounts of water. Eyes should be flushed for at least 15 minutes. All contaminated clothing should be removed at once. No ointments should be applied to the affected areas under any conditions. Persons who have been overcome by inhaling ammonia should be given artificial respiration if breathing has stopped. Persons overcome by ammonia or those receiving ammonia burns to the eyes should be taken to a physician immediately. Complete familiarity with ammonia safety is essential for everyone working with the material.

7.1.5 Modern Ammonia Application

Knife Design. Development of applicators for ammonia application has led to the design of several different types of knives or shanks. Basically, these various types of knives differ only in the rigidity or flexibility of the shanks and the size of opening that they leave in the soil. Agronomically, applications with most of these types of equipment produce identical results as long as good ammonia retention by soil occurs. A common design in ammonia application equipment is indicated in Fig. 7-8. The rigid shanks on this applicator are frequently equipped with drag paddles to mound soil over the knife opening to aid in sealing in the ammonia.

The applicator pictured in Fig. 7-9 employs flexible knives or tines with hardened tips. Lateral movement of these tines in the soil is said to aid in sealing of ammonia. This particular applicator is hinged to allow road movement.

Fig. 7-8. A conventional, rigid shank field ammonia applicator. (Courtesy Lely Independence Equipment Co.)

Fig. 7-9. A flexible shank ammonia applicator. Such large applicators are frequently equipped with a positive displacement pump although that is not mandatory. (Courtesy Tennessee Valley Authority.)

Spacing of the knives on ammonia applicators is usually modified for the type of cropping system planned. Wider spacings are possible with row crops. Grasses or small grains require narrower spacings to avoid strip effects in the nitrogen response such as those noted in Fig. 7-10. The ammonia retention zones match the wheat growth patterns in Fig. 7-10. The ideal spacings between points of release are difficult to generalize because of differences in row spacing for various crops. Small grains seem to respond well to spacings in the vicinity of 40 cm (16 in.) whereas row crops respond well to spacings as great as 75 cm (30 in.).

A considerable amount of effort has been devoted to the development of ammonia knife designs that will allow ammonia use on established stands of grasses and/or small grains. In the mid-1960s, a rolling coulter design was studied extensively and employed in the field to some extent. That design employed a rolling coulter that cut a narrow opening through the sod or small grain with minimal disturbance. A narrow knife followed the coulter and injected ammonia to a depth slightly shallower than the depth of the coulter. Excellent nitrogen responses in grasses and wheat were noted with the use of this type of equipment. Press wheels were sometimes used to seal the knife slot in fine-textured soils. Unfortunately, the high cost of this type of equipment and some difficulty in penetration on packed soils with trailer-type applicators did not allow wide adaptation.

Fig. 7-10. Distribution of ammonia in the soil is not uniform due to rapid adsorption of ammonia by soil colloids. Both figures indicate the effect of spacing between points of ammonia release and growth of winter wheat. Note that the wider spacing (left) produced uneven growth that corresponded to the ammonia retention pattern. Narrower spacings (right) produced better plant growth and improved yields. (Courtesy L. S. Murphy.)

Additives. Addition of nitrification inhibitors such as nitrapyrin to ammonia have led to the development of additional equipment for ammonia application. This particular nitrification inhibitor is soluble in anhydrous ammonia and has been added successfully to the applicator or nurse tank just prior to application. However, some problems with corrosion of float gauges and other aluminum alloy parts and problems with elimination of all traces of the inhibitor from tanks led to the development of systems that inject metered amounts of the inhibitor directly into the ammonia stream following metering through the flow control apparatus and prior to entering the manifold. This new type of system (Fig. 7-11) is much more flexible, does not involve problems with cleaning tanks, eliminates any possible corrosion problems with float gauges, and allows the operator to apply the inhibitor to only certain parts of the field. The type of metering pump for the inhibitor pictured in Fig. 7-11 is ground-driven. Another system is electrically powered from the tractor's electrical system.

Tillage Implement Applications. One of the most significant developments in ammonia application in the last several years has been the adaptation of tillage equipment for ammonia application. This practice, although not totally new, has had its most rapid development in the last 7 to 8 years. It has been estimated that the average rate for ammonia application in the Midwest is about 4 ha/hr (10 A/hr), which is considerably below that for application of solid or liquid fertilizers.[3] Development of new applicators incorporated into

[3] F. P. Achorn, H. L. Kimbrough, and L. S. Murphy, "Latest Techniques for Applying Anhydrous Ammonia," Proc. TVA Fertilizer Conf., Kansas City, Mo., July 26–27, 1977, pp. 36–45.

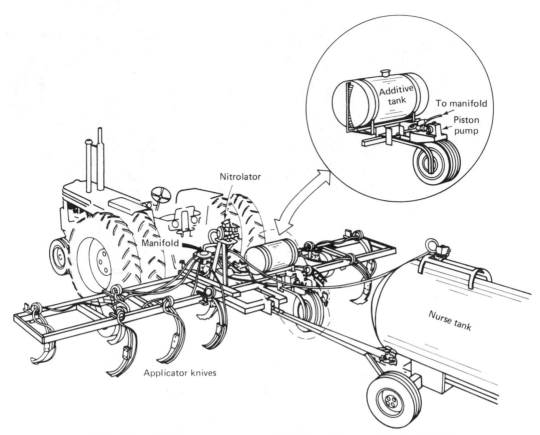

Fig. 7–11. Equipment is now available for adding nitrification inhibitors to ammonia without having to inject the inhibitor into the nurse tank or applicator tank. In this diagram, the inhibitor is injected into the ammonia stream between the flow control mechanism and the flow-dividing manifold. (Courtesy TVA.)

tillage equipment has led to rates of application as high as 16 ha/hr (40 A/hr). Swath widths as great as 20 m (65 ft) are not uncommon in some areas.

In the Great Plains region of the United States, adaptation of tillage equipment for ammonia application has found ready acceptance. Undercutting blades or sweeps are commonly used in the winter wheat and grain sorghum-producing areas of the Plains for weed control while leaving a surface mulch. These sweeps have been successfully adapted for ammonia application by placing tubes beneath the blades for ammonia distribution (Fig. 7–12). Ammonia is released from the delivery lines on approximately 40-cm centers (Fig. 7–13) with two or three delivery holes in the ammonia line beneath either side of the blade depending on blade width. Maintaining a back pressure of about 1.8 kg/cm^2 (25 lb/in.^{2}) in the delivery lines aids in uniformity of distribution. Wider spacings between points of release and large openings in the delivery lines (greater than 2–3 mm) may lead to uneven distribution

Fig. 7-12. An undercutting blade equipped for direct application of ammonia during a tillage operation. Note that a nurse tank is trailing the applicator and is serving as an applicator tank. The type of equipment shown here is common in the Plains states of the United States. (Courtesy L. S. Murphy.)

and poor plant growth patterns at relatively low rates of ammonia application. Note the uneven plant growth that occurred in Fig. 7-10 when 100-cm spacings were used with undercutting blade ammonia applications for wheat versus more conventional 40-cm spacings. This uneven growth was also reflected in wheat yields.[4]

Ammonia retention has been no problem with this type of application due to low pressure of the ammonia at the point of application. Excellent retention has been the rule with applications as shallow as 10 cm (4 in.). The loose mass of soil flowing over the blade gives a wider ammonia retention zone in the soil due to greater ease of lateral movement of the ammonia. Retention zones as wide as 30 to 40 cm (12 to 16 in.) are not uncommon with blade ammonia applications. In the same soil, knifed applications would be expected to produce bulb-shaped retention zones about 15 to 20 cm (6 to 8 in.) in width.

Tillage implement adaptation for ammonia application has frequently eliminated the applicator tank and employs instead a nurse tank trailing the tillage implement as the ammonia carrier. Heavier equipment of this type dictates greater horsepower in tractors but usually this is not a problem with the large tractors in use in the Plains states. Ammonia tanks can be mounted directly on the tillage equipment if that is desirable.

Besides the undercutting blade adaptation to ammonia application, chisel

[4] C. L. Swart, L. S. Murphy, and C. W. Swallow, "Retention Patterns and Effectiveness of Ammonia Applied with an Undercutting Blade," *Agron. J.*, 63:881, 1971.

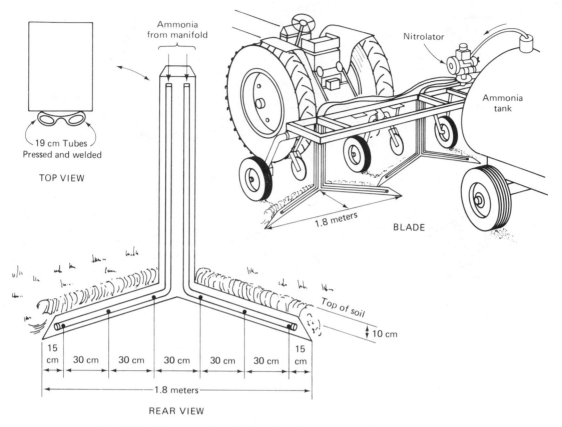

Fig. 7–13. Diagram of method used in adapting an undercutting blade for application of ammonia. (Courtesy TVA.)

plows, field cultivators, moldboard plows, disks, and rod weeders among other implements have been successfully adapted for ammonia application.

One of the most recent developments in the area of equipping tillage tools for ammonia application is the application of *cold* ammonia. In this type of system, depicted in Fig. 7–14, ammonia is allowed to boil in a converter lowering its pressure to 1 atm and decreasing the depth to which ammonia must be injected for good retention. Two types of converters are currently being marketed. The large converter in Fig. 7–15 delivers cold liquid ammonia to flow dividers that in turn lead to individual knives or release points. Boiling ammonia produces about 15% vapor at 10°C (50°F) and this vapor must also be injected. Separate hose systems are needed to handle the vapor that is frequently applied at the same point as the liquid. Figure 7–16 indicates the type of arrangement necessary on shanks handling both liquid and gaseous ammonia. Frequently, alternate shanks are used for both liquid and gas injection. Liquid flow from these large converters is by gravity so slope must be maintained on all liquid lines to obtain good distribution.

Fig. 7–14. New developments in ammonia application include the use of chambers that allow the ammonia to boil, cooling it and allowing shallower application. Such equipment does require handling of vapor separate from the liquid. Advantages include less draft for application and use of tillage implements for ammonia application. (Courtesy USS Agrichemicals, Division of United States Steel Corp.)

Another system of cold ammonia application involves small individual converters (Fig. 7–17) that are mounted on individual shanks. Slightly higher pressures are associated with the liquid ammonia with these smaller converters than is the case with the larger units.

The possible advantage associated with the use of such equipment is the shallower depth of application that can be compatible with good soil retention. Heavy rates of ammonia application for row crops in the Midwest may allow application at depths of 10 to 15 cm (4 to 6 in.), whereas conventional warm ammonia applications might necessitate applications at 20 to 25 cm (8 to 10 in.). Shallower application depth also cuts the draft required for the ammonia application allowing the use of lower horsepower tractors or wider applicators. Tillage operations that normally do not involve deep penetration of the soil such as field cultivators and disks are more adaptable to ammonia application with this cold ammonia equipment. Agronomically, little difference would be expected between conventional ammonia and cold ammonia applications unless one system produced unsatisfactory retention. Shallow application could be disadvantageous in areas of limited rainfall by placing the ammonia higher in the soil profile where moisture for nitrogen absorption by plant roots could be more limiting.

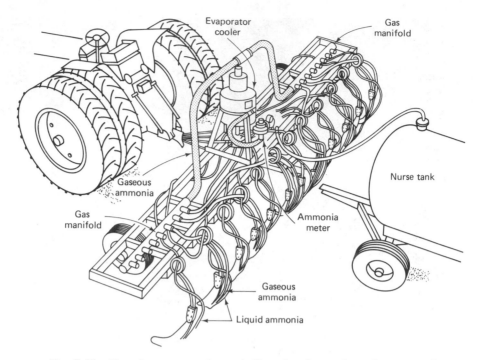

Evaporator cooler

Gas manifold

Gaseous ammonia

Gas manifold

Ammonia meter

Nurse tank

Gaseous ammonia

Liquid ammonia

Fig. 7–15. Use of an evaporation or boiling chamber such as the one on the applicator in Fig. 7–14 requires a flow control device just like conventional ammonia applicators but the flow control device must be placed before the evaporation chamber. The approximately 15% vapor from the chamber must be split by a manifold and injected into the soil for uniformity of application. The liquid ammonia is at a temperature of $-33°C$ $(-28°F)$. (Courtesy TVA.)

Dual Applications. Dual application of ammonia and other fertilizer materials has developed over the last several years as a means of providing all necessary elements for a crop in a single preplant application. In the northwestern part of the United States (Idaho, Washington, Oregon), ammonia applicators have been equipped with a second delivery system to add sulfur at the time of nitrogen application. Ammonium polysulfide (20% N, 45% S) is carried in a separate tank (Fig. 7–18), metered by a positive displacement pump, and delivered to the soil by a second line on the back of the ammonia knife. Other fertilizer solutions have been applied successfully in this manner including nitrogen-phosphorus-potassium liquids and suspensions. As indicated in Chapter 3, Section 3.5.4, phosphoric acid has also been applied in combination with ammonia through a dual delivery system. Agronomic advantages of this method of dual application of nitrogen and phosphorus may occur because of accelerated availability and greater plant growth.

The presence of large quantities of ammonium ions in the soil along with phosphorus has been identified as a positive factor in phosphorus absorption

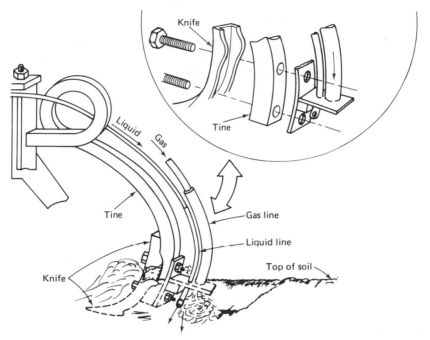

Fig. 7–16. Knives adapted for application of cold ammonia must handle both liquid and gaseous ammonia or either form separately. (Courtesy TVA.)

for over 20 years. Olson and Drier[5] noted positive effects of ammonium nitrogen on phosphorus absorption in Nebraska. Riley and Barber[6] pointed out that ammonium nitrogen may influence phosphorus absorption positively through an induced decrease in soil pH in the vicinity of the roots as the plant absorbs ammonium ions.

Results of several field studies conducted in Kansas since 1971 have indicated agronomic superiority of dual application of ammonia and liquid ammonium polyphosphate (APP) knifed into the soil preplant for wheat. Under-cutting blades and chisel plows adapted with two lines below each wing of each blade for delivery of liquid APP (10–34–0) and ammonia produced superior growth, leaf composition, and yield in wheat and grain sorghum as compared to a mixture of nitrogen solution and liquid APP sprayed on the soil surface at the same time and incorporated by disking[7] (Table 7–1). Later

[5] R. A. Olson and A. F. Drier, "Nitrogen—A Key Factor in Fertilizer Phosphorus Efficiency," *Soil Sci. Soc. Amer. Proc.*, 20:509, 1956.

[6] D. Riley and S. A. Barber, "Effect of Ammonium and Nitrate Fertilization on Phosphorus Uptake as Related to Root-Induced pH Changes at the Root-Soil Interface," *Soil Sci. Soc. Amer. Proc.*, 35:301, 1971.

[7] L. S. Murphy, K. W. Kelley, P. J. Gallagher, and C. W. Swallow, "Tillage Implement Applications of Anhydrous Ammonia and Liquid Ammonium Polyphosphate," *Proc. 8th Intnl. Fert. Cong., Moscow, USSR, June 21–26, 1976, pp. 293–305.

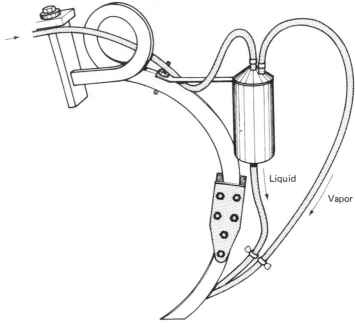

Ammonia knife equipped with individual cold
ammonia converter

Fig. 7-17. Smaller evaporative coolers for ammonia application have also been devised and are attached to individual shanks of an applicator. Both liquid and vapor must be injected but the unit has the advantage of fewer problems with flow distribution compared with the larger models. (Courtesy L. S. Murphy.)

studies in 1975-1980 involving dual delivery of ammonia and liquid APP down separate lines on an ammonia knife and applied preplant for wheat continued to indicate general superiority of placement of ammonia and liquid APP in the same soil zone (Fig. 7-18). Knifed applications of a mixture of nitrogen solution (UAN) and APP also produced higher wheat grain yield and higher leaf concentrations of phosphorus when compared to methods of nitrogen and phosphorus application that separated most of the two nutrients (Table 7-2, Fig. 7-19).

Deeper placement of nitrogen and phosphorus for dryland crops in the Great Plains can involve positional effects on nutrients availability due to greater soil moisture at lower depths. However, advantages do seem to exist for such placement aside from moisture relationships arising from deeper placement. In addition to the pH effects pointed out by Riley and Barber in Footnote 6, the presence of high concentrations of ammonium ions either from ammonia or from nitrogen solution may alter the phosphorus fixation reactions in both calcareous and high iron soils to the extent that greater plant availability of the phosphorus may result (refer to Chapter 3).

Both tillage implements and existing ammonia applicators can be

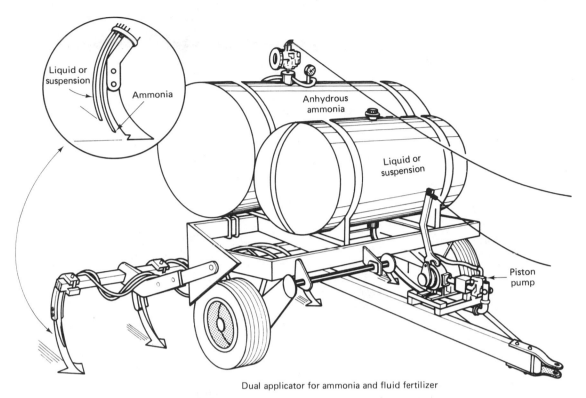

Liquid or suspension

Ammonia

Anhydrous ammonia

Liquid or suspension

Piston pump

Dual applicator for ammonia and fluid fertilizer

Fig. 7–18. Applicators for anhydrous ammonia can be adapted for dual application of ammonia and liquids. Usually the liquid or fluid fertilizer line is placed behind and above the ammonia line to avoid freezing the liquid line by vaporizing ammonia. Phosphoric acid or liquid mixed fertilizers have successfully been applied by this technique. (Courtesy TVA.)

equipped for dual applications of ammonia and liquid fertilizers. In order to have an effective system, however, the ammonia and liquid lines must be kept separate to prevent the vaporizing ammonia from freezing the liquid line. Separation by 2 or 3 cm (about 1 in.) at the delivery points of both lines should prevent freezing. Vertical separation of the release points is also suggested (2–3 cm) on shank-type applicators in order to avoid the accumulation of solid APP compounds around the holes in the liquid line. These solids result from reactions of the ammonia and the liquid APP.

In summary, both tillage implement applications of ammonia and dual applications of ammonia and liquid fertilizers pose some obvious agronomic and economic advantages. Lowered costs of ammonia application result from the combination of ammonia application with a tillage operation. Because the field is to be cultivated anyway prior to seeding, doubling up on operations will cut fuel costs, save time, and perhaps produce some agronomic advantages when compared to surface nitrogen applications. Dual applications of

Table 7-1. Tillage Implement Applications of Nitrogen and Phosphorus Have Been Demonstrated to Frequently Be More Effective than Surface Applications of the Same Amounts of Nutrients, Even When Surface Applications Were Incorporated by a Later Tillage Operation.

WINTER WHEAT

Treatments	Carwile Loamy Sand (Typic Argiaquoll) Yield (kg/ha)	% Grain Protein	Goessel Silty Clay (Udic Pellustert) Yield (kg/ha)	% Grain Protein	Detroit Silty Clay (Pachic Argiustoll) Yield (kg/ha)	% Grain Protein
Control (O-N, O-P)	2486	10.6	2285	10.8	2390	12.8
Undercutting blade[a]	3242	11.8	3125	10.9	2957	13.4
Chisel plow	3066	10.4	2872	10.6	2738	13.2
Surface	2906	11.1	2839	10.8	2789	12.9
LSD.05	134	0.5	202	NS	NS	NS

GRAIN SORGHUM

Treatments	Grundy Silty Clay Loam (Aquic Argiudoll) Yield (kg/ha) 1973	Leaf[b] % N	% P	Yield (kg/ha) 1974	Leaf % N	% P
Control (O-N, O-P)	3073	1.66	0.20	1944	2.69	0.31
Undercutting blade	4892	2.32	0.24	3562	2.90	0.27
Chisel plow	4641	2.44	0.24	3976	3.08	0.29
Surface	4328	2.06	0.22	2666	2.73	0.31
LSD.05	251	0.14	0.01	326	0.12	NS

[a] Mean N rate 50 kg/ha, phosphorus (P) rate 8 kg/ha.

[b] Plants sampled at heading, youngest fully-emerged leaf. Mean N rate 100 kg/ha; mean P rate 8 kg/ha.

SOURCE: L. S. Murphy, K. W. Kelley, P. J. Gallagher, and C. W. Swallow, "Tillage Implement Applications of Anhydrous Ammonia and Liquid Ammonium Polyphosphate," Proc. 8th Intl. Fert. Congress, Moscow, USSR, June 21–26, 1976, p. 293.

ammonia and other nutrients further add to the savings through elimination of another operation. Because dual applications can also be readily adapted to tillage equipment, further advantages may result from placement of ammonia and phosphorus in the same soil zone. Undoubtedly, more emphasis will be afforded such multiple operations as the costs of fuel continue to increase. Savings can partially offset the lower application costs of solids and liquids. At the present time, however, it is much more common for ammonia to be applied as a separate application in combination with broadcast applications of solid

Table 7-2. Comparative Effectiveness of Methods of N-P Application on Grain Yield and Leaf Concentrations for Wheat (Kansas)

	Reno Co.			Ellsworth Co.			Labette Co.	Dickinson Co.	
	bu/A	% N	% P	bu/A	% N	% P	bu/A	% N	% P
No N	21	3.80	.17	38	3.91	.20	26	3.58	.22
Knifed NH₃	22	3.44	.16	43	3.90	.16	38	4.14	.23
Knifed NH₃-knifed APP	39	3.74	.25	58	4.74	.30	45	4.50	.28
Knifed NH₃-b'cast APP	38	3.78	.20	53	3.92	.19	43	4.06	.22
Knifed NH₃-band APP	35	3.80	.19	50	4.18	.21	41	4.12	.24
Knifed UAN	25	3.84	.16	40	4.02	.19	37	4.07	.23
Knifed UAN-knifed APP	43	3.87	.23	55	4.62	.29	45	4.02	.28
Knifed UAN-b'cast APP	39	3.82	.20	52	4.02	.20	45	3.92	.21
Knifed UAN-band APP	35	3.84	.20	45	4.06	.21	45	4.07	.23
Knifed NH₃-knifed APP	39	3.74	.25	58	4.74	.30	45	4.50	.28
Knifed NH₃-knifed APP plus N-SERVE®	47	4.02	.23	45	4.49	.30	49	4.36	.33
Soil pH		7.3			6.0			5.5	6.1	
Soil-test P, ppm		4			5			4	11	

NOTE: APP supplied as 11–37–0 from Tennessee Valley Authority at rate of 19 kg/ha of P (40 lb/A of P_2O_5). N constant at 84 kg/ha N (75 lb/A N). UAN is non-pressure, 28% urea-ammonium nitrate solution. N-SERVE is a nitrification inhibitor product of Dow Chemical Company. Band means placement of P with seed.

SOURCE: D. R. Leikam, R. E. Lamond, P. J. Gallagher, and L. S. Murphy, "Improving N-P Application," *Agrichemical Age,* 22(3)6, 1978.

or liquid fertilizers. Disadvantages of the dual application method include a slower rate of application because of the need to fill two separate tanks on the applicator.

7.1.6   Other Compressed Gases

Some examination of the effectiveness of soil injection of compressed sulfur dioxide has been undertaken in the Northwest where sulfur is frequently deficient. Although sulfur dioxide (SO_2) can be handled with ammonia equipment, sulfur dioxide is not compatible with ammonia and separate tanks must be used if the materials are to be injected into the soil simultaneously. Sulfur dioxide is a liquid under pressure very much like ammonia and has a vapor pressure of 1.6 kg/cm² (23 lb/in.²) at 15°C. Despite some favorable results from its use, it has never gained wide acceptance.

Some use of sulfur dioxide other than as a plant nutrient has occurred in the San Joaquin and Imperial Valleys in California. Sulfur dioxide injected into tile drains has succeeded in dissolving and cleaning the drains of iron and manganese compounds.

Fig. 7-19. Dual-knifed application of ammonia and liquid ammonium polyphosphate (bottom) produced better plant growth than did surface applications of an all liquid mixture (top) or a combination of ammonia knifed and liquid ammonium polyphosphate banded in direct seed contact (center). (Courtesy L. S. Murphy.)

Although elemental sulfur dissolved in ammonia can hardly be considered as another compressed gas, the characteristics of this unique solution were examined for a time as a possible source of nitrogen and sulfur. TVA examined a solution containing 74% nitrogen and 10% sulfur but encountered insurmountable equipment problems. Sulfur dissolved in ammonia tends to precipitate as the ammonia boils and is deposited in the metering equipment, manifolds, and distribution lines.

7.2 APPLYING FERTILIZERS THROUGH IRRIGATION WATER

The process of application of nutrients through irrigation systems is not new but is still developing. This technique has been named *fertigation,* a contraction of fertilization and irrigation. This term is now finding wide use in the popular literature and will be used throughout the remainder of this section. Interest in this technique has stemmed from a desire to reduce fertilizer application costs through elimination of an operation and possibly to improve the efficiency of nutrient use by applications close to the time of actual plant need. Applications closer to the time of plant need could conceivably reduce leaching or denitrification losses of nitrogen and lower the possibility of luxury uptake of nutrients by plants.

The most common nutrient applied by fertigation is nitrogen. Other nutrient elements applied more or less frequently include phosphorus, sulfur, potassium, zinc, and iron. Because nitrogen fertigation developed first, it will be considered first in this discussion.

7.2.1 Nitrogen Application

Applying Ammonia in Water. The first use of anhydrous ammonia for direct application as a fertilizer involved injection of ammonia into irrigation water in California.[8,9] These early efforts demonstrated that ammonia was a feasible source of nitrogen but also revealed that precipitation can occur when ammonia is injected into water containing large quantities of dissolved calcium and magnesium salts.

When ammonia is injected into irrigation water, an equilibrium is set up that generates hydroxyl (OH^-) ions along with ammonium (NH_4^+) ions.

$$\underset{\text{Ammonia}}{NH_3} + H_2O \rightleftharpoons \underset{\text{Ammonium Hydroxyl}}{NH_4^+ + OH^-}$$

The hydroxyl ions increase the pH of the water causing the solubility of dissolved salts to decline, particularly calcium salts. Subsequently, these salts precipitate, nozzles are clogged in sprinkler systems, pipe weight increases substantially, and water flow is restricted.

Ammonia injection into ditch or siphon tube irrigation systems is somewhat less troublesome. However, siphon tubes can become encrusted with the precipitated calcium salts and are not easily cleaned. If the precipitation occurs in the irrigation ditch, no particular problems are encountered (Fig. 7-20).

Some relief from problems of ammonia injection into hard water can be ob-

---

[8] D. D. Waynick, "Anhydrous Ammonia as a Fertilizer," *Calif. Citrog.,* 19(11):295, 1934.
[9] L. Rosenstein, "Increased Yields Obtained from Shell Agricultural Ammonia in Irrigated Agriculture," *Shell Chem. Co. Bul.* No. 1, p. 23, 1936.

Fig. 7-20. Injection of ammonia into irrigation water containing large quantities of dissolved calcium carbonate or bicarbonate can produce immediate precipitation of the calcium compounds due to increase in the pH of the water. This picture shows precipitation in a siphon tube from the injection of 22 kg N/ha (20 lb N/A) as ammonia. This is a serious problem in underground systems and sprinklers. (Courtesy L. S. Murphy.)

Table 7-3. Suggested Anhydrous Ammonia and Calcium Precipitation Inhibitor Rates

Irrigation Water Hardness (ppm)	Maximum Ammonia Application per 4000 l/min H_2O (kg/hr)	Inhibitor Required[a] per 4000 l/min H_2O (g/hr)
35–120	18	120
120–170	18	180
170–425	18	300
425–850	13	360

[a] Sodium hexametaphosphate, "Calgon."

SOURCE: H. R. Mullinier, *Applying Anhydrous Ammonia in Irrigation Water,* University of Nebraska Guide G 74-129, 1974.

tained by using an inhibitor such as sodium hexametaphosphate that tends to sequester the calcium, reducing precipitation. Agricultural engineers[10] at the University of Nebraska have published guidelines for precipitation inhibitor use based on amount of ammonia being applied and the quality of the water. That information is summarized in Table 7-3.

Another major problem in applying anhydrous ammonia or free ammonia-containing solutions through sprinkler irrigation systems is the fact that volatilization losses of ammonia occur from the time that the ammonia-water mixture leaves the sprinkler until the ammonia reaches the soil surface. Ad-

[10] H. R. Mullinier, *Applying Anhydrous Ammonia in Irrigation Water,* University of Nebraska Guide G 74-129, 1974.

Applying Fertilizers and Soil Amendments

mittedly, some of the nitrogen is present as ammonium ions but much more is in the ammonia gas form. Subsequently, as water vaporizes, ammonia is also lost to the atmosphere as a gas.

There has been much controversy over the actual amount of ammonia lost through a sprinkler irrigation system. Most of the work in that area was carried out in the mid-1950s. Scott[11] suggested that the net loss might range from 5 to 40%. Jackson and Chang[12] estimated losses to range as high as 58%, and Henderson et al.[13] presented data showing losses as high as 60%. Recently, University of Nebraska research has corroborated these findings. Conditions governing such losses include initial pH of the water, temperature at the time of application, wind velocity, and concentration of ammonia in the water.

Ammonia or any nitrogen source applied in irrigation water is distributed in about the same manner as the water. Poor sprinkler patterns are not too much of a problem with ammonia application because of the already overwhelming problem of volatilization losses. Application through gated pipe or siphon tube systems must consider the fact that more water percolates into the soil at the head of the run than at the tail. Similarly, more ammonia would be expected to be adsorbed by colloids at the head of the run. To offset this problem, ammonia is frequently withheld from the water during the first part of the irrigation thus avoiding part of the accumulation of nitrogen at the head of the run.

Fischbach[14] recommends that nitrogen injection begin as soon as the water stream is started down the furrow and continue until the water and fertilizer reach the end of the run. Fertilizer application should then be shut off and water application continued until the proper amount of water has been applied.

Leavitt[15] commented on his early experience with ammonia application in irrigation water for contour irrigation of rice and noted that severe problems were encountered due to adsorption of all the ammonia by soil and organic matter during the early stages of migration of water through the contour system. All ammonia was adsorbed in the first 2 to 3 km of travel.

Some of the problems encountered in the application of ammonia in sprinkler irrigation systems have been successfully countered in work reported from the University of Nebraska. Researchers there have buffered

[11] V. H. Scott, "Sprinkler Irrigation," Calif. Agr. Exp. Sta. Ext. Serv. Circ. 456, 1956.

[12] M. L. Jackson and S. C. Chang, "Anhydrous Ammonia Retention by Soils as Influenced by Depth of Application, Soil Texture, Moisture Content, pH Value and Tilth," *Agron. J.*, 39:623-33, 1947.

[13] D. W. Henderson, W. C. Bianchi, and L. D. Doneen, "Ammonia Loss from Sprinkler Jets," *Agr. Eng.*, 36:398-99, 1955.

[14] P. E. Fischbach, "Irrigate, Fertilize in One Operation," *Nebraska Quarterly,* Summer, 15-17, 1964.

[15] F. H. Leavitt, "Agricultural Ammonia Equipment, Development and History." In *Agricultural Anhydrous Ammonia Technology and Use,* M. H. McVickar et al. (Ed.), American Society of Agronomy, Madison, Wis., pp. 125-142, 1966.

Table 7-4. Volatilization Losses From Ammonia Applied in Sprinkler Irrigation System

Applied mgm N/l	Acid Applied	Water pH when NH_3 Added	% Loss
48	None	9.0	24
49	Equivalent	7.4	4
179	None	9.6	51
189	Equivalent	7.1	5

NOTE: Relative humidity, 45%; wind 20 kph; temperature, 29°C.

SOURCE: R. R. Bock, R. A. Olson, D. H. Sander, and K. D. Frank, "Effective Sprinkler Irrigation Application of Anhydrous Ammonia," *Soil Sci. Prog. Report,* Dept. of Agronomy, University of Nebraska, 1976, pp. 27-1 to 27-6.

the pH change induced in irrigation water by application of ammonia through the introduction of equivalent amounts of sulfuric acid. By keeping the pH from rising, precipitation of the calcium carbonate and bicarbonate in the water can be avoided. In addition and perhaps more importantly, the flash (volatilization) loss of ammonia was cut substantially by making the water acid (Table 7-4). Admittedly, this process involves substantial use of a corrosive agent but it does have potential and will continue to be researched.

Applications of Nitrogen Solutions in Water. In irrigated agriculture, water application of nonpressure urea-ammonium nitrate (UAN) solutions has become increasingly popular as a means of improving nitrogen use efficiency. The University of Nebraska has led in the development and study of fertigation. Nebraska research has shown that nitrogen use efficiency can be improved by applications through the water. Losses due to leaching are minimized by split or multiple nitrogen applications as compared to applications prior to seeding, particularly on sandy soils. Data in Table 7-5 indicate the

Table 7-5. Comparative Effects of Mechanical and Nitrogen Fertilizer Application in Irrigation Water on Corn

Method of Application	Yield (kg/ha)
Normal mechanical[a]	7275
Normal (mechanical and irrigation split)	7526
Normal irrigation (all in one irrigation)	7777
Normal irrigation (split into two irrigations)	7840
Normal mechanical plus 20 kg/ha of nitrogen in irrigation water	8216

[a] Recommended amount of nitrogen determined by soil test and past cropping history.

SOURCE: P. E. Fischbach, "Irrigate, Fertilize in One Operation," *Nebraska Quarterly,* Summer, 15-17, 1964.

Fig. 7-21. The development of center pivot irrigation in the Great Plains of the United States led to rapid development of fertilization by injection of liquids into these sprinkler systems commonly known as fertigation. (Courtesy Swanson, Rollheiser and Holland, Inc. and Valmont Industries.)

effect that fertigation can have on corn yields. Widest adaptation for this technique has occurred in sprinkler-irrigated sandy soils (Figs. 7-21 and 7-22). Estimates indicate that about 60 to 70% of the sprinkler irrigation systems in Colorado, Kansas, Nebraska, and Oklahoma involve application of nitrogen and/or other nutrients through the irrigation water. This technique is also adaptable to gated pipe or ditch irrigation, but development there has not been so widespread as with sprinkler systems.

Fischbach[16] examined the distribution pattern of nitrogen solutions applied through sprinkler and gated pipe irrigation systems. Because nitrogen applications could be affected by improper or incomplete mixing in the water, tests were conducted for uniformity of nitrogen concentrations at various points on a gated pipe system and at points along a center pivot sprinkler system. Data reported in Tables 7-6 and 7-7 show good uniformity indicating good mixing.

Timing nitrogen solution application in water under furrow irrigated conditions is important in maintaining a good distribution of nitrogen. Fischbach[14] suggested that acceptable nitrogen distribution could be achieved by using as large an irrigation stream in each furrow as possible without causing

[16] P. E. Fischbach, "Applying Fertilizers Uniformly Through Irrigation Systems," *Fertilizer Solutions*, 14(6):92, 1970.

Fig. 7-22. An installation for injection of liquid fertilizers (tank) into a center pivot irrigation system. The pump (foreground) is powered from the driveshaft for the well pump. If the engine stops, fertilizer injection stops. A check valve prevents the fertilizer solution from running into the well. (Courtesy Allied Chemical Co.)

Table 7-6. Concentration of Nitrate-Nitrogen in Irrigation Water from Sequential Gate Openings at Various Intervals of Time in a Gated Pipe Irrigation System[a]

Elapsed Time Minutes	Gate Number			
	Fifth ppm NO_3-N	Eleventh ppm NO_3-N	Twenty-Third ppm NO_3-N	Twenty-Ninth ppm NO_3-N
10	27.2	27.2	27.2	27.2
20	27.2	28.8	27.2	27.2
30	28.8	28.8	28.8	27.2
40	27.2	28.8	28.8	27.2

[a] A 4000 l/min system, using 20-cm gated pipe 384 m long. The water pressure at the beginning of the gated pipe was 528 g/cm².

SOURCE: P. E. Fischbach, "Applying Fertilizers Uniformly Through Irrigation Systems," *Fertilizer Solutions*, 14(6):92, 1970 by permission of the American Society of Agronomy.

Table 7-7. Concentration of Nitrate-Nitrogen in Irrigation Water Caught Halfway Between Towers of a 12-Tower Center-Pivot Sprinkler System[a]

Location of Sample	NO_3-N (ppm)
Pivot Point and Tower 1	35
Tower 1 and 2	35
Tower 2 and 3	35
Tower 3 and 4	32
Tower 4 and 5	36
Tower 5 and 6	34
Tower 6 and 7	33
Tower 7 and 8	34
Tower 8 and 9	35
Tower 9 and 10	34
Tower 10 and 11	37
Tower 11 and 12	34

[a] A-3600 *l*/min system, 392 m long using 17-cm diameter OD pipe, 4.22 kg/cm^2 pressure at the pivot. Fertilizer solution (UAN) injected by positive displacement pump.

SOURCE: P. E. Fischbach, "Applying Fertilizers Uniformly Through Irrigation Systems," *Fertilizer Solutions*, 14(6):92, 1970 by permission of the American Society of Agronomy.

serious erosion on the particular slope being irrigated. He suggested that the stream should reach the far end of the furrow in a maximum recommended time for various soil textures (Table 7-8). If it does not reach the end of the furrow in the time indicated, the run may be too long, water penetration is uneven, and nitrogen distribution suffers. If the water flows through the field in less than the recommended maximum time, the distribution will be still better. For efficient use of fertilizer nitrogen, Nebraska researchers suggest that the nitrogen solution should be injected into the water as soon as the stream

Table 7-8. Approximate Maximum Length of Time for Water to Flow to the End of the Furrow on Various Soil Textures[a]

Soil Texture	Hours
Loamy sands	2–3
Sandy loams	3–4
Fine sandy loams	4–5
Silt loams	5–6
Silty clay loams	6–7

[a] Length of furrows should not exceed 185 m (600 ft) on sandy soils and 400 m (1300 ft) on clay soils.

SOURCE: P. E. Fischbach, "Irrigate, Fertilize in One Operation," *Nebraska Quarterly*, Summer, 15–17, 1964.

is started down the furrow and continue until the water with the fertilizer reach the far end of the field in all furrows. The nitrogen should then be shut off and irrigation continued until water penetrates to the desired depth.

Use of nitrogen solutions or other nitrogen sources in irrigation water should include a reuse system for tail water (surface runoff) in order to protect against contamination of surface water with nitrogen and to make efficient use of both nitrogen and water.

Nonpressure nitrogen solutions are most commonly used for fertigation because they avoid calcium salt precipitation produced by ammonia and they avoid ammonia loss through sprinklers. Pressure solutions would be most compatible with ditch-siphon tube irrigation systems.

Applications of Solid Nitrogen Sources in Irrigation Water. Any soluble nitrogen fertilizer can be applied in irrigation water by converting the solid to a liquid and placing it in the irrigation water as already described. Techniques also exist for direct application of solids such as urea to irrigation water. When solids are added directly to the water, there should be ample distance to the first turnout or junction in the pipe/ditch to allow for good mixing. No reaction problems would be encountered when solids are used. The system would operate very much like a system involving UAN.

7.2.2 Phosphorus Application

Injection of phosphorus fertilizers into irrigation water is less common that injection of nitrogen. As long as phosphorus can be maintained in the water stream totally dissolved, no major problems in distribution should occur. However, if no reactions with the water occur, phosphorus must still be applied very early in the growing cycle, preferably before seeding, and for most crops it should be incorporated into the plant root zone.

Problems in application of phosphorus in irrigation water are at least threefold: (1) precipitation may be encountered when ammonium polyphosphate-containing liquids are injected into high calcium/magnesium water; (2) phosphorus must be applied very early in the growing cycle for most crops particularly if there is a definite phosphorus need in order to prevent any early season stunting of yield potentials; and (3) phosphorus applied for row crops through irrigation water may remain on or near the soil surface if not incorporated and be less effective than phosphorus applied preplant and incorporated by a tillage operation. The latter problem may be insignificant in the irrigation of crops such as alfalfa and grasses.

Precipitation from the injection of ammonium polyphosphate liquids into irrigation water results from the presence of calcium and magnesium in the water. Duis and Burman[17] studied this problem and determined that the precipitates that form at certain concentrations of ammonium polyphosphate

[17] J. H. Duis and K. A. Burman, "Polyphosphates in Irrigation Systems," *Fertilizer Solutions,* 13(2):46, 1969.

were calcium ammonium pyrophosphates. These precipitates are quite insoluble and represent both a clogging hazard to the sprinkler system and a potentially poor source of phosphorus for plants. Gallagher[18] researched the availability of these precipitates as phosphorus sources and found them to be less available than ammonium polyphosphates, particularly on high pH soils.

Duis and Burman did report, however, that success can occur in the mixing of ammonium polyphosphates and irrigation water. This can be tested simply by calculating the ratio of liquid fertilizer to be applied per hectare and the amount of water that will be applied to the same area in a given amount of time. By merely mixing in a glass jar the liquid APP and the water on the same proportionate basis and observing for a few minutes, one can quickly tell if precipitation will occur, because, if it forms, the precipitate forms quickly.

Injecting orthophosphates into high calcium irrigation water is less likely to produce precipitation. Duis and Burman also examined the use of an all-orthophosphate liquid and noted that precipitation was less pronounced. This fact may be of limited practicality, however, because most liquid fertilizers contain some polyphosphate.

Hergert and Reuss[19] studied sprinkler irrigation application of phosphorus and zinc in northeastern Colorado using 10-34-0 liquid ammonium polyphosphate as the phosphorus source. Phosphorus movement in a Nunn clay loam (Aridic Argiustoll) was only 4 to 5 cm (about 2 in.) following water application. Later work on a Haxtun loamy sand (Pachic Argiustoll) showed that sprinkler-applied phosphorus moved to a depth of approximately 18 cm (7 in.) indicating the greater fixing capacity of the finer-textured Nunn soil. At the same time, sprinkler-applied zinc moved to a depth of 5 cm on both soils. This study demonstrated the problems involved in water application of phosphorus. For adequate movement of phosphorus into the root zone (except on very sandy soils) mechanical incorporation must take place following irrigation.

Greater ease of movement of a phosphorus source into soil was noted in work in California by Rolston et al.[20] They studied organic phosphorus compounds including glycerophosphate, methyl ester phosphate, glycol phosphate, ethyl ester phosphate, glucose-1-phosphate, and glucose-6-phosphate. They theorized that organic compounds might move further into soil than inorganic phosphorus compounds due to the necessity for organic phosphate hydrolysis to orthophosphate before soil reactions would hinder further movement. Generally, their work indicated that glycerophosphate, the compound chosen for most intensive study, moved downward approximately 12 cm in a Panoche clay loam whereas a control involving potassium dihyrogen phosphate (KH_2PO_4) moved about 3 cm. Not all soils responded in the same

[18] P. J. Gallagher, "Reactions of Ammonium Pyrophosphates with High Calcium Irrigation Waters," Ph.D. Thesis, Dept. of Agronomy, Kansas State University, 1976.

[19] G. W. Hergert and J. O. Reuss, "Sprinkler Application of P and Zn Fertilizers," *Agron. J.*, 68:5, 1976.

[20] D. E. Rolston, R. S. Rauschkolb, and D. L. Hoffman, "Infiltration of Organic Phosphate Compounds in Soil," *Soil Sci. Soc. Amer. Proc.*, 30:1089, 1975.

fashion but this study did demonstrate some potential for organic phosphates as phosphorus sources in irrigation water if their prices were economically comparable.

Rauschkolb and others of the California group further examined the possibilities of glycerophosphate as a phosphorus source that could be effectively applied in a drip irrigation system.[21] They compared phosphorus movement in soil and uptake by tomatoes when orthophosphate and glycerophosphate were applied through a drip irrigation system and also included a comparable rate of phosphorus banded below the seed at planting. Phosphorus applied in the drip irrigation system on a Panoche clay loam moved much farther in general than had previous surface applications. Glycerophosphate moved 5 to 10 cm farther through the soil at application rates of 6 and 13 kg/ha of phosphorus than did orthophosphate. At the relatively high phosphorus rate of 39 kg/ha, orthophosphate moved 25 cm horizontally and 30 cm vertically in the soil. Such movement of orthophosphate resulted from saturation of the soil reaction sites near the point of application and subsequent mass flow with the water. Distance of phosphorus movement was proportional to the application rate. As a further indication of the possibilities of this technique, their data indicated a significantly higher phosphorus content in seedling leaves when 26 kg/ha of phosphorus were applied in irrigation water compared to the same amount banded.

In summary, applications of phosphorus in irrigation water are subject to precipitation and uneven distribution in the soil. The practice can work with very good-quality water but nutrient distribution will be no better than water distribution. Compatibility of liquid ammonium polyphosphate fertilizers with irrigation water should be checked prior to injection. This is obviously very important on drip and sprinkler irrigation systems.

7.2.3 Potassium and Sulfur Application

Much less research has been conducted into the effectiveness of potassium and sulfur applied in irrigation water. These techniques are receiving some attention particularly in irrigated areas that have developed with center pivot irrigation in the last 10 to 15 years. Interest in fertigation application of both of these elements has developed where sandy soils are sprinkler irrigated.

Application of potassium has usually involved small treatments in each of several irrigations (10 kg/ha of K). Usually, a nitrogen-potassium solution has been used as the potassium source with a 20-0-4 analysis being popular. Farmers and fertilizer dealers have believed that this process may improve plant absorption of potassium and result in higher yields where soil solution potassium is low due to low exchangeable potassium, leaching, or heavy plant use. Very little if any field data are available at the time of this writing,

[21] R. S. Rauschkolb, D. E. Rolston, R. J. Miller, A. B. Carlton, and R. G. Burau, "Phosphorus Fertilization With Drip Irrigation," *Soil Sci. Soc. Amer. Proc.,* 40:68, 1976.

however. Corn, grain sorghum, soybeans, and potatoes are some of the crops receiving potassium in irrigation water research.

Sulfur applications in irrigation water have been more common than applications of potassium. Injection of sulfur is quite easy and usually involves the use of ammonium thiosulfate (12% N, 26% S) or solutions of ammonium sulfate. These sulfur carriers already contain nitrogen but can be blended with nitrogen solutions quite easily. Interest in this technique has been intense in sulfur-deficient areas where sandy soils are common and where soil organic matter levels are low. Addition of sulfur in irrigation water is a common natural phenomenon on large acreages particularly in the Great Plains area of the United States where groundwater contains appreciable quantities of sulfate. Sulfur determinations should be requested along with irrigation water quality determinations so that the amount of this element applied will be known. The sulfur present in the water arises from gypsum and is available to the plants. Like the application of potassium, very few research projects have addressed the efficiency of sulfur applied in this manner. Generally, potassium and sulfur should encounter no detrimental reactions in the water even if the water is quite high in soluble calcium and/or magnesium. Timing and placement would largely determine the effectiveness of the applied nutrients.

7.2.4 Micronutrient Application

Micronutrient applications for crops have usually been relegated to preplant applications in order to supply the needed nutrients when plants are small and beginning to develop roots. However, successful application of micronutrients through irrigation systems has been effected usually through sprinkler systems. Application of micronutrients through a sprinkler irrigation system poses a possible method of treatment of a crop when deficiencies are detected midway through the season. Zinc and iron are two elements that have been successfully applied in this manner although others have been used in some cases with success, notably boron and copper.

The entire subject of efficiency of micronutrient applications in irrigation water has been only lightly researched in row crops. More practical experience has been gained by horticulturists who have successfully injected micronutrients into irrigation water or into sprays for tree crops. One should not assume, however, that just because a nutrient is injected into irrigation water and applied through a sprinkler that all the nutrient will be absorbed by the leaves of the plants. Much of the nutrient will eventually contact the soil surface with the irrigation water. If micronutrient application is a major objective in a plant nutrient program, then a small water application could be used to assure a modest concentration of nutrients on leaf surfaces where they may be rapidly absorbed. Low-pressure, drop-delivery sprinkler irrigation systems are particularly well adapted to micronutrient application.

Some evidence of the effectiveness of sprinkler irrigation application of nutrients is presented in some work with citrus in California reported by

Table 7–9. Effects of Iron Ligninsulfonate Applied in Irrigation Water on Grain Sorghum Yield and Leaf Composition

Fe Applied (kg/ha)		Leaf P (%)	Leaf Fe (ppm)	Grain Yield (kg/ha)
Soil	Irrigation H$_2$O			
0	0	0.23	274	5582
2.2	0	0.22	219	5582
4.5	0	0.22	242	5896
9.0	0	0.22	272	5896
0	4.5	0.22	902	6021
2.2	4.5	0.18	915	6272
4.5	4.5	0.20	872	6272
9.0	4.5	0.19	827	6523

SOURCE: L. S. Murphy, M. C. Axelton, and P. J. Gallagher, "Iron Fertigation of Grain Sorghum," *1970 Kansas Fert. Res. Report of Progress*, p. 133.

Wallace and North.[22] They suggested the feasibility of applying iron as the chelate FeEDDHA for orange trees at rates of 0.28, 0.66, and 1.35 kg/ha of iron (Fe). Another study conducted in Kansas with sprinkler-irrigated grain sorghum gave some encouragement to further evaluation of this technique. Applications of iron to the soil to correct iron deficiency in grain sorghum in Texas, New Mexico, Oklahoma, Colorado, and Kansas are notably ineffective. Foliar applications are costly and must be repeated several times. The study reported in Table 7–9 included applications of 4.5 kg/ha of iron as an iron ligninsulfonate through a sprinkler irrigation system. Iron was injected into the irrigation system when the plants were in the 10- to 12-leaf stage and was applied only once. Plant tissue samples were analyzed to determine if the iron was actually absorbed. Iron applications in the water increased yields over soil-applied treatments and produced a striking increase in the iron concentrations in the leaves. The leaf analyses should be viewed with some skepticism, however, because the higher values could have been due to some residues on the leaves even though they were thoroughly washed prior to analysis.

Some successful experience with applications of zinc fertilizer in irrigation water has been reported from Nebraska.[23] Injection of an inorganic nitrogen-zinc combination into water applied through a low-pressure sprinkler system produced consistent increases in the zinc concentration in corn leaves even though application only totaled 0.2 kg/ha of zinc. The effects of applied zinc on tissue zinc concentrations declined in magnitude as the amount of water applied decreased. Apparently the most effective zinc application occurred when only 6 mm (0.25 in.) of water was applied. Other rates of water application were 19 and 32 mm (0.75 and 1.25 in.). Yields trended up but differences

[22] A. Wallace and C. P. North, "Supplying FeEDDHA to Orange Trees in Irrigation Water." In *Current Topics in Plant Nutrition*, A. Wallace, (Ed.), Ann Arbor, Mich.: Edwards Bros., 1966.
[23] Personal communication from Don Johnson, Allied Chemical Company.

were variable. Much more research is needed in this area to confirm the effectiveness for both zinc and iron.

7.3 APPLYING SOLID FERTILIZERS

Applications of solid fertilizers range from spreading of single element materials such as urea and ammonium nitrate (Fig. 7-23) to applications of granular materials containing several elements in each granule to spreading bulk blends that are mechanical mixtures of several different nutrient sources. Methods of application for solid fertilizers are numerous and include broadcast, banded near the seed, banded in direct seed contact (pop-up) and sidedress methods.

In the following sections, we examine problems relative to the various materials and methods of application.

7.3.1 Broadcast Applications

Broadcast, preplant application of solid and liquid fertilizers has grown rapidly in the past 20 years due to a need to reduce the time involved in the seeding of crops and subsequently a desire to lessen the time involved in handling fertilizers. The development of bulk blends of solid fertilizers has occurred in response to the need for faster and less labor-intensive methods of fertilization at the time of seeding.

Fig. 7-23. Uniform broadcast application of solid fertilizers are essential for most efficient use of the applied nutrients. Irrigation following this application will improve the efficiency of the applied nutrients. (Courtesy TVA.)

Applying Fertilizers and Soil Amendments

A considerable amount of discussion has been afforded the relative efficiencies of broadcast and band or seed placement of nutrients. Phosphorus, potassium, and micronutrient use efficiencies have been the major reason for concern over method of application. Recommendations in many states recognize the superior efficiency of band-applied phosphorus close to the seed where emerging seedlings can contact the fertilizer source early in the growing season. Usually, areas that encounter the greatest improvement in efficiency from band placement *versus* broadcast treatments are the northern states in the United States that experience colder conditions during spring seeding of row crops. Recommendations in those areas frequently call for a doubling of application rate if broadcast applications are to be used *versus* band treatments. Very acid soil conditions and very low soil-test values also combine to improve the efficiency of banded applications *versus* broadcast treatments. In warmer geographical areas with soil conditions other than those favoring phosphorus banding, broadcast treatments have usually performed quite well and recommendations do not take into account the method of application.

Movement toward higher rates of nutrient application in an attempt to improve the nutrient supplying ability of the soil has also led to more equality in efficiency between banded and broadcast applications. However, whatever method of application is used, nutrients in fertilizers are used more efficiently by plants when the fertilizer has been properly blended.

Bulk blending is generally defined as the physical mixing, without chemical reaction, of two or more dry fertilizer materials to produce a mixture including two or more nutrients. A typical bulk blend is formulated from at least two materials. The earliest vestiges of bulk blending and broadcast applications can be traced back to the 1940s in northern Illinois where fall plowdown of rock phosphate was a common recommended practice. Some vendors who used trucks to spread rock phosphate began to dump in potassium fertilizers with the rock phosphate creating a poorly mixed but never-the-less a bulk blend. Later ammonium sulfate was added to the mixture and a "complete" blend resulted.

Recognizing the problems in mixing all the materials together, some individuals began to use feed mill mixers for a better blend. This practice grew slowly in the late 1940s and early 1950s but rapid growth began in the mid-1950s and by 1959 there were 186 bulk blend facilities in the North Central region of the United States. As an indication of the speed of development of the practice in the late 1950s and early 1960s, there were more than 1200 bulk blend facilities in that same region by 1964. By 1966 there were more than 3000 nationwide and by 1978 about 6000 bulk blend facilities (Fig. 7-24).

Many factors contributed to the development of bulk blending. Major factors included the rapid expansion in nutrient use, technological developments, convenience, growth of custom application by dealers and others, and favorable economics.

Convenience alone was a strong force in the acceptance of bulk blends and broadcast applications of nutrients. Trends toward larger farms combined

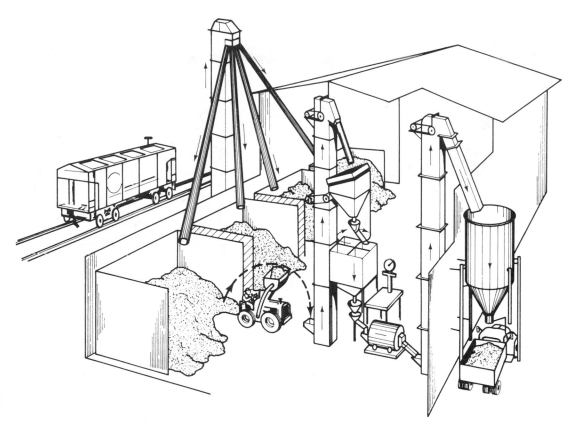

Fig. 7-24. A diagram of a facility for the production of bulk blended fertilizers. (Courtesy TVA.)

with a smaller number of farm workers put more emphasis on custom application particularly by fertilizer dealers with a strong customer service orientation. Another convenience factor in the development of bulk blends and broadcast application was the flexibility in formulating to specific recommendations through the use of several materials rather than the restrictive ratios imposed by granulated, bagged materials that had only limited flexibility. Costs of bulk blends also competed favorably with other types of solid fertilizers due to the higher costs of bagged materials, lower freight costs per unit of nutrient with the use of high-analysis materials and lower storage costs for the dealers. These combinations put bulk blends in a very favorable position to capture much of the solid fertilizer market from specific ratio materials.

With all the advantages for bulk blends and broadcast applications of nutrients came some pretty significant problems. One of these that discussed in some detail is the problem of segregation (particle separation) (Fig. 7-25) and poor distribution patterns. Probably the most consistent problem of these two is that of nonuniformity of the blended materials. Fertilizer materials of

Fig. 7–25. Segregation or separation of components in a bulk blend is or can be a serious problem: (left) KCl too coarse, causing segregation; (right) KCl is granular and no segregation results because of the similarity of the two materials' particle size. (Courtesy TVA.)

different particle size and different specific gravity simply do not respond to gravitational forces in the same way.

Segregation of bulk blends may occur immediately after mixing, even though the various materials are thoroughly mixed. Handling, transportation, and spreading following mixing causes separation of the various materials. Three distinct types of separation are recognized: (1) rolling action or coning, which occurs when the mixer is unloaded into a bin, truck, or spreader resulting in accumulation of different-sized particles in different areas of the pile; (2) vibration during transportation and application resulting in stratification of materials by particle size and density; (3) ballistic separation during application due to different trajectories of particles of varying density and particle size.

Excellent reviews of the problems associated with bulk blending have been assembled by TVA and other segments of the fertilizer industry. Proceedings of the 1973 TVA Bulk Blending Conference at Louisville, Kentucky, provide an excellent overview of problems and solutions.

Hoffmeister[24] discussed the importance of size-matching of materials in bulk blending and points out that this technique is essential to production of high-quality blends. Numerous studies have indicated that uniform particle size is the single most important contribution to reduce segregation in mixes; shape and density of particles tend to fade into insignificance by comparison. The practice of including micronutrients or herbicides in bulk blends makes

[24] G. Hoffmeister, "Quality Control in a Bulk Blending Plant," Proc. TVA Bulk Blending Conf., Aug. 1–2, 1973, Louisville, Ky., pp. 59–70.

the problem of segregation due to particle size even more important because only very small quantities of these materials are used in a given formulation. To overcome this problem, some companies have introduced coating of micronutrients on other particles through the use of powdered or liquid micronutrient and herbicide materials and frequently employing some type of binding agent such as a small amount of nitrogen solution or liquid ammonium polyphosphate.

Achorn and Wright[25] pointed out that a great deal of effort has been expended in the past few years to improve the quality of bulk blends so they can be handled and applied uniformly. Still, problems do exist either through poor mixing or poorly adjusted applicators. The experiences of one company that examined the distribution and analysis of a bulk blend spread pattern from a dual-spinner applicator (Fig. 7-26) indicate the types of problems that can occur. Two passes with a slight overlap (estimated to be 10%) were made in several tests. Samples in the catch trays in the applicator's path were analyzed and the results reported in terms of the application rate, amount of nutrients applied per hectare, and the analysis of the mixture (Table 7-10). Results indicate that the application rate varied widely from 176 kg/ha at the center of the swath to about 447 kg/ha in the overlap area. Nitrogen content of the mixture caught in the trays varied from 14.0 to 22.4% with original target formulation being 21.4%. This represents a variation of about 35% in the nitrogen analysis alone and subsequently appeared in the nitrogen application rate as well. Similar variation was noted in the phosphorus and potassium applications.

Considering the variations in analysis due to segregation and rate of fertilizer applied, a poor crop response and a very angry customer would have resulted. The degree of overlap was an important factor in the variation. The obvious problem in this entire operation was the fact that the operator had not recently calibrated his equipment. More attention is given to distribution patterns and calibration of equipment for both solids and liquids in Section 8.

Wells[26] points out that the greatest question in use of bulk blends, particularly broadcast applications of bulk blends, relates to the effects of nonuniformity of application on crop production. Jensen and Pesek at Iowa State University examined theoretical yield losses that might be expected from nonuniform applications of fertilizer spreaders. Their model[27,28,29]

[25] F. P. Achorn and E. B. Wright, Jr., "Fertilizer Field Distribution Problems and How to Solve Them," Proc. Iowa Fert. and Ag Chem. Dealers Conf., Des Moines, Jan., 1973.

[26] K. L. Wells, "The Importance of Timely and Accurate Application," Proc. TVA Bulk Blending Conf., Aug. 1-2, 1973, Louisville, Ky. TVA Bul. Y-62, pp. 85-93.

[27] D. Jensen and J. Pesek, "Inefficiency of Fertilizer Use Resulting from Nonuniform Spatial Distribution: I Theory," Soil Sci. Soc. Amer. Proc., 26:170, 1962.

[28] D. Jensen and J. Pesek, "Inefficiency of Fertilizer Use Resulting from Nonuniform Spatial Distribution: II Yield Losses Under Selected Distribution Patterns," Soil Sci. Soc. Amer. Proc., 26:174, 1962.

[29] D. Jensen and J. Pesek, "Inefficiency of Fertilizer Use Resulting from Nonuniform Spatial Distribution: III Fractional Segregation in Fertilizer Materials," Soil Sci. Soc. Amer. Proc., 26:178, 1962.

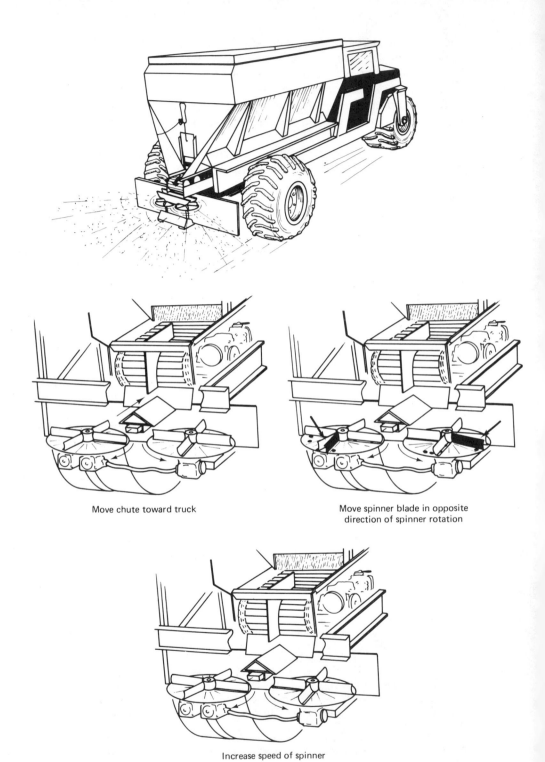

Move chute toward truck

Move spinner blade in opposite direction of spinner rotation

Increase speed of spinner

Fig. 7-26. Dual spinner solid fertilizer applicator showing adjustments for corrections of an M distribution pattern characterized by low delivery in the center and high delivery on the edges of the swath. (Courtesy TVA.)

Table 7–10. A Poorly Calibrated and Adjusted Applicator Combined with Segregation Can Produce Dramatic Variation in Application Rate, Distribution Pattern, and Analysis of the Applied Material at the Soil Surface[a]

		Catch Pan								
		1L	2L	3C	4R	5RR	6R	7C	8L	9L
	Rate (kg/ha)	342	295	176	472	447	362	188	265	226
%N		17.5	21.5	22.4	21.6	14.0	20.8	22.4	20.1	11.5
%P_2O_5		16.2	10.0	9.5	8.6	20.5	9.0	9.4	10.0	23.2
%K_2O		3.8	6.7	5.7	7.3	3.2	7.4	4.7	7.6	4.0
%Zn		2.9	2.3	1.7	2.6	3.2	1.2	1.7	2.1	3.6
N (kg/ha)		60	58	39	102	63	75	42	54	26
P (kg/ha)		25	13	7	18	40	14	8	11	23
K (kg/ha)		11	16	8	28	12	22	7	17	8
Zn (kg/ha)		10	7	3	12	14	8	3	6	8

[a] This test involved application with a double spinner solid applicator spreading a bulk blend material.

NOTE: Nominal analysis: 21.4% N, 4.7% P, 4% K, 2.1% Zn. Application involved two passes with the applicator. C refers to center of the swath. RR is an overlap area. L and R refer to left and right of center of the swath.

SOURCE: L. W. Lohry, "Putting a Dollar Value on Uniform Application," *Proc. National Fertilizer Solutions Association Round-Up,* July 24–25, 1973, St. Louis, Mo., p. 31.

predicted the greatest loss of corn production from fields of very low fertility. Other estimates of yield losses from nonuniform applications have varied considerably in forms of magnitude of the problem. Englestad[30] estimated yield and profit losses of corn and cotton on responsive soils from uneven distribution and concluded that a wide range of flexibility exists in rate and ratio of nutrients that crops will tolerate without serious effects on either yield or profit. Dumenil and Benson[31] analyzed agronomic effects of uneven distribution of corn yields and pointed out that yield losses can vary from appreciable losses to none and are influenced by (1) yield response of the nutrients applied, (2) nutrient rates applied, and (3) type of fertilizer distribution pattern across and along the path of the applicator. Dumenil and Benson noted that uneven growth and color effects are often caused by other cultural practices including compaction and drainage. Nitrogen deficiency symptoms were the most notable cases cited resulting from uneven distribution.

[30] O. P. Englestad, "Effect of Variation in Fertilizer Rates and Ratios on Yield and P Profit Surfaces," *Agron. J.,* 55:263, 1963.

[31] L. Dumenil and G. Benson, "Fertilizer Field Distribution Problems — Agronomic Implications," Proc. 25th Annual Fert. and Ag Chemicals Dealers Conf., Des Moines, Iowa, Jan. 9–10, 1973.

Although the discussion to this point has related primarily to broadcast applications of bulk blends of solid fertilizers, the same concerns for uniform distribution also pertain to applications of single nutrient materials such as topdressed applications of nitrogen materials on grasses.

7.3.2 Pop-Up Applications

Pop-up fertilization refers to placement of small amounts of nutrients in direct seed contact usually for row crops. Kresge[32] pointed out that pop-up placement of nutrients was originally studied in Minnesota and Ontario where cooler soil conditions dominate at the time of seeding. Early Minnesota work indicated yield increases in corn of 1200 kg/ha from application of 56 kg/ha of a 6–10–10 solid fertilizer in direct seed contact when the corn was planted in 100-cm rows. Ontario studies produced a 370 kg/ha grain yield increase from a similar 56 kg/ha application of 6–10–5 liquid fertilizer in direct seed contact. Because colder soil conditions occur with earlier planting dates, this technique is more likely to exert beneficial effects at earlier planting. Root development is slow under those conditions and additional nutrients close to the seed can aid in absorption. Furthermore, in cold soils phosphorus and potassium are not as readily available. Kresge noted that research by Ohlrogge at Purdue produced 800 to 1600 kg/ha increase in grain yield in corn with May 5 planting but produced variable 900 kg/ha decreases to 600 kg/ha increases in yield with a June 3 planting. Inclusion of banded applications close to the seed but not in direct seed contact will usually diminish the pop-up response but may not totally erase it.

Solid fertilizers used for pop-up application should have most of the following characteristics:

1. Contain nitrogen, phosphorus, and potassium with high phosphorus content.
2. High water solubility.
3. Low salt index.
4. Minimal content of compounds that liberate ammonia (DAP, urea).
5. Some nitrate-nitrogen; some ammonium-nitrogen.
6. High nutrient concentration.

One of the critical aspects of pop-up placement of solid and liquid fertilizers is the salt index of the materials in the fertilizer. The salt index is a measure of the salt concentration that the fertilizer induces in the soil solution. The osmotic pressure of the soil solution is actually determined to measure salt index. Salt index is expressed as the ratio of the increase in osmotic pressure of the soil solution produced by the material to the osmotic pressure of the same weight of sodium nitrate, based on a relative value of 100. Higher-analysis fertilizers (higher nutrient concentrations) usually have a

---

[32] C. B. Kresge, "Let's Take a Look at Pop-Up Fertilizer," *Potash Institute Newsletter*, M-143, Jan.–Feb., 1967.

lower salt index because fewer molecules of salt are placed in the soil solution when they dissolve. For example, to apply 10 kg/ha of nitrogen as ammonium sulfate would require about 48 kg of material. On the other hand, ammonium nitrate would require only 29 kg of material. More water would be required to dissolve the ammonium sulfate than the ammonium nitrate. Less requirement of water to dissolve the fertilizer will cause less competition between germinating seedlings and the fertilizer for water present in the soil.

High concentrations of salts and subsequently high osmotic pressures of the soil solution in the vicinity of the germinating seed can cause salt injury (plasmolysis) similar to that which occurs in a natural saline soil (see Chapter 9). Major concern over the use of pop-up fertilizers is seedling injury or fertilizer "burn." Use, then, of a low salt index fertilizer in pop-up placement is mandatory. Kresge noted that stand damage was substantial (50%) when high rates of fertilizer were placed in direct seed contact (pop-up) for corn in Ohio. Applications in that case were 180 kg/ha of 6-8-6 fertilizer. On the other hand, only 57 kg of a 6-10-20 in direct seed contact in Michigan on a sandy soil reduced stands by 30% and reduced corn yield by 1500 kg/ha.

Use of fertilizer formulations for pop-up placement should take into account the materials that are involved in the particular analysis. For instance, it is quite possible that a given analysis of a complete mixed fertilizer could vary widely in salt index between two different producers or even vary in salt index in time at a given point of production. Some idea of the salt indices of various fertilizer materials is given in Table 7-11. Note that the nitrogen and potassium materials have higher salt indices than the phosphorus compounds. Ammonium phosphates, though, have appreciable salt indices and they are much higher than the superphosphates. The computed salt index for a mixed fertilizer containing nitrogen, phosphorus, and potassium is the sum effects of its components. Although the *total* salt index for a high analysis nitrogen-phosphorus-potassium mix may be *greater* than that for a lower-analysis material, the salt index per unit of plant nutrients is lower and subsequent lower use per hectare subjects the seedlings to a lower salt effect.

Liquid fertilizers may produce lower osmotic pressure in the soil solution than solid fertilizers. Because less water is required, the salts already being in solution, generally fewer problems have been encountered in the use of liquids as pop-up fertilizers.

Crop tolerances to osmotic pressure of the soil solution in the vicinity of the seed varies widely. For instance, wheat is more tolerant of high salt concentrations than is grain sorghum. Corn is intermediate. For that reason, recommendations for placement of fertilizer in direct seed contact vary with the crop. As a very general rule, recommendations for corn and grain sorghum call for no more than 8 to 18 kg/ha of nitrogen plus potassium in direct seed contact. Another general rule is to keep pop-up applications of solid fertilizers around 50 kg/ha. Bandel[33] has recommended that for a specific stand density

[33] V. A. Bandel, "Pop-Up Fertilizers for Corn," Maryland Cooperative Ext. Service Fact Sheet 211, Dec., 1969.

Table 7–11. Salt Indices of Fertilizer Materials[a]

Fertilizer Material	Analysis (%) N-P_2O_5-K_2O	Salt Index	Salt Index/20 lb (9 kg) of Nutrient N + P_2O_5 + K_2O
Nitrogen fertilizers			
Anhydrous ammonia	82-0-0	47.1	0.572
Ammonium nitrate	34-0-0	101.7	2.904
Ammonium sulfate	21-0-0	69.0	3.253
Urea	45-0-0	72.7	1.560
Pressure nitrogen solution	41-0-0	78.3	1.930
Sodium nitrate	16.5-0-0	100.0	6.060
Phosphorus fertilizers			
Normal superphosphate	0-20-0	7.8	0.390
Triple superphosphate	0-45-0	10.1	0.224
Monoammonium phosphate	11-55-0	26.9	0.407
Diammonium phosphate	18-46-0	29.0	0.454
Potassium fertilizers			
Potassium chloride	0-0-60	116.3	1.936
Potassium sulfate	0-0-54	46.1	0.853
Potassium nitrate	13.8-0-46.6	73.6	1.219

[a] Salt index of a fertilizer material is an indication of the relative increase in the osmotic pressure of the soil solution produced by the dissolution of the fertilizer relative to that of the same weight of sodium nitrate.

SOURCE: L. F. Rader, Jr., L. M. White, and C. W. Whittaker, "The Salt Index: A Measure of the Effect of Fertilizers on the Concentration of the Soil Solution," *Soil Science*, 55:201, 1943 by permission of the American Society of Agronomy.

the rate of application of pop-up solids be closely tied to row spacing for crops. Wider row spacing would allow the use of very limited amounts of nutrients in direct seed contact, whereas more rows per unit of width results in a dilution effect and larger amounts would be allowable. He recommends that about 9 kg/ha of nitrogen plus potassium is the maximum allowable for 100-cm (40-in.) rows, whereas that amount could be increased to 18 kg/ha of nitrogen plus potassium for 46-cm (18-in.) rows.

7.3.3 Band Applications

The terminology for methods of fertilizer placement for crops tends to vary somewhat depending on the region or country. Banded placement of nutrients is intended here to refer to placement of the fertilizer in bands on one or both sides of the seed or seedlings (Fig. 7–27). Usually, or at least frequently, the fertilizer material is placed at a depth greater than that of the seed in order to separate the fertilizer from the drier surface soil and to allow interception of the fertilizer band as the roots penetrate outward and downward. The terminology "side-banded" is also used interchangably with

Applying Fertilizers and Soil Amendments

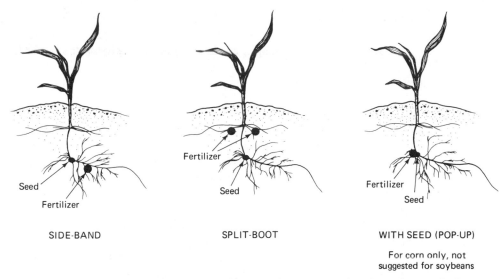

SIDE-BAND SPLIT-BOOT WITH SEED (POP-UP)

For corn only, not
suggested for soybeans

Fig. 7-27. Various methods of applying fertilizer at time of seeding. (Source: R. D. Voss and J. C. Herman, "Three Ways to Place Row Fertilizer," Iowa State Univ. Pamphlet 361, Feb. 1967.)

"banded" in many areas of the United States. U.S.S.R. and east European agronomists consider "banded" to be "localized" placement as this term is frequently translated in the literature. Some further confusion occurs because certain areas of the United States equate banded placement of nutrients for some crops with placement in direct seed contact, similar to "pop-up" described in the preceding section. Banded placement of starter fertilizer for winter or spring wheat in the Great Plains of the United States results in direct seed contact of amounts that may be at or above the limits spelled out for pop-up in Section 7.3.2. Interestingly, in that same region, banded application of starter fertilizers for row crops implies placement to the side and below the seed.

Regardless of the terminology, band placement of nutrients, particularly phosphorus, potassium, and some micronutrients has received a great deal of attention over the years as the most efficient method of fertilizer placement. Studies in many states in the United States have shown some advantage in phosphorus use efficiency when the phosphorus was placed close to or with the seed as compared to "broadcast" applications or "topdressed" applications. Broadcast usually implies applications prior to seeding that are incorporated by a tillage operation. Topdressed applications of nutrients usually implies postemergence applications. Topdressing is usually used to describe applications of nitrogen for small grain. "Side-dressed" treatments, by contrast, usually refer to postemergence nitrogen applications for row crops such as corn or sorghum.

Evidence for the efficiency of banded applications of phosphorus was accumulated by several authors in a publication edited by Richards.[34] Individual articles in that publication cited reasons for the generally accepted theory of superior performance of phosphorus in bands *versus* broadcast applications. Among the factors cited or related to superior performance of banded phosphorus were the following:

1. Adaptable to root interception in most crops.
2. High concentration in limited soil zone conducive to early contact by plant roots particularly those growing in cold, wet soils.
3. Inclusion of ammonium nitrogen in the phosphorus fertilizer and subsequent greater phosphorus absorption.
4. Reduced rate of reaction (fixation) of phosphorus with soil due to high concentrations in a limited amount of soil.
5. Greater depth of placement of phosphorus into soil zones that remain moist longer (positional availability).

Environmental and cultural conditions that tend to favor band application of phosphorus (the element generally considered to respond best to placement) include the following:

1. Low phosphorus soil test levels.
2. Low soil pH.
3. Large amounts of soluble aluminum and iron in the soil.
4. Highly calcareous soil conditions.
5. Crops planted or growing in cold soil (planting date, residue management).
6. Poor soil aeration (wetness or compaction).
7. Use of very small amounts of fertilizer (usually goes hand in hand with short term lease arrangements).
8. Poor incorporation of broadcast applications with existing tillage equipment.

For the sake of brevity, discussion of comparative effectiveness of methods of fertilizer application will focus on phosphorus.

Olson and Sander[35] pointed out that rooting habit and activity strongly influence the responsiveness of a certain crop to band applications of nutrients, particularly phosphorus. They pointed out that corn and grain sorghum will show greater early growth responses to band applications of phosphorus than will soybeans, and quoted data from the University of Nebraska showed that crops vary in their abilities to recover phosphorus from broadcast or band placement (Table 7–12).

Olson and Sander further pointed out that there is good evidence that soybeans absorb significant quantities of nutrients from greater soil depths than

[34] G. E. Richards, "Band Application of Phosphatic Fertilizers," Olin Corp., Agric. Div., Little Rock, Ark., 1977.
[35] R. A. Olson and D. H. Sander, "Corn, Grain Sorghum and Soybeans." In *Band Application of Phosphatic Fertilizers*, G. E. Richards (Ed.), Olin Corp., Little Rock, Ark., 1977, p. 53.

Table 7-12. Utilization of Fertilizer Phosphorus by Various Crops in Greenhouse Pots[a]

Crop	Utilization of Applied P in $80 + 50^a + 0$	
	Mixed	Banded
	% P from Fertilizer	
Soybeans	30.8	42.4
Other legumes		
Alfalfa	52.6	62.1
Red clover	56.6	67.7
Sweetclover	54.2	73.0
Nonlegumes		
Corn	46.6	48.0
Wheat	38.9	56.7
Oats	43.3	60.4
Bromegrass	52.4	53.1
Grain sorghum	63.1	44.8

[a] Fertilizer tagged with ^{32}P was mixed with all the soil or placed about 2.5 cm (1 in.) to the side and below the seed at planting; assay effected at flowering for each crop. Data are mean values for these soils: Nemaha (acid), Moody (neutral), and Crofton (calcareous).

SOURCE: R. D. Green, M. S. Thesis, Univ. of Nebraska, Lincoln, Neb., 1957.

do cereals due to a tap rooting pattern in soybeans. Consequently, a substantial portion of the nutrients absorbed by that crop is taken up late in the growing season. Corn roots on the other hand grow as far laterally as they do vertically early in the growing season and consequently can intercept band-placed nutrients early in their development. Early corn root growth is temperature-controlled and lateral growth of the radicle root of corn exceeds downward penetration until subsoil temperatures warm to an acceptable level. Cold, wet soil conditions would tend to favor corn plant utilization of nutrients placed in the surface soil near the seed. Further contrast in crops results from corn-sorghum comparisons. Later planting dates for sorghum tend to result in rapid downward root penetration due to warmer subsoil conditions and less responsiveness to banded applications of nutrients.

Early planting dates and large amounts of residue on the soil surface could also affect plant responsiveness to methods of fertilizer application. Both of these factors would subject plants to colder conditions early in the growing season and could enhance the effectiveness of band placement, particularly of phosphorus.

Soil conditions exert dramatic effects on efficiency of methods of nutrient application. Welch and co-workers at Illinois[36] determined that banded ap-

[36] L. F. Welch, D. L. Mulvaney, L. V. Boone, G. E. McKibben, and J. W. Pendleton, "Relative Efficiency of Broadcast versus Banded Phosphorus for Corn," *Agron. J.*, 58:283, 1966.

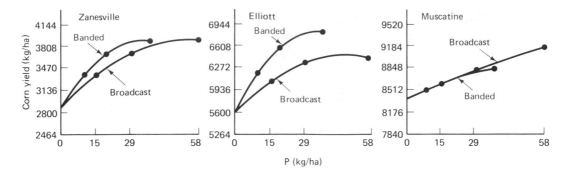

Fig. 7–28. Band placement of phosphorus for corn can lead to more efficient utilization of phosphorus under some conditions. These data from Illinois show greater efficiency of banded phosphorus on some soils than on others. Low available phosphorus, low rates of application and low soil pH tend to favor band applications. (Source: L. F. Welch et al., *Agron. J.,* 58:283, 1966 by permission of the American Society of Agronomy.)

plications of phosphorus for corn were more effective than broadcast phosphorus applications on one medium phosphorus testing soil in Illinois, but another soil with similar surface soil phosphorus levels exhibited essentially no difference in effectiveness of banded and broadcast phosphorus. Obviously, some other factor such as subsoil phosphorus concentration was affecting the plants' nutrition (Fig. 7–28). When methods of phosphorus application were compared for irrigated corn in a cold, wet spring on a sandy loam soil in Kansas, method of phosphorus application had little effect on grain yield but did affect plant phosphorus concentrations early in the growing season (Figs. 7–29 and 7–30). Banded applications of monoammonium phosphate as well as dry and liquid forms of ammonium polyphosphate produced the highest plant concentrations of phosphorus early in the season but, by tasseling, differences in concentrations due to method of application had disappeared. Form of applied phosphorus had no consistent effect on either yield or phosphorus concentrations in the leaves. Nebraska soil test recommendations call for a doubling of phosphorus rates if broadcast application is to be used. Kansas, on the other hand, does not distinguish between methods of phosphorus application in its recommendations.

Although less emphasis has generally been placed on the efficiency of band applications of potassium for various crops, research in Tennessee and Illinois has indicated some improvement from this method of placement. Parks and Walker[37] showed higher efficiency for banded potassium for corn versus comparable broadcast applications but these differences diminished as the soil-test level of potassium increased. Welch and co-workers at Illinois[38]

[37] W. L. Parks and W. M. Walker, "Effect of Soil Potassium, Fertilizer Potassium and Method of Fertilizer Placement Upon Corn Yields," *Soil Sci. Soc. Amer. Proc.,* 33:427, 1969.

[38] L. F. Welch, P. E. Johnson, G. E. McKibben, L. V. Boone, and J. W. Pendleton, "Relative Effectiveness of Broadcast versus Banded Potassium for Corn," *Agron. J.,* 58:618, 1966.

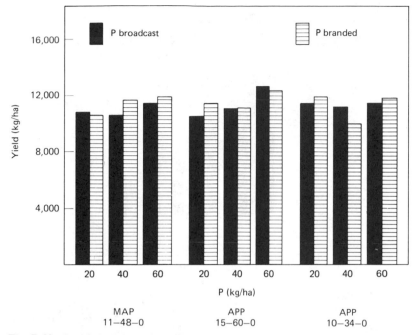

Fig. 7-29. Banded placement of phosphorus for corn does not always produce better yields as indicated by these data from Kansas. Soil pH in this case was only slightly acid but soil test phosphorus was very low. (Source: B. B. Webb, "Field and Growth Chamber Comparisons of Ortho and Polyphosphates," Ph.D. Thesis, Dept. of Agronomy, Kansas State University, 1970.)

also examined method of application effects on potassium efficiency for corn and reported that the broadcast applications ranged from 33 to 88% as efficient as banded applications when soils tested low to medium in available potassium.

Band placement of nutrients for cereal crops has been quite effective but some limitations on amounts applied are necessary. Smith[39] discussed the effectiveness of banded applications of phosphorus for wheat and other small grains and noted that band placement has frequently been more efficient than broadcast placement. In the 1920s, methods of phosphorus placement for small grain were examined with indications of superior performance of band placement on moderately acid soils low in available phosphorus (Table 7-13).

Later data from the Plains States have also indicated greater efficiency of banded applications of phosphorus for wheat both on very acid (pH 4.8) and on calcareous soils (Fig. 7-31).

To complicate the picture further, soil conditions and growing conditions shortly after seeding wheat can also have an important effect upon the efficiency of band or broadcast phosphorus applications. The data in Table 7-14

[39] F. W. Smith, "Wheat and Other Small Grains." In *Band Applications of Phosphatic Fertilizers,* G. E. Richards (Ed.), Olin Corp., Little Rock, Ark., 1977, p. 65.

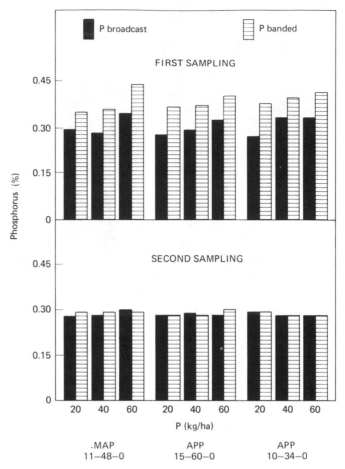

Fig. 7-30. Banded applications of phosphorus usually increase plant concentrations of phosphorus early in the growing season. Note that the differences between methods of placement had disappeared by the second sampling date, tasseling in this case. (Source: B. B. Webb, "Field and Growth Chamber Comparisons of Ortho and Polyphosphates," Ph.D Thesis, Dept. of Agronomy, Kansas State University, 1970.)

are a good example of what can happen on two successive years at the same site where treatments were repeated on exactly the same area. The 1976 data indicated some superiority of banded phosphorus applications versus broadcast at the lowest rate of applied phosphorus but very little difference at the other rates or between means for method average across all rates. The fall of 1975 was warm and dry with good growing conditions. The same study was seeded in the fall of 1976 followed by very cold wet weather. Results of the 1977 harvest showed highly significant and consistent differences favoring banded applications. This indicates that efficiency of placement is strongly related to growing conditions, greater efficiency from band applications being expected when growing conditions are adverse.

Table 7-13. Winter Wheat Yield Responses to Broadcast and Drilled (Banded) Fertilizer

Method of Fertilizer Application	Wheat Yield (kg/ha)	
	0-16-0	2-12-2
Control	867	
Broadcast	1237	1008
Drilled (banded)	1821	1532

SOURCE: F. L. Duley, Kansas Agr. Exp. Sta.

More recently, data have suggested that banded and broadcast applications of phosphorus for winter wheat may both be inferior to preplant applications that place the entire nitrogen and phosphorus fertilizer supply approximately 15 cm below the soil surface prior to seeding. Data from these studies indicate that the deeper placement and combination of nitrogen with the phosphorus improves plant phosphorus concentrations and subsequently increases yields (Table 7-15). Very similar information has been generated in Sweden where placement of nitrogen and phosphorus below the soil surface prior to seeding was superior to banded placement (Fig. 7-32). As noted earlier in this chapter, beneficial aspects of such placement may arise both from positional effects, placing the nutrients in a soil zone that will remain moist longer, and from possible disruption of phosphorus fixation reactions by the presence of high concentrations of ammonium ions. More efficient utilization of banded applications of phosphorus reported by Thompson in

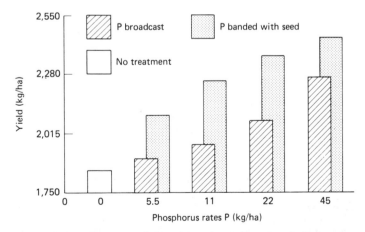

Fig. 7-31. Banded applications of phosphorus have also given better performance for winter wheat on calcareous soils under some conditions. Applications of banded phosphorus using a hoe-drill under dryland conditions could also provide some positional availability benefits by placing the phosphorus in more moist soil. (Source: C. A. Thompson, "Fertilizing Winter Wheat in a 16–24 Inch Rainfall Area," Kansas Exp. Sta. Bul. 276, 1973.)

Table 7-14. Comparative Effectiveness of Broadcast and Banded Applications of Phosphorus for Winter Wheat in Two Successive Years — Kansas[a]

| | Grain Yield (kg/ha) Labette County, Kansas | |
Treatment	1976	1977
Control, no P	1485	1868
Broadcast P	2258	2318
Banded P	2271	2843
LSD$_{.05}$	NS	430

[a] Cold conditions prevailed after seeding in fall of 1976 (1977 crop). Conditions in the fall of 1975 (1976 crop) were warm and dry.

NOTE: All plots received 84 kg/ha of N (75 lb/A). Results are the mean of three rates of P (12, 24, 36 kg/ha P or 25, 50, and 75 lb/A P_2O_5) and three P sources.

SOURCE: K. W. Kelley et al., "Wheat Responses to P Materials, P Rates and Methods of Application," *Kansas Fert. Res. Report of Progress, 313,* 1977, p. 26.

Table 7-15. Summarized Results of N-P Placement Studies with Wheat[a]

| | Grain Yield (kg/ha) | | | | | | | |
Treatment	Harper Co. 1975	Harper Co. 1976	Reno Co. 1977	Ellsworth Co. 1977	Labette Co. 1977	Dickinson Co. 1978	Osage Co. 1978	Ellsworth Co. 1978
Knifed N	2285	1620	1492	2903	2533	2379	3972	3172
Knifed N-knifed P	3024	2261	2641	3884	2997	4684	5074	4220
Knifed N-b'cast P	—	—	2580	3535	2621	3851	—	4052
Knifed N-banded P	2587	2560	2365	3346	2775	3817	—	—
B'cast N-b'cast P	2722	1744	2450	4005	2621	3898	2883	3044
B'cast N	2150	1304	1391	2795	2110	2957	2493	2500
LSD$_{.05}$	538	403	410	558	302	638	625	719

[a] Kansas studies have indicated that dual, knifed preplant applications of N and P for winter wheat are frequently more effective than P applications with the seed or broadcast, surface applications of N and P.

NOTE: N applications were 84 kg/ha N (75 lb/A N). Knifed N applications were anhydrous ammonia. Broadcast N was applied as 28% N nitrogen solution. P applications involved use of 10–34–0 liquid ammonium polyphosphate at rates of 18 kg/ha P in 1975 and 76 and 19 kg/ha P in 1977–78 (37.5 lb/A P_2O_5 and 40 lb/A P_2O_5). N-P knifed treatments were injected into the soil on 46-cm centers at a depth of 15 cm (18-in. centers at a depth of 6 in). Banded P was placed in direct seed contact at planting. Broadcast treatments were incorporated into the soil by disking. See Figs. 7-18, 7-19 for a drawing of application equipment and photos of field plot results.

SOURCES: D. F. Leikam, R. E. Lamond, P. J. Gallagher, and L. S. Murphy, "Improving N-P Application," *Agrichemical Age,* 22(3):6, 1978.
D. F. Leikam et al., "Comparisons of Methods of N and P Application for Winter Wheat," *Kansas Fert. Res. Report of Progress 343,* 1978, p. 20.

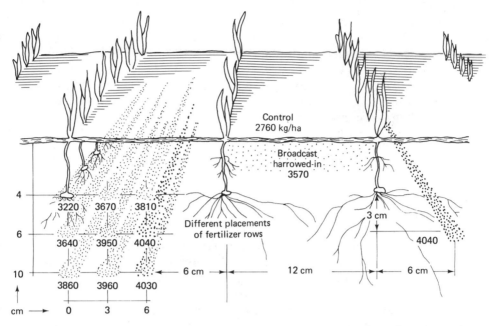

Fig. 7-32. Nitrogen-phosphorus placement studies in Sweden have demonstrated that applications of fertilizer to the side and below the wheat seed have been more effective than shallower applications closer to the seed. The numbers within the figure are yield data in kg/ha for the various placement positions. The highest yield resulted from fertilizer placement 3 cm below and 6 cm to one side of seed. (Source: Reijo Heinonen and Ake Huhtapalo, "Fertilizer Placement for Small Grains," Mimeographed Report, Dept. of Soil Sciences, Swedish University of Agricultural Sciences, Uppsala, 1978. Reproduced with the authors' permission.)

Fig. 7-31 might have arisen by shallow incorporation of the broadcast treatments or by deep placement of the banded phosphorus through the use of a hoe-type grain drill. This seems particularly likely because differences between broadcast and banded methods of phosphorus application for wheat persisted at rates as high as 45 kg/ha of phosphorus (92 lb/A of P_2O_5).

Vegetables represent another class of crops that respond well to banded applications of nutrients. Lucas and Vittum[40] noted that all vegetables require a relatively large percentage of the total nutrients needed early in the growing cycle. High requirements during the seedling stage have produced experimental results suggesting that the best method of supplying nutrients is to place most of the phosphorus and some of the nitrogen and potassium about 5 cm (2 in.) to the side of the seed. They also point out that micronutrients are sometimes needed on vegetables, particularly on organic soils, and banded placement of these nutrients near the seed at planting time is an effective means of application. Boron is an exception, however, because placement of

[40] R. E. Lucas and M. T. Vittum, "Vegetable Crops." In *Band Applications of Phosphatic Fertilizers,* G. E. Richards (Ed.), Olin Corp., Little Rock, Ark., 1977, p. 75.

this element near the seed may result in toxicity. Lucas and Vittum also caution about using too large quantities of all nutrients when placed too close to the seed. Nitrogen and potassium are most likely to contribute to seedling damage through salt injury. General recommendations for most vegetable crops except for potatoes call for a limit of 336 kg/ha (300 lb/A) of fertilizer *materials* (not nutrients). They reinforce the earlier statements of danger in banding large quantities of urea or diammonium phosphate close to the seed, particularly on high pH soils.

For all crops, please refer to your state extension service for recommendations on the most effective methods of nutrient application.

7.3.4 Specialty Applications

Recently, fertilizer technology has been modified to the needs of special uses of some nutrients. Forest fertilization has required the development of new forms of conventional materials such as "large-particle" urea. Fertilizer manufacturers working in cooperation with timber producers have developed very large particles of urea that can be applied by air with minimum drift and with enough ballistic effect to penetrate the foliage of standing timber. Some of the pioneering work in this area has been carried out by the National Fertilizer Development Center of TVA. Urea briquets and

Fig. 7–33. Urea in "mudballs", urea briquets, and large urea granules for special applications including forest fertilization and subsurface nitrogen placement for rice. Root zone placement studies of nitrogen fertilization of rice with such large particles of urea have proven effective (see Fig. 7–34). (Courtesy International Rice Research Institute.)

large urea granules suited for this and other purposes are pictured in Fig. 7–33.

Research at the International Rice Research Institute (IRRI) in the Philippines has indicated superior performance for subsurface placement of nitrogen for rice. Experiments at IRRI have shown that deep placement of fertilizer increases efficiency in lowland rice by minimizing losses due to volatilization and microbial oxidation. In a nitrogen isotope experiment at IRRI in 1966, 68% of the fertilizer nitrogen applied 10 cm (4 in.) deep as a basal treatment was used by the rice, whereas only 28% was used when the nitrogen was broadcast and incorporated by harrowing. Subsequently IRRI personnel studied a technique of applying urea and ammonium sulfate nitrogen in "mudballs" that approximate the size of the large urea particles and are pictured in Fig. 7–33. Nitrogen fertilizer was formed into mudballs by hand labor and dropped between every four rice hills when the rice was transplanted. This method of application of nitrogen has been tested extensively in Asia including locations in the Philippines, Laos, Thailand, and Korea (Fig. 7–34). Higher yields were obtained with 60 kg/ha (54 lb/A) of nitrogen applied in mudballs in the root zone of rice than when 100 kg/ha of nitrogen was applied (89 lb/A) by the conventional topdressing technique (Table 7–16). Because of the high labor requirement for making the mudballs, TVA manufactured the large granules and briquets pictured in Fig. 7–33. Studies of the efficiencies of these particles *versus* hand-formed mudballs indicated equal performance (Table 7–17).

In cooperative studies with IRRI, Indian scientists have tested fertilizer placement for rice at several locations using techniques similar to those described previously. Agronomists at the All India Coordinated Rice Improvement Project (AICRIP) found that 56 kg/ha of nitrogen (50 lb/A) applied

Fig. 7–34. Root zone placement of "mudball" urea on rice in the Philippines and other countries in southeast Asia has been superior to topdressed applications. Similar results were noted when urea briquets were used (see Fig. 7–33). (Courtesy International Rice Research Institute.)

Table 7-16. Higher Increases in Yield Were Obtained with 60 kg/ha of Nitrogen When Concentrated in Mudballs and Applied to the Rice Root Zone than with 100 kg/ha Applied by the Topdressing Method Used by Farmers

Fertilizer Rate (kg/ha N)	Yield (metric tons/ha) when N Applied as		Efficiency of N (kg rice/kg N)	
	Mudball	Topdressing	Mudball	Topdressing
60	8.0	5.8	53	23
100	8.4	6.6	38	21

SOURCE: S. K. De Datta, "Root Zone Placement Stretches Scarce Agricultural Chemicals," *The Intl. Rice Res. Inst. Reporter,* May 1976, p. 3.

Table 7-17. Yields of IR28 Rice When Nitrogen Was Applied at 56 and 112 kg N/ha as Urea-Briquets, as Mudballs, and in Split Doses

Application Method	Yield (metric tons/ha)	
	56 kg/ha N	112 kg/ha N
Urea-briquets	5.4	4.7
Mudballs	5.1	5.7
Split application	4.4	5.2
No nitrogen	3.4	

SOURCE: S. K. DeDatta, "Root Zone Placement Stretches Scarce Agricultural Chemicals," *The Intl. Rice Res. Inst. Reporter,* May 1976, p. 3.

as briquets produced yields about equal to those of 80 kg/ha of nitrogen (71 lb/A) broadcast in three split applications. The savings of 24 kg/ha of nitrogen was worth 115 rupees (about $13) per hectare at the current market value for nitrogen. Extra labor to place the briquets cost about 30 rupees ($3.30), a savings of 85 rupees (about $9) per hectare.

7.4 APPLYING FLUID FERTILIZERS

Much of the discussion in the preceding sections on solid fertilizer application is also relative to the application of fluid (liquid and suspension) fertilizers. Major differences in application considerations relate to the equipment used and to uniformity of distribution. Segregation problems in solid bulk blends do not occur in fluids but problems with uniform distribution can still occur in fluids due to equipment failure and improper formulations. Applications of fluid fertilizers as pop-up treatments or in bands beside the seed require the same considerations of rate and location as solid fertilizers.

7.4.1 Application Equipment

Application equipment for liquid fertilizers has changed dramatically in the past 15 years with development of larger applicators capable of operating in moist soils as well as under dry conditions. Most nonpressure nitrogen solutions (see Chapter 2) and clear liquid mixed fertilizers are applied by self-propelled applicators or by individual, self-contained tank units. Recently, the self-propelled applicators have been commonly equipped with flotation tires to aid in reducing soil compaction (Fig. 7–35). The self-contained tank units can be pulled with tractors and represent a common means of farmer applications of liquids. Self-propelled units are used mainly for custom application.

In most cases, these applicators are equipped with spray booms and nozzles. Delivery of liquids to the nozzles is controlled by a centrifugal pump equipped with a bypass system or by a positive displacement pump. Application rate on equipment having a centrifugal pump driven by a power source on the applicator is controlled by the speed of the applicator, the size of the nozzles, and the pressure on the bypass line. When positive displacement piston pumps (Fig. 7–36) are used, the pump is driven through chains and sprockets from the wheels of the applicator and, subsequently, application rate is largely independent of ground speed.

Adaptation of planting equipment for application of liquid fertilizers is also common in the United States, Canada, and western Europe. Little modification of the existing planting equipment is necessary except for the installation of some type of pump or constant-head delivery system for the liquids. One type of pump commonly used for planters is the squeeze or hose pump. This pump is relatively inexpensive and involves a roller system that traps and forces liquid through a series of hoses. The pump is usually ground-driven

Fig. 7–35. Large flotation-type liquid fertilizer applicator commonly used in the United States and Europe. (Courtesy Fred Myers, TVA.)

Fig. 7-36. A positive displacement pump for metering fluid fertilizers. (Courtesy Dempster Manufacturing Co.)

and application rate is independent of speed. Application rate is modified by changing the size of the drive sprocket.

In some areas liquid or fluid fertilizer is injected into the soil. This is necessary when pressure nitrogen solutions (see Chapter 2) are used or when subsurface placement is desired. Some applicators are equipped to handle both liquids and anhydrous ammonia but these pieces of equipment require two separate delivery systems for obvious reasons. When injection of liquids is planned, a positive displacement piston pump is normally used to meter the solution. The ground-driven pumps deliver solution at a rate that is largely independent of ground speed. Liquid delivery is controlled by the length of stroke of the pump.

Suspensions, the most recent addition to the field of fluid fertilizers, are applied by essentially the same equipment as for clear liquids with one major exception. Equipment used for application of suspensions should be equipped with some means of agitation of the suspension. Suspensions, properly manufactured, are quite stable for relatively long periods of time but higher-analysis materials require continued mixing to assure uniformity of the applied fertilizer. Frequently, agitation is provided by a compressed air sparging system on larger applicators (Fig. 7-37). Compressed air is fed into the bottom of the suspension tank through a multiple-ply rubber hose reinforced with cloth fabric. The hose has slits in it that open when air is applied and close when air shuts off. In some systems, expanding air from the sparger creates enough pressure to force the suspension from the applicator tank through the nozzles. Other systems involve high-volume liquid bypass lines to agitate the

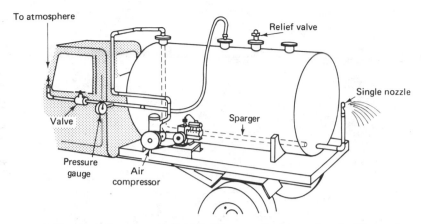

Fig. 7-37. A truck-mounted, fluid fertilizer applicator equipped with an air compressor and air sparging system. Note that a single flood nozzle is used. Proper pressure and speed must be maintained for correct application. (Courtesy TVA.)

suspension within the tank. The liquid returned to the tank is directed across the tank bottom to keep settling solids in suspension.

Nozzles used with suspension applicators are normally larger in capacity than those used for clear liquid applications in order to accommodate the larger amounts of fertilizer material. Suspension nozzles are commonly known as flooding nozzles or flood jet nozzles. Examples of those types of nozzles are shown in Fig. 7-38. A single flooding nozzle can give swath widths up to about 12 m (40 ft) when mounted at the proper height. Uniform application occurs over about 95% of the swath width. Application rate is controlled by nozzle height, attitude, and applicator speed if air or centrifugal pump systems are used and by pump setting in the case of positive displacement pumps. High-capacity flood nozzles are occasionally used without a spray boom due to the wide swath width. A boom with multiple nozzles is sometimes preferred and is more common on the newer high-flotation type of equipment. Use of booms and high-capacity nozzles can increase swath widths up to about 30 m (100 ft).

Suspensions can also be applied through planting equipment by the use of squeeze pumps or positive displacement pumps, the former being more common. Hose pumps are usually quite trouble-free because they normally do not allow the plugging that is frequently the result of attempting to apply suspensions through nozzles of small capacity.

Suspensions can also be knifed into the soil just as can clear liquids. Essentially the same pump systems are used but suspension applications frequently include the entire nutrient supply for a crop and consequently more volume of material is involved. Larger pumps are necessary in those cases.

Application of liquids through irrigation systems is discussed in Section 7.2. Normally, only clear liquids (true solutions) are used for irrigation applica-

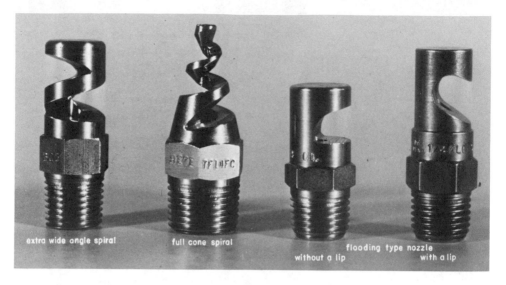

Fig. 7-38. Several types of nozzles for application of fluid fertilizers. Manufacturers' specifications must be closely followed for proper application and spray pattern. (Courtesy TVA.)

tion particularly through sprinkler systems. Suspensions applied in irrigation water may cause problems by uneven distribution of nutrients with time due to settling out of suspended salts in the irrigation water and slow dissolution of those materials.

Foliar applications of liquid fertilizer is discussed in Section 7.6.

7.4.2 Uniform Application

Interpretation of the term *uniform* application varies depending on the type of applicator being used. Obviously, uniform application of fluid fertilizers through a grain drill or row crop planter is quite different from uniform application through a boom-type applicator that is producing a broadcast application. In this discussion, comments are directed primarily at broadcast application of fluid fertilizers but the same concepts that control uniform delivery through a nozzle or orifice on a broadcast applicator also refer to applications through planters or shank applicators.

Several factors affect the uniformity of application of fluid fertilizers including pressure, height, and attitude of nozzles; speed of ground travel; size; pump characteristics; nozzle spacing; nozzle matching; specific gravity of the fluid; fluid viscosity; agitation; swath marking; and uniformity of the fluid formulation itself.

Specific Gravity of the Material. The primary effect of specific gravity is on the flow rate, which varies inversely as the square root of the density of the fluid. Basically it is the velocity of the fluid flow that is affected, which in turn affects the flow rate. The change in flow rate with change in density can be calculated from the following formula:

Density Correction

$$\text{Density of fluid} = \frac{\text{kg}/l \text{ of fluid}}{\text{kg}/l \text{ of water}} \text{ or } \frac{\text{lb/gal of fluid}}{\text{lb/gal of water}}$$

$$\text{Conversion factor} = \frac{1}{\sqrt{\text{density of fluid}}}$$

Nozzle capacity, water (l/min or gal/min) × conversion factor $=l$/min or gal/min of fluid. The spray angle of a given nozzle is not too adversely affected by changes in specific gravity and only slight effect is noted on the distribution of fluid within the spray pattern.

Viscosity. Viscosity effects on uniform distribution depends on the type of nozzle used. Hollow cone sprays will show an increase in capacity when there is an increase in viscosity. This is due to a slowing of the whirling action within the nozzle which decreases the friction of fluid on nozzle material and which *increases* the flow rate. Delivery capacity *decreases* when viscosity increases with *flat spray* and *straight stream* nozzles and is due to an increase in the friction within the nozzle and its orifice. Spray angles also decrease with an increase in viscosity.

When a viscous fluid is sprayed through flooding nozzles, there is a tendency toward a heavier concentration of fluid on the edge of the pattern that is often referred to as "horning." Suspension fertilizers are a good example of materials that, although somewhat similar in terms of weight per unit volume (usually about 1.44 kg/l or 12 lb/gal), vary in viscosity depending on materials used to produce the fluid. Ratio of liquids to solids, clay content, and quantity and type of micronutrients and pesticides in a suspension can affect viscosity. Kimbrough and Balay[41] have prepared data for determining flow rate correction factors to compensate for suspension viscosity (Table 7-18). This correction factor is used in addition to the correction factor for density. For instance, water flow through a nozzle times density correction factor times the viscosity correction factor equals the approximate suspension flow for a material with a given viscosity. A 20-10-10 grade of suspension may have an apparent viscosity of about 150 centipoises, whereas grades such as 15-15-15 or 7-21-21 have viscosities in the 250- to 350-centipoise range. Other grades such as 3-10-30 or 5-15-30 will have viscosities above 350 centipoises.

[41] H. L. Kimbrough and H. L. Balay, "Evaluating Spray Patterns," Proceedings 1975 National Fertilizer Solutions Association Round-up, St. Louis, Mo., July 23-24, 1975, pp. 30.

Table 7–18. Flow Correction Factors for Different Viscosities of Fluid Fertilizers[a]

Viscosity Correction	
Brookfield Apparent Viscosity (centipoises)	Flow Correction Factor
150	0.90
250	0.75
350	0.70
400	0.65
500+	0.60

[a] To obtain correct flow rate, multiply water flow rate by factor for given viscosity.

SOURCE: H. L. Kimbrough and H. L. Balay, "Evaluating Spray Patterns," Proc. National Fertilizer Solutions Association Round-Up. St. Louis, Mo., July 23-24, 1975, p. 30.

The correction factors in Table 7–18 are approximate, but they are accurate enough to calculate approximate nozzle orifice size required for desired application rates. These data are intended to be supplementary to actual nozzle calibration in the field that is mandatory for accurate applications.

Pressure. Pressure affects droplet size, rate, and application pattern. Droplets must be sufficiently large to restrict drift and still small enough to give adequate distribution. The latter is particularly important when pesticides are included in the fluid formulation. Equipment operators frequently attempt to correct rates by varying pressure. This will modify the application rate but doubling the pressure *does not* double application rate. One of the reasons that application rate is not directly proportional to pressure is that the velocity of a liquid stream is proportional to the square root of the pressure. The nozzle manufacturer's operating manuals usually show fluid flow rate for 10, 20, 30, and 40 psi (pounds per square inch) or 0.70, 1.41, 2.11, and 2.81 kg/cm.^{2} Table 7–19 shows calculated factors that can be used by operators to determine fluid flow rates between these reference pressures.

The width of spray pattern of a single nozzle is determined in part by pressure. Figure 7–39 indicates the effects of changing pressure from 10 psi (0.70 kg/cm^2) to 40 psi (2.81 kg/cm^2). Width of the swath, originally 21 ft (6.5 m) increased to 31 ft (9.5 m). Another common arrangement of nozzles is to use two flood-type nozzles on an applicator. This does not eliminate the pressure effects, however, because with operation at specified pressure no overlap is planned between the two nozzles. Decreasing pressure would then allow a skip to exist between the two strips creating 100% error in application rate in that strip *versus* the covered area and producing a higher than intended rate in the smaller strip covered at the lower pressure. Use of three or more nozzles with 100% overlap reduces possible error in application through

Table 7-19. Flow Correction Factors for Fluid Fertilizers[a]

Pressure		Flow Rate		Pressure		Flow Rate	
psi	kg/cm²	(gallons or liters)	Factor	psi	kg/cm²	(gallons or liters)	Factor
10	0.70	X	—	30	2.10	X	—
12	0.84		1.10	32	2.24		1.033
14	0.98		1.18	34	2.38		1.065
16	1.12		1.26	36	2.52		1.095
18	1.26		1.34	38	2.66		1.125
20	1.40		1.41	40	2.82		1.155
20	1.40	X	—	40	2.82	X	—
22	1.55		1.05	42	2.96		1.025
24	1.69		1.10	44	3.10		1.049
26	1.83		1.14	46	3.24		1.072
28	1.97		1.18	48	3.38		1.095
30	2.12		1.23	50	3.52		1.118

[a] Nozzle flows at intermediate pressures can be approximated by multiplying the manufacturer's calibrated flow (X) by the appropriate factor.

NOTE: Correction factor = $\sqrt{\text{intermediate pressure/calibrated pressure}} = \sqrt{12/10}$ = 1.10

SOURCE: H. L. Kimbrough and H. L. Balay. "Evaluating Spray Patterns," Proc. National Fertilizer Solutions Association Round-Up. St. Louis, Mo., July 23-24, 1975, p. 30.

skips between nozzles but pressure drop still creates errors because of relatively heavy rates of application in the reduced zones of overlap.

Many of the problems encountered with large flooding nozzles are reduced with flat spray booms where nozzles are placed much closer together (40 in. or 100 cm). If there is a nozzle every 100 cm and the pattern is changed 10% due to pressure drop, there would be only a 10-cm (4-in.) skip if the nozzles normally operated without overlap. Small skips of that size would probably not be noticed but large skips with the larger nozzles would create problems both in plant nutrition and in weed control if herbicides were included in the formulation.

False pressure readings are another common cause of error in fluid applications. Normal operating pressure for nozzles listed in the manufacturer's charts are pressures at the nozzle. If the plumbing on the equipment is not of sufficient size, the pressure indicated may be due to restrictions in the system rather than pressure at the delivery point. This drop in pressure between a remote reading position and the nozzle may range from 15 to 30 psi (1.05–2.11 kg/cm²). This problem can be corrected by reducing the pressure requirements of the system, replumbing the system, or recalibrating the gauge. Plugged screens may also be a cause of abnormal pressures and these should be checked regularly, particularly when a wettable powder herbicide is being applied in the fluid.

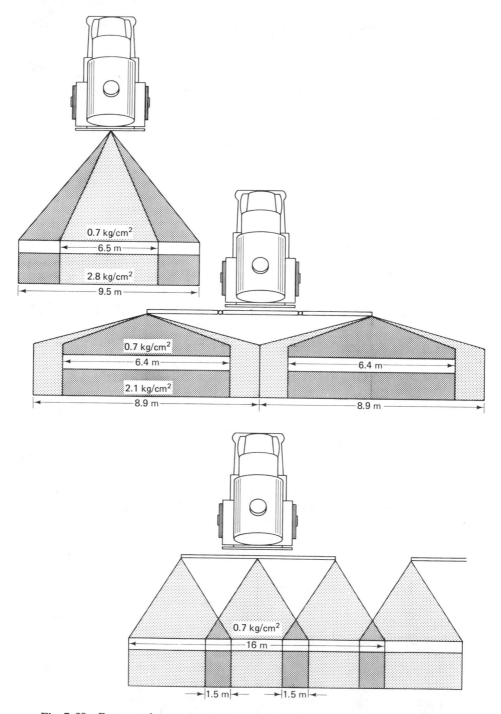

Fig. 7–39. Pressure changes in a fluid fertilizer application system can create non-uniformity of application. Note that pressure drops result in narrower application swaths and can lead to 100% errors in application by creating skips. Use of more nozzles tends to diminish the effects of pressure changes but they are still important even in such a system. (Source: K. W. Bradfield, 1969 Round-Up Proceedings, National Fertilizer Solutions Association, p. 58. Used with permission.)

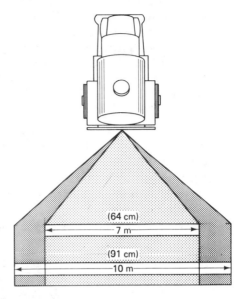

Fig. 7-40. Changes in height of nozzles on a fluid fertilizer applicator result in changes in the distribution pattern. Note that as the height of the nozzle(s) is raised (from 64 to 91 cm), the swath width increases. (Source: K. W. Bradfield, 1969 Round-Up Proceedings, National Fertilizer Solutions Association, p. 58. Used with permission.)

Speed of Ground Travel. With all other factors constant such as orifice size, pressure, and nozzle spacing, ground speed has an inverse effect on amount of fluid delivered when the system is powered by a power-takeoff centrifugal pump or by an air-sparged system. If the applicator is normally working at 6 km/hr (3.6 mph) and the speed was increased to 12 km/hr (7.2 mph), the application rate would be reduced by 50%. Ground-driven pumps vary pressure with changes in speed. If an applicator traveling 6 km/hr has a pressure of 1.4 kg/cm^2 (20 psi) at the pump and the speed is doubled, the effect is to quadruple approximately the pressure using the same nozzle. In strip application of fluids, this would not pose a particular problem, however.

Height and Attitude of Nozzles. The height of a nozzle from the soil surface with most nozzles used for fluid applications has an effect on width of the swath. Figure 7-40 demonstrates the effect of increasing height of a given nozzle from 64 cm (25 in.) to 91 cm (36 in.) when all other operating conditions were constant. Swath width increased from 7 m (23 ft) to 10 m (32 ft). Swath width also can be decreased by lowering the angle between an upright nozzle and the soil surface.

Orifice Size. When the orifice (opening) in a nozzle is used as part of the calibration package, it is one of the major determinants of application rate. The larger the hole in a nozzle at a given speed and pressure, the greater the application rate. When the orifice size is doubled on most nozzles and fluid fertilizers, the application rate is increased by a factor of 4. Also, eroded (worn) nozzles have higher flow rates. Orifice size also controls the droplet size. As pressure is increased, droplet size decreases. This can pose a problem in application as wind drift increases with smaller droplet size. Application of fertilizers alone may dictate the use of larger orifices to increase droplet size to minimize foliage desiccation such as in the topdressing of small grains or grasses with nitrogen solution. Larger droplets will not cling as readily to the leaves of the plants. On the other hand, inclusion of a pesticide in a fertilizer formulation may require the use of smaller orifices with smaller droplets to increase foliage or soil contact.

Pump Characteristics. Many field application problems are due to a lack of knowledge of pump capabilities and operation. One of the most common problems affecting application is pump capacity. Use of older equipment designed to apply say 90 *l*/ha (10 gal/A) in 6-m (20-ft) swaths for applications three or four times that great on swaths twice as wide leads to erroneous applications. Lack of pump capacity results in less than anticipated application with less than intended swath width.

Nozzle Spacing. Nozzle spacing according to manufacturer's specifications is an important factor in precision application of fluids any time multiple nozzles are being used. A common error is to *assume* that a certain nozzle is designed for some spacing without referring to its specifications. Spacing nozzles on 100-cm (40-in.) intervals when the units are designed for 91 cm (36 in.) or 107 cm (42 in.) is a common problem that can be overcome by familiarization with specifications. Obviously, nozzles spaced too wide will produce skips.

Nozzle Matching. Nozzle tips and orifices that look alike are not necessarily identical. Many different orifice sizes exist in the line of any manufacturer even within a given shape of nozzles. Units that are difficult or impossible to calibrate in the field have frequently been found to be equipped with many different-sized nozzles all delivering different amounts depending on the pressure and speed of operation. It is a common and good practice to remove nozzles and tips during the off season to facilitate cleaning. Replacement must be by number and not by appearance.

Agitation. Agitation is not a problem in application of true solution fluid fertilizers but assumes major importance in fluid fertilizer-herbicide mixtures and in application of suspension fertilizers. Any suspended agent in a fluid fertilizer settles slowly, the speed of settling depending on the particle size of the suspended material and the viscosity of the fluid through which the particle is settling. Mechanical, air, or hydraulic (liquid) agitation can keep these suspended materials adequately dispersed until application is completed. Improper agitation allows separation of some fertilizers and herbicides, depending on the materials involved, with subsequent uneven distribution of the components even if all other aspects for the system are workable. The amount of agitation necessary varies with the materials being used. Some suspensions are quite stable and actually require little if any agitation during application.

On the other hand, some mixtures of fluid fertilizers and liquid herbicides require continuous, vigorous agitation to prevent separation.

Arrangements of the agitation system within the tank of the applicator can also affect the uniformity of application. Misdirected streams of air or liquid can miss the material that tends to settle out on the bottom of the tank and are of little help in maintaining uniformity of the suspension or emulsion. The best correction for this type of problem is for permanent installation of the agitation or sparging equipment in the applicator tank by the manufacturer with a subsequent sweeping action of the air or liquid jet across the tank bottom.

Physical Uniformity (Composition) of the Fluid Fertilizer. This point sounds like further discussion of the importance of agitation and in fact is closely related. However, in the not too distant past, problems with uniform application of fluids revolved around problems with formation of gels and precipitates in what was supposed to be a clear liquid formulation. This problem was largely solved by the use of high polyphosphate liquids in formulations containing phosphorus. Polyphosphates sequestered the impurities and added micronutrients, essentially eliminating precipitation problems as long as the polyphosphates did not hydrolyze. Usually, the impurities were due to such normal components of the rock phosphate as iron, aluminum, and magnesium. (See Chapters 3 and 6).

Research has shown that low-polyphosphate liquid fertilizers can be stabilized (guarded against precipitation) by the addition of fluorine compounds. Addition of substantially *more* fluoride to levels greater than 0.7% may improve storage properties of liquids by sequestration of magnesium.[42] Normal fluoride content would have been about 0.2 to 0.3%. On the other hand, decreasing the fluoride content from about 0.35 to about 0.05% also increased the storage life of liquid fertilizers. Procedures for the production of liquid fertilizers free of gels and sludge are available and clear liquids are now of generally excellent quality.

Swath Marking. There is no substitute for good swath marking in correct application of any type of fertilizers, particularly bulk applications of solids and fluids. A common and generally acceptable method of swath marking is to use the rows of the preceding crop. That can fail when point rows occur. Use of flags of stakes is time-consuming and relatively uncommon. More recently, marking has become a more sophisticated technology with the use of foam markers particularly in the application of fluid fertilizers. Special equipment on the applicator drops large puffs of foam at regular intervals marking the outside edge of the swath, which allows the operator easy return to that same mark for the next pass through the field (Fig. 7–41). Proper overlap where required is easily accomplished. The foam eventually evaporates and has no ef-

[42] A. W. Frazier and R. G. Lee, "Stabilizing Liquid Fertilizers with Fluorine Compounds," *Fertilizer Solutions*, 16(4):32, 1972.

Fig. 7–41. Foam makers on fertilizer applicators leave an easily followed trail across the field, thus assuring uniformity of fertilizer application. Courtesy Fred Myers, TVA).

fect on the following crop. This type of marking is most readily adapted to applicators using booms. Mechanical swath marking is also possible except that slightly more difficulty is encountered in moving equipment because of the need to fold the markers.

7.5 FERTILIZER–PESTICIDE COMBINATIONS

The desirability of cutting crop production costs has resulted in the practice of applying pesticides in fertilizers. These dual applications potentially can reduce equipment, labor, and time requirements for crop production and reduce soil compaction by elimination of operations.

The concept of combining fertilizers and pesticides for simultaneous application is not new but research in this area is somewhat limited. Jacob[43] reported that fertilizer-pesticide mixture consumption in the United States for the year ending June 30, 1953 totaled only 87,000 short tons (79,000 metric tons). This amounted to only 0.55% of the total mixed fertilizer consumption. Most of these early combinations were mixtures of solid materials. Most of the emphasis was on fertilizer-insecticide mixtures, although

---

[43] K. D. Jacob, "Status and Problems of Fertilizer-Pesticide Mixtures," *J. Agr. Food Chem.*, 2:970, 1954.

fertilizer-herbicide mixtures were largely limited to the use of 2,4-dichlorophenoxyacetic acid (2,4–D) in solid fertilizers. Murphy[44] reported that the use of fertilizer-pesticide mixtures began in earnest in the United States in 1954 with the introduction of soil insecticides, particularly the chlorinated hydrocarbons. Early combinations of herbicides with fertilizers entailed direct impregnation of solid fertilizers with some of the liquid herbicide formulations.

7.5.1 Fluid Fertilizer-Herbicide Combinations

Early emphasis on liquid fertilizer-herbicide mixtures was an attempt to improve the weed control characteristics of some of the chemicals.[45] This use was characterized by the inclusion of 2,4–D in liquid fertilizers under the assumption that the adhesive properties of liquid fertilizer on plant foliage would substantially increase the effectiveness of the herbicide. Fertilizers may also exert beneficial effects on the phytotoxicity of herbicides because they increase the vigor of plants and actively growing weeds are more susceptible to herbicidal damage. Also, the heavier fertilizer solutions are subject to less wind drift than herbicide formulations in water. Early success of this type of dual application led to the development of the practice that is today known as "weed and feed." Because most of the dual application of fertilizers and herbicides involves combinations of fluid fertilizers and herbicides, that is the primary focus of this discussion.

Not all herbicides are stable in or compatible with fluid fertilizer formulations. Achorn, Scott, and Wilbanks of TVA reported on physical compatibility tests of fluid fertilizer-herbicide combinations.[46] Their results indicated that more pesticides were physically compatible in suspension fertilizers than in clear liquids. It was noted that less agitation was necessary to maintain dispersion of herbicides in suspensions than in clear liquids. On the basis of that work, they suggested constant agitation of clear liquid-herbicide mixtures during application and use of an emulsifying agent to help prevent separation of the herbicide from the liquid fertilizer. Martens[47] and Meyer[48] performed extensive evaluations of the physical and chemical compatibility of fluid fertilizers and herbicides and both concluded that agitation and use of an emulsifying agent was highly desirable in maintaining a homogenous, workable, effective mixture.

[44] D. R. Murphy, "Pesticide/Fertilizer Combinations—Why They'll Continue to Grow," *Fertilizer Solutions*, 15(6):92, 1971.

[45] W. R. Furtick, "A Look at Liquid Fertilizer-Herbicide-Pesticide Mixtures," *Fertilizer Solutions*, 12(3):22, 1968.

[46] F. P. Achorn, W. C. Scott, and J. A. Wilbanks, "Physical Compatibility Tests of Fluid Fertilizer-Pesticide Mixtures," *Fertilizer Solutions*, 14(6):40, 1970.

[47] Alan R. Martens, "Compatibility and Phytotoxicity of Herbicide-Fertilizer Combinations," M.S. Thesis, Department of Agronomy, University of Nebraska, March, 1973.

[48] L. J. Meyer, "Physical and Chemical Compatibility of Herbicide-Suspension Fertilizer Combinations," M.S. Thesis, Department of Agronomy, Kansas State University, 1972.

Achorn and co-workers outlined a simple process for field testing of compatibility of herbicide-fluid fertilizer combinations. That process includes these steps:

1. Place one-half liter (approximately a pint) of the fluid fertilizer in a jar.
2. Add an equivalent amount of pesticide in the same proportion to be applied in the field.
3. Close jar and shake well.
4. Observe the mixture at once and again after 30 minutes.

Testing for the requirement of an emulsifying agent (adjuvant) in the fluid fertilizer-herbicide mixture involves repeating these same outlined steps with two jars. The emulsifying agent should be added to one of the jars at a rate equivalent to the amount of fertilizer to be applied and the recommendations of the emulsifying agent manufacturer. Add the emulsifying agent prior to adding the herbicide. Following good mixing, observe the mixture for the development of an oily layer, large oil globules, gels, large flakes, crystalline precipitates, or sludge (Fig. 7–42). If the mixture without adjuvant shows no sign of incompatibility, then no adjuvant is needed. If the mixture without adjuvant is incompatible but the one with adjuvant is compatible, then use adjuvant in the spray tank. If more than one herbicide is to be used in a mixture, add them separately in order of wettable powders first, flowable materials second, and liquids last. If any of the indications of incompatibility just mentioned occur in the jar containing the adjuvant, the fluid fertilizer-herbicide mixture should not be made on a large scale. Obviously, if a fluid fertilizer-herbicide combination is not stable and separates into two distinct layers with the lower-specific-gravity herbicide solution rising to the surface of the mixture, application of an unagitated load of such a combination would result in only fertilizer being applied for most of the time but essentially straight herbicide application occurring as the tank is emptied. Results of exactly that type of problem have been disastrous to weed control and crop growth.

Tests of this type should be made prior to attempting to formulate fertilizer-herbicide mixtures either in the field or at the dealer's plant. It does little good to take a jar of material out of a tank after the mixture has been made and observe it for the formation of the abnormalities just listed. A nurse or saddle tank full of a material with the consistency of mayonnaise salad dressing is impossible to spread. Remember also that fertilizer materials of identical nutrient content can and frequently do differ considerably in their chemistry. For instance, it is always a good idea to check the compatibility of a *known* mixture whenever the source of the fertilizer is changed. Switching from 28% N urea-ammonium nitrate solution to a 32% N solution of nominally the same composition can create incompatibility that might not have been observed earlier. The presence of free ammonia in a nominal nonpressure nitrogen solution can cause incompatibility although relatively little is known on the subject. Remember too that you can never really be too careful and make too many checks on compatibility of fertilizer-herbicide mixtures. Label

Fig. 7-42. Pesticide-fertilizer mixture that was unagitated for 30 minutes shows definite separation of the components. Agitation is essential for uniform mixing and application of both pesticide and fertilizer. (Courtesy TVA.)

directions for herbicides and pesticides of all types should be closely followed in preparing fertilizer mixtures with these materials. Even then, a jar test is good insurance for success.

Meyer[46] investigated the effects of extended contact time between mixtures of fluid fertilizers and herbicides on the phytotoxicity of several classes of herbicides. Of the materials examined, contact with a suspension fertilizer for periods up to 230 days exerted no detrimental effect on phytotoxicity. This type of long-term storage of fertilizer-herbicide formulations is not likely because most combinations are made up immediately prior to application. Still, these results do indicate that if application is delayed for a few days due to rain, the combination is still effective in weed control. One should not leap to the conclusion that *all* herbicides are unaffected by contact with fluid fertilizers, either clear liquids or suspensions. Some evidence exists that *some* herbicides are strongly adsorbed by the clay in suspensions and phytotoxicity is subsequently greatly reduced. Familiarity with manufacturer's recommendations will avoid formulation of such mixtures.

Timing and Placement of Fertilizer-Herbicide Mixtures. Timing of fertilizer-herbicide applications are not always optimal for one or both of the components to exert maximum effectiveness. Surface placement of herbicides in some cases may be desirable for control of a certain weed, but if phosphorus is applied in the same manner and is not incorporated, plant absorption of the applied nutrient may be low. No general recommendation fits all conditions but most preplant herbicide-fluid fertilizer applications should provide adequate efficiency for both.

Much of the field work and experience with combinations of nitrogen solutions and herbicides has been concerned with *preplant* applications of the two materials. Actual field application of such combinations occasionally has been delayed until a *postemergence* date due to inclement weather. Such combinations and late applications frequently are associated with herbicide injury in row crops such as corn and grain sorghum. Uncertainty over the effects of such foliar burn has occasionally been the cause for litigation due to the supposition that the crop was ruined by the herbicide. Studies of the effects of dual nitrogen solution-herbicide applications at growth stages of corn and sorghum ranging up to the six-leaf stage were reported by Russ et al.[49] Russ and co-workers indicated that the applications of herbicide-nitrogen solution combinations as late as the six-leaf stage for both corn and sorghum provided a feasible means of nitrogen application as well as an acceptable means of herbicide application for weed control. They concluded that applications as late as the six-leaf stage should be viewed as a salvage operation, particularly from the standpoint of weed control and recognition should be given to the fact that some leaf injury may be produced particularly in dryland grain sorghum. Such treatments can hardly be construed as producing total crop destruction despite visible leaf injury from both herbicide and nitrogen solution. Yield reduction from the lack of both nitrogen and adequate weed control is much more likely under identical conditions than is yield depression that might result from delayed applications of nitrogen solution-herbicide combinations.

7.5.2 Fertilizer-Insecticide Combinations

Development of the low-cost chlorinated hydrocarbon insecticides was largely responsible for rapid growth in the use of insecticides by cereal farmers. Early in the development of insecticide use on a large scale, the advantages of inclusion of insecticides in fertilizers involved blending of liquid or solid forms of insecticides with solid fertilizer materials either at the point

[49] O. G. Russ, R. F. Sloan, C. W. Swallow, P. J. Gallagher, and L. S. Murphy, "Effects of Application Date on Crop Responses to Herbicide-Nitrogen Solution Combinations," *Fertilizer Solutions*, 18(4):24, 1974.

of production or at the local dealer level. Local blending developed more rapidly due to the flexibility of the system. The greatest advantage and impetus in the use of these dual applications was the savings in time through elimination of an operation, the same advantage that led to the development of dual herbicide-fertilizer combinations.

Development and release of aldrin and heptachlor for the control of soil insects in corn was one of the strong factors for development of dual applications. Watts and Nettles[50] estimated that 50% of all the corn planted in South Carolina as early as 1949 was treated with a combination of chlordane-fertilizer.

Application of insecticides in fluid fertilizers has developed to a greater extent in recent years but has never reached the popularity of herbicide-fluid fertilizer combinations. Generally, the use of insecticides in fluid fertilizers has been relegated to applications at the time of seeding. Problems may be encountered in the residual effectiveness of insecticides applied far in advance of the actual need for plant protection. Insecticides in fall preplant or early spring preplant applications may be much less effective than applications at seeding time. Usually, the insecticide-fluid fertilizer mixture is placed close to or beneath the seed. Banding to the side of the seed conventionally used for starter fertilizers is also acceptable for these mixtures. Applications on both sides of the seed may even be more effective in the case of some insecticides. Some manufacturers suggest that their insecticides not be placed in direct seed contact such as would occur with a pop-up fertilizer application.

One of the disadvantages that must be considered in using insecticide-fluid fertilizer mixtures is that the time of application and placement of one of the components may not result in the most effective response of the other. Also, the same problems in compatibility that were discussed for fertilizer herbicide mixtures must be recognized for insecticides in fluid fertilizers. Use of adjuvants and batch testing *before* formulation should be practiced.

Label registrations must be followed for mixing insecticides into fluid fertilizers. Flexibility of fluid fertilizer-insecticide formulations that can be produced in the field for a given situation hold more promise for continued use than premixed solid materials. Premixed formulations of dry fertilizers and insecticides were outlawed in Great Britain several years ago because unnecessary use of insecticides was encouraged by its presence in fertilizers. Applications according to need must govern the use of such mixtures.

7.6 FOLIAR FERTILIZATION

Foliar applications of nutrients, conceptually over 100 years old, involves the use of soluble, liquid sources of fertilizer. Foliar fertilization results in rapid nutrient absorption and utilization and has the advantage of allowing

[50] J. G. Watts and W. C. Nettles, "Control of the Sand Wireworm with Chlorinated Hydrocarbons," *J. Econ. Entomol.*, 44:619, 1951.

immediate correction of deficiencies determined by observation or plant analysis. On the other hand, foliar fertilization normally does not adapt itself to the application of large amounts of nutrients and should be considered as supplementary to a sound soil fertility program. Due to the small amounts of nutrients that are usually applied in foliar fertilization, benefits are frequently temporary and repeated applications may be necessary to correct the deficiency.

The most common use of foliar fertilization has been to supply micronutrients to crops, particularly tree fruit crops (see Chapter 6). Some successes have also been reported in correction of deficiencies in field crops. The expense of repeated sprayings dictates that this procedure be utilized primarily on relatively high-value crops. Problems with soil fixation of needed nutrients or soil availability under unusual conditions such as very low temperatures provides another possible advantage to the practice of foliar fertilization. On the other side of the coin, severe leaf burn (plasmolysis) can and does occur when large amounts of nutrients are applied or when certain forms of nutrients are used in the foliar fertilization.

Foliar fertilization allows a combination of operations, namely application of solutions containing both pesticides (insecticides or fungicides) and fertilizer. Where repeated sprayings are normally undertaken because of pest problems, the disadvantage of high application costs for foliar fertilization (foliar feeding) is reduced.

Because of the relatively small amounts of nutrients applied in foliar fertilization, application techniques vary from ground sprayers to aircraft. The fact that only small applications of fertilizers are required does lend this method to the widespread use of aircraft.

Foliar absorption of urea nitrogen has been recognized in several crops. Apparently, nitrogen absorption in this case is influenced by the urease enzyme that hydrolyzes urea to carbon dioxide and ammonia (ammonium) either at the leaf surface or in its immediate vicinity. Urea absorption is believed to occur through stomata[51] although uptake has been shown to occur in several other structures.[52] Apparently, absorption is most rapid during the first hours after application and, subsequently, conditions that are associated with stomatal closure may negatively affect the efficiency of the operation. Most recommendations call for foliar fertilization to take place during periods of low temperature and relatively high humidity such as early morning or late evening. Applications during this period are also likely to cause less leaf burn.

In a recent review of foliar fertilization, Ham et al.[53] pointed out that foliar fertilization has recently changed from a means of providing a recovery

[51] K. Horie, "The Permeation of Urea into Leaf Cells. II. Absorption of Urea in the Dark," *Sci. Rep. Hyogo Univ. Agr. Series: Agr.*, 2:183, 1956.

[52] M. Yatazawa and M. Namiki, "Direct Evidence of Foliar Absorption of Urea: Synthesis and Utilization of ^{15}N–Rich Urea," *J. Sci. Soil Manure*, 26:219, 1955.

[53] G. E. Ham, W. D. Poole, and G. W. Randall, "Soybean Yield Improvement with Foliar Fertilization," Proceedings of 1978 National Fertilizer Solutions Association Round-Up, July 20–21, 1978, Kansas City, Mo., p. 54.

mechanism from deficiencies to a technique of providing large amounts of nutrients for some crops. Ham noted that pineapple receives as much as 75 to 80% of the nitrogen supplied to the crop as urea sprays and received from 40 to 50% of phosphorus and potassium by foliar application.

Nutrient uptake tends to parallel dry matter accumulation and is particularly high during flowering and pod filling (seed set). The protein content of wheat has been shown repeatedly to respond to applications of urea solutions nearing the flowering stage of development (see Chapter 2). Hanway[54] recognized the occurrence of processes in the soybean plant that might be conducive to foliar fertilization of that crop. He noted that during the seed-filling process, soybean seed becomes the dominant sink for carbohydrates produced in the leaves. Subsequently, the soluble carbohydrate content of the stems and roots decreases with a parallel reduction in the activity of nitrogen-fixing bacteria in root nodules due to a lack of carbohydrate. Root growth also slows with a simultaneous reduction in nutrient absorption. Nutrients needed by the developing seeds are translocated from the leaves and other vegetative parts of the plant. The detrimental cycle is completed with a reduction of photosynthesis because of nutrient shortages in the leaves.

Recognizing this problem, Garcia and Hanway[55] began a series of investigations into the efficiency of foliar fertilization of soybeans in an attempt to maintain photosynthetic activity in the leaves, supply adequate carbohydrate for the roots and seeds, and consequently increase yields. They initiated a series of field experiments of foliar fertilization of soybeans with a mixture of nitrogen, phosphorus, potassium, and sulfur compounds formulated to approximate the ratio of these four elements in soybean seeds. That ratio of N:P:K:S was 10:1:3:0.5, respectively. Nutrients were supplied as urea, potassium polyphosphate, and potassium sulfate. Very significant yield increases were obtained (Table 7–20) from two to four sprayings at several locations. These applications involved relatively heavy applications of nutrients, a radical departure from most investigations involving foliar fertilization of soybeans. Results of those studies suggested that foliar fertilization of soybeans during the seed-filling period can be a practical means of increasing soybean yields.

Following the work by Garcia and Hanway, a very large number of foliar fertilization investigations were begun by universities and the fertilizer industry in North and South America. Results of summarized investigations in 16 states in 1977 (Table 7–21) using soybeans as a test crop and involving heavy rates of nutrient application indicated no consistent trend in soybean yield effects from foliar fertilization. Although research continues with soybeans and other crops, only guarded optimism is expressed for this technique to exert major influence on yields.

Mixed results from heavy applications of nitrogen-phosphorus-potassium-

[54] J. J. Hanway, "Interrelated Development and Biochemical Processes in the Growth of Soybean Plants." In *Proc. World Soybean Res. Conf.,* Univ. of Illinois, Urbana, Ill., 1975, p. 172.

[55] Ramon Garcia and J. J. Hanway, "Foliar Fertilization of Soybeans During the Seed-Filling Period," *Agron. J.,* 68:653, 1976.

Table 7-20. Effects of Foliar Fertilization on Yield of Two Soybean Cultivars in Iowa[a]

Soybean Cultivar	Yield Not Sprayed (kg/ha)	Yield Increase from Spraying (kg/ha)	Seed Size	
			Not Sprayed	Sprayed
			(g/100 seeds)	
Corsoy	3540	1570	15.2	15.8
Amsoy	3850	1490	16.7	16.0

[a] A total application of 96 kg N, 9.6 kg P, 28.8 kg K, and 4.8 kg S per hectare was applied between developmental stages R5 and R7.

NOTE: Yield increases were significant at the 1% level.

SOURCE: R. Garcia and J. J. Hanway, "Foliar Fertilization of Soybeans During the Seed-Filling Period," *Agron. J.*, 68:653, 1976 by permission of the American Society of Agronomy.

Table 7-21. Summarized Results of Soybean Foliar Fertilization Studies in the United States in 1977[a]

Fertilizer Analysis	Number of Applications	Number of Comparisons	Approximate Total Nutrients Applied $N+P_2O_5+K_2O$		Yield			
			(kg/ha)	(lb/A)	Control	Treated	Control	Treated
					(bu/A)		(kg/ha)	
10-2.4-4.0-0.6S	1	32	28+7+11	25+6+10	37.0	36.4	2486	2446
10-2.4-4.0-0.6S	2	31	56+13+22	50+12+20	36.9	36.4	2480	2446
10-2.4-4.0-0.6S	3	43	84+20+34	75+18+30	39.8	39.8	2674	2674
10-2.4-4.0-0.6S	4	13	112+27+45	100+24+40	41.4	41.4	2782	2782
12-6-6-0.5S	1	40	13+7+7	12+6+6	41.2	41.0	2769	2755

[a] Summary includes results from 68 replicated studies in 16 states.

SOURCE: R. C. Gray, "Results of Foliar Fertilization of Soybeans in 1977," Mimeographed report, Tennessee Valley Authority, Muscle Shoals, Alabama.

sulfur mixtures on soybeans should not be taken as a condemnation of foliar fertilization, however. Strong recommendations exist for the use of foliar fertilization with micronutrients. Micronutrients are more adaptable to foliar application because of the small amounts of nutrients involved. This method of nutrient application is discussed further in Chapter 6.

7.7 TILLAGE SYSTEMS AND FERTILIZATION

The role of the tillage system in determining the method of fertilizer application for a given crop varies depending on several factors. Climate and soil are the two most important variables in determining how efficiently fertilization can be incorporated into a tillage system, such as applying fertilizers during tillage operations. In Section 7.1.5 ammonia application with tillage implements is discussed at length. Advantages to combined operations noted include savings in labor, time, and fuel through elimination of an operation. Additional advantages to deep placement of nitrogen and other elements have been noted in the drier areas of the United States (positional availability). The latter point would be less important where surface moisture is usually not a limiting factor.

Special consideration in this section is given to fertilization practices for reduced tillage or minimum tillage cropping systems. Most of the research into fertilization practices for no-till or reduced tillage cropping has been carried out in the eastern Cornbelt, particularly the southern areas of the Cornbelt, where no-till farming has been most successful. Introduction of reduced tillage systems has been successful through the elimination of costly tillage operations, better control of erosion by soil surface protection, and moisture conservation through reduction of evaporation (see Chapter 10).

When little or no tillage is practiced, incorporation of fertilizer into the soil from preplant applications is essentially impossible. If climatic conditions, soil texture, and crop residue amounts are conducive to maintenance of moist soil conditions at the soil surface, then unincorporated broadcast applications of nutrients will probably be effective. On the other hand, if climatic conditions are such that surface moisture is limiting even with large amounts of crop residue, then some technique of subsurface placement of the needed fertilizer has to be used. Obviously, root activity at the soil surface would be necessary to allow good utilization of nonincorporated nutrients.

Phillips and Young[56] discussed fertilization practices in their book on no-tillage farming and concluded that, under Kentucky conditions or at least conditions of the southern Cornbelt and eastern seaboard states, method of fertilizer application for no-tillage crop production was "astonishingly simple" and no more critical than with other, conventional tillage systems. In fact, much of the fertilizer placement research for that region shows good utilization of surface-applied nitrogen, phosphorus, and potassium. Banded applications of potassium may be somewhat more effective but the conditions that dictate greater effectiveness of banded applications of phosphorus are the same that control phosphorus-use efficiency under conventional tillage conditions.

Probably more attention has been given to nitrogen and phosphorus management under reduced tillage crop production than has been afforded most of the other nutrients. Recent research has shown that more denitrification may occur under reduced tillage conditions due to more moist soil conditions and high concentrations of easily oxidizable organic matter. Moist, warm conditions with readily oxidizable carbon present are exactly the conditions which promote denitrification (see Chapter 2). Use of a nitrification in-

[56] S. H. Phillips and H. M. Young, Jr., *No-Tillage Farming*, Milwaukee, Wis.: Reiman Associates, 1973, p. 151.

hibitor may prove effective in improving nitrogen-use efficiency under those conditions. Higher nitrogen rates may be required to offset denitrification if not controlled.

Applications of anhydrous ammonia for no-till or reduced tillage crops are easily adaptable with some attention possibly necessary to seal the ammonia in the soil. Use of special sealers on the ammonia knives can solve this problem, however. Also, use of a rolling coulter in front of the ammonia knife may be necessary under heavy residue conditions to cut the residues and keep them from hanging on the knife.

Use of urea-ammonium nitrate (UAN) nonpressure solutions in nitrogen fertilization of no-till or reduced tillage crops is fairly common but application of this form of nitrogen where residues are very heavy may result in poor nitrogen-use efficiency due to volatilization losses of ammonia (see Chapter 2). This problem can become severe under some conditions and tends to make this method of nitrogen fertilization rather uncertain. Higher temperatures and relatively low relative humidity seem to increase the likelihood of ammonia volatilization. For that reason use of UAN solutions for nitrogen fertilization of reduced tillage crops in the Great Plains of the United States is even more uncertain than this method of nitrogen fertilization in the southern Cornbelt. Poor nitrogen responses have occasionally been noted when preplant, preemergence, or postemergence applications of UAN solution have been used on no-till grain sorghum in the Kansas-Nebraska area. Inclusion of an herbicide in the UAN had no effect on the performance of the nitrogen source. Raines[57] reported poor utilization of surface applications of UAN for no-till irrigated corn under Kansas conditions.

Moschler and Martens[58] studied soil surface applications of nitrogen, phosphorus, and potassium for conventional and no-till corn culture. Their results indicated plants in the no-till system were efficient in utilizing the applied nutrients, in fact, more efficient than comparable applications under conventional tillage conditions. Their results and those of Lutz[59] suggested that phosphorus and potassium broadcast on the surface for no-till corn were as available as those elements incorporated into the soil under the relatively high rainfall conditions of the eastern United States. Triplett et al.[60] noted that most of the phosphorus applied to the soil surface remained in the surface 2.5 cm (1 in.) of soil. Raines[57] recorded somewhat deeper movement (5 cm) of phosphorus applied to the surface of a sand cropped with no-till irrigated corn. Raines also noted much poorer utilization of phosphorus from surface applications *versus* banded or knifed treatments.

[57] G. A. Raines, "Conventional and No-Tillage System Effects on Plant Composition and Yield," M.S. Thesis, Department of Agronomy, Kansas State University, 1978.

[58] W. W. Moschler and D. C. Martens, "Nitrogen, Phosphorus and Potassium Requirements in No-Tillage and Conventionally Tilled Corn," *Soil Sci. Soc. Amer. Proc.*, 39:886, 1975.

[59] J. A. Lutz and J. H. Lillard, "Effect of Fertility Treatments on the Growth, Chemical Composition and Yield of No-Tillage Corn on Orchardgrass Sod," *Agron. J.*, 65:733, 1973.

[60] G. B. Triplett and D. M. Van Doren, "Nitrogen, Phosphorus and Potassium Fertilization of Non-Tilled Maize," *Agron. J.*, 61:637, 1969.

In conclusion, fertilizer application techniques for no-till or reduced tillage cropping systems have been fairly well developed in the southern Cornbelt and eastern seaboard areas of the United States. However, the need for additional research into fertilizer application techniques for reduced tillage cropping systems in the western Cornbelt and the Great Plains of the United States remains a high-priority item. Wide adaptation of eastern United States techniques to the lower rainfall areas simply is not verified and such adaptation should be undertaken with caution. Refer to local recommendations for fertilization practices for reduced tillage systems available through the local extension service offices or available from industry service groups in the area.

7.8 CALIBRATING APPLICATION EQUIPMENT

Uniform and accurate application of fertilizer, lime, other soil amendments, and pesticides is essential for optimal utilization of these production inputs. Failures in the performance of agricultural chemicals can frequently be directly related to improper application, particularly uneven application caused by equipment failure; careless application, or separation of components of the materials being applied. The last point deals specifically with segregation in bulk blends of solid fertilizer materials (see Section 7.3). Proper swath marking is possible and practical (see Section 7.4) so carelessness in application and poor swath overlap is inexcusable. Because the subject of proper mixing of pesticides and fertilizers is discussed in detail in Section 7.5, this section addresses proper calibration and operation of application equipment. The net goal of proper application is of course higher economic yield.

Uneven distribution of lime and fertilizer can reduce crop yields. Stewart and Bandel[61] noted that research conducted in Virginia indicates that when one area of the field is overfertilized and another is underfertilized, the total yield is less than if the correct amount of fertilizer were spread evenly over the entire field. The effects of variability in fertilizer distribution are most notable on areas that are low in soil fertility because response to applied nutrients is greatest on such areas. In the data reported by Stewart and Bandel, yields were reduced by 20 to 25% when the fertilizer had been applied in a skewed or nonuniform pattern. Yield losses due to lower than recommended rates of application far exceed the slight increase in yield obtained from excess application over the recommended rate. If applications far in excess of the recommended rate continue to produce dramatic yield increases, then the original recommendations were too low. The reason for the greater depression in yield from inadequate application of nutrients *versus* slight yield increase from excessive applications is the law of diminishing returns. That means that after a certain rate, there is a decreasing return for each additional increment of applied nutrient. The last increment of fertilizer applied will increase the yield

[61] Larry Stewart and V. A. Bandel, "Uniform Lime and Fertilizer Spreading," University of Maryland, Coop. Ext. Ser. Bul. 254, undated.

by a smaller amount than the preceding increment. Eventually, no further increase occurs and, eventually, yield depression may result (see Chapter 11).

7.8.1 Calibration and Adjustment of Solid Application Equipment

Uneven or inaccurate application of solid fertilizers, lime, gypsum, or sulfur can be caused by segregation of materials as well as by improper adjustment or improper calibration of application equipment. Calibration techniques differ depending on the type of applicator being used. Obviously, techniques for determination of the amount of solid fertilizer delivered by a grain drill or row crop planter unit are much different than determination of the amount of fertilizer spread by a large flotation spreader covering an 18-m (60-ft) swath.

Calibration of fertilizer spread by planting equipment is relatively simple. Distribution between planter units on a multirow planter admittedly may vary in amount but the pattern of delivery is fairly uniform. These units can easily be calibrated by driving the unit a specified distance and catching the fertilizer delivered in containers such as plastic buckets. Because the row spacing between the planter units is known and because a particular unit is "responsible" for a given width, area "covered" can be calculated by the product of width and distance traveled, say 100 m (325 ft). The following example indicates how to check the fertilizer delivery for a planter unit after the manufacturer's settings have been met.

Intended application rate = 100 kg/ha of material (89 lb/A)

Row width = 70 cm (27.5 in.)

Length of trial run = 100 m (325 ft)

Area of trial = 100 m x 0.7 m = 70 m²

70 m²/10,000 m²/ha = 0.007 ha

Fertilizer delivered in trial run = 810 g (1.78 lb)

Fertilizer required for desired rate = 700 g (1.54 lb)

Result: Overdelivery by 15.7%. Try next lower setting.

Planter units for row crops will have to be calibrated individually. Usually, manufacturer's settings will be close but for best operation this process should be repeated on each separate unit. The process is time-consuming but well worth the effort. Excessive rates of application, particularly in pop-up placement (see Section 7.3.2), could result in a reduction in germination and should not be left to chance.

Grain drills operate essentially as a unit and, for proper calibration, fertilizer delivered from each opener should be collected in an individual bucket for determination of both uniformity of distribution and rate of application. Because openers on most drills are adjusted collectively, there may be little that can be done to change the distribution pattern but the rate of application can be easily changed by following the same procedure outlined for the row crop planter. Calibration during inclement weather can be carried out by jacking up the implement off the floor in the shop, determining the diameter of the wheel driving the fertilizer section, and merely simulating movement of the drill by turning the wheel the number of times necessary to produce the predetermined distance. The disadvantage of this technique (which can also be used on row crop planters) is that the smooth operation inside may deliver slightly less than would be the case if the machine were operating in the field on a rough surface. Still the procedure is a good one and should give a very good idea of how much fertilizer the drill or planter will deliver.

One other factor that should be kept in mind (and perhaps this is more for the researcher than the farmer) is that the rate of delivery of fertilizer materials can change significantly among different sources of material. Say for instance you are using brand X of diammonium phosphate for wheat and you calibrate your drill for that material that is a certain particle size. Changing to brand Y, which may not be as uniform in particle size, may result in a higher rate of application. Recalibration may be necessary to utilize that bargain material you bought if you note nonuniformity of the product. Practically speaking, however, it is unlikely that anyone will go to the trouble to recalibrate a planter during the busy planting season just because you are afraid that the bulk blend you are using may not be made out of exactly the same materials as the last truck load. The fertilizer dealer will be cognizant of the need to use good quality, uniform materials in his bulk blends so that they will spread accurately through his bulk applicators and your planter with a minimum of segregation.

The problem of obtaining a uniform application of fertilizer or lime from a bulk spreader is magnified by wide ranges in swath width; field conditions including shape, slope, and soil conditions; weather conditions (wind and rain); intended rate of application; equipment design; and composition of the material being spread (particle density, particle size, particle shape, and moisture content).

Spread patterns for a twin-disk solid fertilizer applicator can be classified into about six different types (Fig. 7–43). The flat top, oval, and pyramid patterns are most desirable because they lend themselves to uniform overlapping of swaths. The most common undesirable patterns are the M, W, and offside (skewed or lopsided) patterns. Achorn and Kimbrough[62] suggested improvements in the M pattern by one or more of the following adjustments:

1. Move the delivery chute toward the applicator to change the point of delivery of the material closer to the outer edge of the spinners.

[62] F. P. Achorn and H. L. Kimbrough, "Uniform Application of Granular Fertilizers," Proc. TVA Fert. Conf., July 29–31, 1975, Louisville, Ky., p. 25.

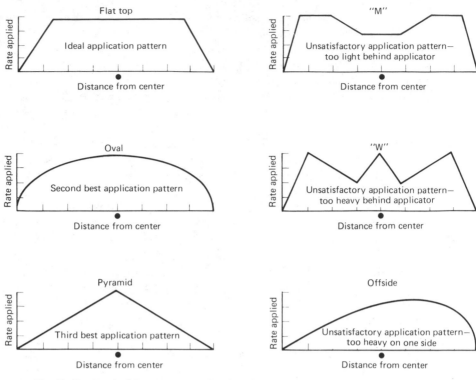

Fig. 7-43. Typical fan-type spreader distribution patterns, some good, some bad. (Courtesy TVA.)

2. Move spinner blades in the opposite direction of the spinner rotation.
3. Increase the spinner speed. A spinner speed of 550 to 650 rpm (revolutions per minute) is recommended. Higher spinner speeds shatter the granules and can lead to segregation and uneven distribution.

The W pattern may result from applicator conditions similar to those causing the M pattern. A heavy band of material occurs in the center of the swath in addition to concentrations on both the right and left sides. The heavy center concentration may be caused by an improperly adjusted delivery chute or leaks that permit material to fall immediately behind the applicator. Wet material that sticks to the conveyor belt or chain and falls later would also cause a heavy application in back of the spreader. In correcting this problem, operations will be essentially opposite of those to correct the M patterns, as follows:

1. Move delivery chute away from the applicator to move the point of material delivery closer to the center of the spinner.
2. Move blades of the spinner in the direction of spinner rotation.
3. Decrease spinner speed.

One or more of the changes should alter the pattern to an acceptable flat top or oval pattern.

Lopsided patterns either right or left may result from twin spinner applicators because of uneven delivery of fertilizer material to the spinners. An improperly adjusted flow divider is usually the cause. Operations on steep slopes can also produce heavier flow to the downhill side if an effective flow divider is not included in the system. This problem can be overcome to some extent by proper overlap between swaths and application in a circular pattern.

Single spinner solid applicators can also produce a lopsided pattern when the delivery of the material to the spinner is not positioned properly. If the right half of the pattern from a clockwise rotating spinner is heavier than the left half, any of the following adjustments will delay release of the fertilizer from the spinner and improve the pattern (Fig. 7–44):

1. Adjust the delivery chute to deliver closer to the outside edge of the spinner.

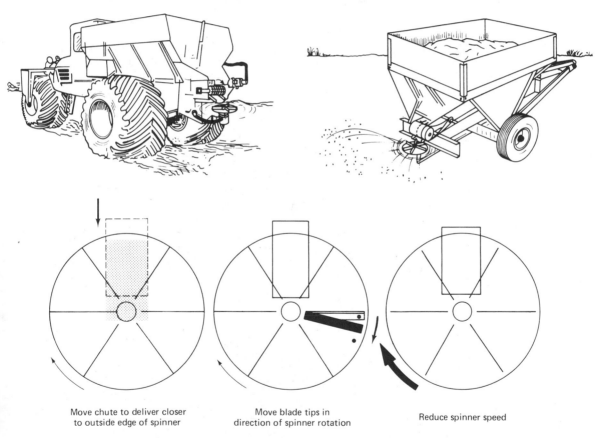

Move chute to deliver closer to outside edge of spinner

Move blade tips in direction of spinner rotation

Reduce spinner speed

Fig. 7–44. Three adjustments to correct "heavy-right" distribution pattern from a single spinner (fan) solid fertilizer applicator. Corrections for heavy-left patterns are just the opposite. (Courtesy TVA.)

Applying Fertilizers and Soil Amendments

2. Move spinner blade tips in the direction of spinner rotation.
3. Reduce spinner speed.

Heavy applications to the left side call for adjustments in the opposite direction.

Achorn and Kimbrough[62] further note that one of the main problems in adequate application of solid fertilizers, lime, and similar materials is maintenance of the equipment. Specific problems requiring attention include accumulation of fertilizer material on spinner blades and on flow dividers, worn disk blades, broken hold-down bolts for spinner blades, accumulated fertilizer material on the flow control gate and on conveyor chain or belt, and loose drive belts for the conveyor. These problems result in modified angles of the spinner blades or inadequate rate of delivery to the spinners.

Hydraulic equipment including hydraulic powered spinners should be checked regularly according to manufacturer's instructions. Low oil levels and excessive valve wear can significantly affect operation of the equipment.

Calibration of Equipment. Smith[63] of the Department of Agricultural Engineering at Virginia Polytechnic Institute and State University points out that calibration checks for solid fertilizer equipment should show the type of spread pattern, including the degree of uniformity obtained across the swath, the effective swath width, and the rate of application. Obviously, this type of operation will have to be carried out to determine the adjustments needed to improve the characteristics of the applicator.

Smith has outlined a simple system of calibrating fertilizer and lime spreaders involving the use of collecting trays or pans with grid baffles, test tubes, a test-tube rack, a funnel, and a measuring tape. A small scale helps improve the accuracy of the calibration test. The size of the pans and test tubes should be in proportion to the rates of application intended. Generally, a shallow 2.5- to 6-cm (1- to 2.5-in.) deep tray or pan about 30 × 40 cm (12 × 16 in.) is adequate. A removable plastic baffle should be placed in each pan to avoid bounce and ricochet of the particles. A satisfactory baffle system can be made from the plastic diffusing lenses of fluorescent light fixtures cut to fit the pans.

Test tubes, the same number as number of pans to be used in the test, usually about 15, are used to evaluate the distribution pattern visually and to estimate the amount of material delivered. Tubes about 1.25 cm (0.5 in.) × 11.5 cm (4.5 in.) are recommended for applications up to about 1200 kg/ha (1100 lb/A). Larger tubes are required for heavier rates of application.

The 15 pans should be spaced across the path of the applicator on 1.5- to 3-m (5-to 10-ft) intervals, depending on swath width estimated. The applicator, adjusted for an intended rate of application, is then driven at normal

[63] E. S. Smith, "Proper Use and Maintenance of Application Equipment," Proc. TVA Bulk Blending Conf., Louisville, Ky., Aug. 1–2, 1973, p. 94.

            Applying Fertilizers and Soil Amendments

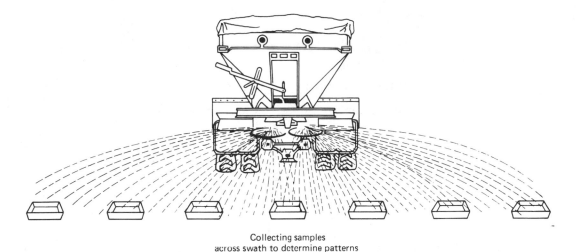

Collecting samples
across swath to determine patterns

Fig. 7-45. Applications of solids and bulk blends must be closely calibrated not only for delivery of the correct total amount of material but also for the correct pattern of application. The diagram demonstrates a method of determining both amount and distribution of a solid fertilizer. (Courtesy Tennessee Valley Authority.)

field speed straddling the middle pan (Fig. 7–45). Material collected in each pan is then poured into the corresponding test tube in a test-tube rack.

The material in the tubes allows a quick determination of the spread pattern. Typical flat top, pyramid, oval, M, W, or lopsided patterns can be easily detected against a black background. Constructing a small case/rack for the test tubes with painted or black felt background greatly aids in repeated tests of this type. Smith points out that the effective swath width can be determined by locating the points on the right and left of center when the amount of material in the tube is about half that in the center tube. This procedure would be acceptable for swath widths in pyramid, flat top, or oval patterns but would of course be in error when an M or W pattern has resulted. These patterns would require further pattern adjustment before the amount of material delivered is finally determined, however.

The test tubes can be calibrated to estimate the amount of material delivered per unit of area by preweighing material calculated for a given area (the area of the pan) and pouring into the tube and marking the height for some unit such as 100 kg/ha. When round-bottomed tubes are used, the first 100-kg/ha unit will be a little higher in the tube than the following 100-kg units. This system is only approximate because particle size and density of fertilizer and lime materials vary. The best procedure is to empty the contents of the tubes individually or collectively onto the pan of a small scale and weigh them.

Let's consider the calculations and procedures for determining the *rate* of application assuming that a satisfactory pattern has been determined.

Effective swath width = 18 m (58.5 ft)

Number of pans = 13

Pan spacing = 1.5 m (4.88 ft)

Pan area = 1200 cm² (0.12 m²) (186 in.²)

Intended application = 1000 kg/ha (892 lb/A)

Accumulative weight of material in all pans = 156 g (0.344 lb)

Accumulative area of all pans = 13 × 0.12 m² = 1.56 m² (16.79 ft²)

Net application of fertilizer = $\dfrac{10{,}000 \text{ m}^2/\text{ha}}{1.56 \text{ m}^2 \text{ of test}}$ × 0.156 kg applied = 1000 kg/ha

or

$\dfrac{43{,}560 \text{ ft}^2/\text{A}}{16.79 \text{ ft}^2 \text{ of test}}$ × 0.344 lb applied = 892 lb/A (1000 kg/ha)

7.8.2 Calibration of Fluid Application Equipment

Calibration of fluid application equipment requires essentially the same attention to detail as calibrations for solids. The type of applicator used dictates the type of calibration technique. For instance, if an applicator equipped with several nozzles on a boom is powered by a centrifugal pump powered by the applicator's power takeoff system, then the system can be calibrated while stationary. The approximate rate of application can be determined from the manufacturer's specifications for pump and nozzles. Assuming that the proper pressure for correct nozzle operation can be maintained by the pump, calibration of the system merely involves collection of the fluid that is delivered in a given period of time. The fluid can be collected in any container as long as it has adequate capacity. Small plastic buckets are ideal. Measurement of the fluid delivered can be either by volume or by weight. Weight per unit volume of the fluid mixture must be known to translate volume into weight per unit of land covered in the application. Let's look at a sample calibration test:

Effective swath width = 18 m (58.5 ft)

Number of nozzles = 12

Effective width of each nozzle = 1.5 m (4.88 ft)

Intended application = 267 kg/ha (238 lb/A)

Fluid characteristics = 1.44 kg/l (12 lb/gal)

Intended application speed = 16 km/hr (9.6 mph)

Application capacity = 28.8 ha/hr (71.1 A/hr)

Test time = 60 sec

Test delivery per nozzle = 7.42 l (1.96 gal)

Test delivery all nozzles = 89.04 l (23.52 gal) = 89.04 × 1.44 kg/l
= 128.22 kg (282.1 lb)

Field delivery 1 hour = 7693 kg (16,925 lb)

Application/hectare (acre) = 7693 kg/hr ÷ 28.8 ha/hr
= 267 kg/ha (238 lb/A)

If the pump is ground-driven with positive displacement and operates essentially independently of ground speed, then calibration must involve either actual movement of the applicator over a course of determined length or rotation of the ground drive wheel the proper number of times to simulate travel of a certain length. Fluid delivery is controlled by the length of stroke of the pump and the number of strokes. Calibration of amount delivered still involves collection of fluid from all nozzles both for the purpose of comparing delivery uniformity and calculation of total amount delivered. Let's go through a sample calibration test for a system of this type.

Effective swath width = 12 m (39 ft)

Number of nozzles = 8

Effective width of each nozzle = 1.5 m (4.88 ft)

Intended application = 267 kg/ha (238 lb/A)

Fluid characteristics = 1.44 kg/l (12 lb/gal)

Length of test run = 100 m (325 ft)

Area of test = 100 m x 12 m = 1200 m² (12,675 ft²)

Test delivery per nozzle = 2.78 l (0.73 gal)

Test delivery all nozzles = 22.24 l (5.88 gal) = 22.24 l × 1.44 kg/l
= 32.02 kg (70.46 lb)

Relative delivery = 32.02 kg ÷ 0.12 ha = 267 kg/ha (238 lb/A)

Visual calibration or determination of spread pattern characteristics for liquids when the entire amount of fluid is caught over the test strip is not so easy as for solids unless the size of the test is greatly diminished. In other words, a test-tube-visual-delivery-per-nozzle system similar to that for solids would require very large containers or a very short test, the latter being undesirable due to large errors. Kimbrough and Balay[42] used a strip of wrapping paper spread across the applicator path to evaluate the uniformity of the application visually. This is a quick and cheap way but the same effect could be achieved by merely running the applicator with water on an area of smooth concrete. Dyed water would provide an even better means of examining the general spray pattern. Don't assume, however, that the pattern with water would be exactly like that of some heavier fertilizer suspension.

The problems associated with proper application of fluid fertilizer were discussed in Section 7.4.

7.9 APPLYING SOIL AMENDMENTS

Problems in spreading fluid or solid fertilizers also pertain to applications of soil amendments. Particularly unique problems in application of finely divided materials such as gypsum or elemental sulfur have not been discussed, however. Applications of finely divided materials such as gypsum are extremely subject to wind drift in the field, to say nothing of problems encountered in getting the material out of a hopper car or truck. This problem is being attacked by formulation of gypsum suspensions much like lime suspensions (see Chapter 8). Application of fluids are much less subject to wind drift than are finely-ground particles approximately 0.01 mm in diameter (200 mesh). This technique involves preparation of suspensions containing about 50% solids, 47% water, and around 2 or 3% attapulgite clay. If this technique solves the problem of drift, it creates another, that of extreme logistical problems in applying enough suspended gypsum to combat high sodium in soils effectively. Refer to Chapter 9 for examples of the amount of gypsum required.

Applications of elemental sulfur as a soil amendment are subject to wind drift and also subject to the additional hazard of fire. Sulfur in the finely divided elemental form is extremely flammable. Sulfur suspensions have been successfully formulated in California and may provide an alternative to conventional application techniques. Use of larger particle-size sulfur reduced both the hazard of fire and eliminates drift but greatly slows the reaction time of sulfur in the soil.

SUMMARY

Fertilizer application techniques are as important in the performance of this crop production input as the analysis of the materials themselves. Proper application requires knowledge of the materials to be used, their chemical characteristics, their physical properties, and how they interact with each other and with soil and water. Improper application techniques can lead to poor utilization of the applied nutrients, uneconomic crop response, possible germination damage in the fertilized crop, and damage to irrigation systems through which fertilizer is sometimes applied.

Proper calibration and operation of application equipment is mandatory for efficient use of fertilizers, lime, and other soil amendments. Uneven application can lead to serious yield losses. Attention to adjustment of the equipment and to proper overlap in broadcast applications of solid fertilizers and lime may increase yields significantly. Calibration of equipment is time-consuming but is essential for uniform application, proper rates of application, and proper placement of fertilizers.

PROBLEMS

1. Describe the problems associated with the injection of anhydrous ammonia into sprinkler irrigation systems. Indicate how the presence of dissolved calcium bicarbonate in the irrigation water can be a factor in the use of anhydrous ammonia in either sprinkler or gated-pipe irrigation systems.

2. Outline a procedure for the calibration of fertilizer application rate of (a) a grain drill, and (b) a row crop planter. Use a solid material such as DAP in your problem.

3. Describe how a lopsided (left) distribution pattern in a single-spinner solid fertilizer applicator can be corrected.

4. Outline a calibration process for a bulk fertilizer applicator.

5. Indicate your familiarity with the advantages and potential disadvantages of the use of pop-up fertilizers.

6. Fertilizer-pesticide dual applications have some special characteristics that must be considered in the efficient use of such a mixture. Describe problems to watch out for in the use of these mixtures including the initial formulation of the mixture.

REFERENCES

Anonymous, "Application of Fluid Fertilizer," TVA Circular Z-6, 1970.

Anonymous, "Application of Granular Fertilizer," TVA Circular Z-12, 1970.

Bandel, V. A., S. Dzienia, G. Stanford, and J. O. Legg, "N Behavior Under No-Till vs. Conventional Corn Culture," *Agron. J.*, 67 (1975) 782.

Barel, D., "Foliar Application of Phosphorus Compounds," Ph.D. Thesis, Dept. of Agronomy, Iowa State Univ., Univ. Microfilms, Ann Arbor, Mich., 1975.

Bates, T. E., "Response of Corn to Small Amounts of Fertilizer Placed with the Seed, III. Relation to P and K Placement and Tillage," *Agron. J.*, 63 (1971) 372.

Blevins, R. L., D. Cook, S. H. Phillips, and R. E. Phillips, "Influence of No-Tillage on Soil Moisture," *Agron. J.*, 63 (1971) 593.

Blevins, R. L., G. W. Thomas, and P. L. Cornelius, "Influence of No-Tillage and Nitrogen Fertilization on Certain Soil Properties," *Agron. J.*, 69 (1977) 383.

Buehrer, T. F., and R. F. Reitemeier, "The Inhibiting Action of Minute Amounts of Sodium Hexametaphosphate on the Precipitation of Calcium Carbonate from Ammoniacal Solutions," *J. Phys. Chem.*, 44(5) (1940) 552.

Czahkowski, A. C., "Compatibility: Herbicides/Liquid Fertilizers," *Agrichemical Age*, 22(5) (1978) 6.

Estes, G. O., "Elemental Composition of Maize Grown Under No-Tillage and Conventional Tillage," *Agron. J.*, 64 (1972) 733.

Fink, R. J., "Corn Yield as Affected by Fertilization and Tillage," *Agron. J.*, 66 (1974) 70.

Jackson, T. L., H. J. Mack, and R. M. Bullock, "Methods of Fertilizer Application for Vegetable Crops," *Oregon's Vegetable Digest,* 17(1) (1968) 4.

Jones, J. B. Jr., and W. H. Schmidt, "Pop-up Fertilizer," *Ohio Report,* 53(2) (1968) 21.

Jung, G. A., J. A. Balasko, and G. E. Toben, "The Response of Hillside Pastures to Fertilizer Applied by Airplane," W. Virginia Agr. Exp. Sta. Bul. 545, 1967.

McVickar, M. H., W. P. Martin, I. E. Miles, and H. H. Tucker, "Agricultural Anhydrous Ammonia, Technology and Use," American Society of Agronomy, Madison, Wis., 1966.

Murphy, L. S., D. R. Leikam, R. E. Lamond, and P. J. Gallagher, "Dual Application of N and P—Better Agronomics and Economics?" *Fertilizer Solutions,* 22(4) (1978) 8.

Murphy, L. S., K. Winter, and R. Lamond, "Lime Suspension Update," Proc. National Fertilizer Solutions Assn. Round-Up, Kansas City, Mo., July 20–21, 1978, p. 41.

Nelson, G. S., "Aerial Application of Granular Fertilizer and Rice and Lespedeza Seed," Arkansas Agr. Exp. Sta. Bul. 671, 1963.

Shear, G. M., and W. W. Moschler, "Continuous Corn by the No-Tillage and Conventional Tillage Methods: A Six-Year Comparison," *Agron. J.*, 61 (1969) 524.

Triplett, G. B., C. A. Osmond, and P. Sutton, "Fertilizer Application Methods for No-Till Corn," *Ohio Report,* 57(3) (1972) 39.

CHAPTER EIGHT

Lime

8.1 INTRODUCTION

Farmers have been educated to expect immediate and spectacular results from the use of fertilizers. "You can see where you put it," is one of fertilizer's strong, frequent, and usually true sales pitches.

This is not true for lime. Yet on most soils with a pH below 5.0, lime must be applied *before* fertilizers to achieve maximum fertilizer efficiency in plant nutrition and economic crop production. Plant response to lime is usually slow, of long duration, and not very conspicuous. Lime applications on acid soils result in higher yields of most grasses and of almost all legumes and a more vigorous root growth of all plants. In temperate regions, the duration of *one* application is usually for a 5-year period; in the tropics, smaller rates applied more frequently are usually more efficient. By contrast, a small amount of nitrogen, the most universally needed fertilizer, makes plants darker green within a week but lasts for only a few weeks before another application is needed. Furthermore, nitrogen fertilizers are one of the dominant causes of greater soil acidity and the need for more lime. But for most efficient plant production on acid soils the correct answer is *not either* fertilizers or lime *but both* — based on a soil test. Newer technology permits the simultaneous application of fluid lime with fluid nitrogen fertilizer.

8.2 BENEFITS OF LIME

The scientific use of lime on acid soils can double the plant uptake of nitrogen, phosphorus, potassium, molybdenum, and other essential nutrients and thus enhance plant nutrition and growth. Lime also aids both symbiotic and non-symbiotic fixation of atmospheric nitrogen (N_2), hastens the release of essential nutrients from organic matter and clay colloids, reduces the buildup of "thatch" in lawns, improves desirable soil structure, and retards the plant uptake of harmful radioactive strontium 90. Furthermore, lime increases the calcium percentage of plants on acid soils and thus increases the nutritional value of foods and feeds for people and animals. Lime applied to acid lake waters clarifies the water, increases its hardness (a measure of more Ca or Mg or both in relation to Na), increases plant uptake of fertilizer and other nutrients from water, and hastens the growth of fish by increasing their food supply.[1]

Lime helps to make the environment cleaner and safer for people and animals. This statement is especially true where sewage sludges are applied to acid soils. Liming acid soils to a pH above 6.5 reduces the solubility and plant uptake of potentially toxic heavy metals such as lead, zinc, copper, nickel, and cadmium (see Chapter 10).

Information developed in Section 8.6 supports the statement that farmers should be applying 200 million metric tons (220 million tons) but are applying about 15% of this optimum amount on only 1% of the cropland and pastureland in the United States.

In one sense, agricultural lime is a fertilizer as well as a soil amendment because it contains the essential plant nutrient, calcium, and sometimes magnesium (see Chapter 5).

8.3 RELATIVE PRODUCTIVITY OF ACID SOILS

Strongly acid soils are relatively low in productivity because of essential nutrient deficiencies and/or because of toxicities of essential or nonessential elements. Essential plant nutrients that are often so slowly available as to be deficient in strongly acid soils include nitrogen, phosphorus, potassium, calcium, magnesium, and/or molybdenum. Toxicities to certain plants growing in strongly acid soils have been caused by highly soluble aluminum, iron, and/or manganese (Fig. 8-1).

Figure 8-1 portrays the general relationship between soil pH and the relative availability of the elements essential to plants. There are two exceptions to this statement: Chlorine is an essential element but is not shown because its availability is not pH-dependent, and nonessential aluminum is indicated because it reduces the solubility of phosphorus and its toxicity to plants at low soil pH's. The wider the bar for each element in Fig. 8-1, the

---

[1] Claude E. Boyd, "Liming Farm Fish Ponds," Auburn University Leaflet No. 91, 1976.

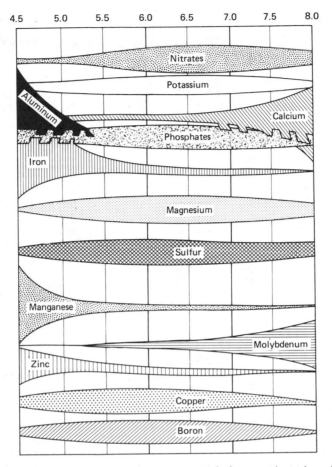

Fig. 8–1. The relative availability of 12 essential plant nutrients in well-drained *mineral* soils in temperate regions in relation to soil pH. A pH range between 6.0 and 7.0 (between heavy lines) is considered ideal for most plants. The thirteenth essential plant nutrient from the soil, chlorine, is not shown because its availability is not pH-dependent. Aluminum is not an essential nutrient for plants but it is shown because it may be toxic below a soil pH of 5.2. (Source: "Liming Acid Soils," Leaflet AGR-19, 1978, University of Kentucky.)

greater the relative availability, assuming that the total supply of each soil nutrient is adequate. The heavy vertical lines in Fig. 8–1 drawn at pH 6 and 7 indicate that in this range most plants on well-drained mineral soils in the temperate region will produce best.

On Histosols (peat and muck soils) in all regions and on Oxisols and Ultisols (mineral soils) in the tropics and subtropics, pH–plant growth relationships are different. This contrast is characterized by the drawing in Fig. 8–2. Although made specifically for Histosols in Michigan, the information is the "best fit" available for Oxisols and Ultisols in the tropics and subtropics.

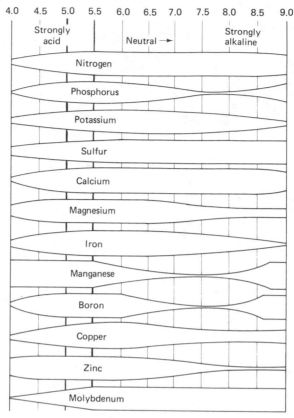

Fig. 8-2. The relative availability of essential plant nutrients from *organic* soils (Histosols). A pH range of 5.0 to 5.5 is considered ideal for the optimum growth of most plants (between heavy lines). Deduction: The more the organic matter in a *mineral* soil, the lower the "ideal" soil pH. *NOTE:* Until more data are available, the authors are suggesting that this figure can also be used for pH-plant growth relationships on Oxisols and Ultisols in the tropics and subtropics. (Source: R. E. Lucas and J. F. Davis, Michigan State University.)

Heavy lines drawn vertically through pH 5.0 and 5.5 are used to indicate the range of normal growth for most plants. This confirms the statement that the same plants tolerate a lower pH (more acidity) on Histosols, Oxisols, and Ultisols than on mineral soil orders in the temperate region. It is also true that the higher the organic matter in a mineral soil, the greater the plant tolerance for acidity. For example, at an organic matter level of < 10%, a desirable pH for most crop plants is 6 to 7; at a level of 15%, it is pH 5.8; and at 20% organic matter, it is pH 5.2.[2]

[2] Anonymous, "Agronomy," Coop. Ext. Serv. Bul. 472, Ohio State University, Feb., 1980.

8.4 PLANT TOLERANCE OF SOIL pH BY SOIL ORDERS

From scores of published information, moderated and amended by the authors' research throughout the world, the approximate plant tolerance by soil pH range and by soil orders has been developed. These data are displayed in Table 8–1. Seventy-one plants have been so characterized, including those of temperate, subtropical, and tropical zones.

Most comparable published data use the terms "desirable pH," "pH preference," "best pH," or "pH requirements for crop production." The authors have chosen to use the term "plant tolerance" because the meaning is broader than those terms traditionally used. Plant tolerance connotes survival under specified conditions of the environment such as the relationships among soil pH, essential plant nutrient availability (e.g., phosphorus), and the presence of toxic substances such as aluminum.

Plants with the greatest range in pH tolerance are barley, cotton, oats, and wheat. These four popular crops tolerate a pH range of 5.5 to 8.5. Two other plants tolerate a pH range of 4.0 to 7.5: cowpeas and redtop grass.

In Section 8.3 the statement is made that the same plants will tolerate a lower pH on an organic soil (Histosols) than on a mineral soil. It will be of interest to test the accuracy of this statement from data given in Table 8–1, derived from a synthesis of research results from many scientists.

Table 8–1. Approximate Plant Tolerance of Soil pH Range by Soil Orders

	Soil Orders [a]				
	Histosols, Oxisols, and Ultisols	Alfisols, Aridisols, Entisols, Inceptisols, Mollisols, Spodosols, and Vertisols			
		pH Range			
Plant Species	4.0–5.5	4.0–5.5	5.5–6.5	6.5–7.5	7.5–8.5
Alfalfa				*	*
Asparagus				*	
Azalea		*			
Barley			*	*	*
Bean, field			*		
Bean, green	*	*	*	*	
Bean, lima			*	*	
Beet, table, garden				*	
Bermudagrass		*	*	*	
Birdsfoot trefoil			*		
Blackberry				*	*

(Table 8–1 cont.)

Table 8–1.(cont.)

Plant Species	Histosols, Oxisols, and Ultisols	Alfisols, Aridisols, Entisols, Inceptisols, Mollisols, Spodosols, and Vertisols			
	pH Range				
	4.0-5.5	4.0-5.5	5.5-6.5	6.5-7.5	7.5-8.5
Blueberry		*	*		
Buckwheat		*	*		
Cabbage			*	*	*
Cantaloupe			*	*	
Carrot			*	*	
Centrosema (tropical legume)	*				
Clover					
Alsike		*	*	*	
Crimson			*	*	
Ladino			*	*	
Red				*	
White			*	*	
Coffee, arabica	*				
Corn, field	*		*	*	
Cotton	*		*	*	*
Cowpea	*	*	*	*	
Cranberry	*	*			
Crownvetch			*	*	*
Dallisgrass			*	*	
Fescue grasses			*		
Kentucky bluegrass				*	*
Kudzu (tropical legume)	*				
Lespedezas, common, Korean		*	*		
Lespedeza sericea		*	*		
Lettuce			*	*	
Millet					
Finger	*				
Pearl			*	*	
Napiergrass	*				
Oat			*	*	*
Onion				*	*
Orchardgrass			*	*	
Pea, English				*	*
Peanut	*		*		

Table 8–1.(cont.)

Plant Species	Histosols, Oxisols, and Ultisols	Alfisols, Aridisols, Entisols, Inceptisols, Mollisols, Spodosols, and Vertisols			
	pH Range	pH Range			
	4.0–5.5	4.0–5.5	5.5–6.5	6.5–7.5	7.5–8.5
Pineapple	*				
Potato, Irish			*		
Potato, sweet	*		*	*	
Redtop grass		*	*	*	
Reed canarygrass			*	*	
Rhodesgrass	*				
Rice	*		*	*	
Rye, cereal		*	*		
Ryegrass, annual			*		
perennial			*		
Smooth bromegrass			*		
Sorghums			*	*	
Soybeans	*		*	*	
Spinach				*	*
Strawberry			*	*	
Stylosanthes (tropical legume)	*				
Sudangrass			*	*	
Sugar beet				*	*
Sugarcane	*				
Sweetclover				*	*
Tea	*				
Timothy			*	*	
Tobacco			*		
Tomato	*			*	
Vetches, annual			*	*	
Wheat			*	*	*
Watermelon	*		*	*	
Weeping lovegrass	*	*			

[a] Histosols are organic soils; Oxisols are *extremely* weathered mineral soils in the humid tropics; Ultisols are *highly* weathered mineral soils in the humid and subhumid tropics and subtropics; the other seven soil orders are not as highly weathered mineral soils that may occur anywhere in the world where the factors of soil formation are favorable for their development.

This table lists seven plants grown in both Histosols as well as in minerals soils in the temperate region. These are green beans, field corn, cotton, peanuts, rice, soybeans, and tomatoes. In the Histosols all plants listed tolerate a pH range of 4.0 to 5.5. In mineral soils in the temperate region, peanuts have a tolerance of the narrowest and most acid pH range, 5.5 to 6.5. Green beans and tomatoes also have a one-unit pH range of tolerance — 6.5 to 7.5. The widest pH range in tolerance, 5.5 to 7.5, is exhibited by field corn, cotton, rice, and soybeans.

The discussion so far has been centered on tolerance of soil acidity by *plant species*. There are also differences in tolerance among varieties and cultivars of the same species. This discovery is a fortunate fact for plant breeders who dream of developing plants adapted to the unamended soils of the world. When this day arrives, lime, fertilizers, and other soil amendments (so the dream continues) will not be needed to produce abundantly for the people of the world. Or will they?

Examples of massive efforts to breed and select high-yielding cultivars or varieties adapted to soil pH levels, seedling vigor, resistance to cutworms and stalk borers, and resistance to foliar and ear diseases are reported by the International Center for Tropical Agriculture in Cali, Colombia, South America. They have discovered so far that differential responses *within the same variety/cultivar* exist for these plant species: cowpeas, black beans, corn, and field beans.

In a similar manner, rice varieties/cultivars at the International Rice Research Institute near Manila, Philippines, have been bred/selected for resistance to alkali injury, salt injury, iron toxicity, aluminum toxicity, manganese toxicity, drought, deep water, phosphorus deficiency, zinc deficiency, and iron deficiency. In addition, success has been achieved to breed into certain rice varieties/cultivars resistance to lodging; resistance to these diseases: blast, bacterial blight, bacterial leaf streak, grassy stunt, and tungro; and resistance to the following insects: green leaf hopper, brown plant hopper, and stem borer.

Concluding question: With such success in breeding/selecting plants to resist so many enemies of plants and to be adapted to existing soil conditions, will the time ever come when no pesticides and no lime or fertilizers will be needed?

The answer is NEVER!

When resistance to a specific disease or insect is built into the plant, two more pests appear from "nowhere" to plague plants. And as soon as plants are bred to be resistant to low fertility or toxic elements, yields are usually so low that high rates of lime and fertilizer are needed to increase them. Continuous high production results only when all scientists work together for long periods of time under stable agencies that supply adequate support.

Research on the yield-limiting effects of acid soils is now receiving worldwide attention but must be intensified if food production is to keep pace with population increase. Examples of plant response to lime in both temperate and tropical regions follow.

8.5 PLANT RESPONSE TO LIME

Plants on acid soils will *not* respond to lime if the first limiting factor is one or more of the following yield-limiting factors: nutrient shortage, insects, diseases, weeds, drought, a cultivar/variety not adapted, the wrong planting date, and/or plant tolerance of the existing soil acidity. If none of these first-limiting factors exists, plant yields can be expected to increase with lime applications if the soil pH is below the optimum for that specific species/cultivar/variety.

Based on many field research and demonstration plots in the temperate region, estimates have been made of crop yield increases resulting from liming soils to specific pH ranges. These data are given in Table 8–2. Note in the table that for alfalfa, 75%, 50%, and 25% yield increases, respectively, can be expected as the soil is limed from pH 5.0 to 5.5, 5.5 to 6.0, and 6.0 to 6.5, respectively. From other data it has been confirmed that on some soils alfalfa responds economically to a soil that has been limed to a pH of 7.5. From these data in Table 8–2, the more responsive crops after alfalfa are annual lespedezas, barley, cotton, soybeans, and wheat. Least responsive are corn, grain sorghum, and tobacco.

Soils and plants in the tropics on Oxisols have a different pH-lime interaction. Whereas, in the temperate region, an application of about 10 metric

Table 8–2. Estimated Yield Increases Resulting from Liming Soils to Specific pH Ranges

| | Percent Yield Increase Predicted from the Harvested Crop | | |
| | Soil pH Increased by Lime Applications from | | |
Crop	5.0 to 5.5	5.5 to 6.0	6.0 to 6.5
Alfalfa	75	50	25
Lespedezas, annual	50	25	1
Barley	20	10	1
Cotton	20	10	1
Soybeans	20	10	1
Wheat	20	10	1
Corn	15	5	1
Grain sorghum	15	5	1
Tobacco	15	5	1

PRIMARY SOURCE: H. Allen and J. M. Soileau, "Increasing Lime Use in the Tennessee Valley Region," Tennessee Valley Authority, Muscle Shoals, Alabama, 1971.

tons/ha (4.5 tons/A) every 5 years on acid clay loam soils is common, in the tropics about 5 metric tons/ha (2.2 tons/A) every 2 years on acid Oxisol clays is closer to the recommended rate. Examples of typical responses to small and frequent lime applications to crops on Oxisols and Ultisols in the tropics are as follows:

1. Green beans and black beans are very sensitive to (not tolerant of) soil acidity and respond readily to lime.
2. Cereal grains respond to lime when soil pH is below 5.0 and exchangeable aluminum is above 15% saturation.
3. Cotton responds to lime in all regions where climatically adapted. Furthermore, cotton is not sensitive to overliming.
4. Cowpeas are tolerant of low pH.
5. Forage legumes respond to lime, especially when soil manganese occurs in toxic quantities.
6. Peanuts have a high calcium requirement but are highly tolerant of aluminum. For this reason, gypsum ($CaSO_4$), a neutral salt, is usually as effective as lime ($CaCO_3$).
7. For rice, the International Rice Research Institute applied 10 metric tons/ha (4.46 tons/A) of lime on a flooded Luisiana clay on a farmer's field in the Philippines and increased the soil pH from 4.7 to 5.9 and the yield from 2460 kg/ha (2194 lb/A) to 2550 kg/ha (2275 lb/A).[3]
8. Soybeans are tolerant of high aluminum but are sensitive to toxic manganese and low molybdenum. Soybeans should respond to lime when the soil pH is about 5 or lower *and* manganese is high and molybdenum is low.
9. Sugarcane responds to lime when soil pH is 5.0 or below, calcium and magnesium levels are low, and exchangeable aluminum exceeds 15% saturation.

Soil scientists and agronomists have researched soil-plant nutrition relationships for more than 100 years. There is general agreement that, in the principal crop-producing areas of the world (the humid regions), excess soil acidity is a major cause of low yields. However, the soils are becoming more strongly acid and less productive, even though apparently adequate fertilizer has been used. For this reason the origin of soil acidity is discussed in specific detail.

8.6 ORIGIN OF SOIL ACIDITY

8.6.1 Introduction

Soils become acid because excess hydrogen ions replace basic cations such as calcium, magnesium, potassium, and sodium (action to the right in Fig. 8–3). The replaced basic ions are then leached below the root zone (rhizo-

[3] International Rice Research Institute, 1976. Annual Report for 1975. Los Baños, Philippines, p. 292.

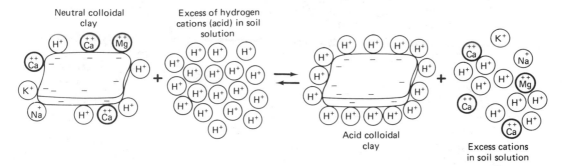

Neutral colloidal clay

Excess of hydrogen cations (acid) in soil solution

Acid colloidal clay

Excess cations in soil solution

Fig. 8–3. Neutral colloidal clay becomes acid by a replacement of basic ions with hydrogen ions (action to the right). Conversely, an acid soil becomes neutral by the application of calcium (Ca) (lime), magnesium (Mg) (lime), and sodium (Na) (action to the left). (Courtesy Roy L. Donahue.)

sphere). The process is reversed (action to the left) when lime (calcium and magnesium) is added that by mass action replaces hydrogen ions and thereby make the soil less acid (toward neutrality).

On soils used for modern crop production, there are at least seven reasons to account for the fact that soils are becoming more acid.

1. Acid parent materials from such rocks as granites, sandstones, and shales.
2. Plant root exudation of hydrogen (upon absorption of other cations) and the formation of carbonic acid from CO_2 excreted by the roots.
3. Crop removal in harvested crops of such basic-forming elements as calcium and magnesium.
4. Leaching from the root zone of bases by the downward percolation of rain and irrigation waters.
5. Erosion by water and wind removes the humus and clay from surface soils. These fine particles carry away adsorbed calcium and magnesium, leaving more acid silt and sand. Sometimes, however, the surface few inches of soil is the most acid and slight sheet erosion will result in exposing less acid horizons.
6. Acid precipitation caused by sulfuric acid and nitric acid in the atmosphere over industrialized and heavily populated areas.
7. Acid-forming fertilizers such as the most common nitrogen fertilizers, monoammonium phosphate, and diammonium phosphate.

Some pesticides used in agriculture are acid but the amount used is so small that their influence on soil acidity is negligible.

Summarized in Table 8–3 is the estimated metric tons of 85% pure calcium carbonate (average agricultural lime) required in the United States each year to neutralize the acidity generated by these mostly anthropogenic (man-induced) activities. Lime used in the United States in 1977 was 28

Table 8–3. Extent of Agricultural Lime Equivalent Needed to Neutralize Soil Acidity Generated by Mostly Anthropogenic Activities[a]

	Agricultural Limestone Equivalent (85% pure $CaCO_3$) Required to Neutralize Acidity Generated	
Origin of Soil Acidity	(million metric tons/yr)	(million tons/yr)
Erosion, accelerated	118	130
Acid fertilizers used	37	41
Crop removal	13	14
Acid precipitation	4	5
Total	172	190

[a] Not estimated is the production of acidity in surface soils by *natural* causes that include geologic erosion, acid parent materials, plant root exudations by native plants, organic matter decomposition, and leaching of basic ions.

million metric tons (31 million tons), only 18% of the 172 million metric tons (190 million tons) required just for maintenance. The rationale used to arrive at this conclusion is detailed in the following sections.

8.6.2 Acid Parent Materials

Acid parent materials include unconsolidated glacial materials as well as acid consolidated rocks low in bases such as granites, sandstones, and many shales. Acid glacial materials occur abundantly in all the northeastern states of Maine, New Hampshire, Vermont, Massachusetts, Connecticut, Rhode Island, New York, northeastern and northwestern Pennsylvania, the northern half of Ohio, northern Michigan, northern Wisconsin, eastern Minnesota, central and southern Iowa, and northern Missouri. Acid crystalline rocks such as granites and granite gneisses occur abundantly in central California, Idaho, Washington, Oregon, and central Arizona. Acid sandstones and shales are common in Pennsylvania, West Virginia, eastern Kentucky, eastern Tennessee, northern Alabama, Arkansas, Oklahoma, central northern Texas, central Kansas, southwestern Colorado, northwestern New Mexico, northeastern Arizona, and southeastern Utah (Fig. 8–4). No estimate is made of acidity from acid parent materials.

8.6.3 Plant Root Exudation of Hydrogen

In the vicinity of actively growing plant roots (rhizosphere), exudations and excretions of carbon dioxide during respiration, plant sap exosmosis, and sloughing of dead cells provide an ideal environment for reproduction of bacteria and fungi. In addition, most root hairs excrete hydrogen ions that are exchanged for nutrient cations on the surface of colloids and in soil solution. Hydrogen ions also arise from root secretions of carbon dioxide as it reacts with water to produce carbonic acid (H_2CO_3). Some of this carbon dioxide is used again by the plant in photosynthesis, some escapes into the atmosphere, and some remains to make the soil more acid by as much as one pH unit. If all crops were perennial, the influence on acidifying the soil by root exudation of hydrogen from carbonic acid could be estimated. However, with most crops occupying the soil for only a few months, no year-around estimates of the acidifying action of crops on soils are made.

8.6.4   Crop Removal

Soils become more acid as hydrogen replaces the calcium and magnesium removed with harvested crops.

Although not a constant value, each crop species varies in calcium and magnesium percentage within specified limits. Because crop yields also vary from soil to soil, management to management, and year to year, the total amount of calcium and magnesium fluctuates, as indicated in Table 8-4.

In the 50 states during the year 1975, there were 136 million ha (336 million A) of crops harvested. Assume that all of these crops contained the *median* amount of calcium and magnesium as the 15 crops shown in Table 8-4, namely, 14 and 11 kg/ha of calcium (Ca) and magnesium (Mg), respectively. The total amount of calcium removed by harvested crops would be 14 × 136 million = 1.904 billion kg of calcium (Ca) with a calcium carbonate equivalent of 4.76 billion kg (4.76 million metric tons or 5.25 million tons). The total amount of magnesium (Mg) removed by all harvested crops in 1975 was 11 × 136 million = 1.496 billion kg with a calcium carbonate equivalent of 6.16 billion kg (6.16 million metric tons or 6.79 million tons).

Summary. 4.76 + 6.16 = 10.92 million metric tons (12.03 million tons) of pure calcium carbonate equivalent. Because agricultural limestone averages 85% pure, this is equal in neutralizing value to 12.85 million metric tons (14.16 million tons) of lime removed each year by harvested crops (rounded to 13 million metric tons per year, summarized in Table 8-3).

8.6.5   Leaching

Rainwater or irrigation water that moves downward through the soil dissolves nutrients that move with the water. These include calcium and magnesium, the "lime" nutrients. The more acid the water, the more effective the leaching. Waters combine with carbon dioxide from the atmosphere (0.03% CO_2) and from decaying organic matter to form carbonic acid (H_2CO_3). This acid reacts with the carbonates of calcium and magnesium, changing them to bicarbonates that are much more soluble. Waters also are better solvents when combined with the oxides of sulfur and nitrogen from the burning of coal and fossil fuels (see Section 8.6.7).

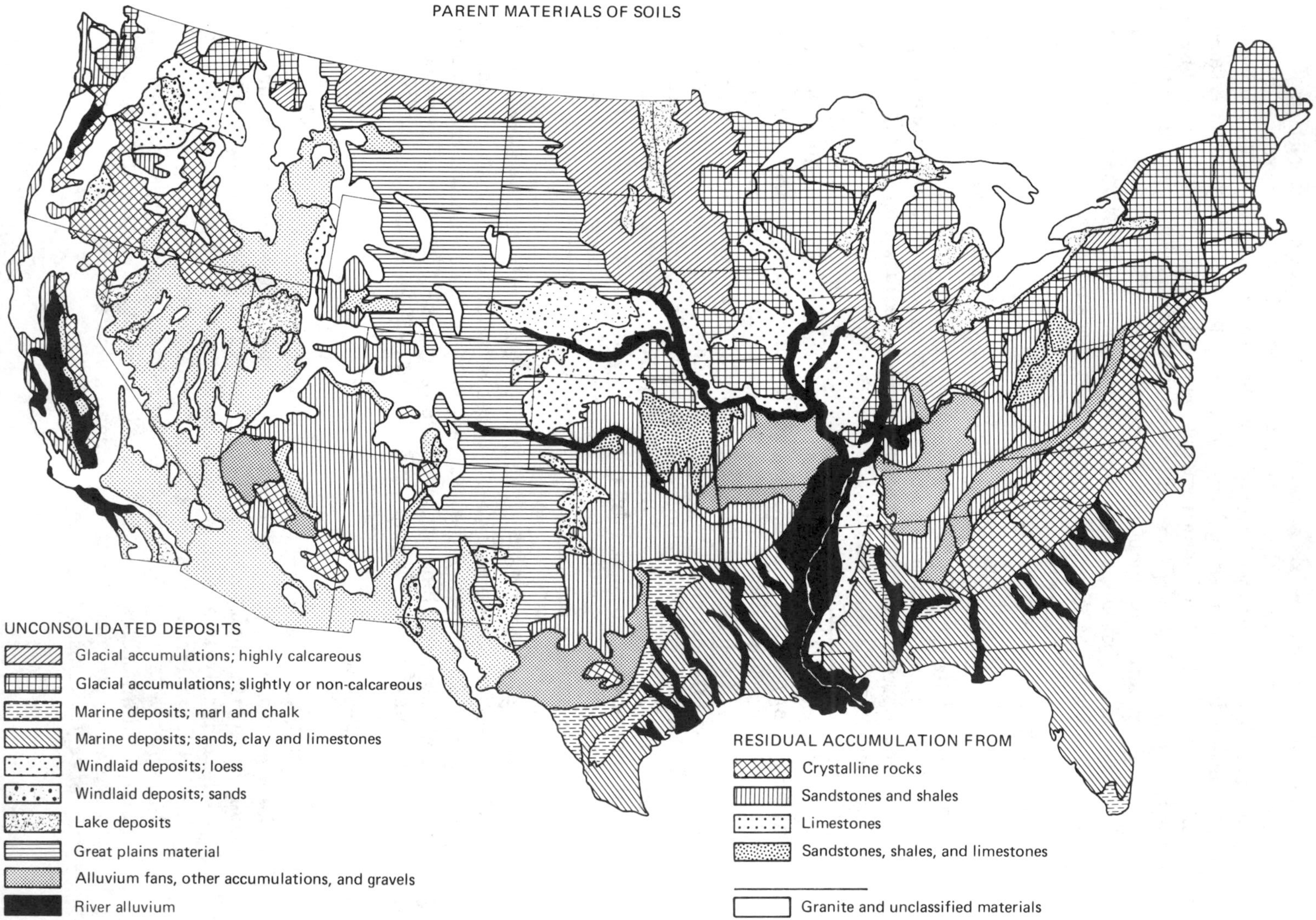

PARENT MATERIALS OF SOILS

RESIDUAL ACCUMULATION FROM

Crystalline rocks
Sandstones and shales
Limestones
Sandstones, shales, and limestones
Granite and unclassified materials

UNCONSOLIDATED DEPOSITS

Glacial accumulations; highly calcareous
Glacial accumulations; slightly or non-calcareous
Marine deposits; marl and chalk
Marine deposits; sands, clay and limestones
Windlaid deposits; loess
Windlaid deposits; sands
Lake deposits
Great plains material
Alluvium fans, other accumulations, and gravels
River alluvium

Table 8-4. Mean Calcium and Magnesium Content of the Harvested Portion of Common Crops

Crop	Yield (kg/ha)	Calcium (Ca) (kg/ha)	Magnesium (Mg) (kg/ha)
Alfalfa	6726	112	22
Barley			
Grain	2152	1	2
Straw	2242	3	1
Corn			
Grain	3139	1	3
Stover	3363	11	6
Cotton			
Lint	560	1	1
Seed	1121	2	6
Oats			
Grain	1794	1	2
Straw	2802	6	2
Rye			
Grain	1256	1	1
Straw	2242	3	1
Soybeans			
Grain	1682	11	11
Hay	4484	34	17
Tobacco	2242	56	11
Wheat			
Grain	2018	1	3
Straw	3363	6	2

SOURCE: "Liming Soils: An Aid to Better Farming," USDA–Farmers' Bul. No. 2124, 1966, p. 28.

Rainwaters leach lime from soils and make them more acid. This occurs when precipitation is greater than evapotranspiration, as portrayed in Fig. 8–5. The authors have not, however, attempted to quantify how leaching has contributed to acidification of soils.

8.6.6 Erosion

The total land area of the 50 states is 930 million ha (2.3 billion A) of which 195 million ha (482 million A) (21%) are usually used for cropland.

Total erosion sediment generated each year from the 50 states is estimated at 3.6 billion metric tons (4 billion tons). About 45% of this sedi-

---

Fig. 8-4. Acid parent materials are primarily those delineated here as (1) glacial accumulations, slightly or noncalcareous, (2) marine deposits; sands and clays (excluding limestones), (3) crystalline rocks, and (4) sandstones and shales. (Redrawn from USDA—"Atlas of American Agriculture, Part III, Soils of the U.S.," by C. F. Marbut, Washington, D. C., 1935, p. 15.

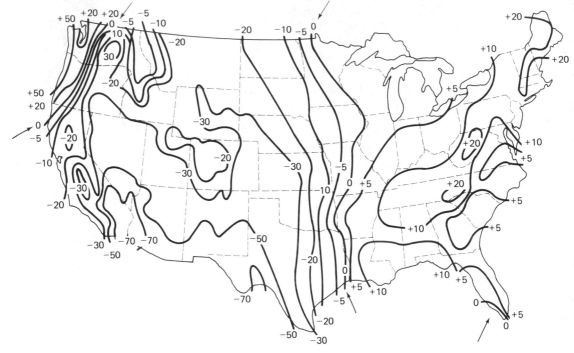

Fig. 8-5. Potential leaching of soils in the United States. Lime will be leached from soil profiles and thus make soils more acid when mean annual precipitation is greater than mean annual evapotranspiration. A plus (+) sign indicates inches of precipitation in excess of evapotranspiration and a minus (−) sign the excess of evapotranspiration over precipitation. Where these two parameters are equal, a zero (0) is shown indicated by arrows. *NOTE:* 1 in. = 2.54 cm. (Source: K. W. Flack, "Land Resources." In *Recycling Municipal Sludges and Effluents on Land*, University of Illinois, July, 1973.)

ment (1.62 billion metric tons, 1.78 billion tons) originates from cropland where most of the lime is applied.

The authors have used scattered sources in the literature to estimate that the average sediment from cropland in the United States contains about 1.5% calcium (Ca) and 0.7% magnesium (Mg). With these assumptions, each year erosion sediments carry away from cropland soils in the United States about 24.3 million metric tons (26.8 million tons) of calcium (Ca) and 11.3 million metric tons (12.5 million tons) of magnesium (Mg). This is equal to 124 kg/ha (111 lb/A) per year of calcium (Ca) and 58 kg/ha (52 lb/A) per year of magnesium (Mg).

Because farmers and scientists are interested primarily in the amount of agricultural limestone that must be applied to cropland in the United States each year to replace that lost by erosion, further calculations are necessary. To convert 24.3 million metric tons (26.8 million tons) of calcium (Ca) lost to calcium carbonate ($CaCO_3$) equivalent, multiply $24.3 \times 2.50 = 60.8$ million metric tons (67.0 million tons). To convert to agricultural lime equivalent (85% pure calcium carbonate ($CaCO_3$)), $60.8 \div 0.85 = 71.5$ million metric tons (78.8 million tons). In a similar way, 11.3 million metric tons (12.5 million tons) of magnesium (Mg) $\times$ 4.12 = 46.6 million metric tons (51.4 million tons) of calcium carbonate ($CaCO_3$) equivalent $\div$ 0.85 = 54.8 million metric tons (60.4 million tons) of agricultural lime equivalent (see Section 8.10).

The total agricultural lime equivalent of calcium (Ca) and magnesium (Mg) lost each year from cropland in the United States through erosion is, therefore:

$$\begin{array}{r} 71.5 \\ + \ 46.6 \\ \hline \end{array}$$
118.1 million metric tons (130.1 million tons)

(rounded to 118 and 130).

8.6.7 Acid Precipitation

The new thrust in the United States to use less nuclear energy and more coal will accelerate the acidification of precipitation, surface waters, and the soil. Sulfur in coal is the principal producer of acid sulfur gases such as sulfur dioxide (SO_2); other acidifying agents are nitrogen oxides (NO_x) gases from gasoline combustion in automobiles and trucks and from the burning of coal. Between 1960 and 1970, sulfur dioxide (SO_2), the principal cause of *acid* precipitation, increased by 44% and nitrogen oxides (NO_x), the secondary cause of acid precipitation, increased by 94%. The data are presented in Table 8–5.

Sulfur in the atmosphere occurs primarily in these acid-producing forms: sulfates (SO_4^{2-}) (aerosols), sulfur trioxides (SO_3) (liquid), sulfur dioxides (SO_2) (gas), and hydrogen sulfides (H_2S) (gas). Atmospheric nitrogen compounds that contribute to acid rainwaters are the gases nitrous oxide (N_2O) nitric oxide (NO), and nitrogen dioxide (NO_2), especially the latter. Nitrogen dioxide (NO_2) reacts with water vapor to form nitric acid (HNO_3). An alkaline-forming compound also may occur in rainwater. As much as 90% of the urea [$CO(NH_2)_2$] excreted by animals may be changed by bacteria to alkaline am-

Table 8–5. Nationwide Total Emissions of Sulfur Dioxide (SO_2) and Nitrogen Oxides (NO_x) in 1960 and 1970

Acid Pollutant	1960	1970	Increase in 1970 over 1960 (%)
	(million metric tons)		
Sulfur dioxide (SO_2)	21.0	30.3	44
Nitrogen oxides (NO_x)	10.3	20.0	94

SOURCE: "Fifth Annual Report of the Council on Environmental Quality," Washington, D.C., Dec., 1974, pp. 267, 276.

monia (NH_3). However, bacterial genera *Nitrosomonas* and *Nitrobacter* in the soil soon change alkaline ammonia into acid nitrates.

Actual field studies with lysimeters in Connecticut have confirmed that each year the acidity generated by acid precipitation requires 36 kg/ha (32 lb/A) of pure calcium carbonate to neutralize it. Total annual precipitation in that area averages 114 cm (45 in.). However, the *acid* precipitation in Connecticut is higher than the average for all states because of (1) 50% more precipitation than the United States average and (2) location of the state in a heavily populated, highly industrialized region.[4]

To generalize from these data to the entire United States is admittedly not highly accurate. However, for this estimate the authors assume that (1) acid soils exist only in areas receiving 76 cm (30 in.) or more of annual precipitation and (2) the amount of acid is proportional to annual precipitation, i.e., in areas receiving from 76 to 114 cm (30 to 45 in.).

Using these assumptions, the amount of 100% pure calcium carbonate equivalency of the acid precipitation would be

$$\frac{30}{45} \times 36 \text{ kg/ha} = 24 \text{ kg/ha}$$

24 kg/ha × 136 million ha (cropland) = 3.264 billion kg/yr
= 3.264 million metric tons/yr

Converted to 85% agricultural limestone, 3.264 ÷ 0.85

= 3.84 million metric tons/yr (rounded to 4 million metric tons/yr)

Acid precipitation has been studied and mapped by the U.S. Environmental Protection Agency and other agencies. This information is shown in Fig. 8–6).

Acid precipitation is not confined to the Unted States; it is common in all industrialized countries, such as Norway.[5]

8.6.8 Acid-Forming Fertilizers

Because they are cheaper per unit (kilograms or pounds) of nutrient delivered to the user, year by year almost all fertilizers except potassium are rapidly becoming more concentrated *and more acid.* For example, from about the 1920s to the 1940s the most commonly used nitrogenous fertilizers were *alkaline.* These included sodium nitrate, calcium nitrate, calcium cyanamide, and guano (wild bird and bat manure). This list also included one *strongly acid*

[4] C. R. Frink and G. K. Voigt, "Potential Effects of Acid Precipitation on Soils in the Humid Temperate Zone." In *Proceedings of the First International Symposium on Acid Precipitation and the Forest Ecosystem,* U.S. Forest Service General Technical Report NE-23, 1976, USDA-Forest Service, Upper Darby, Pa., pp. 685–709.

[5] Finn H. Braekke, ed., "Impact of Acid Precipitation on Forest and Freshwater Ecosystems in Norway," Research Report Fr 6/76, Oslo, Norway, 1976. Agricultural Council of Norway and the Norwegian Council for Scientific and Industrial Research.

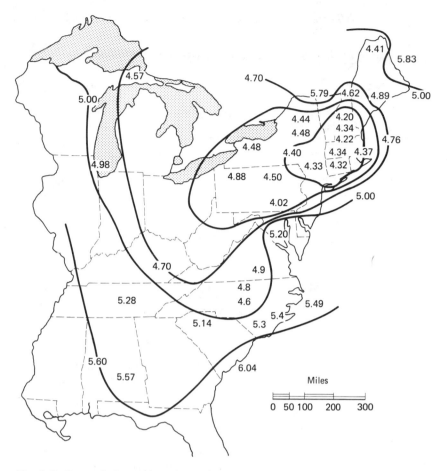

Fig. 8-6. In a pristine environment rainwater has a pH of about 5.7; in a highly industrialized environment rainwater is made more acid by sulfur, nitrogen, and chlorine gases. This 29-state region has acid rainwater with pH values mostly below 5.0 (1965–66 field data). (Source: M. A. Berry and J. D. Bachmann, "Developing Regulatory Programs for the Control of Acid Precipitation," U.S. Environmental Protection Agency. In *Proceedings of the First International Symposium on Acid Precipitation and the Forest Ecosystem,* U.S. Forest Service General Technical Report N3-23, 1976. USDA-Forest Service, Upper Darby, Pa., p. 1064.)

nitrogen fertilizer — ammonium sulfate. As recently as the 1960s, normal superphosphate (18 to 20% P_2O_5) and concentrated superphosphate (42 to 50% P_2O_5) (*both neutral*) were very popular. They have now been almost completely replaced by *acid* diammonium phosphate (16 to 21% N and 48 to 53% P_2O_5) and *acid* monoammonium phosphate (11 to 13% N and 48 to 52% P_2O_5) (Table 8-6).

Table 8-7 sets forth tonnages of and the calcium carbonate equivalent acidity of the seven most popular fertilizers produced in the United States in

Table 8–6. Common Fertilizer Materials and Their Equivalent Acidity or Basicity

NITROGEN FERTILIZERS

Material	Nitrogen (N) (%)	Amount Needed to Supply 20 lb of Nitrogen (N) (lb)	Calcium Carbonate Equiv. per 20 lb of Nitrogen[a] (N) (lb)
Anhydrous ammonia	82	24	− 36
Urea	45	44	− 36
Nitrogen solutions	16–49	125–41	− 36
Ammonium nitrate	33.5	60	− 36
Cyanamide	21	95	+ 57
Ammonium sulfate	20.5	98	− 107
Ammonium nitrate lime	20.5	98	0
Diammonium phosphate	18–21	111–95	− 71
Sodium nitrate	16	125	+ 36
Potassium nitrate	14	143	+ 36
Monoammonium phosphate	11	182	− 107

PHOSPHORUS FERTILIZERS

Material	Phosphate (P_2O_5) (%)	Amount Needed to Supply 20 lb of Phosphorus (P_2O_5) (lb)	Calcium Carbonate Equiv. per 20 lb of Phosphorus (P_2O_5) (lb)
Monoammonium phosphate	48	42	− 21
Diammonium phosphate	46–54	43–37	− 28
Superphosphate, concentrated	42–50	48–40	Neutral
Superphosphate	20	100	Neutral

POTASSIUM FERTILIZERS

Material	Potash (K_2O) (%)	Amount Needed to Supply 20 lb of Potassium (K_2O) (lb)	Calcium Carbonate Equiv. per 20 lb of Potassium (K_2O) (lb)
Potassium chloride	60–62	33–32	Neutral
Potassium sulfate	50–53	40–38	Neutral
Potassium nitrate	44–46	45–43	+ 11
Potassium magnesium sulfate	22	91	Neutral

[a] A minus sign indicates the number of pounds of pure calcium carbonate needed to neutralize the acidity created by 20 lb of N, P_2O_5, or K_2O. A plus sign indicates that the material is basic and is equivalent to the number of pounds of pure calicum carbonate indicated.

NOTE: To convert from pounds to kilograms, multiply pounds by 0.454.

PRINCIPAL SOURCE: James R. Miller, *Fundamental Facts About Soils, Lime, and Fertilizer,* University of Maryland Fact Sheet 186, 1971, 4 pp.

Table 8-7. Production of the Most Popular Fertilizers in the United States in 1978 and the Calcium Carbonate Required to Neutralize Their Acidity

Fertilizer Materials	Percent Nutrient	U.S. Production (1978) (thousand metric tons)		Calcium Carbonate Equivalent (Acidity) (metric tons per metric ton of nutrient)	Calcium Carbonate Equivalent (Total) (thousand metric tons)
		Material	Nutrient		
Anhydrous ammonia	82	15,867	13,011	1.48	19,256
Diammonium phosphate (average analysis, N + P_2O_5)	34.5	3,369	1,162	0.70	813
Phosphoric acid (average analysis, wet process)	56	—[a]	7,404	0.60	4,442
Nitrogen solutions (average analysis)	30	—[b]	2,342	0.63	1,475
Urea	46	4,556	2,096	0.84	1,761
Ammonium sulfate	21	2,116	444	1.10	488
Ammonium nitrate (solid)	33.5	2,768	927	0.59	547
				Total	28,782

[a] Phosphoric acid reported only as P_2O_5.
[b] Nitrogen solutions reported only as nutrient N.

NOTES:
1. *Neutral* fertilizers that are popular include normal superphosphate, concentrated superphosphate, and muriate of potash. *Alkaline* fertilizers and soil amendments used as fertilizers that are sold in lesser quantities include dried animal manures, basic slag, calcium nitrate, cyanamide, potassium nitrate, rock phosphate, sodium nitrate, and bone meal.
2. Additional fertilizers are acid but insufficient data are available for calculating their effect on soil acidity.
3. Conversion factors: metric tons × 1.102 = tons
   kilograms × 2.205 = pounds

PRIMARY SOURCES: "1979 Fertilizer Situation," USDA-Economics, Statistics, and Cooperative Service, FS-9, Washington, D.C., 1978, p. 9; *The Fertilizer Handbook,* The Fertilizer Institute, Washington, D.C., 1976, pp. 62, 63.

1978. The last column in Table 8-7 details the total *pure* calcium needed to neutralize the acidity generated by the use of these fertilizers. This total is 28,782,000 metric tons (31,717,764 tons).

State law specifications of agricultural limestone average about 85% purity for calcium carbonate. The number of tons of 85% pure calcium carbonate needed to neutralize the acidity generated by fertilizers produced in 1978 in the United States totaled 33,852,000 metric tons (37,315,000 tons). Because this is an estimate it can be rounded to 34 million metric tons (31 million tons).

During the last few years, the actual amount of agricultural lime used by farmers each year in the United States has been about 30 million metric tons (33 million tons). All experts on the subject agree that this is not enough.

According to the authors' in-depth analysis of the origin of soil acidity,

172 million metric tons (190 million tons) of agricultural grade lime are needed each year just to replace and neutralize the lime "lost" by erosion, acid fertilizers, crop removal, and acid precipitation. For efficient crop production, the authors conclude that farmers are using only 15% as much lime as they should be using for the most economic production of crops. Farmers would be wise to use about 200 million metric tons (220 million tons) of lime each year.

8.7 LIME REQUIREMENT

When a soil sample is collected and sent to a laboratory for determining lime requirement, it helps in interpreting the results if the soil map unit is known. The unit name includes soil series, slope class, and erosion class. From the soil series name the complete soil taxonomy is known. The taxonomy includes texture, mineralogical characteristics (type of clay), and soil temperature regime (an indication of the amount of organic matter and its rate of mineralization). In the absence of a soil map unit name, these parameters must be estimated before a rational lime requirement can be made: soil textural class, type of clay, organic matter content, and exchangeable aluminum. In addition, the lime requirement of the plant to be grown must be known.

Section 8.4 lists the major crop plants in relation to the widest soil pH range in which they produce satisfactory yields. Note in Table 8-1 that for Histosols (organic soils), Oxisols (well-drained soils on old land surfaces in the humid tropics), and Ultisols (less weathered than Oxisols in tropical and subtropical regions), the crops that are climatically adapted yield best at a pH of about 5.5. Most crops on the other seven soil orders (mineral soils not highly weathered) produce best at a pH of about 6.5. The greatest soil pH tolerance is exhibited by redtop grass (4.0–7.5) and by cabbage, cotton, oat, and wheat (5.5–8.5). Even though these crops exhibit a wide tolerance to soil pH, it probably would not be economical to lime the soil to a pH of more than 6.5 to 7.0.

Soil textural class is a second determinant in estimating lime requirement because of differential buffering capacity. The more clay present in a soil the more lime is required to raise the pH. For example, on a loamy sand only 1 ton of lime may be required to raise soil pH from 4.5 to 6.5, but 4 tons may be needed on a clay loam. More lime is needed on soils high in clay because larger amounts of adsorbed hydrogen must be replaced.

Type of clay also modifies the lime requirement. Other factors being equal, the lime requirement of the principal clay minerals are, from high to low, vermiculite, montmorillonite, illite, kaolinite, and the sesquioxide clays.

Organic matter content has a direct relationship to lime requirement—the more the organic matter (humus), the more the lime needed to raise the soil pH. In general, for each 1% increase in soil organic matter the lime requirement should be increased by about 10%. However, the more the organic matter in soil, the lower the ideal pH for plant growth (see Section 8.3).

Toxic aluminum is so prevalent in Oxisols and Ultisols that *exchangeable aluminum* is being used by many soil scientists as a better criterion for determining lime requirement than soil pH or exchangeable hydrogen. Rates of lime equal to 1.5 to 3.0 times the exchangeable aluminum often give the best results on Oxisols in the tropics.[6,7]

8.8 LIMING MATERIALS

Lime used on acid soils to increase their productivity is mostly ground limestones. Limestones consist of impure calcic limestone ($CaCO_3$) and sometimes impure dolomitic limestone [$CaMg(CO_3)_2$].

The calcium carbonate equivalent of the most common materials used as agricultural lime and their usual chemical composition and purity are exhibited as Table 8–8. The standard for comparison on Line 1 is calcic limestone, 100% purity. No such product exists in nature except crystalline calcite, an uncommon mineral. Other parameters specified are close to the chemical composition of common sources of liming materials found in nature or are variable. New sources of lime should always be tested before they are declared acceptable for use. When using an established source of lime, ask to see the chemical and physical analyses.

To be effective quickly, lime must be finely ground — the finer the grind, the more rapidly lime reacts with the soil. However, a very finely ground product costs more and is easily blown away by the wind in the process of spreading. In addition, a fine product leaches more readily and will not be as effective for as many years as a coarser lime. (However, see *Note 8–1(a)*, p. 419).

All 30 states with laws regulating the sale of agricultural lime specify a minimum chemical analysis and fineness based on sieve analyses. For example, the fineness specification for Georgia is fairly typical of that for most states:

90% through a 2-mm (10-mesh) sieve[8]

25% through a 0.15-mm (100-mesh) sieve

In scientific language, this is written: 90%–2mm + 25%–0.15mm (see Table 8–9 for lime conversion factors).

8.9 APPLYING LIME

Because it is more economical, almost all lime is applied by private contractors from a V-shaped truck bed with a spinner-type propeller in the back. However, this is not the most accurate implement for spreading lime. Inac-

---

[6] Robert W. Pearson, "Soil Acidity and Liming in the Humid Tropics," Cornell International Agriculture Bulletin 30, 1975, Ithaca, New York, pp. 52, 55.

[7] F. L. Long, G. W. Longdale, and D. L. Myhre, "Response of an Al Tolerant and an Al Sensitive Genotype to Lime, P, and K on Three Atlantic Coast Flatwoods Soils," *Agron. J.*, 65:30, 1973.

[8] The 10-mesh (2-mm) sieve has 10 openings per linear inch (100 openings per square inch) and the 100-mesh (0.15 mm) sieve has 100 openings per linear inch (10,000 openings per square inch).

Table 8–8. Approximate Relative Value in Neutralizing Acidity (Calcium Carbonate Equivalent) of Liming Materials

Liming Material	Parameter	Calcium Carbonate Equivalent
Limestone, calcic	CaCO₃, 100% purity (occurs as mineral calcite)	100
Limestone, dolomitic (high magnesian lime)	65% CaCO₃ + 20% MgCO₃, 87% purity[a]	89
Ashes, wood	Fresh, from hardwood (deciduous) species	80
Blast furnace slag	Finely ground	90
Ground shells	Finely ground, 85% purity	85
Hydrated lime	Ca(OH)₂, 85% purity	115
Marl	CaCO₃, 50% purity	50
Quick lime	CaO, 85% purity	151
Waste lime products	Extremely variable in composition Test each source before use	?

NOTE: The "Model Agricultural Liming Material Bill" proposes these minimum calcium carbonate equivalents:[b]

Material	Calcium Carbonate Equivalent (%)
Burnt lime	Not less than 140
Hydrated lime	Not less than 110
Limestone	Not less than 80
Slag	Not less than 80
Shells	Not less than 80

[a] The 30 state laws specify a calcium carbonate equivalent averaging about 85%.

[b] Abstract of State Laws and ACP Specifications for Agricultural Liming Materials," National Limestone Institute, Fairfax, Va., 3rd ed., 1977, p. 75.

curacies in quantities occur because of improper adjustments, "bridging" of lime in the truck bed, losses by "dusting" away in the wind, uneven truck speed, and overlaps and skips in spreading. Much more accurate but slower and more costly is a lime spreader that is pulled by a tractor and operates like a grain drill (Fig. 8–7).

Lime can be applied at any season that the cropping sequence permits. It can even be applied with satisfactory results on the *surface* of the soil in no-till cultivation.[9] Ideally, however, lime should be plowed under, chiseled in, or disked in about once in 5 years. Even in no-till farming, persistent weed

[9] R. L. Blevins, and L. W. Murdock, "Effect of Lime on No-Tillage Corn Yields," *Agronomy Notes*, 12(1), 1979, Univ. of Kentucky.

Table 8–9. Lime Conversion Factors [a]

To Convert from Column A	To Column B	Multiply Column A by
Calcium (Ca)	Calcium oxide (CaO)	1.40
Calcium (Ca)	Calcium carbonate (CaCO₃)	2.50
Calcium (Ca)	Magnesium (Mg)	0.61
Calcium (Ca)	Magnesium oxide (MgO)	1.01
Calcium (Ca)	Magnesium carbonate (MgCO₃)	2.10
Calcium oxide (CaO)	Calcium (Ca)	0.71
Calcium oxide (CaO)	Calcium carbonate (CaCO₃)	1.78
Calcium oxide (CaO)	Magnesium (Mg)	0.43
Calcium oxide (CaO)	Magnesium oxide (MgO)	0.72
Calcium oxide (CaO)	Magnesium carbonate (MgCO₃)	1.50
Calcium carbonate (CaCO₃)	Calcium (Ca)	0.40
Calcium carbonate (CaCO₃)	Calcium oxide (CaO)	0.56
Calcium carbonate (CaCO₃)	Magnesium (Mg)	0.24
Calcium carbonate (CaCO₃)	Magnesium oxide (MgO)	0.40
Calcium carbonate (CaCO₃)	Magnesium carbonate (MgCO₃)	0.84
Magnesium (Mg)	Magnesium oxide (MgO)	1.66
Magnesium (Mg)	Magnesium carbonate (MgCO₃)	3.47
Magnesium (Mg)	Calcium (Ca)	1.65
Magnesium (Mg)	Calcium oxide (CaO)	2.31
Magnesium (Mg)	Calcium carbonate (CaCO₃)	4.12
Magnesium oxide (MgO)	Magnesium (Mg)	0.60
Magnesium oxide (MgO)	Magnesium carbonate (MgCO₃)	2.09
Magnesium oxide (MgO)	Calcium (Ca)	0.99
Magnesium oxide (MgO)	Calcium oxide (CaO)	1.39
Magnesium oxide (MgO)	Calcium carbonate (CaCO₃)	2.48
Magnesium carbonate (MgCO₃)	Magnesium (Mg)	0.29
Magnesium carbonate (MgCO₃)	Magnesium oxide (MgO)	0.48
Magnesium carbonate (MgCO₃)	Calcium (Ca)	0.48
Magnesium carbonate (MgCO₃)	Calcium oxide (CaO)	0.67
Magnesium carbonate (MgCO₃)	Calcium carbonate (CaCO₃)	1.19

[a] Calculated, using the following atomic weights: calcium, 40.08; magnesium, 24.30; oxygen, 16.00; carbon, 12.01; and hydrogen, 1.00. For example, in Line 1 above, to convert from Ca to CaO equivalent, divide the molecular weight of CaO by the atomic weight of Ca $= \frac{56.08}{40.08} = 1.40$. Therefore, to convert Ca to CaO, multiply the kilograms (lb) of Ca by 1.40 to obtain the equivalent kilograms (lb) of CaO.

populations usually increase and can be controlled best by cultivation — and this is the *best* time to apply lime.

Caution: Lime should not be applied on steep slopes when there is snow or the ground is frozen because of erosion hazard; neither should heavy trucks be allowed on wet clay soils because they cause soil compaction.

Although it has been traditional to apply agricultural limes as dry solids, the rapid increase in the use of *fluid fertilizers* has sparked research on *fluid lime* (Note 8–1).

Fig. 8–7. Lime is being applied accurately and evenly on a day with low wind velocity. (Courtesy National Limestone Institute, Inc.)

Fig. 8–8. An example of the fineness of lime most suitable for making fluid lime. Any coarser and the lime would not make a stable suspension; any finer and it would cost more and be more difficult to suspend without forming lumps. (Courtesy Gerald D. McCart, Virginia Polytechnic Institute and State University.)

Note 8-1. Fluid Lime[10,11,12]

The rapid increase in popularity of *fluid* fertilizers has stimulated interest in *fluid* applications of lime. To save expense, fluid lime may be mixed with fluid nitrogen fertilizer. This combination makes economic and agronomic sense. Not only will it save one trip over the field, it will also neutralize acidity created by the nitrogen fertilizer and neutralize it more uniformly and more quickly than conventional applications of dry lime. However, there is no "magic" in using fluid lime. For the same amount of calcium carbonate applied, the total amount of soil acidity neutralized is the same whether dry or fluid lime is used. Using smaller applications of fluid lime means more frequent applications. Farmers with a *1-year* lease are pleased with this newer system because of speed, economy, and convenience. Fluid lime is also ideal for soils under minimum tillage to neutralize acidity from crop residues.

Caution: Fluid lime must not be mixed with fluid phosphorus fertilizers because of the hazard of reversion to less soluble forms of phosphorus.

Usual fluid lime suspensions consist of about 50% water or nitrogen fluid, 48% lime solids, and 2% clay. The lime should be of sufficient fineness that all of it will pass a 250-micron (60-mesh) sieve and have a purity of at least 85% calcium carbonate equivalent (Fig. 8-8). Usual applications of fluid lime are about 500 kg/ha (446 lb/A). Using these parameters, the amount of lime suspension to apply to obtain 500 kg/ha of pure effective calcium carbonate equivalent (E.C.C.) would be

$$\frac{500 \text{ kg/ha E.C.C.}}{0.85 \text{ (\% purity as decimal)} \times 0.48 \text{ (\% solids)}} = 1225 \text{ kg/ha}$$

NOTE 8-1(a): If the fineness is less than "all through a 250-micron (60-mesh) sieve," more lime must be added to supply 500 kg/ha of E.C.C. Because state laws vary in prescribing the fineness parameters, no generalized example is given (Fig. 8-9). [See problem 6 for E.C.C. calculation.] NOTE 8-1(b): Expressions for quality of agricultural limestone vary from state to state as follows: E.C.C. = effective calcium carbonate; E.C.C.E. = effective calcium carbonate equivalent; E.N.P. = effective neutralizing power; E.N.V. = effective neutralizing value; E.N.M. = effective neutralizing material; N.I. = neutralizing index.[13]

[10] Edgar W. Sawyer, "Lime in Suspensions: Part 1. Stable Liquid Lime Slurries Produced With Colloidal Attapulgite Clay," *Fertilizer Solutions,* May–June, 1976.

[11] L. S. Murphy, P. D. Shoemaker, and Carl Fabry, "Lime Suspension Update: Agronomics, Economics and Common Sense," *Fertilizer Solutions,* July–Aug., 1977.

[12] J. R. Wolford and E. W. Sawyer, "Preparation and Application of Fluid Lime Suspensions for Soil Neutralization," *Fertilizer Solutions,* Nov.–Dec., 1977.

[13] Regis D. Voss, "What Constitutes an Effective Liming Material?" Proceedings of the National Conference on Agricultural Limestone, Nashville, Tennessee, Oct. 16–18, 1980.

Fig. 8–9. Finely ground fluid lime is being applied at the rate of 2.2 metric tons/ha (1 ton/A). Fluid line gives a rapid increase in pH that lasts about a year. (Courtesy Big Wheels, Inc.)

SUMMARY

Lime on strongly acid soils reacts slowly to make the major plant nutrients, nitrogen, phosphorus, and potassium, more readily available, adds calcium and sometimes magnesium, increases the availability of molybdenum, but decreases the toxicity of aluminum, iron, and manganese.

The soil pH of greatest nutrient availability lies between 6.0 and 7.0 for well-drained mineral soils in temperate regions but between pH 5.0 and 5.5 for Histosols (organic soils), Oxisols (in humid tropics), and Ultisols (in humid tropics and subtropics).

Not only do soils vary in pH, but plant cultivars and species also are variable in their tolerance to soil acidity. Plant breeders are using this diversification to develop crops that produce optimum yields on soils with specific pH's.

Contributing to soil acidity are acid parent materials of soils, plant root exudations, nutrient removal in harvested crops, leaching, erosion, acid precipitation and acid-forming fertilizers. Of these causes of soil acidity, erosion ranks first, followed by acid-forming fertilizers, crop removal, and acid precipitation. Based upon these estimates, farmers in the 50 states are using each year only 15% of the proposed optimum of 200 million metric tons (220 million tons) of agricultural limestone.

The amount of lime to apply should always be based on a soil test. The soil test is then interpreted into metric tons per hectare (tons per acre) based on the degree of acidity, plant requirements, and soil taxonomy inferred from soil map units. In the absence of a soil map unit, these parameters should be known: soil textural class, type of clay, organic matter content, and presence of toxic aluminum.

The principal liming material is ground limestone (high calcic or high magnesic). Secondary sources include hardwood ashes, blast furnace slag, ground shells, hydrated lime, marl, quick lime, and waste lime products. A satisfactory chemical guarantee is "85% calcium carbonate equivalent." A common physical guarantee is "90% through a 2-mm (10-mesh) sieve *and* 25% through a 0.15-mm (100-mesh) sieve."

Lime can be applied almost anytime that a spreader can operate in the field. Usually one application in each 5-year rotation is satisfactory in the temperate region; in the tropics, smaller amounts about every 2 years are satisfactory. A newer technique includes annual applications of fluid lime with fluid nitrogen or nitrogen-potassium fertilizers.

Plant response to lime in the temperate region is usually more pronounced with alfalfa, annual lespedezas, barley, cotton, soybeans, and wheat. Least responsive is corn, grain sorghum, and tobacco. In the tropics, most crops respond to lime when the soil pH is below 5.0.

PROBLEMS

1. If the need for lime is as critical as the authors have explained in this chapter, why don't farmers use as much as is necessary to obtain maximum plant nutrient efficiency?

2. Why is the pH of greatest nutrient availability between 5.0 and 5.5 for Histosols, Oxisols, and Ultisols and between 6.0 and 7.0 for other soil orders?

3. Plant breeders can develop plants that are either tolerant or resistant to nearly all plant physiological conditions existing in strongly acid soils. Is this a certain means of eliminating the need for lime and fertilizers?

4. Rank the causes of soil acidity, from highest to lowest in importance. Now discuss in a practical way how to reduce soil acidity and increase soil productivity.

5. Convert 5 metric tons of dolomitic limestone (65% pure $CaCO_3$ + 20% pure $MgCO_3$) to $CaCO_3$ equivalent, 85% purity (see Table 8–9).

6. A certain agricultural limestone has been analyzed and found to contain 75% $CaCO_3$ equivalent. That same sample was also subjected to a fineness analysis (sieve analysis) and the following data were recorded:

    5% on an 8-mesh (3.17-mm) sieve, 0% effective
    33% on a 60-mesh (0.42-mm) sieve, 50% effective
    62% through a 60-mesh (250-micron) sieve, 100% effective

    What is the percent E.C.C. (effective calcium carbonate equivalent) in that agricultural lime sample?

    Solution: E.C.C. = fineness factor times purity

Fineness Factor:

5% of lime *on* an 8-mesh sieve × 0% effectiveness = 0.0% effective
33% of lime *on* a 60-mesh sieve × 50% (0.5) effectiveness = 16.5% effective
62% of lime *through* a 60-mesh sieve × 100% (1) effectiveness = 62.0% effective
Total fineness factor is sum of these three classes = 78.5% effective

Purity: 75% $CaCO_3$ equivalent (by laboratory analysis)

% E.C.C. (effective calcium carbonate) = 78.5% × 0.75 = **58.88%**

7. Laboratory analyses indicate a need for 3000 kg/ha E.C.C. for a particular soil. If the agricultural lime available has a 70% E.C.C. content, how many metric tons (1000 kg) of this limestone will have to be applied per hectare?

Solution:

3000 kg/ha E.C.C. needed
Ag lime available 70% E.C.C.
Amount of ag lime per hectare to give

$$3000 \text{ kg E.C.C.} = \frac{3000 \text{ kg E.C.C.}}{0.7 \text{ E.C.C.}} = 4286 \text{ kg/ha}$$

REFERENCES

"Abstract of State Laws and ACP Specifications for Agricultural Liming Materials," 3rd ed., National Limestone Institute, Fairfax, Va., 1977.

Boyd, Claude E., "Liming Farm Fish Ponds," Auburn University Leaflet 91, 1976, 7 pp. Auburn, Alabama.

Braekke, Finn H. (Ed.), "Impact of Acid Precipitation on Forest and Freshwater Ecosystems in Norway," 1976. Oslo, Norway. Agricultural Council of Norway and the Norwegian Council for Scientific and Industrial Research.

Chiu, S. Y., J. W. Nebgen, A. Aleti, and A. D. McElroy, "Methods for Identifying and Evaluating the Nature and Extent of Nonpoint Sources of Pollutants," EPA–430/9–73–014 and Midwest Research Institute, 1973.

"Control of Water Pollution from Cropland," Vol. I. *A Manual for Guideline Development,* USDA-Agricultural Research Service and U.S. Environmental Protection Agency, Washington D.C., 1975.

"Fertilizer Situation, 1979," USDA-Economics, Statistics, and Cooperative Service, FS-9, 1978.

Follett, R. H., and R. F. Follett, "The Basics of Soil Acidity and Liming," National Conference on Agricultural Limestone, Nashville, Tennessee, Oct. 16–18, 1980. Sponsored by the National Crushed Stone Association, U.S. Dept. of Agriculture, Tennessee Valley Authority and the Southern Aglime Committee.

A Handbook for the Ag Lime Salesman, National Limestone Institute, Inc. Fairfax, Va., 1973.

Hoyt, P. B., M. Nyborg, and D. C. Penney, "Farming Acid Soils in Alberta and Northeastern British Columbia," Canada Dept. of Agriculture, Ottawa, Canada, 1974.

"Lime and Fertilizer Recommendations, 1977," University of Kentucky, Lexington, 1976.

"Limestone Purifies Water," National Limestone Institute, Inc., Fairfax, Va., 1977.

McCart, G. D., G. W. Hawkins, G. R. Epperson, and D. C. Martens, "Soil Acidity: Its Nature, Causes, Effects, and Remedy," Virginia Polytechnic Institute and State University, Blacksburg, 1972.

Nesmith, J., and E. W. McElwee, "Soil Reaction (pH) for Flowers, Shrubs, and Lawn," Institute of Food and Agricultural Sciences, University of Florida, Circular 352A, 1974.

Olson, O. E., R. L. Emerick, and L. Lubinus, "Nitrates in Livestock Waters," FS-603, South Dakota State University and U.S. Dept. of Agriculture, 1973.

Pearson, Robert W., "Soil Acidity and Liming in the Humid Tropics," Cornell University, Ithaca, N.Y., 1975.

"Proceedings of the First International Symposium on Acid Precipitation and the Forest Ecosystem," USDA-Forest Service General Technical Report NE-23, Upper Darby, Pa., 1976.

Sanchez, Pedro A. (Ed.), "A Review of Soils Research in Tropical Latin America," North Carolina Agr. Exp. Sta. in Cooperation with the U.S. Agency for International Development. 1973.

"Soil Acidity and Liming," American Society of Agronomy Monograph 12, Madison, Wis., 1967.

"Soils for Management of Organic Wastes and Waste Waters," Soil Science Society of America, Madison, Wis., 1977.

Swoboda, Allen, R., "The Control of Nitrate As a Water Pollutant," EPA-600/2-77-158, Aug., 1977. Robert S. Kerr Environmental Research Laboratory, U.S. Environmental Protection Agency, Ada, Okla.

Wallingford, G. W., L. S. Murphy, W. L. Powers, and H. L. Manges, "Denitrification in Soil Treated With Beef-Feedlot Manure." In *Communications in Soil Science and Plant Analysis,* 6(2) (1975) 147–161.

Whitney, David A., and R. Hunter Follett, "Water Quality and Fertilizer Usage," Fact Sheet 6, "Water Quality Through 208 Planning," Coop. Ext. Serv., Kansas State University, Manhattan, Kan., 1978.

CHAPTER NINE

Reclamation and Management of Saline and Sodic Soils

9.1 INTRODUCTION

Saline and sodic soils reduce the value and productivity of a considerable area of land throughout the world. It has been estimated that salt and sodium affect over 20,000,000 ha (49,420,000 A) on a worldwide basis. More than one-fourth of the irrigated farmland in these United States is affected to some extent by soil salinity. In Canada, salt-affected soils occur on about one-tenth of the irrigated land. In general, soil salinity and its related problems occur in the arid and semiarid climates and in coastal regions in all climates where ocean tides salinize soils and groundwaters. Rainfall leaches salts out of soils in humid regions, making salt problems important but rare and often transitory in such areas.

The ions that contribute to soil salinity include Cl^-, SO_4^{2-}, HCO_3^-, Na^+, Ca^{2+}, Mg^{2+}, and (rarely) NO_3^- and K^+. The salts of these ions occur in highly variable concentrations and proportions. They may be indigenous from geologic formations, but more commonly they are brought into an area in the irrigation water or in waters draining from adjacent areas. Natural surface and subsurface drainage are often so poorly developed in arid regions that salts collect in inland basins rather than being discharged into the sea.

The term *alkali* is often used to refer to soils that are light in color with snow white surface crusts (Fig. 9–1). The term *alkali* is also used to imply that the affected soil is high in exchangeable sodium. Soils affected by salts have been given descriptive names like white alkali, black alkali, gumbo, or slick

Fig. 9–1. Toxic salt accumulation (white layer on ridge at arrow 1) has prevented the growth of all plants on this irrigated cotton field in Texas. The accumulation of excess salt could have been caused by a slight surface depression or a concentration of clay in the soil profile that reduced the infiltration of irrigation water, as seen in the insert (at arrow 2). (Courtesy Tex. Agr. Exp. Sta., El Paso.)

spots. These names come from the land surface appearances as soils become salt contaminated. If the soil pH is high enough, dispersed organic colloids color the surface water black, so that they appear like puddles of oil. Upon drying, the soil has black crusts over its surface. These terms, as commonly used, usually include several kinds of salt-affected soils that not only differ in composition and soil properties but also differ in use suitability, productivity, ease of reclamation, and management needs. Arbitrary limits of salt content and exchangeable sodium percentage are established in soil science to define *saline, sodic,* and *saline-sodic* soils, which are terms now used for salted soils.

9.2 CHARACTERIZATION OF SALINE AND SODIC SOILS

Salt-affected soils can be divided into three groups, depending on the kinds and amounts of the various salts present: saline, sodic, and saline-sodic. The classification depends on the total soluble salts (as measured by conductivity), soil pH, and the exchangeable sodium percentage, as summarized in Table 9–1. It is important to understand these differences because they determine, to a great extent, how these soils should be reclaimed and managed.

Table 9-1. Summary of Salt-Affected Soil Classification

Classification	Conductivity (mmhos/cm)	Soil pH	Exchangeable Sodium Percentage	Soil Physical Condition
Saline	> 4.0	< 8.5	< 15	Normal
Sodic	< 4.0	> 8.5	> 15	Poor
Saline-sodic	> 4.0	< 8.5	> 15	Normal

9.2.1 Saline Soils

Most soils contain some water-soluble salts. Soils containing water-soluble salts in amounts sufficient to be harmful to seed germination and plant nutrition and growth are called *saline soils*. Saline soils are the most common type and are usually the easiest to reclaim. Their structure is generally good, and their permeability to water and tillage characteristics are like those of nonsaline soils. Saline soils are recognized by spotty growth of crops (Fig. 9-2) and when dry often by white crusts of salt on the soil surface (Fig. 9-3). The salts dissolved in the soil moisture move to the surface where they are left as a crust when the water evaporates. These salts are mostly sulfates and chlorides of calcium, magnesium, and/or sodium.

Soils are classified as *saline* if the solution extracted from a saturated soil paste (soil-saturation extract) has an electrical conductivity value of 4 or more millimhos per centimeter (mmhos/cm). The ability of a soil solution to carry a current is called *electrical conductivity* and is usually measured in mmhos/cm. The lower the salt content of the soil, the lower the mmhos/cm rating and the less effect on plant growth. Ordinarily the pH of such soils is below 8.5. Saline soils are generally flocculated (friable), and their permeability to water is similar to that of nonsaline soils of similar texture.

Fig. 9-2. Excess salt has killed large patches of wheat on this saline soil in Kansas. (Courtesy R. H. Follett.)

Reclamation and Management of Saline and Sodic Soils

Fig. 9–3. This farm is suffering serious loss of production because of large saline spots in the field. Whiteness comes from salt carried to the surface through moisture evaporation (Kansas). (Courtesy R. H. Follett.)

9.2.2 Sodic Soils

Sodic soils are relatively low in soluble salts but are high in exchangeable sodium. These soils tend to remain in a dispersed condition, almost impermeable to both rain and irrigation water. They are of poor tilth — plastic and sticky when wet and prone to form hard clods and crusts upon drying. When wet they have a characteristic smooth, slick look caused by the dispersed condition of clay and humus. These areas are commonly referred to as *slick spots* because, when the soil is plowed slightly wet, it turns over in slick, rubbery furrow slices.

These sodic soils are very poor for the growth of most plants. The exchangeable (adsorbed) sodium exceeds 15% of the cation exchange capacity. The high sodium causes the soils to be dispersed and strongly alkaline (soil pH is more than 8.5). Reclamation is necessary before crops can be grown, and final recovery to normal, friable, well-aerated conditions is often a slow process, requiring several years and much expense. Sodic conditions tend to develop in low areas of fine-textured soils periodically subjected to flooding with high sodium waters (Fig. 9–4). This condition can also develop from continued application of waters high in sodium and low in calcium and magnesium.

9.2.3 Saline-Sodic Soils

Saline-sodic soils contain large amounts of total salts as well as more than 15% of exchangeable sodium. The pH is less than 8.5. As long as an excess of all salts is present, the physical properties of these soils are good and similar to those of saline and nonsaline soils.

When rains occur or after irrigation with low-salt (good-quality) water, most of the soluble calcium and magnesium are leached out of the surface soil

Fig. 9-4. This is a good example of a soil that has developed under sodic conditions. Note the columnar structure, with columns that have rounded tops. Subsoil is very slowly permeable and apparently dispersed (Texas). (Courtesy USDA-SCS.)

and the sodium remains attached to the clay and humus. The pH rises above 8.5, the soil becomes dispersed, and permeability to water virtually ceases. Rainwater often stands on these soils for many days until it evaporates. The danger of excess sodium often goes unrecognized until these events occur. Such soils require amendments and leaching to remove the excess sodium.

9.3 EFFECT OF SALTS ON PLANT GROWTH

Plants differ widely in their ability to tolerate salts in the soils. Salt tolerance ratings are based on the yield reduction on salt-affected soils when compared with yields on similar soils not affected by salt. The significance of the soil salinity class is best understood in its relation to its effect on plant growth (Table 9-2).

Plants absorb water and dissolved nutrients from the soil through the root hairs partly by the physical process of osmosis. Water can move from the soil into the root hairs only as long as the osmotic pressure within cells of the root hairs is greater than that of the soil water. Salts, sugars, nutrients, and other soluble materials contribute to the osmotic pressure or solute content of the cell sap, not only in the roots but in all plant tissue. Any increase in the soluble salt content of the soil raises the osmotic pressure of the soil solution. As a result, less water flows from the soil into the plant, thus reducing the amount of soil water available to the plant. Plants growing in three soils — one nonsaline, one moderately saline, and one highly saline — are depicted in Fig. 9-5. In the nonsaline soil, about half of the total soil water could be used before the plant wilts. At moderate and high salinity, smaller and smaller

Table 9-2. Effect of Various Degrees of Soil Salinity on Crop Plants

Soil Salinity Class	Conductivity (mmhos/cm)	Influence on Crop Plants
Nonsaline	0–2	Salinity effects negligible
Slightly saline	2–4	Very sensitive plants restricted
Moderately saline	4–8	Many crops restricted
Strongly saline	8–16	Only tolerant plants yield satisfactorily
Very strongly saline	> 16	Only a few very tolerant plants grow

amounts of soil water are available to plants; therefore, plants would wilt sooner.

9.3.1 Selecting Plants Tolerant of Salinity

Proper selection of crops is an essential part of successful management of salt-affected soils. Some species of plants are very sensitive to salts, many have moderate tolerance, and a few types seem to thrive on saline soils. The mechanisms involved in salt tolerance appear to be related to either one or both of two conditions: (1) the ability of the plant to restrict the entrance of certain salt ions into its roots or (2) the ability to tolerate or adjust to salts after they are taken into the plant.

The salt tolerance of certain crops varies with stage of development (Table 9-3). For example, the salt tolerance of alfalfa is poor during the germination stage but is good after the crop is established. With rye and corn the reverse is true.

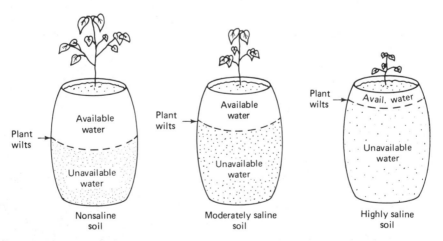

Fig. 9-5. Diagram showing that as the salt content of the soil increases, the availability of water to plants decreases. In a nonsaline soil about one-half of the water is available, whereas in a highly saline soil perhaps only one-tenth is available. (Courtesy USDA-SEA-AR.)

Table 9–3. Tolerance of Crops to Salts at Two Stages of Growth

Crop	Germination Stage	Established Stage
Barley	Very good	Good
Rye	Good	Poor
Corn	Good	Poor
Wheat	Fairly good	Fair
Alfalfa	Poor	Good
Sugar beets	Very poor	Good
Beans	Very poor	Very poor

SOURCE: R. A. Milne and E. Rapp, "Soil Salinity and Drainage Problems—Causes, Effects and Management," Canada Department of Agriculture, Research Station, Lethbridge, Alberta, Publication 1314 (1968).

Soluble salts and nutrients enter the plant with water as it moves through the living tissue of root hairs. The living tissue of root hairs of different plant species has varying degrees of capacity to restrict the entrance of certain ions — the reasons are not yet fully understood. As examples, bermudagrasses and wheatgrasses grow well on saline soils; yet they do not absorb harmful amounts of soluble salts. Their mechanism of salt tolerance could be called *restricted absorption*. Other grasses such as rhodesgrass and alkali sacaton can absorb large amounts of salt and yet still grow well on saline soils. This mechanism could be called *salt tolerance*.

If salinity is extremely high for a crop, a progressive burning or firing (plasmolysis) of the older leaves may result in eventual death of the plants. The effects of salt appear similar to those of drought. However, salinity and drought differ in one important respect; drought-affected plants develop characteristic symptoms such as a wilted appearance; whereas salt-affected plants do not demonstrate a characteristic wilted appearance. A plant that is highly resistant to drought is usually salt-tolerant.

Some crops are affected more by salinity than other crops (Fig. 9–6). Of the field crops, barley is known to have the greatest tolerance. Sugar beets are fairly tolerant once the plants are well established. Rye, wheat, oats, flax, and lettuce are moderately tolerant provided that fertility levels are suitable and adequate moisture is available during germination. Beans and peas are very sensitive to soil salinity.

Forage crops tend to tolerate salt better than most cereal crops. Because most forage crops are perennials, they shade the soil longer than cereals do. The shade reduces evaporation from the soil surface; therefore less salt rises with capillary water to the surface. Furthermore the fibrous roots of forage crops improve the physical condition of the soil because they add organic matter to it; thus, increased permeability aids water infiltration to keep salts leached below the rhizosphere.

Fruit trees are more sensitive to salinity than most field crops. Many

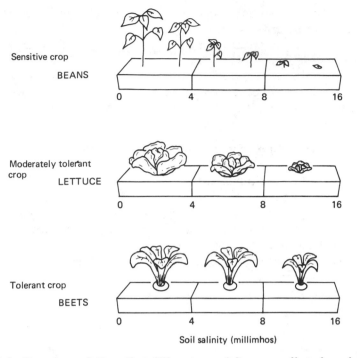

Sensitive crop
BEANS

0   4   8   16

Moderately tolerant
crop
LETTUCE

0   4   8   16

Tolerant crop
BEETS

0   4   8   16

Soil salinity (millimhos)

Fig. 9-6. How does salinity affect different crops? Some are affected much more than others. (Courtesy USDA-SEA-AR.)

trees are sensitive to salts, especially seedlings and transplants of evergreen and deciduous trees. Because trees are extremely sensitive, they should be planted only in areas that are comparatively salt-free.

The relative tolerance of various plant species are rated into what are called *salt tolerance groups:* that is, sensitive, moderately tolerant, tolerant, and highly tolerant. These tolerance ratings are rather general because other factors such as management practices, quality of irrigation water, environmental conditions, crop variety, and type of rootstocks used in propagation of fruit trees can affect tolerance. The tolerance ratings (Table 9–4) should be used only as a general guide. In this respect they can be quite useful.

It is important to select grasses that are salt tolerant for roadside use, particularly where salt is used for ice removal. For example, about 6.8 metric tons/km (12 tons/mi) of sodium chloride has been used on a four-lane highway per year for ice removal on some sections of Interstate 80 in Iowa. Although salt tolerance of mature grass is important, seedling development is critical in areas where new highways are being built. Highways may be open for traffic and pavements salted before roadsides are seeded. Grasses that have been used effectively on roadsides are Kentucky 31 (tall) fescue grass, western wheatgrass, slender wheatgrass, intermediate wheatgrass, Russian wildrye, and Reed canarygrass. Of these grasses, Kentucky 31 fescue grass and western wheatgrass are the most salt tolerant. The salinity tolerance of various turf-grass species is given in Table 9–5.

Precise judgments of plant tolerance to salinity are not simple to obtain because some plants are more affected by salinity at one stage of development than another. Even highly tolerant plants may be acutely affected at some particular stage of development or stress. In general, well-established plants are more tolerant to salinity than new transplants.

9.3.2 Selecting Plants Tolerant of Sodium

The nutritional problems in sodic soils are usually related to the relative amounts of calcium and sodium accumulated by the plants. The tolerance of crops is not so closely related to the absolute amount of exchangeable sodium (ES) in the soil as to the exchangeable sodium percentage (ESP) (the percentage of the total exchangeable cations that are sodium ions). Some crops are much more tolerant of a soil low in calcium or high in sodium than other crops.

At a given ESP, the physical condition of a soil depends on the soil tex-

Table 9–4. General Salt Tolerance Ratings of Various Crops

Sensitive (0–4 mmhos/cm)	Moderately Tolerant (4–6 mmhos/cm)	Tolerant (6–8 mmhos/cm)	Highly Tolerant (8–12 mmhos/cm)
		Field Crops	
Field bean	Soybean	Wheat (grain)	Barley (grain)
	Castorbean	Oats (grain)	Rye (grain)
	Sesbania	Safflower	Sugar beet
	Rice	Cotton	
	Flax	Sunflower	
	Guar	Triticale	
	Sorghum (grain)		
	Corn (field)		
		Forage Crops	
White clover	Reed canarygrass	Hardinggrass	Bermudagrass
Dutch clover	Oats (hay)	Kleingrass	Crested wheatgrass
Alsike clover	Orchardgrass	Alfalfa	Barley (hay)
Red clover	Bromegrass	Birdsfoot trefoil	Rye (hay)
Ladino clover	Big trefoil	Hubam clover	Panicgrass
Crimson clover	Gramagrasses	Dallisgrass	Alkali sacaton
Meadow foxtail	Sour clover	Tall fescue grass	Rhodesgrass
	Milkvetch	White sweetclover	Saltgrass
	Timothy	Yellow sweetclover	
	Sudan-sorghum hybrids	Perennial ryegrass	
	Sorghum (forage)	Wheat (hay)	
	Corn (forage)	Johnsongrass	

Table 9–4. (cont.)

Sensitive (0-4 mmhos/cm)	Moderately Tolerant (4-6 mmhos/cm)	Tolerant (6-8 mmhos/cm)	Highly Tolerant (8-12 mmhos/cm)
		Vegetable Crops	
Carrot	Lettuce	Tomato	Asparagus
English pea	Corn (sweet)	Beet	
Radish	Potato	Kale	
Celery	Squash	Spinach	
Green bean	Onion	Broccoli	
Lima bean	Sweet potato	Cabbage	
Kidney bean		Cauliflower	
Cucumber	Bell pepper	Watermelon	
Rhubarb	Blackeyed pea		
	Muskmelon		
		Fruit, Nut, and Vine Crops[a]	
Grapefruit	Pecan	Pomegranate	Date palm
Orange	Peach	Fig	
Lemon	Apricot	Olive	
Avocado	Grape		
Pear	Quince		
Apple			
Cherry			
Plum			
Walnut			
Blackberry			
Raspberry			
Strawberry			
Boysenberry			
		Ornamental Shrubs	
Viburnum	Spreading juniper	Oleander	Purple sage
	Arborvitae	Bottlebrush	Saltcedar
	Lantana		
	Pyracantha		
	Privet		
	Japonica		

[a] Ratings may vary somewhat depending on the particular rootstock used for propagation.

PRIMARY SOURCE: Publications of the U.S. Salinity Laboratory, Riverside, Calif.

ture. For example, the structure of a high-sodium, fine-textured (clay) soil is usually worse than that of a high-sodium, coarse-textured (sandy) soil. This difference in structure is related to the greater number of colloids in a clay soil that are dispersed.

The retarded growth of sensitive and even moderately tolerant crops is due in part to the nutritional problems associated with sodic soils and in part

Table 9–5. Salinity Tolerance of Selected Turf Grasses

Soil Salinity Class	Conductivity (mmhos/cm)	Adapted Species
Nonsaline	0–2	All turfgrass species
Very slightly saline	2–4	Ky. bluegrass, red fescue grass, colonial bentgrass
Moderately saline	4–8	Alta fescue grass, perennial ryegrass
Strongly saline	8–16	Zoysiagrass, Saint Augustinegrass, bermudagrass, seaside bentgrass
Very strongly saline	> 16	Saltgrass, switchgrass, alkaligrass

SOURCE: R. H. Follett, "Salinity Problems," Proceedings of the 42nd Annual International Turfgrass Conference, 1971, pp. 21–24.

to the adverse structural characteristics of such soils. The most tolerant crops are apparently not affected nutritionally at moderately high ESP values, but growth is retarded as a result of poor physical condition of the soil. Therefore, the reduced growth of crops grown on sodic soils may be due to adverse nutritional factors, adverse physical conditions, or a combination of both. The crops listed in Table 9–6 are arranged approximately in order of increasing tolerance to exchangeable sodium.

Toxicity of sodium soils can occur with almost any crop if concentrations are high enough. Sodium toxicity often accompanies and is complicated by soil salinity or reduced permeability. Plant symptoms of sodium toxicity occur first on the oldest leaves because a period of time (days or weeks) is normally required before plant accumulations reach toxic concentrations. Symptoms usually appear as a burn or drying of tissue at the outer edges of the leaf, progressing inward between veins toward the leaf center as severity increases.

Sodium toxicity can be confirmed by chemical analysis of the leaf tissue, comparing sodium content of damaged leaf blades with that of normal leaf blades of the same species nearby.

Sodium in leaf tissue in excess of 0.25 to 0.50% (dry weight basis) is typical of sodium toxicity for many tree crops. A combination of soil analysis, water analysis, and plant tissue analysis will greatly improve the chances of a correct diagnosis.

9.4 RECLAMATION OF SALT-AFFECTED SOILS

The reclamation of any soil first requires an assessment of the situation: how bad it is, the causative factors, how it can best be reclaimed, and whether it is worth the expense of reclaiming. The first two are relatively simple to determine by analysis of soil and water samples, by a careful examination of the soil profile, and by establishing the depth to the water table. Some salt-affected soils cannot be economically reclaimed because of fine soil texture to

Table 9-6. Tolerance of Selected Crops to Exchangeable Sodium Percentage (ESP)

Crop	Tolerance to ESP and Range at Which Affected	Growth Response Under Field Conditions
Decidous fruits Nuts Citrus Avocado	Extremely sensitive (ESP = 2–10)	Sodium toxicity symptoms even at low ESP value
Beans	Sensitive (ESP = 10–20)	Stunted growth at low ESP values even though the physical condition of soil is good
Clovers Oats Tall fescue grass Rice Dallisgrass	Moderately tolerant (ESP = 20–40)	Stunted growth due to both nutritional factors and adverse soil physical conditions
Wheat Cotton Alfalfa Barley Tomato Beets	Tolerant (ESP = 40–60)	Stunted growth usually due to adverse soil physical conditions
Crested wheatgrass Tall wheatgrass Rhodesgrass	Most tolerant (ESP = more than 60)	Stunted growth usually due to adverse soil physical conditions

SOURCE: George A. Pearson, "Tolerance of Crops to Exchangeable Sodium," USDA Agric. Information Bul. No. 216, 1960.

an excessive depth (60 cm or more), the lack of good-quality irrigation water, or the impossibility of establishing adequate drainage.

The analyses should determine the kinds and amounts of salts present and whether the soil contains gypsum and/or free carbonates. The test will establish whether the soil is saline, sodic, or saline-sodic in nature (refer to Table 9-1). The profile examination should determine the soil permeability characteristics and the relative difficulty of leaching the salts.

If it has been determined that a soil can be economically reclaimed, the next step is to examine and, if feasible, establish drainage facilities for carrying away the leached salts. If sand, silt, or gravel layers are encountered at depths of less than 3 to 4 m (10 to 13 ft), the most common practice is to dig large, deep parallel open drains through the area to be reclaimed, with some means of disposing of the drainage water at the lower end of the irrigated area. Distance between drains and their width and depth should be determined by an irrigation specialist. It is important to eliminate, if possible, any

Reclamation and Management of Saline and Sodic Soils

impermeable or restrictive zones above this 3- to 4-m depth before leaching is attempted. Drainage is an expensive operation requiring technical knowledge and should be performed only after consultation with agricultural irrigation/drainage experts.

9.4.1 Reclaiming Saline Soils

Saline soils are relatively easy to reclaim for crop production if adequate amounts of low-salt irrigation water are available and if internal and surface drainage are feasible. Saline soils cannot be reclaimed by any chemical amendment, conditioner, or fertilizer. Reclamation of these soils consists simply of applying enough high-quality water to leach the soil thoroughly (Fig. 9–7). The water applied should be low in sodium but can be fairly saline (1500 to 2000 ppm total salt), as this helps in keeping the soil permeable during the leaching process. Preferably the water should be applied in several applications, allowing time for the soil to leach and drain well after each application. After reclamation, only good-quality water should be used for irrigation.

The quantity of water required to remove salts from the soil rhizosphere

Fig. 9–7. Continuous leaching by ponding has been developed through research to benefit farmers with soil salinity in the Imperial Valley, California. After leaching is completed, the land must be leveled again before irrigation can continue. (Courtesy USDA-SEA-AR.)

depends on several factors, such as the initial salt level in the soil, the final salt level desired, the quality of the irrigation water, how deep in the soil profile are the salts to be leached, and how the leaching is done (ponding constantly or intermittent flooding). A general guide is that with ponded water, about 30 cm (12 in.) of water are required to remove 70 to 80% of the salt for each 30 cm (12 in.) depth of soil to be leached of salt.[1] Intermittent water additions are more efficient and reduce the water quantities needed to about 70% of that needed with continuous ponding or leaching methods.

9.4.2 Reclaiming Sodic and Saline-Sodic Soils

The reclamation of sodic soils is a different matter than the reclamation of saline soils. In sodic soils, the exchangeable sodium may be so great that the resulting dispersed soil is almost impervious to water. The sodium must first be replaced by another cation and then leached downward and out of reach of plant roots. Sodium soils are treated by replacing the adsorbed sodium with a soluble source of calcium. Native gypsum, calcium in irrigation water, or commercial amendments can supply the calcium. Adequate drainage must also be present.

The purpose of an amendment is to provide soluble calcium to replace exchangeable sodium. The fundamental reaction may be illustrated as follows:

$$\text{Na-clay} + Ca^{2+} \text{ (in solution)} \longrightarrow \text{Ca-clay} + Na^+ \text{ (in solution)}$$

The Ca-clay is the normal and desirable state. Calcium solubilized from gypsum replaces sodium, and sodium is leached out. Of all calcium compounds, calcium sulfate (gypsum, $CaSO_4 \cdot 2H_2O$) is considered the cheapest and most soluble calcium source for this purpose.

If the soil contains lime ($CaCO_3$), then sulfur or sulfuric acid may be added to the soil to furnish calcium indirectly. When sulfur is added to the soil, bacteria slowly oxidize the sulfur to sulfuric acid (H_2SO_4). The hydrogen ions of the sulfuric acid can replace sodium ions or the sulfuric acid may react with the soil lime to form gypsum, which then has the same effect as applied gypsum. Sulfuric acid reacts quickly with free carbonates (lime) to produce gypsum according to the following chemical equation:

$$CaCO_3 + H_2SO_4 + H_2O \longrightarrow CaSO_4 \cdot 2H_2O + CO_2$$

The carbon dioxide (CO_2) escapes immediately as a gas. One advantage of using sulfuric acid is that the gypsum formed is in very fine particles. The fine gypsum particles react more quickly to replace sodium because they are more soluble than coarse gypsum usually applied. Sulfuric acid is a very corrosive

---

[1] J. D. Rhoades, "Drainage for Salinity Control." In *Drainage for Agriculture*, (Ed.) J. V. Schilfgaarde, Number 17 in the Agronomy Series, American Society of Agronomy, Inc., Madison, Wis., 1974, pp. 433–61.

liquid and is difficult and dangerous to handle; thus, extreme care is necessary in using it.

If the sodic soils contain no source of calcium (gypsum or free carbonates), then gypsum or a soluble calcium source should be applied. If the soils contain free carbonates, then acid or acid-forming materials can be used. Examples of acid or acid-forming materials are sulfuric acid, elemental sulfur, ferric sulfate, and lime sulfur (Table 9–7). Sulfuric acid reacts immediately with free carbonates in the soil to release soluble calcium for exchange with sodium and thus is considered a rapid-acting amendment. Elemental sulfur takes several years to oxidize completely into sulfates and gypsum and is the slowest acting of all amendments. Lime sulfur contains some soluble calcium, but its chief effect is due to the oxidation of sulfur to sulfate and eventual production of gypsum, a slow process. Ferric sulfates are relatively expensive; therefore, they are not used unless they offer some advantage, such as a local low-cost source of supply.

The reclamation procedures for sodic and saline-sodic soils usually consist of a series of stages — first reclaiming the surface soil and then adding more amendment to reclaim the soil to greater depths. This improves the physical

Table 9-7. Amendments for Water and Soil and Their Relative Effectiveness in Supplying Calcium

Amendment	Tons Equivalent to 1 Ton of 100% Gypsum [a]
Gypsum ($CaSO_4 \cdot 2H_2O$)[b]	1.00
Sulfur (S)[c]	0.19
Sulfuric acid (H_2SO_4)[b]	0.61
Ferric sulfate [$Fe_2(SO_4)_3 \cdot 9H_2O$][c]	1.09
Lime sulfur (9% Ca + 24% S)[b]	0.78
Calcium chloride ($CaCl_2 \cdot 2H_2O$)[b]	0.86
Calcium nitrate [$Ca(NO_3)_2 \cdot 2H_2O$][b]	1.06

[a] These are based on 100% pure materials. If not 100%, make the following calculation to find tons (X) equivalent to 100% material:

$$X = \frac{100 \times \text{tons}}{\% \text{ purity}}$$

Example: If gypsum is 80% pure,

$$X = \frac{100 \times 1.00}{80} = 1.25 \text{ tons}$$

Answer: 1.25 tons of 80% gypsum is equivalent to 1 ton of 100% gypsum.

[b] Suitable for use as a water or soil amendment.
[c] Suitable only for soil application.

SOURCE: M. Fireman and R. L. Brauson, "Gypsum and Other Chemical Amendments for Soil Improvement," Calif. Expt. Sta. and Ext. Serv. Leaflet 149, 1965, 4 pp.

Fig. 9–8. Total soluble salts and sodium on this soil (Toy soil series) were so high that nothing would grow until 44.8 metric tons/ha (20 tons/A) had been applied (see Fig. 9–9). (Courtesy L. E. Dunn, University of Nevada.)

condition of the surface soil within a period of time and permits crop growth (Figs. 9–8 and 9–9).

In some cases, attempts are made to seed a salt-tolerant crop such as barley or bermudagrass after the first or second stage of the operation. These crops are grown for a year or two and disked in. The roots and added organic matter increase permeability and aid in the reclamation process. As reclamation proceeds to greater depths, salt-tolerant crops can be grown to maturity. However, the reclamation process is not complete until most of the sodium is removed from at least 1- to 1.5-m (3.3- to 4.9-ft) depth of soil. Even then, more time is required for restoration of good soil productivity because soil structure, once completely destroyed, is slow to return to a desirable condition.

Nonirrigated saline-sodic soils have shown definite improvement with the application of gypsum as a soil amendment (Fig. 9–10). Gypsum applied at the rate of 33.6 metric tons/ha (15 tons/A) to a saline-sodic soil increased dryland wheat yields an average of 673 kg/ha (10 bu/A) per year over a 5-year period of time. Annual precipitation in the area averages 71 cm (28 in.).[2]

[2] D. A. Whitney, W. L. Powers, W. A. Moore, and G. W. Wright, *Improving Non-irrigated Saline-Alkali Soils with Gypsum,* unpublished manuscript, Kansas State University.

Fig. 9–9. This saline-sodic soil (Toy soil series) produced a fair crop of barley after 44.8 metric tons/ha (20 tons/A) of gypsum had been applied (see Fig. 9–8). (Courtesy L. E. Dunn, University of Nevada.)

9.4.3 Using Soil and Water Amendments

Improved permeability should result if either the sodium in the irrigation water is reduced or the calcium and magnesium are increased (Fig. 9–11). At present there is no process available for removing salts such as sodium from irrigation water that is low enough in cost for use in general agriculture. Chemicals, however, can be added to the water or soil to increase the calcium and improve the sodium-to-calcium ratio. Under favorable conditions this may improve water penetration into and through the soil. The chemicals used either supply calcium directly (as from gypsum) or supply an acid or acid-forming substance (sulfuric acid or sulfur) that dissolves calcium from lime ($CaCO_3$) in the soil or reduces the bicarbonates in the water. Trials should always be conducted to determine if results are sufficiently beneficial to justify the cost.

Gypsum, sulfur, or sulfuric acid are the most commonly used soil amendments, whereas gypsum and sulfuric acid are used as water amendments. Granular gypsum has been applied broadcast to soils at rates of 2 to 20 metric tons/ha (0.9 to 8.8 tons/A). For land reclamation where sodium problems are extreme, rates as high as 44.8 metric tons/ha (20 tons/A) have been used. Where slow permeability is primarily in the soil surface, granular gypsum

Fig. 9-10. This is an example of the effect of gypsum on a nonirrigated saline-sodic soil (Farnum-Slickspot complex) in Reno County, Kansas. The gypsum was applied in 1965 (left) and the picture on the right was taken in 1971. Dryland wheat yields were increased an average of 673 kg/ha (10 bu/A) per year during that period of time. (Courtesy D. A. Whitney, Kansas State University.)

may be more effective if left on the soil surface or mixed with soil to a shallow depth rather than worked deeper into the soil.

Water applications usually require considerably less gypsum per hectare than soil applications. Water applications are particularly effective with low-salinity water. They are less effective with high salinity water that allows too little calcium to dissolve to balance the high sodium effectively. For water applications, finely-ground gypsum (0.25 mm, or finer — not granular or rock gypsum) is added more or less continuously and at a constant rate during the entire irrigation period. Improvement will depend on the soil situation as well as the water's calcium and salinity content.

Amendments should be used only when needed and demonstrated results justify their use. They may be useful where soil permeability is low due to low salinity, excess sodium, or high carbonate/bicarbonate in the water. They will not be useful, however, if poor permeability is due to problems of soil texture, soil compaction, restrictive layers (hardpans, claypans), or high water tables. If the crop is receiving adequate water for near-maximum yields, amendments may not increase yield but may make water management a little easier but at an additional cost for amendments, handling, and application.

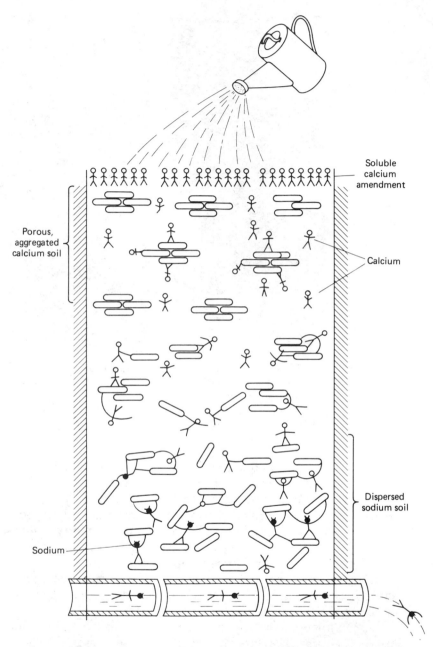

Fig. 9–11. When irrigation water and an amendment that supplies soluble calcium are applied to a sodic (high sodium) soil having dispersed clay and humus particles and small pores, the calcium replaces the adsorbed sodium and permits the particles to aggregate so as to form large pores. The aggregation of calcium soil is facilitated by alternate wetting and drying and by the action of growing plant roots. (Source: C. A. Bower, "Chemical Amendments for Improving Sodium Soils," Agr. Inf. Bul. No. 195, 9 pp., 1959.)

Fig. 9-12. A saline seep in northeastern Montana. The saline seep area grew a normal crop 3 years before but is now a nonproductive land area. The primary salts are sodium, calcium, and magnesium sulfates. If the lateral movement of water can be reduced and the water table in the seep recedes, then reclamation proceeds naturally because of the gypsum that precipitated during the salinization process. (Courtesy USDA-SEA-AR.)

9.5 SALINE SEEPS

Saline seep areas have developed in several semiarid areas of the Northern Plains of the United States and the Prairie Provinces of Canada. In recent years, they have been occurring more frequently and in numerous new areas.

The Soil Conservation Service estimated that Montana had 32,805 ha (81,000 A) of saline seeps in 1971 and 61,560 ha (152,000 A) in 1973. Annual farm income losses would total somewhere between $5 and $10 million, not including the value of the land lost.[3]

Saline seeps are formerly productive nonirrigated soil areas that became too wet and saline for economical crop production as a result of seepage of saline subsurface water to the soil surface (Fig. 9-12). A saline seep develops when soil water infiltrates below the rooting depth of crops and comes in contact with an almost impermeable soil layer, which causes water to move laterally. Eventually, at some lower elevation on slopes, the water will seep out at the soil surface and/or move to the soil surface by capillary action. As the water evaporates, it leaves a surface soil contaminated by salts. Seep areas vary in size from very small to more than 6 ha (15 A). An additional loss results from the odd shapes and sizes of the seep areas that make mechanized seeding and harvesting difficult.

Summer fallow has been identified by many as the principal cause of

---

[3] A. D. Halvorson, A. L. Black, and C. A. Rule, "Saline Seeps — Continued Bad News for Many Farmers," *Montana Farmer-Stockman* 61(18):34–38, March 21, 1974.

saline seeps. When precipitation exceeds the needs of a crop, water percolates below the root zone of annual crops and accumulates as a perched water table above a dense, nearly impermeable layer of clay.

Farmers can use a variety of cultural practices to contain and reclaim saline seep areas. These include reducing fallow acreage, more intensive cropping, and the use of grasses or deep-rooted perennials so that water is used up in areas above the recharge areas, thereby reducing seepage movement. Alfalfa has been particularly effective in controlling saline seeps when grown on the recharge area. Research and farmer experience indicate that if recharge areas could be cropped more frequently and if conventional crop-fallow methods were used less, much of the water that contributes to the formation of saline seeps could be intercepted and used. Drainage is usually not feasible on saline seeps because of the high concentration of salts and no place to dispose of the drainage water.

9.6 MANAGING SALINE AND SODIC SOILS

Salts accumulate in arid and semiarid regions mainly because rainfall is insufficient to leach them out of the soil rhizosphere and natural drainage is not adequate to remove salts from the soil. Another cause of salt accumulation in soil is the use of irrigation water high in salt content. Planting methods and irrigation practices need to be modified for salt-affected soils.

9.6.1 Irrigation and Water Quality

Many salt problems develop directly from salts added in the irrigation water. This is usually a gradual effect because large amounts of salt must accumulate before effects become visible. Plants absorb and transpire much water. Additional water is evaporated from the soil surface. In both processes, salt-free water is removed, and the salts are left behind to accumulate. If only enough water is added each time to replace moisture that has been lost by consumptive use (that is, by the amount of water that has been absorbed by plants and lost by evaporation — also called evapotranspiration), then each irrigation will introduce additional amounts of salt. This salt will be added to that already present in the soil and soil salinity will increase with each irrigation.

Salts added to the soil must be removed before accumulation becomes serious. Thus, leaching, deep percolation, and lateral drainage become essential in control of salinity. Leaching and deep percolation of salts can be done by applying extra water with every irrigation. The extra water will leach the salts downward and prevent salt accumulation in the plant root zone.

Maintaining a high water content in the soil, near field capacity, dilutes salts and lessens their toxic and osmotic effects. If a soil is irrigated lightly and more frequently during the salt-sensitive germination and seedling stages, most plants may be able to survive to the more tolerant mature stage of growth (see Table 9–3).

Level land is essential for good irrigation water management. Uniform distribution and uniform penetration of water is necessary for prevention of localized salt accumulations. Land should be "touched up" frequently with a land plane to increase water use efficiency and to reduce salt accumulations.

Where a choice of different quality waters is available, it is a good practice to use the best quality (lowest salt) water for the first (or preplant) irrigation. Once plants are past the seedling stage, waters of higher salt content can be used with less risk of adverse effects because root systems become deeper and more extensive, and plant species have had time to adjust somewhat to saline conditions.

9.6.2 Determining Potential Salinity and Sodium Hazard of Water

To evaluate an irrigation water's potential to cause plant toxicity, a water analysis is needed that includes boron, sodium, calcium, magnesium, chlorine, carbonate, bicarbonate (BO_3^{3-}, Na^+, Ca^{2+}, Mg^{2+}, Cl^-, CO_3^{2-}, HCO_3^-), and total salt content (conductivity). However, it is not necessary to include all these analyses to evaluate the salinity and sodium hazard of irrigation water.

The potential salinity and sodium hazard of a certain irrigation water can be determined with a laboratory test for electrical conductivity and soluble sodium percentage of the irrigation water. These two laboratory tests coupled with the soil texture are used in determining water quality by the State Soil Testing Laboratory at Kansas State University (Fig. 9–13). Irrigation water quality charts are then used for ranking the potential salinity and sodium hazard of the water. The charts (Figs. 9–14, 9–15, and 9–16) are based on experiments conducted by the Kansas Agricultural Experiment Station and utilize the electrical conductivity (millimhos per centimeter) and the soluble sodium percentage of the irrigation water. These results coupled with the soil texture are used in determining irrigation water quality.[4]

The soil conductivity and the percent exchangeable sodium values given in the charts at the top and on the curved lines, respectively, represent statistical estimates of the maximum values that will develop in any soil horizon when irrigated with a given water quality.

Prior to using the water quality charts (Figs. 9–14, 9–15, 9–16), the proper chart based on soil texture must be selected. Sandy (light-textured) soils include sands, loamy sands, and sandy loams. Loamy (medium-textured) soils include sandy loams, loams, silt loams, clay loams, and silty clay loams. Clayey (heavy-textured) soils include clay loams, silty clay loams, and clays. With the following exception, the texture used is the average texture of the root zone in the soil profile and not that of the surface soil only. Where a

---

[4] H. S. Jacobs and D. A. Whitney, "Determining Water Quality for Irrigation," Extension Service Cir. No. C-396, Kansas State University, 1975, 8 pp.

State Soil Testing Laboratory
Department of Agronomy, KSU
Manhattan, Kansas

REPORT OF TEST FOR IRRIGATION WATER QUALITY

Name _John Farmer_ _____ Laboratory sample Nos. _25_ _____

Address _Abilene KS 67410_ _____ County _Dickinson_ _____

Source of water _Well_ _____ Soil texture (profile) _____

Legal description of irrigated land: _____ $\frac{1}{4}$, _____ $\frac{1}{4}$, Sec. _____ ; T_____ ; R_____

Sample No.	Electrical Conductivity Millimhos/cm	Soluble Sodium Percentage	Soil Texture	Irrigation Hazard	
				Salinity	Sodium
			Light	Low	Low
			Medium	Low	Low
25	1.0	25 %	Heavy	Medium	Medium
			Light		
			Medium		
			Heavy		

Analyst _____   Date _____

Fig. 9–13. The potential salinity and sodium hazard of irrigation water can be determined by a laboratory test for electrical conductivity and soluble sodium percentage. This is reported to irrigators on a water quality test report. *NOTES:* (1) Soluble sodium percentage $= \dfrac{\text{soluble sodium (meq}/l)}{\text{total cation content (meq}/l)} \times 100$; (2) total cation content (meq/l) = 10 x electrical conductivity; (3) electrical conductivity (millimhos/cm) = EC x 10^3. (Source: Kansas State University.)

buried, clayey layer occupies sufficient depth in the root zone to impede water penetration, use the water quality chart for clayey soils. Water movement through such soil can be very slow.

As an example, a water sample is ranked on the test report from Fig. 9–13 for salinity and sodium hazard under the three soil classifications. These values also are marked on the water quality charts (Figs. 9–14, 9–15, 9–16).

The interpretation and significance of the salinity and sodium hazards are summarized as follows.

Salinity

Low Salinity Water. Water can be used for irrigation with most crops with little chance that salinity will develop. Some leaching is required but this occurs under normal irrigation and rainfall, except in soils characterized by low permeabilities.

Medium Salinity Water. Water is acceptable for irrigation, but moderate amounts of leaching are necessary to control salt accumulation. Irrigators normally apply sufficient water to provide the necessary leaching but some irrigators will need to increase the amount of water applied. A salt and sodium

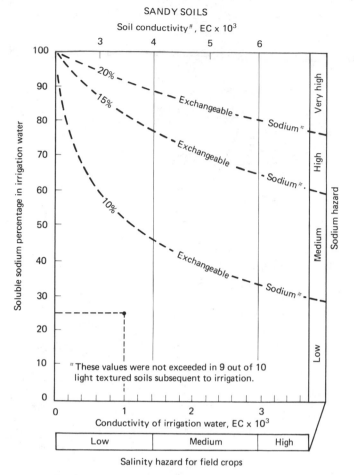

Fig. 9–14. Salinity and sodium hazard classification of irrigation water for *sandy soils* based on the conductivity and soluble sodium percentage in the water. The dot marks the water test report in Fig. 9–13. (Source: Kansas State University.)

test every 3 to 5 years should be used to assess the salt balance in the soil. Most common field crops can be grown without difficulty if these precautions are followed.

High Salinity Water. Water is of questionable quality for irrigation due to the likelihood of salt accumulation. Adequate leaching and growing salt-tolerant crops may allow the use of this water but favorable drainage conditions are essential. Salinity tests every 3 years should be used to follow the salt balance in the soil.

Very High Salinity Water. Water is not suitable for irrigation under most conditions but may be used occasionally under special circumstances where the soil is permeable and the subsurface drainage is excellent. Considerable

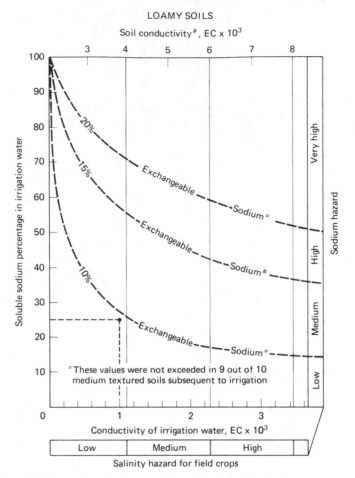

Fig. 9–15. Salinity and sodium hazard classification of irrigation water for *loamy soils* based on the conductivity and soluble sodium percentage in the water. The dot marks the water test report in Fig. 9–13. (Source: Kansas State University.)

leaching is necessary and salt-tolerant crops should be selected. Tests for salinity should be used without fail every year to monitor the salt balance in the soil.

Sodium

Low Sodium Water. Can be used for irrigation of virtually all soils with little danger that the soil will disperse due to harmful sodium accumulations.

Medium Sodium Water. Water is of permissible quality for most soils. Precautions should be taken so that excess exchangeable sodium does not develop because some soil horizons, in certain soils, will disperse when irrigated with this water. Chemical amendments may be necessary. Soil tests

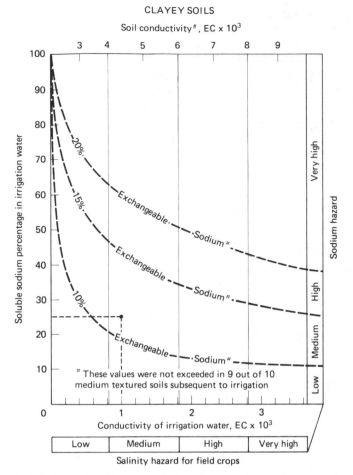

Fig. 9-16. Salinity and sodium hazard classification of irrigation water for *clayey soils* based on the conductivity and soluble sodium percentage in the water. The dot marks the water test report in Fig. 9–13. (Source: Kansas State University.)

for sodic (alkali) hazard every 3 to 5 years should be used to follow the sodium balance in the soil.

High Sodium Water. May produce harmful sodium levels. Chemical amendments, permeable soils, good subsurface drainage, and careful soil management would be required for sustained use of this water. Soil tests for sodium hazard should be used without fail every 3 years to follow the sodium balance. Once the soil disperses, it is very difficult to reclaim.

Very High Sodium Water. Generally not acceptable for irrigation. A major portion of the soils irrigated with this water may disperse unless the water is low or medium in salinity and gypsum or other chemical amendments are added to reduce the sodium hazard to low or medium.

Reclamation and Management of Saline and Sodic Soils

9.6.3 Tillage

It is known that salts tend to accumulate closer to the soil surface in nonirrigated drylands. Tillage speeds up the desalinization by mixing the easily soluble salts deeper in the soil and loosening the dense subsoil. Subsoil tillage operations — such as chiseling or moldboard or disk-plowing land with compact layers, hardpans, or cemented layers — frequently will improve infiltration and uniformity of water and root penetration.

Sodic claypan soils have responded to deep plowing of 61 to 76 cm (24 to 30 in.). In 9 years of testing, deep-plowed plots averaged 605 kg/ha (9 bu/A) more wheat compared to conventional plowed tests in North Dakota.[5] Deep plowing does at least two things that improve the soil. It breaks up the sodic layer and mixes in native soil gypsum that often occurs below the dense sodic layer. The broken sodic layer, mixed with the gypsiferous subsoil, lets more water enter the soil. Leaching can now take place, tending to move sodium below the root zone. When no restrictive layers are present, subsoil operations are not effective.

9.6.4 Planting

Generally, crop seeds should be planted in such a way as to avoid a salt buildup in the immediate zone of the seed placement. When a dry soil is wetted, the soluble salts already in the soil move with the water because they dissolve and are carried along with it. Germinating seed and seedlings are usually sensitive to salt. Farmers in saline areas avoid salt accumulation by planting seeds on the shoulders of the beds (Figs. 9–17 and 9–18).

9.6.5 Organic Matter and Soil Structure

Many factors are involved in the development and maintenance of soil structure, but the most important is growing salt-tolerant crops that add fresh organic matter. The regular addition of organic matter, whether from crops, manure, sludge, or compost, is the best possible insurance of good soil physical conditions and a safeguard against the adverse effects of sodium.

When soils contain plenty of rapidly decomposing organic matter, their desirable structure is less readily broken down by high sodium or compacted by heavy farm equipment. They remain open and permeable to water and air because the organic matter physically prevents the dispersed soil from compacting.

Crop residues left on the soil or worked into a rough cloddy soil surface often improves water penetration. The more fibrous crop residues such as from cereals and sudangrass and deep-rooted crops such as alfalfa increase

5 "Sodic Soils Respond to Deep Plowing," *Agricultural Research*, pp. 12–13, July 1976. USDA–ARS, Washington, D.C.

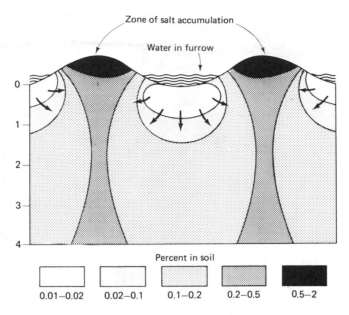

Fig. 9-17. Direction of salt flow and salt accumulation during irrigation. The zone of maximum salt accumulation is on the tops of the beds. (Courtesy Texas Vocational Instructional Services.)

water penetration. Presumably, the cereal and sudangrass straw along with crop roots physically keep the soil porous by maintaining channels and voids that improve water penetration. To be very effective, however, relatively large quantities of crop or other organic residues are usually needed. Rice hulls, cotton gin wastes, sawdust, shredded bark, and many other waste products in large volumes (10 to 20% by volume in the upper 15-cm depth) have been used with varying degrees of success. From a long-term standpoint the regular return of crop and animal residues to the soil is considered to be beneficial in that this helps to maintain granular soil structure as well as returning needed nutrients to the soil (see Chapter 10).

9.6.6 Managing Fertility While Controlling Salinity

Under conditions of high salinity there may be little or no response to applied fertilizers because excess salts may be the first factor limiting growth. When excess salts are removed by leaching, much of the soluble fertilizer is leached out with them. The problem then becomes how to remove soluble salts and still maintain necessary fertility levels. This is not easy to do, particularly on sandy soils.

The best way is to leach out most of the accumulated salts before fertilizer

Fig. 9–18. The small lettuce plants planted on the shoulder of the beds (at arrows) avoided the zone of high salt concentration at the top of the bed. The lettuce continued to grow and produce a good crop (Colorado). (Courtesy R. H. Follett.)

is applied. Most irrigating farmers do this with a heavy preplant water application in the spring. After planting, nitrogen (and potassium, if needed) is topdressed or sidedressed before the first summer irrigation. Phosphate can be applied any time because it is not leached from soils. *Do not overfertilize soils that are saline. Fertilizers are also salts and will increase the salinity of the soil. Follow a conservative soil testing program.*

Another effective way to maintain fertility is to depend on crop residues, manure, or compost to supply part of the fertilizer needed. These materials decompose slowly in soils and can supply ample plant nutrients, if enough material is incorporated. However, avoid heavy applications of manure on saline soils because of the presence of high concentration of soluble salts in the manure. Guidelines have been worked out for the application of feedlot manure to soils (see Chapter 10). The salinity hazard from the manure depends on the salt content of the irrigation water, the salt content of the manure, and the soil texture.[6] In general, farmers must depend largely on the return of crop residues to maintain and improve the organic matter content of the soil.

9.7 SAMPLING SOILS AND WATERS FOR SALINITY ANALYSIS

If analyses are to be meaningful, samples must be typical of the soil or water they are intended to represent. If a field contains both productive and un-

---

[6] W. L. Powers, G. W. Wallingford, L. S. Murphy, D. A. Whitney, H. L. Manges, and H. E. Jones, "Guidelines for Applying Beef Feedlot Manure to Fields," Kansas State Univ. Ext. Cir., C-502, 1974, 11 pp.

productive areas, an analysis of one soil sample containing a mixture of soil from the entire field would have little meaning. Salt content varies widely within the soil profile — usually it is highest in the centers of beds and tops of ridges and lowest under the furrows (see Fig. 9–18). The magnitude of this variation depends on many factors, such as type of planting bed, management practices, soil permeability, and salt concentration of the irrigation water. A tenfold to twentyfold increase in salt concentration from furrow to the top of the bed is not unusual.

9.7.1 Soil Sampling

The proper procedure for sampling soils for salinity analysis depends on the specific purpose for which the samples are taken. If the purpose is to determine the general salt or sodium content of the soil, the entire soil profile should be represented; however, it is best to avoid the tops of ridges where there are abnormally high-salt concentrations. Where crops are flat-planted (not on beds), sampling sites should be selected away from the borders. High spots and depressions should also be avoided. With irrigated row crops, sampling of either beds or furrows could give misleading information because of the great variations in salt content between furrows and tops of beds. The best time to sample probably is after land preparation and just before listing or preplant irrigating.

It is usually desirable to sample both surface soil and subsoil to a depth of at least 1 m (3.3 ft). Salinity or low productivity is often a result of restricted permeability in deep horizons in the soil. Dense or impervious layers 2 to 3 m (6.6 to 9.8 ft) below the surface may be responsible for salt accumulation in the plant root zone, sometimes resulting in total crop failure.

All soil samples submitted for analysis should be composite samples. Individual samples often vary widely in salt content; therefore, a number of subsamples from each depth at different locations should be composited and mixed together before extracting a representative sample for analysis. Composite samples should be composed of 5 to 10 subsamples taken from one general area. Do not attempt to composite samples from too large an area. Where differences in soil texture or crop productivity exist, separate composites should be made from each area. In some instances it may be advisable to obtain the advice of a soil specialist before sampling (Fig. 9–19).

9.7.2 Four-Electrode Technique for Measuring Soil Salinity

The four-electrode technique for measuring bulk soil electrical conductivity has been developed for use on irrigated soils and on dryland saline seeps in the field. This is a relatively new technique that has great potential for measuring soil salinity in the field without soil sampling and subsequent laboratory analyses. The technique employs a direct measure of the soil

State Soil Testing Laboratory
Department of Agronomy, KSU
Manhattan, Kansas

Form ST6, 9-59

REPORT OF SOIL TEST FOR SALTS AND ALKALI

Name John Farmer Sample No. 1

Address Alkali Flats, Kansas County

Soil type or symbol Laboratory Sample Nos. 1850

Legal description: ____ 1/4, ____ 1/4, Sec. ____; T ____; R ____

SOIL TEST RESULTS:

(1) Soil Layer	(2) Soluble Sodium meq/100 gms	(3) Exchangeable Sodium meq/100 gms	(4) Approximate Exchange Capacity meq/100 gms	(5) Exchangeable Sodium Percentage
0-12"	19.4	11.8	36.8	32.1

(6) Soil Layer	(7) Elec. Cond. Sat. Extract millimhos/cm	(8) pH	(9) Salt Rank	(10) Alkali Rank
	21.8	8.6	V. Excess	V. Excess

SOIL AMENDMENT RECOMMENDED:

Analyst Date

Fig. 9–19. Report of soil test for salts and alkali (sodium). (Courtesy Soil Testing Laboratory, Kansas State University.)

properties using four electrode conductivity cells inserted into the soil at a certain depth and spacing.[7]

9.7.3 Water Sampling

Water samples are comparatively simple to take, but several precautions should be observed. When sampling pump water, samples always should be taken at or near the well and only after the pump has been operating for ½ hour or more. Samples never should be taken of stagnant water or from ditches where the water is not flowing.

Surface waters should be sampled from the running stream as near the point of field application as possible. Containers should be approximately 1 *l* (1.1 qt) in size and thoroughly clean. Plastic bottles with tight lids are ideal for water sampling because breakage is eliminated.

[7] A. D. Halvorson, J. D. Rhoads, and C. A. Revle, "Soil Salinity — Four-Electrode Conductivity Relationships for Soils of the Northern Great Plains," *Soil Sci. Soc. Amer. Proc.*, 41(5):966–971, 1977.

Because most salinity problems are caused by salts dissolved in the irrigation water, a knowledge of what constitutes good, fair, or poor water quality is very important. A water analysis is generally used to evaluate the relative salinity hazard and sodium hazard of a certain water for irrigation purposes. Also, the occurrence of boron in toxic concentrations in certain irrigation waters makes it necessary to consider this element when assessing the quality of water. The classification of irrigation water requires a chemical analysis that can be performed by either private or state laboratories. Contact your local County Extension Agent for information on how to take a sample and where to send irrigation water for chemical analysis.

SUMMARY

Salt problems usually develop in one or more of three ways: (1) from salts already present in the soil, (2) from high groundwater tables, and/or (3) from salts added in the irrigation water.

Soil salinity and its related problems are most likely found in (though not necessarily restricted to) the arid and semiarid climates where rainfall is not sufficient to dissolve and leach these salts below the root zone. Salinity often develops from the presence of high groundwater tables. Inadequate drainage is probably the most important factor responsible for soil salinity. Groundwaters in many areas are quite salty. Water containing dissolved salts moves upward into the soil above a water table through the fine pores or capillaries. More efficient use of soil water in recharge areas by crops is an effective means of controlling saline seeps. Salinity that develops from salts added in the irrigation water is a gradual process. Large amounts of salts must be left behind to accumulate before the effects become visible.

Salt-affected soils are classified into three categories: saline, sodic, and saline-sodic. The classification depends on the total soluble salts (conductivity), soil pH and the exchangeable sodium percentage (refer to Table 9-1).

Corrective practices for saline and sodic conditions include both chemical and physical methods. The suggested practices that maintain or bring about a beneficial change in the soil or water chemistry include the following:

1. Establish drainage facilities, if feasible.
2. Use soil or water amendments such as gypsum, sulfur, or sulfuric acid, if needed.
3. Irrigate more frequently.
4. Cultivate and till deeply.
5. Use organic residues.

A soil test determines the kind of soil salinity problem, and a water analysis determines whether the water that is to be used for irrigation is satisfactory for the problem soil. Included in the reclamation is the selection of the crops most tolerant to salinity and sodium.

PROBLEMS

1. List the categories of salt-affected soils, and explain the classification of each category.

2. List a few field crops and forage crops that have a high tolerance to salinity.

3. Explain how a saline seep develops.

4. Explain how poor-quality irrigation water can result in a salt-affected soil.

5. Soil amendment problem: A soil sample from a salt-affected area in a field was sent to a soil testing laboratory for analysis. The report of the soil test is shown in Fig. 9–19.
   (a) Would you classify this soil as a saline soil, sodic soil, or saline-sodic-soil? Explain.
   (b) Calculate the tons of gypsum needed to reclaim the soil to a depth of about 30 cm (1 ft). Calculate both tons per acre and metric tons per hectare.

   Solution:

   We can use the following equation:
   Tons of gypsum/A = 1.7 [meq Na/100 g − (CEC x 5%)]

   In order to calculate the amount of gypsum needed for this example, we first take a look at the exchange capacity, which is 36.8 meq/100 g. We will recommend an amount of gypsum so that the calcium will replace sodium on all but 5% of the CEC. (5% × 36.8 = 1.8). We next consider the exchangeable sodium level of 11.8 m.e./100 g. If we then subtract the 1.8 from 11.8, we end up with about 10.0 meq/100 g of sodium that we want to replace with calcium.

   Answer:

   Tons of gypsum/A-ft = 1.7 [11.8 − (36.8 × 0.05)]
   Tons of gypsum/A-ft = **17 tons/A**

   Convert from tons per acre-foot to metric tons per hectare
   Metric tons/ha = tons/A × 2.242
   Metric tons gypsum/30 cm soil/ha = 17 × 2.242
   Metric tons gypsum/30 cm soil/ha = **38.1 metric tons**

General Comment: The common practice is to apply sufficient amendment to remove most of the adsorbed sodium from the top 30 cm (12 in.) of soil. This improves the physical condition of the surface soil within a short time and permits plant growth. By the continued use of good-quality irrigation water, good irrigation methods, and good cropping practices, further removal of adsorbed sodium, especially in the subsoil, usually takes place. In some instances, it may be necessary to reclaim to greater depths in order to obtain adequate drainage and desirable root penetration.

REFERENCES

Ayers, R. S., and D. W. Westcot, "Water Quality for Agriculture," Irrigation and Drainage Paper No. 29, FAO Rome (1976).

Bernstein, Leon, "Salt Tolerance of Grasses and Forage Legumes," USDA Agric. Information Bul. No. 194 (1958).

Bernstein, Leon, "Salt Tolerance of Vegetable Crops In the West," USDA Agric. Information Bul. No. 205 (1959).

Bernstein, Leon, "Salt Tolerance of Field Crops," USDA Agric. Information Bul. No. 217 (1960).

Bernstein, Leon, "Effects of Salinity and Sodicity on Plant Growth," *Annual Review of Phytopathology,* 13 (1975) 295–312.

Branson, R. L., "Gypsum for Soil Improvement," University of California Leaflet No. 2149 (1976).

Branson, R. L., P. F. Pratt, J. D. Rhoades, and J. D. Oster, "Water Quality in Irrigated Watersheds," *J. Environ. Qual.* 4 (1975) 33–40.

Doering, E. J., and F. M. Sandoval, "Hydrology of Saline Seeps in the Northern Great Plains," *Transactions of the ASAE,* 19(5) (1976) 856–865.

Fuller, W. H., "Water, Soil and Crop Management Principles for Control of Salts," Univ. of Arizona Agric. Exp. Sta. Bul. A-43 (1965).

Halvorson, A. D., and A. L. Black, "Saline-Seep Development in Drylands of Northeastern Montana," *Jour. of Soil and Water Conservation,* 29(2), (1974) 77–81.

Johnsgard, G. A., "Salt Affected Problem Soils in North Dakota," North Dakota State Univ. Ext. Bul. No. 2 (1970).

Longnecker, D. E., and P. J. Lyerly, "Control of Soluble Salts in Farming and Gardening," The Texas Agric. Exp. Sta. Bul. B-876 (1974).

Ponnamperuma, F. N., R. S. Lantin, and M. T. C. Cayton, "Boron Toxicity in Rice Soils," *International Rice Research Newsletter* 4(6), IRRI, Manila, Philippines (Dec., 1979) 8.

Rhoades, J. D., and A. D. Halvorson, "Electrical Conductivity Methods for Detecting and Delineating Saline Seeps and Measuring Salinity in Northern Great Plains Soils, ARSW-42, 45 pp, USDA–ARS, March, 1977.

Sandoval, F. M., J. J. Bond, and G. A. Reichman, "Deep Plowing and Chemical Amendment Effect on a Sodic Claypan Soil," *Transactions of ASAE,* 15 (1972) 681–687.

Soils Bulletin 31, "Prognosis of Salinity and Alkalinity," FAO, Rome, Italy (1976).

U. S. Salinity Laboratory Staff. Diagnosis and Improvement of Saline and Alkali Soils, *USDA Handbook 60* (1969).

CHAPTER TEN

Organic Amendments: Plant and Animal Residues

10.1 INTRODUCTION

On July 20, 1969, Neil A. Armstrong and Edwin E. Aldrin, Jr., were the first persons to walk on the surface of the Moon and to collect soil and rock samples. When these were later analyzed on Earth, no organic matter was present, but the moon soil grew plants.

Plants *do not* need organic matter to grow and reproduce, but they produce more efficiently when it is present. Animals and people *do* need organic matter for all functions of life.

Progressive farmers realize the plant-growth-enhancing benefits of organic matter in soils. They also are aware that it takes more knowledge and skill to make maximum use of plant and animal organic residues than to use chemical fertilizers. Ranchers have known for years the wisdom of grazing only the *top* half of the grass blades, leaving the *bottom* half for growth, reproduction, and organic residue.

About 95% of the dry weight of all organic matter originates from green plants through photosynthesis. The other 5% is the nutrients from the soil, water, and air. Energy for photosynthesis comes from the sun. Of the total solar energy on earth, only about 0.15% is utilized in photosynthesis. On any one corn field, the percentage utilized is 1.6%; but of this amount 0.4% is used in plant respiration, making the efficiency of solar energy by corn about 1.2%. Effective use of fertilizers and soil amendments can greatly increase the capability of plants to capture "free" solar energy.

Each year throughout the world an estimated 27×10^9 metric tons (30×10^9 tons) (dry weight basis) of organic matter are produced by photosynthesis. About one-third of this organic matter eventually comes in contact with the soil and the remaining is fixed in perennial vegetation, such as in tree trunks and in bodies of animals and people.[1] Most of these residues will be eventually returned to the soil to maintain soil productivity.

10.1.1 Functions of Organic Matter in Soils

Organic residues from plants and animals that are on or in the soil are beneficial in the following ways:

1. Serving as the principal storehouse for anions essential for plant growth such as nitrates, phosphates, sulfates, borates, molybdates, and chlorides.
2. Increasing the cation exchange capacity of soils by a factor of 5 to 10 times that of clay. This is true for humified organic matter (humus). More available nutrient cations such as ammonium (NH_4^+), potassium (K^+), calcium (Ca^{2+}), and magnesium (Mg^{2+}) are thus adsorbed by humus.
3. Buffering the soil against rapid changes due to acidity, alkalinity, salinity, pesticides, and toxic heavy metals.
4. Protecting the surface soil against erosion by water and wind by reducing the impact of raindrops on soil peds and clods, increasing infiltration, reducing water runoff, increasing the soil's total and available water-holding capacity, and increasing surface wetness. In the Great Plains where hail is fairly common, a surface organic mulch also reduces water and wind erosion by protecting soil peds and clods against destruction by hail stones.
5. Supplying food for beneficial soil organisms such as earthworms, symbiotic nitrogen-fixing bacteria, and mycorrhizae (beneficial fungi).
6. Reducing extremes of soil surface temperature. This is especially true of organic residues used as a surface mulch.
7. Decreasing surface crust formation by decreasing the soil-dispersing action of beating raindrops.
8. Reducing the crystallization and hardening of the plinthite (laterite) layer of soils in the humid tropics that are rich in soluble iron and aluminum. Humate complexes with iron and aluminum reduce crystallization. Organic matter reduces hardening also by maintaining more uniform soil temperature and soil moisture.
9. Supplying to growing plants, as organic residues decompose, small quantities of *all* essential plant nutrients, usually in time-sequence harmony with the needs of plants.
10. Making phosphorus and micronutrients more readily available over a wider pH range. This is a function especially of soil humus, a resistant decomposition product of soil organic matter.

[1] J. Balogh, "Biogeographical Aspects of Soil Ecology." In *Ecology and Conservation*, Proceedings of the Paris Symposium, UNESCO, 1970, p. 33.

11. Increasing the application rates of selected herbicides. For example, in Michigan the recommended rate of postemergent herbicides on Histosols (peat and muck soils) is approximately 50% greater than on mineral soils.[2] For some herbicides, application rates are determined partly by the percentage of organic matter, sand, silt, and clay in mineral soils. An example is cited in Kansas with the recommendation of Sencor (metribuzin) 50% wettable powder when used as a preemergent broadcast application for soybeans. On coarse-textured soils with less than 2% organic matter, Sencor is *not recommended*. With increasing clay and increasing organic matter, the rate gradually increases to 2.2 kg/ha (2 lb/A).[3] A mixture of Sencor (metribuzin) 50% wettable powder and Surflan (oryzalin) 75% wettable powder is even more soil-dependent. It should not be used on a soil with a pH greater than 7.4 nor on loamy sand or sandy loam soils nor on soil with *less* than 2% organic matter nor *more* than 3% organic matter.[4]
12. Decreasing the bulk density of the soil. Tillage pans or naturally indurated horizons in soils may be so dense as to reduce the rate of infiltration, decrease water storage capacity, and restrict normal root development. Compacted layers in fine-textured soils may be as dense as 2 g/cc (125 lb/ft^3). Chiseling to stir the dense layers followed by an organic residue management system can reduce this bulk density to 1.4 g/cc (90 lb/ft^3). At this soil density the soil environment enhances plant nutrition, growth, and yield.

The composition of soil organic matter as well as ambient temperature and moisture determine its rate of microbial decomposition and transformation to available nutrients.

10.1.2 Mineralization and Immobilization

Mineralization and immobilization are antonyms. Mineralization is the microbial transformation of an element from organic to its inorganic form; for example the conversion of proteinaceous nitrogen to *ammonium* carbonate [$(NH_4)_2CO_3$] or calcium *nitrate* [$Ca(NO_3)_2$]. Immobilization is the microbial conversion of a mineral element or radical from the inorganic to the organic form. An example is the uptake of ammonium (NH_4^+) or nitrate (NO_3^-) by a plant or a microbe and its transformation into proteins. Nitrogen in these proteins will again be mineralized to contribute substantial nitrogen for plant growth (Note 10–1).

When fresh crop residues are added to a soil, the nitrogen in the residues may be decomposed and mineralized by microbes and soon become available

[2] William F. Meggitt, "1977–78 Weed Control Guide for Field Crops," Cooperative Extension Service Bul. E-434, Michigan State University.

[3] Mobay Chemical Corporation, Kansas City, Mo., Directions for Use of Sencor 50% Wettable Powder Herbicide, dated Feb. 23, 1977.

[4] E. B. Nilson, O. G. Russ, W. M. Phillips, and J. L. Condray, "Chemical Weed Control in Field Crops, 1977," Agricultural Experiment Station Report of Progress 284, Kansas State University, 28 pp., p. 16.

Note 10-1.

Mineralization of Plant and Animal Residues
Contributes Substantial Nitrogen for Plant Growth

The storehouses for combined nitrogen in the soil are plant and animal residues that comprise the soil organic matter. Organic residues have been used for centuries to enhance plant growth, but there has been a recent acceleration of awareness of their unique value as a companion to chemical fertilizers for obtaining maximum economic yields.

Although raw residues benefit a soil physically by increasing infiltration, moderating soil temperature, and reducing erosion, plant nutritional values are derived only upon mineralization of the organic matter. All essential elements from the organic matter thus become more readily available, but uniquely so for nitrogen. This is true for nitrogen because mineralization, microbial decomposition, and microbial oxidation are the *only* pathways by which organic nitrogen in organic residues is made available for plant nutrition and growth.

The nitrogen (N) in soils of the United States derived from organic matter was estimated by the National Research Council in 1970 as almost 15% of annual soil N input (Table 10-1).

Table 10-1. Estimated Annual Nitrogen Input to Soil in the United States in 1970

Source	Total Annual Input (million metric tons N)	Percent of Total	Annual Input	
			All Land	Cropland + Pastureland[a] (kg/ha N)
Fertilizers	7.5[b]	35.7	8.3	17.4
Precipitation	5.6	26.7	6.1	6.1
Symbiotic N_2 fixation	3.6	17.1	3.9	8.3
Mineralization of organic matter	3.1	14.8	3.4	3.4
Nonsymbiotic N_2 fixation	1.2	5.7	1.3	1.3
Total	21.0	100.0	22.9	36.5

[a] Calculations based on the assumption that all nitrogen fertilizers and all symbiotic N_2 fixation are on only cropland plus pastureland and that the other inputs are in proportion to total land area.

[b] Fertilizer N input increased to about 10 million metric tons in 1978.

SOURCE: Calculated from data in *Nitrates: An Environmental Assessment,* National Research Council, National Academy of Sciences, Washington, D.C., 1978, 723 pp., p. 214.

The authors estimate that this proportion will increase during the decade of the 1980s because farmers and other land managers will be using more crop residues, animal manures, and sewage sludges. An estimate of the amounts of each residue available each year and the release of available nitrogen ($NH_4^+ + NO_3^-$) from them are given here.

Annual United States production of crop residues, including tops and roots, is estimated at 6 metric tons/ha (2.7 tons/A) (dry basis). From 142 million ha (350 million A) of crops, the total would be 852 million metric tons (939 million tons) a year of crop residues. Assuming an average composition of 1% N, this would equal 8.5 million metric tons (9.4 million tons) of N. If 25% of this N became available to crops by mineralization the first year, the total available N would be about 2.1 million metric tons (2.3 million tons).

Livestock manures produced in the United States each year total about 1.8 billion metric tons (2 billion tons) (dry basis). With an estimated composition of the manures of 0.5% N, the total N would be 9 million metric tons (9.9 million tons). Perhaps half of the total N would become available to plants during the first year; if so, the total would be 4.5 million metric tons (5 million tons).

As stated in Section 10.4.1, sewage sludge production each year in the United States is about 5 million metric tons (5.5 million tons) (dry basis). Average composition is approximately 3.5% total N, or 175,000 metric tons (192,500 tons). Assuming that 75% of the N would mineralize to NH_4^+ and/or NO_3^- the first year, this would be 131,250 metric tons (144,375 tons) of N.

Assuming that in 1980 *all* the crop residues, animal manures, and sewage effluents from municipal plants were used for plant production, the total N would be 17.7 million metric tons (19.5 million tons) that year. The estimated N mineralized into NH_4^+ and NO_3^- the first year is 6.73 million metric tons (7.44 million tons). This compares with the National Research Council estimate for 1970 of 3.1 million metric tons (3.4 million tons) of N a year mineralized from organic matter. The authors' estimates made for 1980 are therefore more than twice those made by the National Research Council in 1970. The explanation of the difference in estimates may be because different parameters were used and/or there may have been actually more total organic residues produced in 1980 than in 1970.

It cannot be assumed that all the N released from organic residues will be available for enhanced plant nutrition and growth. As much as 10 to 50% can be *lost* through the following pathways: As atmospheric N_2 as oxides of N by denitrification, leaching as NO_3^-, runoff and erosion, or fixation as ammonium between clay platelets.

Farmers, gardeners, and other land managers have been using more organic residues in recent years. Predictions are that the trend in use will accelerate for agronomic, economic, and environmental reasons.

for new plant growth, or the nitrogen may be immobilized (not decomposed) and therefore unavailable to plants. Assuming that all other factors that influence microbial decomposition are present in optimum amounts (such as temperature and moisture), the limiting factor in immobilization may be the ratio of organic carbon to total nitrogen in the plant residue.

In arable soils when the carbon:nitrogen ratio of a fresh organic residue is about 30:1, ammonium (NH_4^+) and nitrates (NO_3^-) may be released for plant growth by mineralization or immobilized because the plant tissue resists microbial attack. At this 30:1 carbon:nitrogen ratio, soil moisture or temperature may determine the direction of the mineralization/immobilization threshold. Such a critical C:N ratio of 30:1 is shown in Table 10-2 for bluegrass.

Summary. Bluegrass residues may be mineralized by microbes if all factors of decomposition are ideal, or they may not mineralize and thereby form thatch on a lawn if temperature and soil moisture are *not* ideal for microbial decomposition.

Organic residues with wider C:N ratios such as corn stalks (40:1) and small grain straw (80:1) must have a nitrogenous fertilizer added to prevent nitrogen deficiency if a crop is planted soon after their incorporation with the soil. Crop residues with narrower ratios such as clovers (20:1), alfalfa (13:1), and sewage sludge (10-12:1) can be incorporated in or left on the soil and a crop can be planted immediately thereafter without causing nitrogen deficiency (Fig. 10-1).

The C:N ratio is also given for soil microbes and soil humus.

Table 10-2. Approximate Composition of Organic Carbon, Total Nitrogen, and C:N Ratio of Common Organic Materials and Soil Microbes and Humus on/in Arable Soils (Dry Weight Basis)

Organic Material	Organic Carbon (C) (%)	Total Nitrogen (N) (%)	C:N Ratio
Crop residues			
Alfalfa (young)	40	3	13:1
Clovers (mature)	40	2	20:1
Bluegrass	40	1.3	30:1
Corn stalks	40	1	40:1
Straw, small grain	40	0.5	80:1
Cattle manure	15	0.5	30:1
Sewage sludge	30-50	2.5-5.0	10-12:1
Soil microbes			
Bacteria	50	10	5:1
Actinomycetes	50	8.5	6:1
Fungi	50	5	10:1
Soil humus	2	0.2	10:1

Fig. 10–1. Plant residues from this legume-grass pasture are estimated to have a C:N ratio of about 25:1. When incorporated in the soil they should start microbial mineralization immediately. This means that the following crop can be seeded/planted within a few days with almost no hazard of nitrogen deficiency. (Courtesy USDA-Agricultural Stabilization and Conservation Service.)

10.1.3   Soil Humus Characterization

Humus in arable soils is a residual organic product in equilibrium with the ecological environment. The amount of humus present at any specific time is a function of relative soil wetness, soil temperature, soil texture, the supply and balance of essential elements, salinity, alkalinity, sodicity, tillage, erosion, freedom from toxins, and the amounts and kinds of organic residues added.

On any particular soil, if the soil management practices over the years do not change, the amount of soil humus will reach an equilibrium and become stabilized in a steady state. Contrary to popular belief, if all environmental and management factors remain constant, increasing additions of organic matter such as animal manures will increase the soil humus percentage of the soil only for a year or two; after that period, the percentage of soil humus will increase no further. At this point of equilibrium it can be said of soil organic matter that

$$\text{mineralization} \leftrightharpoons \text{immobilization}$$

In the central United States on a fine-textured, cultivated soil, the amount of organic matter needed to maintain this soil humus equilibrium has been predicted to be about 10 metric tons/ha/yr (4.5 tons/A/yr). These estimates assume no erosion losses of humus.

If all factors of the ecological environment remain constant except temperature, the higher the temperature, the less the accumulation of humus. When only precipitation varies, the more the precipitation, the more the humus accumulation.

Organic residues are the food for these soil microbes: bacteria, actinomycetes, and fungi. All soil algae and a few soil protozoa contain chlorophyll and manufacture their own food. The living and dead microbes plus resistant organic crop residues such as lignin comprise soil humus of arable soils.

Soil humus is very important in plant nutrition. Humus serves as a storehouse (pool) of readily available major nutrients, secondary nutrients, and micronutrients. Humus is the principal pool for micronutrients as well as carbon, nitrogen, phosphorus, and sulfur. The approximate percentages of organic carbon, total nitrogen, organic phosphorus, and organic sulfur in humus in fine-textured, cultivated, humid-region soils are approximately as follows (dry weight basis):

Organic carbon	50.0%
Total nitrogen	5.0%
Organic phosphorus	0.5%
Organic sulfur	0.5%

The nutrients (N, P, and S) become readily available to plants upon microbial decomposition. Based upon the approximate compositions, the ratios of these nutrients in humus are, therefore,

$$C:N = 10:1$$
$$N:P = 10:1$$
$$N:S = 10:1$$
$$C:P = 100:1$$
$$C:S = 100:1$$

The approximate factors for conversion of these nutrients to percentage of organic matter are therefore:

Organic carbon (C) $\times$ 2 = organic matter %

Total nitrogen (N) $\times$ 20 = organic matter %

Organic phosphorus (P) $\times$ 200 = organic matter %

Organic sulfur (S) $\times$ 200 = organic matter %

NOTE: Many exceptions to these generalizations can be found in world research literature especially in noncultivated sandy soils in arid regions. When found, the authors believe that the variations can be explained, based on pedological, anthropological, nutritional, ecological, and climatological criteria.

Organic Amendments: Plant and Animal Residues

10.1.4 Soil Amendments: A Closer Look

There is an economic reason and also a legal reason for taking a closer look at organic fertilizers and soil amendments. Prices of chemical fertilizers are variable and U.S. Public Law 92–500 as amended makes it mandatory to "safely dispose" of all potential pollutants. Both factors will encourage the application of more animal manures and sewage sludge effluents on soils.

Each year there have been increasing amounts of organic materials bought in the 50 states and used as fertilizers/organic amendments. In 1976 the natural organic materials bought totaled 374,864 metric tons (413,208 tons). These materials are listed in Table 10-3 from greatest to least amounts sold, as follows: dried animal manures, compost, sewage sludges, and tankage. Only small amounts of other natural organic materials were sold, but many other kinds are used by gardeners.

For the developing countries throughout the world, a summary was made of all important organic wastes available annually for use, together with its metric tonnage, composition of nitrogen (N), phosphorus (P), and potassium (K) (Table 10-4). No estimates have been made of the current use of these organic residues produced in the developed countries. The amounts of nitrogen, phosphorus, and potassium in the organic wastes used by the developing countries are greater by a factor of 7 or 8 than the nutrients used by the same developing countries in chemical fertilizers. With effective technical assistance, farmers in developing countries are expected to use more organic residues *and* more chemical fertilizers to enhance economic production.

Among farmers of the United States there is a rapid increase in the adoption of more efficient systems of crop residue management. This has led to minimum tillage whose success depends on leaving about 6 metric tons/ha/yr

Table 10–3. Natural Organic Materials Sold in the Contiguous United States During the Fertilizer Year Ending June 30, 1976

Natural Organic Material	Metric Tons	Short Tons
Animal manures, dried	175,284	193,214
Compost	62,804	69,228
Sewage sludge, activated	57,609	63,502
Sewage sludge, other	44,894	49,485
Tankage	13,742	15,147
Other organic materials	13,484	14,863
Cottonseed meal	4,688	5,167
Castor pomace	1,195	1,317
Dried blood	1,166	1,285
Total	**374,864**	**413,208**

SOURCE: Commercial Fertilizers, Final Consumption for Year Ended June 30, 1976, Sp. Cir. 7 (77), USDA, 1977, p. 6.

Table 10-4. Total Annual Production of Primary Plant Nutrients in Organic Wastes in the Developing World for 1971 (Actual) and 1980 (Estimated)

Source	Year	Nitrogen (N)	Phosphorus (P)	Potassium (K)
		Million Metric Tons		
Cattle manure	1971	17.80	4.91	14.12
	1980	22.25	6.14	17.65
"Night soil"	1971	12.25	2.87	2.61
(human wastes)	1980	15.26	3.57	3.25
Compost, farm	1971	9.54	3.34	9.54
	1980	11.93	4.18	11.93
Sewage, urban	1971	1.43	0.29	0.86
	1980	1.79	0.36	1.08
Compost, urban	1971	0.48	0.38	0.57
	1980	0.60	0.48	0.71
Other wastes[a]	1971	6.63	4.44	11.35
	1980	8.29	5.55	14.19
Total wastes	1971	48.13	16.23	39.05
	1980	60.12	20.28	48.81

[a] Other wastes include poultry manure, goat manure, sheep manure, sugarcane wastes, and miscellaneous wastes.

SOURCE: A. Duncan, "Economic Aspects of the Use of Organic Materials as Fertilizers," In *Organic Materials as Fertilizers*, FAO Soils Bul. No. 27, Rome, Italy, 1975, p. 356.

(2.68 tons/A/yr) of crop residues on or near the surface of the soil. These amounts of organic residues include stems, leaves, and roots (see Section 10.2).

There has also been a rapid change in farm practice back to the efficient use of animal manures. There are at least two reasons for this change:

1. Public Law 92-500 as amended gives the U.S. Environmental Protection Agency authority to make all water bodies clean again.
2. Most domestic livestock are now raised/fattened in confinement and high labor costs have made it more economical for farmers to concentrate thousands of head into a small space (see Section 10.3).

Human wastes (solid and liquid sewage effluents) are also being used more in the United States now than formerly. The reason is primarily one of keeping the environment clean. Dried organic fertilizer materials are made from human wastes and sold by several municipalities, but the sales of most of such products are subsidized. Sewage effluents (sewage sludge and sewage waste waters) are also used increasingly for fertilizing and irrigating crops and for enhancing the growth of vegetation to stabilize quickly soils disturbed by urbanization and mining (see Section 10.4).

Composts are on the market as specialty organic materials. They are sold as "potting soil" for house plants. Composting and composts are also promoted as "backyard" enterprises for use on flower gardens or vegetable gardens. Green-manure (cover) crops, once very popular, are being replaced by more efficient and more economical crop residue management systems (see Section 10.6). Wood residues such as sawdust, wood chips, wood fiber, and tree bark have unique and specialized uses to reduce erosion and sediment from soils drastically disturbed by mining and construction. Peat has always been a popular natural organic product, especially for acidifying soil and for increasing the soils' total and available water-holding capacity and cation exchange capacity (see Section 10.7).

10.2 CROP RESIDUES

Efficiency of plant nutrition is increased when the *surface* of the soil is protected by organic amendments. Crop residues are ideal for this purpose. For the benefit of the soil's productivity, crop residues should be used efficiently and not burned or otherwise removed (Fig. 10–2). The trend toward narrower rows, thicker planting, and heavier fertilization has resulted not only in higher

Fig. 10–2. Blowing wheat straw and perennial grass seed on bare roadside cut and fill slopes. The organic residue protects the soil while providing an ideal seed bed for perennial grasses (Kentucky). (Courtesy USDA-Soil Conservation Service.)

yields but also in more fine-stalked/stemmed crops that are easier to manage as surface residues.

Desirable crop residue management means the conservation of all organic material (roots, leaves, and harvest residues) on or near the soil surface. This practice was researched extensively in the decades of the 1930s and 1940s under the name of "stubble mulching."[5] The benefits of such a crop management residue system claimed at that time were as follows:

1. Increased crop yields.
2. Less surface erosion by water and wind.
3. More available soil moisture.
4. More soil humus.

A good residue management system today can claim these same four advantages over clean tillage and at least three additional advantages:

5. Lower tractor fuel costs.
6. Less soil compaction.
7. More desirable bacteria, actinomycetes, and fungi.

Disadvantages of a crop residue management system include a greater hazard of the following:

1. More insects, diseases, weeds, and rodents.
2. More wetness and coldness in the soil surface, resulting in slower germination.
3. More difficulty in getting good seed-soil contact, resulting in a nonuniform stand.

Newer names for modern popular practices similar to stubble mulching include the following:

1. Mulch farming.
2. Conservation tillage.
3. No-tillage.
4. Zero tillage.
5. Strip tillage.
6. Sweep tillage.
7. Plow-plant.
8. Wheel-track planting.
9. Reduced tillage.
10. Minimum tillage.
11. Chemical fallow.
12. Eco-fallow.

[5] F. L. Duley and J. C. Russell, "Effect of Stubble Mulching on Soil Erosion and Runoff," *Soil Sci. Soc. Amer. Proc.,* 7:77–81, 1942.

10.2.1 Quantity of Crop Residues

In the United States each year there are about 142 million ha (350 million A) of crops harvested. Each hectare is estimated to produce an average of 4 metric tons/ha (1.78 tons/A) of aboveground crop residues each year. Roots of crop plants are estimated to add 50% more crop residues, making a total of about 6 metric tons/ha (2.68 tons/A) per year, or a total United States annual production of 852 million metric tons (937 million tons). Under continuous cultivation, these amounts of residues are not enough to maintain soil humus levels. However, with a rapidly increasing hectarage (acreage) of minimum tillage, double cropping, and other desirable residue-management systems, the more progressive farmers *are* maintaining the ecological levels of soil humus (see Section 10.1.3).

10.2.2 Minimum Tillage — A Crop Residue Management System

In 1979 an estimated 22 million ha (56 million A) of minimum tillage was reported in the United States. The success of minimum tillage is due in large part to the fact that it is a *surface crop residue* management system. Research in the midwestern United States Corn Belt confirms this fact. In a cropping system of continuous corn, with conventional plowing in the fall with a turning plow, the corn residues at corn planting time covered about 1% of the soil surface; with double disking in the spring, the residues covered about 13% of the soil surface; with soil preparation with a rotary tiller in the spring only in a narrow planting strip, 62% corn residue cover remained; but with minimum tillage there was 76% of the soil covered by last season's cornstalks.[6]

Some scientists and many farmers remain skeptical about the value of minimum tillage. There is no doubt that fine-textured soils with poor internal drainage are not well adapted because a surface mulch delays the spring warming of an already slow-to-warm soil. For the same reason, it is also true that north-temperate region clay soils are not as well adapted to the use of a surface mulch. It has been estimated that surface crop residues in a system of minimum tillage delay spring planting on well-drained soils by about 1 day in the southern United States to as much as 1 week in the northern United States. Planting on poorly drained soils is delayed even more. Where growing seasons are already short for the crop to be grown, minimum tillage with crop surface residue management may never be popular.

Opponents of minimum tillage claim that phosphorus, calcium, and magnesium will not move fast enough into the rhizosphere when applied on the surface of soils managed under minimum tillage. Such conclusions were valid for bare soil between conventionally tilled crops. Proponents of minimum tillage conducted field trials and concluded that microbial products from crop residue decomposition solubilized phosphorus, calcium, and

---

[6] J. V. Mannering, D. R. Griffith, and C. B. Richey, "Tillage for Moisture Conservation," Paper No. 75–2523. American Society of Agricultural Enginers, St. Joseph, Mich., 1975.

magnesium. These nutrients then moved into the rhizosphere and were absorbed efficiently by plants.[7,8] Another fertilizer-related modification is the recommendation to add 4.5 kg/ha (10 lb/A) *more* nitrogen (N) for each 454 kg/ha (1000 lb/A) of crop residues left on the soil surface. These higher rates of N are used more efficiently by plants.

Leaving all crop residues on the soil surface can be expected to decrease runoff by about 50% but not to increase the nutrient percentage of the runoff water. Soil sediment losses are also reduced by similar amounts. The net result of surface crop residue management in minimum tillage is to conserve nitrogen, phosphorus, and other nutrients as well as to stabilize the soil.[9]

10.2.3 Crop Residues as Mulches on Intensively Managed Areas

Flower gardens and vegetable gardens are examples of intensively managed areas often mulched with small-grain straw or hay to control weeds, conserve moisture, reduce mud and dust on flowers and vegetables, and permit nonmuddy access to the garden even after heavy rains.

Construction sites and mine spoils areas must be managed intensively to reduce erosion sediment losses. Such areas are almost always seeded to adapted vegetation and, prior to seeding, an organic mulch is used sometimes to stabilize the seedbed. Many of the cut-and-fill slopes along the service roads of the Alaskan Pipeline were stabilized with grain straw before being seeded to adapted grasses. In windy location, the straw was stabilized with an emulsified asphalt.[10]

Surface mining in the United States has created "spoils" not yet stabilized by vegetation that now total about 405,000 ha (1 million A). In addition, an estimated 40,500 ha (100,000 A) *each year* will require stabilization. An example of the need for crop residues for stabilizing mining wastes is cited from Idaho.[11] As is true with all mine spoils, the mine spoils in Idaho were extremely variable as to slope, texture, rockiness, acidity, toxic elements, total soluble salts, and available plant nutrients.

A field experiment was conducted in Idaho to determine the best way to establish vegetation on the mine spoils to reduce erosion and sediment yield to streams. The variables were native and introduced grass species, chemical fertilizers, straw mulch, and irrigation. The results are summarized as follows:

[7] C. R. Belcher and J. L. Ragland, "Phosphorus Absorption by Sod-Planted Corn (*Zea Mays*) From Surface Applied Phosphorus," *Agron. J.,* 64:754–756, 1972.

[8] W. W. Moschler, D. C. Martens, and G. M. Shear, "Residual Fertility in Soil Continuously Field Cropped to Corn by Conventional Tillage and No-Tillage Methods," *Agron. J.,* 67:45–48, 1975.

[9] M. H. Frere, "Nutrient Aspects of Pollution from Cropland." In *Control of Water Pollution from Cropland: Volume II, An Overview,* USDA-Agricultural Research Service and U.S. Environmental Protection Agency, 1976, pp. 59–90.

[10] J. C. Stover, "Mulching in Alaska." In *1973 Alaska Revegetation Workshop Notes,* No. RP-239, University of Alaska, 1973, pp. 27–32.

[11] E. E. Farmer, B. Z. Richardson, and R. W. Brown, "Revegetation of Acid Mining Wastes in Central Idaho," USDA-Forest Service Research Paper INT-178, 1976.

1. Introduced grass species produced best, followed by a mixture of introduced + native species. Native species seeded *alone* produced the least stabilizing vegetation.
2. Chemical fertilizer *alone* did not increase yields of any grass species.
3. Mulch *alone* did not increase yields.
4. Irrigation increased yields during the first year but depressed yields the second year. The reason given for the decrease was that reacidification was taking place through the oxidation of sulfides of such metals as copper, iron, and aluminum.
5. Chemical fertilizer + mulch together (but not separately) had a significant and positive interaction on grass establishment and production.

Crop residue mulches, especially when the crops are adequately fertilized, have increased yields in all 50 states as well as in every country of the world. This statement can be confirmed by writing to the Land-Grant Universities and to the International Agricultural Research Centers.

10.2.4 Shifting Cultivation — An Organic Residue Management System

Nye and Greenland, in their preface, state that "Over 200 million people, thinly scattered over 14 million square miles (3.63 billion hectares — 8.96 billion acres) of the tropics, obtain the bulk of their food by the system of shifting cultivation."[12]

Shifting cultivation means partially clearing and burning the forest; growing cultivated crops for 2 to 3 years until yields decline; allowing the forest to sprout, reseed, and grow for 10 to 20 years; and then cutting, burning, and planting cultivated crops again. Cultivation is done with a village-made hand hoe. No commercial lime or fertilizer is used. The wood ashes add lime and fertilizers and the "resting" period rejuvenates the soil physically and makes it more productive. Success of the practice is dependent on the rapid regrowth of *woody* vegetation to supply lime and fertilizer when burnt. If all forest vegetation were killed during the cropping cycle, grasses and forbs would invade the site and the amount of organic matter would be too small to supply adequate lime and fertilizer when burned (Fig. 10–3).

Another vital function of organic matter in the success of shifting cultivation is that of preventing the crystallization of iron and aluminum in the soil. Within a forest, the microbial organic residues actually soften hardened (crystallized) iron and aluminum compounds. Furthermore, on the forest floor uniform soil temperatures and a more uniform soil moisture reduce crystallization.[13]

[12] P. H. Nye and D. J. Greenland, "The Soil Under Shifting Cultivation," Tech. Comm. No. 51, Commonwealth Bureau of Soils, Harpenden, England, 1960.

[13] Lyle T. Alexander and John G. Cady, "Genesis and Hardening of Laterite in Soils," Tech. Bul. No. 1282, U.S. Dept. of Agriculture, 1962, 90 pp.

(a)

(b)

(c)

Fig. 10-3. Shifting cultivation as practiced in Ghana, western Africa. (a) The crop is a tall grain sorghum (Guinea corn). In the left and right backgrounds, trees are growing rapidly and in a few years will crowd out all cultivated crops. Trees will remain for 10 to 20 years and will be cut again to plant field crops. Note the village-made hoe in the farmer's hand. The other two men are agricultural extension officers. (b) A 25-cm (10-in.) chunk of hardened plinthite (laterite, ironstone) dug from a nearby soil that developed by an inflow of iron and aluminum and hardened under *grass* vegetation. (c) A corn field on soil similar to that in (b) that must be declared a total crop failure because of hardened plinthite near the soil surface. An estimated 5% of the tropics has similar soils. (Courtesy Roy L. Donahue.)

473

The International Institute of Tropical Agriculture is conducting field research that supports the hypothesis that the proper use of organic residues under a system of minimum tillage can replace the 10 to 20 years of forest vegetation needed to rejuvenate the soil.[14]

10.3 ANIMAL MANURES

The range in attitudes toward animal manures throughout the world today extends from complete dependence on manures as fertilizers on subsistent farms to negative values on commercial feedlots. A cycle of similar contrasts in attitudes toward manures has existed during the 200 years of United States history. Until the late 1930s, United States farmers placed a high value on manures as fertilizers. With the low cost and plentiful supply of chemical fertilizers, starting in the mid-1940s, animal manures were seldom competitive. We are now in the stage of "What do you do with the surplus animal manures?" As recently as the 1960s, it was officially recommended that the surplus manures should be flushed into a nearby creek. Anyone doing this today would be in violation of State and Federal water quality laws.

The question of whether animal manures are a waste with negative value or a valuable resource also depends on the cost of chemical fertilizers. Using animal manures instead of chemical fertilizers has been calculated to be economical of total energy when the manure is hauled and spread within a radius of about 5 km (3 mi) from its source. However, most large feedlots are in arid and semiarid regions where there are not sufficient croplands, pasturelands, and rangelands for land spreading at environmentally safe rates. Heavy rates in arid regions result in a hazard of excess soil salt. Rational use of manures is almost site-specific.[15]

Total manure produced by *confined* domestic livestock and poultry is estimated at 900 million metric tons (990 million tons). However, only 10% of the tonnage and 5% of the nutrients are estimated to be spread on cropland. (The amount reaching pastureland has not been estimated.) By class of livestock, the manure estimated to be applied on cropland each year in the United States and its nitrogen, phosphorus, and potassium content is presented in Table 10–5.

10.3.1 Disposal Systems

To comply with new State and Federal laws, all animal manures must be confined to the farm or feedlot property. Furthermore, leachings from the manure are not to pollute the water table.

The solid waste system is the old-fashioned method of collecting the

[14] "Annual Report, 1975," International Institute of Tropical Agriculture, Ibadan, Nigeria, 1976.
[15] U.S. Library of Congress, "Opportunities for Energy Savings in Crop Production," Congressional Research Service. U.S. Government Printing Office, 1978, pp. 24–26.

Table 10-5. Manures Estimated to be Applied to Cropland Each Year in the United States (Dry Weight Basis)

Animal	Total Applied (million metric tons)	Composition (%)		
		N	P_2O_5	K_2O
Dairy cattle	15	2.0	1.4	2.0
Beef cattle	12	2.5	1.8	2.0
Swine	10	2.8	2.3	1.8
Broilers	3	3.8	3.0	1.9
Laying hens	2.5	4.5	4.0	2.0

SOURCE: M. H. Frere, "Nutrient Aspects of Pollution From Cropland." In *Control of Water Pollution From Cropland: Volume II, an Overview,* USDA-Agricultural Research Service and U.S. Environmental Protection Agency, 1976, pp. 59–90.

manure and manure plus bedding and spreading it on fields every day the weather, soil, and crop permitted. The liquid waste systems are all of more recent origin. A liquid pit is a fairly deep (> 1 m) pit that is never stirred with electric motors; therefore it is anaerobic. The oxidation ditch may be any size and depth into which manure is collected in slurry form. It is oxidized because it is stirred mechanically. A lagoon is a large receiving basin with more water than either a liquid pit or an oxidation ditch. When a lagoon is stirred, it is aerobic; when not stirred, anaerobic.

Field application consists of the following:

1. Spreading of solid wastes on soils (Fig. 10–4).
2. Injecting a slurry of water and manure directly into the soil or spraying it on the soil surface (Fig. 10–5).
3. Injecting a slurry of water and manure into a sprinkler irrigation system and applying it to the soil. The slurry for systems 2 and 3 may come from an aerobic or an anaerobic animal waste pit, pond, ditch, lagoon, or debris basin.
4. Composting of solid wastes in small (aerobic) piles or large (anaerobic) piles. Aerobic piles have a less offensive odor. Partial decomposition takes place during composting and some nitrogen is lost. A local market near metropolitan areas can usually be developed for aerobic animal manure compost. Compost not sold is more readily applied to soils because of a decrease in bulk during composting (for more on composts, see Section 10.5).
5. Dehydrating. Dehydrated (air-dried) animal manures were the most popular natural organic materials sold in the United States in 1976, representing almost 47% of all such materials. There is an increasing market for dehydrated manures and an increasing need for this type of disposal system.
6. Producing plants by hydroponics. The University of Maryland and the USDA-SEA-AR are researching the feasibility of using animal manure ef-

Fig. 10-4. Spreading animal manures on fields and pastures as fast as it is produced is an old custom that is still considered a good practice. The major precaution, however, is to avoid spreading manure on frozen soils; to do so will cause pollution of the water environment by biochemical oxygen demand (BOD), phosphates, and nitrates. (Courtesy New Idea Equipment Company, Coldwater, Ohio.)

fluent from a lagoon to produce livestock forage. Gravel is applied to serve as a rootbed. Grasses successfully produced include bromegrass, reed canarygrass, tall fescue grass, and grain rye. Neither timothy nor orchardgrass could be so grown successfully.[16]

10.3.2 Composition

The composition of domestic animal manures varies because of the kind and age of animal, feed consumed, bedding used and the waste management system.

A comparison of the dry matter, available N, total N, P_2O_5, and K_2O in manures from beef cattle, dairy cattle, poultry, and swine under the most com-

[16] J. Ronald Miner, (Ed.), "Farm Animal Waste Management," North Central Regional Publication 206, Iowa State University, Special Report 67, 1971.

(a)

(b)

(c)

(d)

Fig. 10–5. Liquid waste management systems. (a) A swine facility with slatted floor. The manure drops directly into the lagoon. (b) Motors with propellers keep the effluent circulating and aerated. (c) A liquid pit to contain all liquid wastes from a modern dairy facility. (d) A tank truck for loading any of the liquid effluents and spraying them on a field. To reduce odors and loss of ammonia the soil should be tilled immediately after application. (Courtesy USDA-Soil Conservation Service.)

Table 10-6. Solid Waste Systems of Managing Animal Manures: Dry Matter and Major Fertilizer Nutrient Percentage Composition at Time of Soil Application

Domestic Animal	Waste Handling System	Dry Matter (%)	Nitrogen (N) (%)		Phosphorus (P_2O_5) (%)	Potassium (K_2O) (%)
			Available[a]	Total[b]		
Beef cattle	Without bedding	15	0.20	0.55	0.35	0.50
	With bedding	50	0.40	1.05	0.90	1.30
Dairy cattle	Without bedding	18	0.20	0.45	0.20	0.50
	With bedding	21	0.25	0.45	0.20	0.50
Poultry	Without litter	45	1.30	1.65	2.40	1.70
	With litter	75	1.80	2.80	2.25	1.70
Swine	Without bedding	18	0.30	0.50	0.45	0.40
	With bedding	18	0.25	0.40	0.35	0.35

[a] Primarily ammonium nitrogen, which is available to plants the first year.
[b] Ammonium nitrogen plus organic nitrogen, available over several years.

NOTES:
lb/ton $\cong$ ½ kg/metric ton
$P_2O_5 \times 0.44 = P$
$K_2O \times 0.83 = K$

SOURCE: A. L. Sutton, J. V. Mannering, D. H. Bache, J. F. Marten, and D. D. Jones, "Utilization of Animal Waste as Fertilizer," Purdue University, 1D-101, 1975.

mon waste management systems is characterized in Tables 10–6 and 10–7. An itemized summary of the comparisons are detailed here:

1. The solid waste system is best for the conservation of organic matter.
2. Available N and total N are highest in the *liquid* waste systems. However, of the three liquid waste systems, the lagoon system has the lowest available and lowest total N.
3. Except for the lagoon system of managing swine manures, all liquid waste managing systems result in a higher P_2O_5 content (dry basis) than the solid waste systems.
4. The lagoon system results in a lower K_2O content of all manures using this system of management; otherwise, the liquid pit and the oxidation ditch system result in a higher amount of K_2O than does the solid waste system.

Summary. To conserve the most dry matter, the *solid* waste system is superior; for the maximum conservation of the three primary nutrients, the *liquid* pit and the oxidation ditch are the superior systems of management.

10.3.3 Crop Response

Except in sodic soils or in fine-textured semiarid soils where salt accumulation is a hazard, the use of animal manures on cropland soils and

Organic Amendments: Plant and Animal Residues

Table 10–7. Liquid Waste System of Managing Animal Manures: Dry Matter Percentage and Fertilizer Nutrient Composition at Time of Soil Application

Domestic Animal	Waste Handling System	Dry Matter (%)	Nitrogen (N)		Phosphorus (P_2O_5)	Potassium (K_2O)
			Available[a]	Total[b]		
				lb/1000 gal of Raw Waste		
Beef cattle	Liquid pit (anaerobic)	11	24	40	27	34
	Oxidation ditch	3	16	28	18	29
	Lagoon	1	2	4	9	5
Dairy cattle	Liquid pit (anaerobic)	8	12	24	18	29
	Lagoon	1	2.5	4	4	5
Poultry	Liquid pit (anaerobic)	13	64	80	36	96
Swine	Liquid pit (anaerobic)	4	20	36	27	19
	Oxidation ditch	2.5	12	24	27	19
	Lagoon	1	3	4	2	4

[a] Primarily ammonium nitrogen, which is available to plants the first year.

[b] Ammonium nitrogen plus organic nitrogen, available over several years.

NOTES:
1000 gal = 4.4 metric tons (4 tons)
27,154 gal (102,778 *l*) = A-in. (0.973 ha-cm)
$P_2O_5 \times 0.44 = P$
$K_2O \times 0.83 = K$

SOURCE: A. L. Sutton, J. V. Mannering, D. H. Bache, J. F. Marten, and D. D. Jones, "Utilization of Animal Waste as Fertilizer," Purdue University, 1D–101, 1975.

pastures enhances plant growth. Many experiments have been conducted comparing manures with an equivalent amount of nitrogen-phosphorus-potassium chemical fertilizers; most often the results favor the manures. Reasons hypothesized for the superiority of manures follow:

1. An increase in decomposing organic matter, resulting in a slow release of major and secondary plant nutrients from soil minerals and from the animal manures.[17]
2. An increase in infiltration rate and thus more rainwater and more irrigation water entering the soil.
3. A decrease in soil bulk density, resulting in a greater capacity for more air and water within the soil.

Current and specific crop responses from using animal manures are summarized from field research in Alabama and Maryland, followed by generalized historical responses, and an economic analysis in Connecticut.

[17] Leon Chesnin and F. N. Anderson, "Manure: Long-Term Study Shows Its Value to Western Soils," *Farm, Ranch, and Home Quarterly*, 22(3), Fall, 1975. University of Nebraska.

Pearl millet and cereal rye were grown on three Ultisols (highly weathered, acid soils) in Alabama, using dairy cattle manure at rates of 0, 22, 45, 90, 179, and 269 metric tons/ha (0, 10, 20, 40, 80, and 120 tons/A). On the zero manure plots, nitrogen-phosphorus-potassium chemical fertilizers were applied that were comparable to the composition of the 22 metric tons/ha (10 tons/A) rate of dairy cattle manure.[18]

Crop responses over a 3-year period were measured for cereal rye and pearl millet on three soil series: Lucedale, Decatur, and Dothan. The results are itemized in summary as follows:

1. Chemical fertilizers on the zero manure plot produced a greater yield increase than the 22-metric ton/ha (10-ton/A) rate of cattle manure.
2. All other manure application rates above 22 metric tons/ha (10 tons/A) resulted in yields greater than those on the plots receiving chemical fertilizer and no manure.
3. Manure rates greater than 90 metric tons/ha (40 tons/A) produced less, the same, or only slightly greater yields than the 90-metric tons/ha (40-ton/A) rate.

The University of Maryland recommends that poultry manure can be used very efficiently in the following ways:[19]

1. Poultry manure from Maryland broiler houses average 4.3% N, 2.8% P_2O_5, and 3.2% K_2O (oven-dry weight basis). The manure also contains measurable quantities of all secondary nutrients and micronutrients.
2. For corn, ryegrasses, small grains, and pastures, topdress 7 to 11 metric tons/ha (3 to 5 tons/A) of poultry manure and let it remain as a surface mulch or incorporate it in the soil by plowing or disking. When the pasture mixture contains both grasses and legumes, applications greater than these favor the grasses, which then crowd out the legumes.
3. Vegetable crops respond efficiently to applications of poultry manure. This is true for warm-season crops of long duration such as tomatoes and squash. Such crops need available nutrients at about the same rate as they become available from manure by microbial decomposition. Short-season and cool-season crops respond best to chemical fertilizers because they supply nutrients that are more quickly available.

Most crops are more efficient in their nutrition when a soil test has been made and the needed plant nutrients come from *both* manures and chemical fertilizers.

[18] B. D. Doss, Z. F. Lund, F. L. Long, and Luke Mugwira, "Dairy Cattle Waste Management: Its Effect on Forage Production and Runoff Water Quality," Auburn University Agricultural Experiment Station Bul. 485, Dec., 1976.

[19] V. A. Bandel, C. S. Shaffner, and C. A. McClurg, "Poultry Manure — A Valuable Fertilizer," Fact Sheet 39, Cooperative Extension Service, University of Maryland, 1972.

Long-term field studies that have compared domestic animal manures with chemical fertilizers include the following (with date of establishment):

1. Broadbalk Field, Rothamsted, England — 1843.
2. Jordon Soil Fertility Plots, University Park, Pa. — 1881.
3. Sanborn Field, Columbia, Mo. — 1888.
4. University of Rhode Island, Kingston, R.I. — 1889.
5. Ohio State University, Columbus, Ohio — 1893.
6. Voorhees Plots, Rutgers University, New Brunswick, N.J. — 1898.
7. Morrow Plots, University of Illinois, Urbana, Ill. — 1904.
8. Ohio State University, Wooster, Ohio — 1904.

Most of these long-term field plots have been discontinued but the results from all of them can be generalized in this way:

1. The lower the level of soil management, the greater the relative crop response from manures over chemical fertilizers.
2. The higher the level of soil management, the more the manures can be replaced by chemical fertilizers without sacrifice in crop production.

A critical analysis of the cost/benefit ratio of using dairy cattle manure on corn and alfalfa in Connecticut was made by C. R. Frink as of May, 1973.[20] An average of 12 experiments indicated that corn yields were increased by 1.5% per ton of manure applied and alfalfa yields by 2.4% per ton. The value of these increases were less than the cost to the farmer of hauling and spreading the manure.

Conclusion: There is no better environmental alternative to manure disposal than spreading on soils; therefore, the consumer must share some of this loss to the farmer by voting higher prices for farm products.

10.4 HUMAN WASTES (SOLIDS AND LIQUIDS)

For centuries Japan and China have maintained the productivity of their soils with "night soil." Many people are questioning why the United States and other countries cannot do the same.

Update Fact 1. Until the end of World War II in 1945, Japan depended mostly on night soil to fertilize its crops. However, Japan soon built chemical fertilizer plants to manufacture fertilizers for domestic use and for export. Today very little night soil is used because Japan is highly urbanized and highly industrialized and therefore chemical fertilizers are environmentally safer to use.

[20] C. R. Frink, "Agricultural Waste Management," Special Soils Bul. 34, Conn. Agr. Exp. Sta., New Haven, Conn., 1973, 15 pp.

Update Fact 2. About 80% of the people in China live in rural areas with no sewage system; "night soil" is therefore relatively easy to collect and the fields where it is used are nearby. This is in contrast to the United States where less than 5% of the people have no municipal or septic tank sewage system.

Update Fact 3. Starting in the 1960s and extending to 1976, China had built many chemical fertilizer plants, using coal as a source of hydrogen. Early emphasis was on establishing an ammonium bicarbonate (NH_4HCO_3 — 17.5% N) factory in nearly every county. Now most of the nitrogen fertilizer will be produced as urea [$CO(NH_2)_2$ — 46% N] in modern large-scale plants. In addition, China has become one of the world's largest importers of chemical nitrogen fertilizer.[21]

During an average day the 4 billion people in the world excrete 600 million kg (1.3 billion lb) of human wastes (night soil) (wet weight basis). This averages about 0.15 kg/day (0.33 lb/day). The night soil has an approximate composition of 0.57% N, 0.13% P_2O_5 (0.06% P), and 0.27% K_2O (0.22% K).[22]

10.4.1 Sewage Effluents (Solids and Liquids)

Sewage effluents from the 18,000 municipal treating plants in the United States include about 5 million metric tons (5.5 million tons) (dry weight) of sewage sludge each year. Calculated at the ratio of 3.4 million l of wastewater per metric ton of sludge (1 million gal/ton) (dry basis), the total amount of wastewater in the United States each year requiring environmentally safe disposal would be 17 trillion l (4.5 trillion gal).

As it enters the municipal sewage treating plants, an estimated 75% of the volume of influent originates from human effluents and 25% from industrial sources. In addition, most industrial plants have their own waste treatment system that also must be discharged safely into the environment.

Sewage treatment systems at the plant may consist of primary, secondary, tertiary, or "other." Effluents are sludges with varying amounts of water and wastewaters from which most solids have been removed. These 4 systems are described as follows:

1. Primary sewage effluent treatment may include flocculation of the solids by lime, iron, or aluminum salts. This removes from 50 to 75% of the suspended solids as sludge.
2. Secondary treatment includes primary treatment and in addition the sewage effluents may be sprayed over rocks or gravel beds (trickling filter)

---

[21] Norman E. Borlaug, "The Role of Fertilizers — Especially Nitrogenous — in Increasing World Food Production," In *Third Regional Wheat Workshop*, Tunis, Tunisia, April 28–May 2, 1975, pp. 218–40.

[22] Tomoji Egawa, "Utilization of Organic Materials as Fertilizers in Japan." In *Organic Materials as Fertilizers*, FAO Soils Bul. No. 27, 1975, pp. 266–67.

or aerated mechanically. This treatment removes as sludge from 80 to 95% of the suspended solids and 20 to 50% of the inorganic salts.

3. Tertiary treatment includes primary and secondary treatment plus floc-culation by a polyelectrolyte, adsorption on activated carbon, and some-times electrodialysis. This removes 98 to 100% of the suspended solids, 95% of the phosphates, and about half of the nitrates. These are all a part of the sludge.

4. "Other" processes sometimes used include sterilization with heat to kill pathogens, neutralization with lime to hasten decomposition and reduce objectional odors, and/or separation into wastewaters and sludges of specific moisture percentage to facilitate disposal by incineration, landfill burial, plant nutrition applications, or "dump" application on submarginal lands or on soils drastically disturbed by erosion, mining, or construction.[23]

All types of solid and liquid sewage effluents are applied to soils because that method of use/disposal is usually most economical. Environmental hazards of soil application include pathogenic organisms; heavy metals; and carbon, nitrogen, and phosphorus to eutrophy surface waters. The treatment systems used and kind of soil chosen for sewage effluent application will dominate the environmental impact (Fig. 10–6).

10.4.2 Soils for Sewage Effluent Disposal

Sewage effluents are an environmental hazard when applied on some soils. If there is a large percentage of fine montmorillonitic clay, infiltration is too slow and surface ponding results. If coarse sand and gravel extend to the surface, infiltration is so rapid as to contaminate the groundwaters beneath. The ideal soil for sewage effluent disposal has these parameters (Fig. 10–7):

1. The soil should be internally well-drained and medium textured, have a pH between 6.5 and 8.2, be high in organic matter, be as deep as possible to bedrock or other restrictive layer, and be level to very gently sloping with a uniformly smooth surface.

2. The surface of the soil should be supporting a dense stand of trees, shrubs, or perennial grasses. Surface and subsoil roots should be dense.

An example of a nonamended soil that approximates the ideal site for sewage sludge "dump" disposal or application as a fertilizer is the Fox series,

[23] Sources: (1) Bernard D. Knezek, and Robert H. Miller, "Application of Sludges and Waste-waters on Agricultural Land: A Planning and Educational Guide," North Central Regional Research Publication 235, 1976 and Ohio Agricultural Research and Development Center, Research Bul. 1090, 1976. Reprinted by U.S. Environmental Protection Agency, MCD-35, 1978.

(2) "Process Design Manual for Land Treatment of Municipal Wastewater," U.S. Environmental Protection Agency, U.S. Army Corps of Engineers, and U.S. Department of Agriculture, EPA 625/1-77-008 (COE EM 1110-1-501) 1977.

(a)

(b)

(c)

Fig. 10–6. During the years 1972 to 1974, the city of Muskegon, Michigan, established a model wastewater treatment system that integrated pollution abatement technologies for effluents from industry and domestic sources. Primary and secondary treatment is standard procedure, but tertiary treatment consists of using the wastewater to irrigate farm crops. The soil acts as a biological medium to adsorb viruses and toxic heavy metals and to nourish bacteria to denitrify excess nitrates. (a) 1. Chlorination plant; 2. Outlet lagoon (anaerobic); 3. Settling lagoon (anaerobic); 4. Three oxidation lagoons (aerobic); 5. Storage reservoir (anaerobic), 344 ha (850 A). (b) 1. Settling lagoon (anaerobic); 2. Three oxidation lagoons (aerobic) each with six aeration motors in operation; 3. Storage reservoir (anaerobic), 344 ha (850 A). (c) A sprinkler irrigation system that sprays all treated wastewater on corn (shown here) or other farm crops. The surplus water not transpired by plants or evaporated from the soil surface percolates below the plant root zone and is discharged by pumping into local streams, where it flows into Lake Michigan. The wastewater so treated meets all water quality standards. (Courtesy Muskegon County and Teledyne Triple R.)

Fig. 10–7. An ideal soil for the disposal of sewage effluent is this Wickham loam, a member of the fine-loamy, mixed, thermic family of Typic Hapludults. It has physical and chemical characteristics similar to the Fox series described in the text. (Courtesy H. C. Porter, Virginia Polytechnic Institute and State University.)

as described and characterized in the "Soil Survey Report of Hendricks County, Indiana."[24]

The Fox series is classified in the fine-loamy over sandy or sandy skeletal, mixed, mesic family of Typic Hapludalf. The soil is on a nearly level to moderately sloping outwash plain. The surface texture is a loam and permeability is moderate under cultivation but high under native hardwood vegetation. Depth to seasonal high water table is greater than 1.8 m (6 ft). The pH of the surface (A horizon) is 6.1 to 6.5, of the subsurface (B horizon) is 5.6 to 7.3, and of the subsoil (C horizon) is 7.4 to 8.4. Organic matter is moderate in the A horizon and low in the B and C horizons. Crop plant roots extend to a maximum depth of about 91 cm (3 ft). The C horizon is stratified sand and gravel.

To make the Fox series truly ideal as an environmentally safe place to dispose of large quantities of sewage sludge would require the following:

1. Liming all horizons of the soil to a depth of 91 cm (3 ft) to bring the pH between 6.5 and 8.2. Because the B horizon (depth of 36–79 cm, 14–31 in.)

[24] John M. Robbins, Jr., Earl E. Voss, Donald R. Ruesch, and Hezekiah Benton, Jr., "Soil Survey of Hendricks County, Indiana," National Cooperative Soil Survey, USDA-Soil Conservation Service in cooperation with the Purdue University Agricultural Experiment Station, 1974, pp. 12, 13.

may be below pH 6.5, a readily-soluble lime such as cement dust slurry, finely ground limestone slurry, or calcium hydroxide should be applied. Periodic tests of pH should be made so as to maintain this pH range (see Chapter 8).

2. Maintaining a dense perennial crop such as trees or pasture grasses on the soil by adequate fertilization and recycling as many plant residues as possible. This will reduce surface soil sealing, increase infiltration, decrease runoff and erosion, and increase crop yields.

3. Applying wastewaters on the Fox soil series because it is somewhat droughty but moderately permeable.

4. Monitoring the groundwaters for excess nitrates. Because the C horizon is sandy and gravelly, there is some hazard of surface waters high in nitrates leaching into the watertable. When this is suspected, water in shallow wells downslope should be tested for nitrates. If nitrates exceed 10 ppm of N (45 ppm of nitrate, apply less sewage effluent until the nitrates are below this critical level.

As a substitute for soil, in some states a few centimeters of sewage sludge are applied as a surface layer to a level area and commercial turfgrass sod is seeded. The turfgrass roots permeate the sludge layer, which can then be cut in strips and easily rolled, transported, and relaid on areas to be sodded. The practice has been judged as economical and environmentally safe. Sludges have also been used for many years as a fertilizer.

10.4.3 Sewage Sludges As Fertilizers

Early field research on the fertilizing value of sewage sludge may be summarized in general by this statement: Crop response to applications of 1 to 90 metric tons/ha (0.5 to 40 tons/A) was about the same as that for animal manures.[25]

Current research on sewage sludge is focused more on *maximum* applications without decreasing crop yields and without permanent heavy metal damage to the soil. Currently there are also two criteria for maximum soil applications; these are based on land use:

1. On submarginal agricultural lands such as mine spoils, gullied lands, noncommercial forest lands, and soils disturbed by construction activities, with suitable environmental protection, sewage sludge can be "dumped."

2. On lands suitable for *commercial* production of crops, pastures, ranges, and forests, application rates of sewage sludge are made to maximize plant nutrition for greater crop yields without a buildup of toxic metals or an imbalance of essential plant nutrients.

[25] (1) H. L. Russell, F. B. Morrison, and W. H. Ebling,"Fertilizer Value of Activated Sludge," Wisconsin Agr. Exp. Sta. Bul. 388, 1926. *Note: Activated* means aerobic.

(2) G. S. Fraps, "The Composition and Fertilizing Value of Sewage Sludge," Texas Agr. Exp. Sta.

The difference between environmentally safe and "dumping" rates of sewage sludge and applications to maximize plant nutrition may be one of composition of the sludge and/or buffering capacity of the soil as measured by cation exchange capacity.

One of the greatest hazards in using sewage sludge is the potential for adsorption of toxic quantities of heavy metal cations on the soil's exchange surfaces. Once adsorbed, these cations are "available" to plants but are not leachable by rainwater or irrigation water. If the metal cations are toxic to plants, animals, or people, there is no viable mechanism for reducing their concentration except to remove the entire soil in which they are concentrated or to cover it with enough soil not so contaminated. Both of these alternatives are too expensive for soil to be used for crop production. The only economic solution to the hazard of toxic heavy metal accumulation is to *avoid* the problem.

The University of Maryland has proposed guidelines designed to avoid such a hazard. Two assumptions are made in establishing the guidelines (Table 10–8):

1. The environmentally safe application rate of sewage sludge on *cropland* is directly proportional to the cation exchange capacity (CEC).
2. The environmentally safe application rate of sewage sludge on *sub-*

Table 10–8. Environmentally Safe Cumulative Sewage Sludge Heavy Metal Application on Soils of Various Textural Classes

| | *Maximum Cumulative Applications (kg/ha)*[a] | | | | | | | | | |
| | *Commercial Agricultural Land* | | | | | *Submarginal Agricultural Land* | | | | |
Heavy Metal	*Loamy Sands*	*Loams*	*Silt Loams*	*Clay Soils*	*Organic Soils*	*Loamy Sands*	*Loams*	*Silt Loams*	*Clay Soils*	*Organic Soils*
Cadmium (Cd)	0.08	0.15	0.3	0.4	1.5	0.16	0.3	0.6	0.8	3.0
Nickel (Ni)	1.1	2.2	4.3	6.5	21.5	2.2	4.2	8.6	13.0	43.0
Copper (Cu)	2.6	5.2	10.5	15.7	52.5	5.2	10.4	21.0	31.4	105.0
Zinc (Zn)	5.3	10.6	21.2	31.8	106.0	10.6	21.2	42.4	63.6	212.0
Lead (Pb)	10.6	21.2	42.4	63.6	212.0	21.2	42.4	84.8	127.2	424.0

[a] Kilograms per hectare × 0.892 = pounds per acre.

NOTES:

1. Based upon these approximate cation exchange capacities in milliequivalents per 100 grams of soil: Loamy sands, 5; loams, 10; silt loams, 20; clay soils, 30; and organic soils, 100. For the same textural class of mineral soil, the CEC will usually be higher in poorly drained soils and in soils of north-temperate regions because of a greater percentage of organic matter and less intense weathering of soil minerals.
2. Heavy metals already in the soil should be subtracted from these values.

SOURCE: "Guidelines for the Application of Sewage Sludge to Land," University of Maryland, Agronomy Mimeo. No. 10, 1976, 4 pp.

marginal land is estimated to be about twice that on commercial agricultural land (cropland).

Although extremely variable in composition, the University of Illinois has presented what they consider typical analyses from fresh, heated, and anaerobically digested sewage sludge (Table 10–9). The sewage sludge produces a plant yield response equal to that of a 5–5–0 chemical fertilizer with a value of about $22/metric ton ($20/ton) (dry basis). However, the sludge contains many more micronutrients and heavy metals.

Table 10–9. Composition of Typical Sewage Sludge

Element	Composition (dry weight basis) (%)
Essential for plant growth	
Organic carbon (C)	50
Nitrogen:	
Ammonium (NH_4^+)	2
Organic N	3
Total N	5
Phosphorus:	
P	3
P_2O_5	6.8
Potassium:	
K	0.4
K_2O	0.5
Calcium (Ca)	3
Magnesium (Mg)	1
Sulfur (S)	0.9
	parts per million (ppm)
Iron (Fe)	40 000
Zinc (Zn)	5 000
Copper (Cu)	1 000
Manganese (Mn)	500
Boron (B)	100
Not essential for plant growth	
Chromium (Cr)	3 000
Sodium (Na)	2 000
Lead (Pb)	1 000
Nickel (Ni)	400
Cadmium (Cd)	150

NOTES:
1. The sewage sludges analyzed were fresh, heated, and anaerobically digested.
2. Sewage sludges may vary by more than a factor of 10, depending on admixtures with factory wastes and methods of treatment. Before use, each batch of sewage sludge should be analyzed.

SOURCE: "Utilization of Sewage Sludge on Agricultural Land," University of Illinois Soil Management and Conservation Series No. SM–29, 1975. 7 pp.

Compared with poultry manure without bedding, sewage sludge contains about the same percentage of organic carbon (50% C), about the same percentage of total nitrogen (5% N), twice as much phosphorus (3% P), but only one fourth as much potassium (0.4% K). About the same calcium and magnesium occurs in both sewage sludge and poultry manure (3% Ca, 1% Mg). The greatest difference in composition, however, as well as the greatest hazard of toxicity are in the metal cations and sodium; sewage sludge contains a much greater concentration of iron, zinc, copper, manganese, and sodium than poultry manure.

Field research on the use of variable rates of sewage sludge in the States of Illinois and Wisconsin are now cited.

Illinois

Near Elwood, Illinois, on a Blount silt loam, classified in the fine, illitic, mesic family of Aeric Ochraqualfs, variable rates of sewage sludge were added to soils for corn and soybeans.[26] The sludge was an anaerobically digested liquid. The check plot received no sewage sludge but did have each year 269 kg/ha (240 lb/A) of N, 303 kg/ha (270 lb/A) of P_2O_2, and 224 kg/ha (200 lb/A) of K_2O. In addition, all sludge plots received 224 kg/ha (200 lb/A) of K_2O each year.

Average corn grain yields reported for the 7-year period, 1968–1974, were as follows:

Cumulative Liquid Sewage Sludge Applied During 7 Years
(metric tons/ha; dry weight basis)

	0	110	221	442
	Corn Grain Yield			
kg/ha	5901	6782	7282	7784
bu/A	94	108	116	124

Summary

1. The more the sewage sludge applied, the higher the yield.
2. Although the check plot received what was assumed to be adequate chemical fertilizers, corn grain yields averaged 13% *below* the lowest rate of sewage sludge.
3. After the first 2 years the soil pH changed from 5.6 to 4.9 and limestone was added to raise the pH to 6.0.

[26] Adapted from "Utilization of Sewage Sludge on Agricultural Land," University of Illinois, Soil Management and Conservation Series No. SM-29, 1975.

Wisconsin

On a Warsaw sandy loam, fine-loamy over sandy or sandy-skeletal, mixed, mesic family of Typic Argiudolls near Janesville, Wisconsin, yields of corn, alfalfa, and sorghum-sudangrass were measured, following variable rates of sewage sludge (Table 10–10).

The highest corn grain yields the first year resulted from chemical fertilizer alone. The highest first-year forage yield of sorghum-sudangrass was shared by chemical fertilizer and 47.1 metric tons/ha (21.0 tons/A) of sewage sludge. Carryover (residual) effects of fertilizer during the second year produced the lowest corn yield and almost the lowest forage yield of sorghum-sudangrass.

Summary

1. Residual effects of sewage sludge during the second year averaged 54% of first year's yields for corn and 80% for sorghum-sudangrass.

Table 10–10. Yields of Corn Grain and Sorghum-Sudangrass Forage Following Sewage Sludge, Irrigation Water, and Fertilizers

Treatment	Corn Grain (kg/ha)		Sorghum-Sudangrass (air-dry forage) (metric tons/ha)	
	First Year	Second Year	First Year	Second Year
Sewage sludge applied (dry weight basis) (metric tons/ha)				
0	3452	2322	4.6	4.1
7.8	5147	2327	6.5	4.8
15.7	5273	2636	6.8	5.0
31.4	6340	2762	6.6	5.5
47.1	5838	3013	7.6	5.9
Water	4017	2825	5.3	3.9
Fertilizer	6403	1444	7.6	4.0

NOTES:
1. On the "water" plot, only water was applied once, at the rate of 3.1 ha-cm (3 A-in.).
2. On the "fertilizer" plot, only chemical fertilizer was applied once, at the rate of 364 kg/ha (325 lb/A) of N, 247 kg/ha (220 lb/A) of P_2O_5, and 112 kg/ha (100 lb/A) of K_2O.
3. The "sludge" plots received all sludge in one application.
4. metric tons/ha × 0.446 = tons/A.
5. kg/ha × 0.892 = lb/A.

SOURCE: Leo Walsh, Art Peterson, and Dennis Kenney, "Sewage Sludge: Wastes That Can Be Resources," Research Report R 2 779, University of Wisconsin, 1976, 9 pp.

2. Chemical fertilizers alone gave highest yields the first year but near-lowest yields as residual carryover during the second year.
3. Irrigation water alone gave low yields for both crops both years.

10.4.4 Sewage Wastewaters

Sewage wastewaters may be a valuable water and nutrient resource to enhance growth of economic plants or a pollutant to lands and waters. Because of large volumes generated within a small space, wastewaters are usually pollutants in search of environmentally safe disposal systems. Potential polluting aspects of wastewaters include the following:

1. Applying wastewaters faster than the steady-state infiltration rate of the soil. Ponding on uplands will kill most vegetation and generate foul odors.
2. Spreading pathogenic organisms.
3. Increasing heavy metal concentrations of soils to levels toxic to plants or to people or animals eating the plants.
4. Increasing carbon, nitrogen, and phosphorus of surface waters and hastening their eutrophication.
5. Increasing the sodium and/or total soluble salt concentrations of soils to levels toxic to plants.

The hazard of spreading pathogenic organisms by land application of effluent wastewater is real. Drying and sunlight, however, are effective control agents for most disease organisms. Many states have laws governing the spreading of substances with a disease-producing hazard. Before applying wastewaters, the State Public Health Department should first be consulted (Table 10–11).

Heavy metal concentrations in wastewaters may also be a health hazard if

Table 10–11. Pathogens Present in Some Wastewaters and the Diseases They Cause in Humans

Pathogen	Disease
Vibrio cholera	Cholera
Salmonella typhi	Typhoid and other enteric fevers
Shigella species	Bacterial dysentery
Coliform species	Diarrhea
Pseudomonas species	Local infection
Infectious hepatitis virus	Hepatitis
Poliovirus	Poliomyelitis
Entamoeba histolytica	Amoebic dysentery
Pinworms (eggs)	Aseariasis
Tapeworms	Tapeworm infestation

SOURCE: "Sludge Treatment and Disposal: Volume I, Sludge Treatment," U.S. Environmental Protection Agency, EPA-625/4-78-012, 1978, 140 pp., p. 47.

Organic Amendments: Plant and Animal Residues

drinking water supplies are contaminated by them. This hazard is apparent in Table 10–12.

Wastewater contains active carbon that supplies energy for bacterial growth and reproduction. An indirect measure of active carbon is the biochemical oxygen demand (BOD). This is the amount of oxygen required for bacteria when they utilize the organic carbon for energy. When wastewater enters surface waters, bacteria utilize the organic carbon in their metabolism and thereby decrease oxygen needed for fish and other aquatic life. This is an example of eutrophication.

Wastewater is also fairly high in nitrogen and phosphorus, which enrich the surface waters and hasten the growth of algae and other aquatic plants. Dead residues from the abundant plants increase the BOD and further deplete oxygen in the water. The result is more eutrophication (Table 10–13).

For typical wastewater and average soils, the application of about 50 cm (20 in.) of wastewater to most nonleguminous field crops should supply about the right amount of carbon, nitrogen, and phosphorus to enhance plant growth. More potassium may be needed in humid regions and perhaps other nutrients to balance the nitrogen and phosphorus. A soil test as well as a test of the wastewater should be obtained to ascertain a balanced application for the particular soil and crop.

One of the oldest examples of the successful use of wastewaters for the purposes of disposal, to replenish the falling watertable, and to enhance plant growth is at Pennsylvania State University, University Park, Pennsylvania.

Starting in 1963 and continuing in 1980, secondary treated sewage effluents have been sprayed year around on field crops, abandoned fields, and

Table 10–12. Range of Concentration of Heavy Metals in Typical United States Wastewaters Compared with Drinking Water Standards

| Element | Sewage Wastewater (mg/l) | | | EPA Recommended Drinking Water Standard (mg/l) |
	Raw (Untreated)	Primary Treatment	Secondary Treatment	
Arsenic	0.003	0.002	0.005–0.01	0.05
Cadmium	0.004–0.14	0.004–0.028	0.0002–0.02	0.01
Chromium	0.02–0.7	0.001–0.3	0.01–0.17	0.05
Copper	0.02–3.36	0.024–0.13	0.05–0.22	1.0
Iron	0.9–3.54	0.41–0.83	0.04–3.89	0.3
Lead	0.05–1.27	0.016–0.11	0.0005–0.2	0.05
Manganese	0.11–0.14	0.032–0.16	0.021–0.38	0.05
Mercury	0.002–0.044	0.009–0.035	0.0005–0.0015	0.002
Nickel	0.002–0.105	0.063–0.20	0.10–0.149	No standard
Zinc	0.030–8.31	0.015–0.75	0.047–0.35	5.0

SOURCE: "Process Design Manual for Land Treatment of Municipal Wastewater," U.S. Environmental Protection Agency, U.S. Army Corps of Engineers, and USDA-EPA 625/1-77-008 (COE EM 1110-1-501) 1977, pp. 3–16.

Organic Amendments: Plant and Animal Residues

Table 10–13. Typical Raw (Untreated) Wastewater Characteristics

Parameter	Concentration (mg/l)
Biochemical oxygen demand (BOD)	200
Suspended solids	240
Nitrogen	
Organic N	15
NH_4^+	25
Phosphorus (P)	10

SOURCE: "Sludge Treatment and Disposal: Volume I, Sludge Treatment," U.S. Environmental Protection Agency, EPA–625/4–78–012, 1978, p. 22.

forests at rates of 2.5 cm (1 in.) or 5 cm (2 in.) a week. The two soils receiving the effluent waters are Typic Hapludults, one a Morrison sandy loam and the other a Hublersburg silt loam. Both soils are acid, well drained, and permeable.

A report based on 9 years of field and laboratory research can be summarized in this way:[27]

1. There are no indications that the sewage effluent had any influences detrimental to most plants.
2. Plots receiving the sewage effluent had significant increases in phosphorus, calcium, magnesium, boron, and sodium and significant decreases in manganese.

From this and related research, it can be concluded that plants most suitable for use with "dump" applications of wastewaters are goldenrod, aster, daisy, wild carrot, reed canarygrass, corn, white pine, and white spruce (Fig. 10–8). Plants sensitive to sewage wastewaters are Kentucky bluegrass, alfalfa, soybeans, small grains, and red pine.

10.5 COMPOSTS

Composting means aerobic mesophillic and thermophillic decomposition of organic residues to a relatively stable humuslike material. Aeration is essential for encouraging specific microbial proliferation of bacteria, actinomycetes, and fungi to achieve rapid decomposition with minimum objectional odors. Heat is generated by microbial activity. At the edges of the compost pile, mesophyllic bacteria are very active at temperatures of 25° to 40°C (77° to

---

[27] James L. Richenderfer, William E. Sopper, and Louis T. Kardos, "Spray-Irrigation of Treated Municipal Sewage Effluent and Its Effect on Chemical Properties of Forest Soils," USDA-Forest Service General Technical Report NE-17, 1975.

Fig. 10–8. White spruce that did not receive sewage wastewater (left) grew 1.2 m (4 ft) in height in 10 years. White spruce that received 5 cm (2 in.) of sewage wastewater a week (right) grew 5.5 m (18 ft) in height in 10 years. (Courtesy William E. Sopper, Pennsylvania State University.)

104°F) and within the pile the thermophiles proliferate at 40° to 65°C (104° to 149°F). Special inoculum can be purchased for hastening microbial proliferation in composts but it is not necessary. Organic residues contain all essential microbes. It is true, however, that the kinds of residues used for composts will determine the speed of decomposition and the quality of the finished product.

10.5.1 Materials for Composting

From popular literature, a person gets the impression that anything that rots will make good compost. This is not true. Materials that should never be put in the compost pile are weeds and grasses with mature seeds, diseased plants, and noxious plants that grow from vegetative sprigs such as bermudagrass and quackgrass.

Leaves/needles that should never be put in a compost include black walnut (toxic), eucalyptus (toxic and oily), red cedar (too prickly), and pine (too slow to decompose).

Desirable materials to put in a compost heap include those with a ratio of organic carbon to total nitrogen as narrow as possible, i.e., high in nitrogen. For example, when fresh organic materials are placed in a compost heap, microbes (bacteria, actinomycetes, and fungi) decompose them. Organic carbon in excess of that required for building microbial tissues is released by respiration as gaseous carbon dioxide (CO_2). Microbial tissues have a composition averaging 50% organic carbon and 5% to 10% total nitrogen (C:N ratio = 10:1 to 5:1) (see Table 10–2).

Conclusion. If organic materials with a C:N ratio wider than 10:1 are applied to the soil, microbes will take the nitrogen and plants will be nitrogen-deficient. There is one principal exception to this general rule: Materials high in *lignin* such as sawdust, bark, and paper have a wide C:N ratio but decompose very slowly because only a few microbes attack lignin. When added to the soil, these products do not deplete soil nitrogen to any great extent.

Table 10–14 lists the organic carbon, total nitrogen, and C:N ratio of materials commonly available for making compost. When the C:N ratio is extremely wide, the slow decomposition can be hastened by adding materials rich in nitrogen including nitrogen fertilizer.

Table 10–15 details the N, P_2O_5, and K_2O content of common organic materials. When rapid decomposition of compost is desired, a material must be used that has 1% or more of nitrogen (N). Alternatively a nitrogen fertilizer

Table 10–14. Organic Carbon (C), Total Nitrogen (N), and C:N Ratio of Materials Commonly Used for Making Compost (Dry Weight Basis)

Material	Organic Carbon (C) (%)	Total Nitrogen (N) (%)	C:N Ratio
Sewage sludge (dry weight basis)			
Aerobic	35	5.60	6:1
Anaerobic	30	1.90	16:1
Garbage, municipal	36	2.40	15:1
Alfalfa hay	43	2.40	18:1
Grass clippings, fresh	43	2.20	20:1
Garbage + paper, municipal	44	1.20	37:1
Leaves, freshly fallen	20–80	0.50–1.00	40:1–80:1
Moss peat	48	0.83	58:1
Corncobs	47	0.45	104:1
Red alder sawdust	50	0.37	135:1
Paper, mostly newspaper	43	0.25	172:1
Hardwood sawdust	50	0.20	250:1
Douglas-fir			
Old bark	59	0.20	295:1
Sawdust	51	0.07	728:1
Wheat straw	45	0.12	375:1
Pine sawdust	51	0.07	729:1

SOURCE: Principal reference: Raymond P. Poincelot, "The Biochemistry and Methodology of Composting," Connecticut Agr. Exp. Station Bul. 754, 1975.

Table 10–15. Major Plant Nutrients in Materials Commonly Available for Making Compost (Dry Weight Basis)

Material	Total Nitrogen (N) (%)	Phosphorus (P$_2$O$_5$) (%)	Potassium (K$_2$O) (%)
Ammonium nitrate	33.50	0.00	0.00
Blood, dried	13.00	1.61	0.84
Bone meal	4.00	23.23	0.00
Cottonseed meal	6.00	2.53	2.76
Fish meal	4.00	23.23	0.00
Hay			
Alfalfa	2.40	0.46	2.04
Timothy	1.00	0.23	1.56
Manure			
Cattle	2.00	2.30	2.40
Hog	2.00	1.78	1.80
Horse	1.70	0.69	1.80
Poultry	4.30	3.68	1.92
Sheep	4.00	1.38	3.48
Muriate of potash	0.00	0.00	60.00
Peat/muck	2.30	0.46	0.72
Sawdust, mixed hardwood and softwood	0.20	0.00	0.24
Seaweed (kelp)	0.60	0.00	1.32
Sewage sludge	5.00	6.80	0.50
Superphosphate	0.00	20.00	0.00
Tankage			
Animal	9.00	10.12	1.56
Garbage	2.50	1.61	2.76
Wood ashes	0.00	2.07	6.00

PRINCIPAL SOURCES: "Let's Take a Closer Look at Organic Gardening," Ohio State University, Ext. Bul. 555, 1973; "Utilization of Sewage Sludge on Agricultural Land," University of Illinois Soil Management and Conservation Series No. SM-29, 1975, 7 pp.

can be added to make the total nitrogen (N) of the organic material equal to 1% nitrogen. Organic materials low in phosphorus and/or potassium should be reinforced with the low nutrient(s). Most organic products of plant origin need dolomitic lime added to equal about 0.3% total calcium carbonate equivalent and slightly less of magnesium.

When microbes complete the composting process and the compost is "ripe" and ready to use, no one in the world can determine whether the ingredients were originally all organic or partly organic and partly inorganic. The reason is because the microbes have changed it *all to organic forms.* When chemical fertilizers are added to organic residues to hasten decomposition and increase plant nutrient percentage, microbes assimilate the fertilizers into their protoplasm. Upon death and decomposition of the microbial tissues, the nutrients from chemical and organic fertilizers and those from the original organic residues are all organic.

10.5.2 Sewage Sludge Compost[28]

Starting in the 1950s, extensive research and practical demonstrations were conducted on composting and use of municipal solid wastes. The pioneering work was done by the University of California and Michigan State University and by federally assisted projects at Johnson City, Tennessee, and Gainesville, Florida. The conclusions reached were *not* favorable toward the practice because of a low cost:benefit ratio, the presence of broken glass, and the variation in biodegradability of the refuse. Such difficulties are not encountered when municipal sewage sludge is composted.

The presently designated agency, USDA–SEA Agricultural Research, started a comprehensive sludge-composting project at the Agricultural Research Center at Beltsville, Maryland. The digested sludge was mixed with wood chips to improve aeration. The mixture consisted of two parts of wood chips to one of sludge, on a volume basis. This mixture was dumped on a 30-cm (12-in.) layer of wood chips to assure adequate aerobic decomposition. Under the layer of wood chips was a network of aeration pipes to which a suction fan and motor were connected. To reduce ambient emission odors, the exhausted air was directed into a pile of "finished" compost capable of absorbing the odors.

With intermittent forced aeration, the compost is "finished" in about 30 days but can remain for many more days without environmental harm. The "ripe" compost is offered free to the public and the demand has always exceeded the supply. As a commercial enterprise, such sludge compost could be sold probably at a profit.

A comparison of the composition of undigested sludge as it comes from the sewage treating plant and finished and screened aerated sludge compost made from it is presented in Table 10–16. Compared with beef cattle manure without bedding, sludge compost contains slightly less nitrogen, three times more P_2O_5, but only one-fifth as much K_2O.

10.6 GREEN-MANURE (COVER) CROPS

A "green-manure" crop is one that is grown and incorporated in the soil to add organic matter and to improve crop yields. A closely related term is "cover crop," which is one planted in orchards or between regular field crops in sequence to reduce erosion and loss of nutrients by leaching. Definitions merge when green-manure crops are known to reduce erosion and leaching and cover crops are incorporated in the soil to enhance crop yields. Both green-manure crops and cover crops may be harvested for seed/grain with very little loss of organic matter if the residues are returned to the soil.

Evidence of decreasing popularity of green-manure crops and cover crops can be found in the records of cost-sharing of the USDA-Agricultural

[28] "Sludge Treatment and Disposal: Volume 2, Sludge Disposal," U.S. Environmental Protection Agency, EPA–625/4–78–012, 1978, 155 pp., pp. 35–55.

Table 10–16. Average Composition of Undigested Sludge and Aerated Sludge Compost Made From It at Beltsville, Maryland

Parameter	Undigested Sludge (%)	Finished and Screened Aerated Sludge Compost (%)
Digested sludge		
Water	80	35
Total solids	20	65
Solid fraction		
Organic matter	50	50
Nitrogen (N)	2.5	0.9
Phosphorus (P_2O_5)	2.7	2.3
Potassium (K_2O)	0.6	0.2
Calcium (Ca)	2.9	2.6
Magnesium (Mg)	1.0	0.3
Sulfur (S)	0.9	0.4
	Parts Per Million (ppm)	
Zinc (Zn)	2,000	1,000
Copper (Cu)	600	250
Lead (Pb)	540	320
Boron (B)	23	27
Cadmium (Cd)	19	9
	Most Probable Number (MPN)/100 g	
Total coliforms	23 billion	97,000
Fecal coliforms	2 billion	3,000
Salmonella	6,000	0

NOTE: Elemental composition is on a *total* (not *available*) basis.

SOURCE: E. Epstein, and G. B. Wilson, "Composting Sewage Sludge," Proceedings, National Conference on Municipal Sewage Sludge Management, Pittsburgh, Pa., June, 1974, pp. 123–28.

Stabilization and Conservation Service.[29] The hectares (acres) of "Interim Cover Crops" decreased steadily from 2.8 million ha (6.8 million A) in 1966 to 405,000 ha (about 1 million A) in 1976.

Primarily, it is a question of changes in terminology as a result of more intensive cropping practices. For example, it is a fairly common practice to grow a warm-season row crop such as grain sorghum, soybeans, or corn, followed by a cool-season crop such as rye, barley, or wheat. Such a double-cropping practice with corn and wheat is portrayed in Fig. 10–9. The wheat, although harvested, serves as a good winter cover crop.

There are still times and places when the use of a green-manure (cover) crop is the best solution to some problems. For the control of the root rot

[29] "Agricultural Statistics," 1976, USDA, p. 532.

Fig. 10-9. One month before this corn was ready to harvest (left), an aircraft seeded wheat (September; Kentucky). (Right) The wheat in February that was seeded into the standing corn 6 months before. After the harvest of wheat, another warm-season crop will be planted in the wheat crop residues. Note that this double cropping resulted in twice the organic residues as in single cropping and the soil surface was protected against raindrop splash erosion every day in the year. (Courtesy Charles W. Martin, USDA-Soil Conservation Service.)

fungus, *Phymatotrichum omnivorium,* on cotton and sweet potato, the *only* recommendation is to grow them in annual rotation with cool-season green-manure (cover) crops such as hubam (annual) sweetclover, rye, wheat, or barley. Crownvetch is a unique perennial legume that can grow on root-rot infested soil. Farmer experience and research have demonstrated that the more fresh organic matter incorporated in the soil and the faster it decomposes, the greater is the fungal root rot control.[30]

No one knows why the practice of incorporating fresh organic matter is effective but there are several hypotheses:

1. Attack of the root rot fungal hyphae by enzymes excreted by certain bacteria and actinomycetes. This is known as *heterolysis.*
2. High concentration of ammonia resulting from organic decomposition.
3. High concentration of tannins in plants.
4. High concentration of soluble aluminum in soils caused by acidifying action during organic decomposition.

A second and unique use of a green-manure (cover) crop is on steep cropland. At Pennsylvania State University, University of Maryland, and Iowa State University, crownvetch has been established on steep and eroding fields and used successfully in minimum tillage management with continuous corn. Crownvetch has perennial roots but the tops die to the ground each winter. When corn-planting time comes, the corn is planted through the dead

---

[30] "Control" means "to reduce the severity of" and not to eliminate.

branches and in among the live crownvetch roots. Other crops grown successfully in among the live roots are winter wheat, winter barley, and spring oats.

Intensively fertilized crops such as peanuts, truck crops, and nursery crops usually need green-manure (cover) crops to reduce losses of nitrates and soil sediments. The most common crops for these land uses include cereal rye, sudangrass, singletary pea, oats, soybeans, sorghums, millets, wheat, vetches, alfalfa, and lupines. In Delaware, winter vetch with no nitrogen fertilizer increased corn yield about 25% more than an application of 112 kg/ha (100 lb/A) of nitrogen (N). Vetch turned under in Alabama resulted in yields of corn and cotton equal to those obtained with 135 kg/ha (120 lb/A) of nitrogen (N) fertilizer.

Green-manure (cover) crops have unique value except in areas of limited precipitation and high-cost irrigation water.

10.7   PEAT AND WOOD PRODUCTS

Peats, mucks, sawdusts, wood chips, pine needles, finely divided wood fibers, and tree bark are used as organic amendments for specialized purposes. All are acid-forming and are therefore used to acidify soil for acid-loving plants such as blueberries, rhododendrons, and azaleas. All of these organic amendments are used as a surface mulch over the top of seeded or planted areas to aid the establishment and growth of small-seeded and difficult-to-establish plants. Wood chips and supersaturated wood fibers are used frequently to stabilize soil slopes that have been eroded and/or disturbed by mining or construction activities until vegetation becomes established. Tree bark is often used as a mulch primarily for its beauty.

10.7.1   Peats

Peats and mucks are both sold in specialty urban markets as peat. However, there are two scientific classification systems currently in use: *commercial* and *soil taxonomic*.

Commercial System

There are five categories of the commercial system:

1. *Moss peat* — mainly sphagnum and hypnum mosses. The fibers are readily identified because they have not been noticeably decomposed. This is the most acid, the most expensive and the most desirable of the peats.
2. *Reed-sedge peat* — a mixture of residues from reeds, sedge grasses, and cattails.
3. *Peat humus* — from advanced decomposition of hypnum moss and reed-sedge peat.
4. *Muck soil* — highly decomposed peat of any source, usually mixed with mineral soil. Often sold as "topsoil."

5. *Sedimentary peat* — highly decomposed, colloidal, organic residues mixed with clay and sometimes marl. Of no value as an organic amendment.

Soil Taxonomic System

There are four suborders of the soil order of Histosols in the soil taxonomic system:

1. *Fibrists* — mostly undecomposed and recognizable organic fibers.
2. *Folists* — mostly recognizable leaf mat accumulations.
3. *Hemists* — "half" of the fibers are recognizable.
4. *Saprists* — unrecognizable plant fibers because of advanced decomposition.

All soil survey reports from the National Cooperative Soil Survey (Soil Conservation Service and a State agency, usually the land-grant university) that have been published since 1965 have classified organic soils according to this soil taxonomic system (see Chapter 1).

10.7.2 Wood Products

Wood products used as soil amendments include sawdust, wood bark, wood chips, and finely divided wood fiber.

From personal experience and from two scientific references, the authors suggest these guidelines on the safe use of sawdust:

1. Sawdust from redcedar and black walnut should not be used around growing plants because some toxins are present.
2. *Fresh* sawdust of any woody species should not be used. Green beans and tomatoes are especially sensitive.
3. Hardwood sawdust is superior to softwood sawdust for use around growing plants.

The average hardwood and softwood sawdust has this composition:[31]

Woody Plant Group	Nitrogen (N) (%)	Phosphorus (P_2O_5) (%)	Potassium (K_2O) (%)
Hardwoods (average of cottonwood and sweetgum)	0.20	0.03	0.15
Softwoods (mostly pines and Douglas-fir)	0.10	0.03	0.10

[31] (1) W. B. Bollen and D. W. Glennie, "Sawdust, Bark, and Other Wood Wastes for Soil Conditioning and Mulching," *Forest Products Journal,* January, 1961, 38–46.
(2) B. G. Blackman, Southern Forest Experiment Station, Stoneville, Miss. Unpublished data supplied to Roy L. Donahue by letter dated April 15, 1977.

Fig. 10-10. Wood chips are being used to mulch a strawberry bed in a home garden in West Virginia. (Courtesy U. S. Department of Agriculture.)

Tree bark, wood chips, or a mixture of the two are nearly ideal for use in stabilizing steep slopes until vegetation is established. (Tree bark, however, is less stable because it resists wetting and will readily wash and blow away.) This practice has been confirmed by the authors in Massachusetts, West Virginia, and Mississippi. The use of a slurry of wood fibers, grass seeds, and water applied under pressure has also been documented in humid areas of California on unstable slopes as a means of establishing adapted grasses.[32]

Wood chips for a garden mulch is a practice that is predicted to spread rapidly. One special reason for this prediction is that hardwood bark helps to control nematodes[33] (Figs. 10-10 and 10-11).

[32] "Methods of Quickly Vegetating Soils of Low Productivity, Construction Activities," U.S. Environmental Protection Agency Bul. No. EPA-440/9-75-006, July, 1975.

[33] (1) R. B. Malek and J. B. Gartner, "Hardwood Bark as a Soil Amendment for Suppression of Plant Parasitic Nematodes on Container-Grown Plants," *Hortscience,* 10(1), Feb., 1975, 33-35.

(2) "Annual Report, 1975," International Institute of Tropical Agriculture, Ibadan, Nigeria, 1976, p. 47.

(3) "PL 480 Research: Soil Amendments Reduce Root-Knots," *Agricultural Research,* Oct., 1974, p. 17.

Fig. 10–11. Wood chips are used extensively in forested areas for spreading on soils severely disturbed by mining or construction activities. The disturbed soils are graded, tilled, fertilized, limed (if needed), and seeded or planted to adapted vegetation. Wood chips are then spread to reduce raindrop splash erosion and to protect the new seedlings. (Courtesy USDA-Forest Service.)

SUMMARY

Plant and animal residues are not necessary for plant nutrition but they are essential for *efficient* plant nutrition. The *most efficient* plant production system, however, consists of an economical combination of using organic residues and chemical fertilizers. Over the decades, this combination is not just complementary (2 + 2 = 4) but synergistic (2 + 2 = 5).

Organic residues on the *surface* of the soil protect the soil against raindrop splash erosion, reduce surface crusting, reduce the extremes of surface soil temperature, and delay spring planting. When the organic residues are in the process of becoming soil humus, they supply some of *all* essential nutrients to plants; serve as the principal source of nitrates, organic phosphorus, organic sulfates, borates, molybdates, and chlorides; increase the cation exchange capacity; increase the buffering capacity; and make phosphorus and most micronutrients more readily available to plants over a wider pH range. Organic residues release essential nutrients faster by microbial decomposition when their ratio of organic carbon to total nitrogen is no wider than about 20:1.

Farmers are spreading more animal manures and sewage effluents every year, although the same increase in yield may often be obtained more economically with chemical fertilizers. U.S. Public Law 92–500 as amended relates to water quality and makes it mandatory for farmers to dispose of animal manures safely. Sewage effluents are being spread by many farmers on a trial basis. Urbanites are also buying more dried animal manures and more dried sludges.

Multiple cropping, minimum tillage, closer spaced crops, and more efficient use of chemical fertilizers have all increased the amount of crop residues per hectare (acre). If possible, all of these residues should be left on the soil surface or incorporated in the soil. Some farmers burn these residues, feed them to domestic animals, or make them into alcohol.

The success of shifting cultivation in the humid tropics is dependent on large quantities of organic residues produced by trees in rotation for 10 to 20 years, with 2 to 3 years of cultivated crops. There is recent field research, however, confirming the fact that surface organic residue management with minimum tillage can substitute for the forest growth cycle as a means of keeping the soil continuously productive.

Principal hazards of using sewage sludges include pathogens, toxic metal buildup, and a physiological imbalance of plant nutrients. Sewage wastewaters, although a slight health hazard, are a practical means of replenishing falling water tables.

Animal manures must be disposed of in an environmentally safe manner. This creates a pressure to spread too much on some *level* areas. Overloading the soil with excess manure can cause salt injury; human diseases from the animals; nitrates in wells, ponds, and streams; and grass tetany in livestock.

Japan has almost quit using night soil as a fertilizer and China is rapidly substituting it with chemical fertilizers.

Compost pits are for people who like to do a large amount of physical labor; for those more efficient of their energy, sheet composting in place is recommended. (The authors use both kinds.)

Green-manure crops and cover crops are being replaced by multiple cropping and close spacing of highly fertilized cash or forage crops. For the control of root rot of cotton and sweet potato, however, there is no substitute for certain cool-season green-manure crops.

The use of composts and peats is increasing, especially among urban flower and vegetable gardeners.

PROBLEMS

1. If organic residues are as beneficial as claimed in this chapter, why are not more of them used?

2. Why are chemical fertilizers used when organic residues are so plentiful?

3. Compare animal manures and sewage sludges as to
   (a) Chemical composition.
   (b) Crop production increases.

(c) Hazards of use in the environment.

(d) Hazards of use in crop production.

4. Write a paragraph on how to make and use compost scientifically.

5. Explain:

(a) How to control the root rot fungus of cotton and sweet potato with fresh organic residues.

(b) How to reduce nematode numbers by the use of sawdust or wood chips.

(c) The role of immobilization and mineralization in relation to nutrient cycling and plant nutrition.

(d) The various methods of treating human wastes prior to recycling in the environment.

REFERENCES

Alexander, M., *Introduction to Soil Microbiology,* 2nd ed., New York: John Wiley & Sons, 1977.

"Application of Sewage Sludge to Cropland: Appraisal of Potential Hazards of the Heavy Metals to Plants and Animals," EPA 430/9-76-013, Nov., 1976, U.S. Environmental Protection Agency.

"Conservation Tillage For Wheat in the Great Plains," Extension Service, USDA Bul. No. PA-1190. U.S. Government Printing Office, Washington, D.C., 1977.

Davies, J. W., (Ed.), "Mulching Effects on Plant Climate and Yield," World Meteorological Organization of the United Nations, Tech. Note No. 136, 1975.

Donahue, Roy L., Roy H. Follett, and Rodney W. Tulloch, *Our Soils and Their Management,* Danville, Ill.: The Interstate, 1976.

Donahue, Roy L., Raymond W. Miller, and John C. Shickluna, *Soils: An Introduction to Soils and Plant Growth,* Englewood Cliffs, N.J., Prentice-Hall, Inc., 1977.

Elliott, L. F., and F. J. Stevenson, (Eds.), *Soils for Management of Organic Wastes and Waste Waters,* American Society of Agronomy, Crop Science Society of America, and Soil Science Society of America, Madison, Wis., 1977.

Hinesly, T. D., R. E. Thomas, and R. G. Stevens, "Environmental Changes from Long-Term Land Application of Municipal Effluents," EPA 430/9-78-003, 1978.

Knezek, Bernard D., and Robert H. Miller, "Application of Sludges and Wastewaters on Agricultural Land: A Planning and Educational Guide," North Central Regional Research Publication 235, 1976 and Ohio Agricultural Research and Development Center Research Bul. 1090, Wooster, Ohio. Reprinted by U.S. Environmental Protection Agency, MCD-35, 1978, variously paged.

Land Application of Waste Materials, Soil Conservation Society of America, Ankeny, Iowa, 1976.

Linn, D. M., and J. W. Doran, "Microbial Changes Associated With Reduced Tillage and Residue Management," *Agronomy Abstracts,* American Society of Agronomy, Madison, Wis., Dec. 3-8, 1978, p. 142.

Midgley, Alvin R., and David E. Dunklee, "The Availability to Plants of Phosphates Applied with Cattle Manure," Agr. Exp. Sta. Bul. 525, 1945, University of Vermont, Burlington, Vt.

Multiple Cropping, American Society of Agronomy, Spec. Pub. 27, 1976.

Organic Materials as Fertilizers, Food and Agriculture Organization of the United Nations, 1975.

Oschwald, W. R., (Ed.), "Crop Residue Management Systems," American Society of Agronomy, Madison, Wis., 1978.

Ponnamperuma, F. N., R. U. Castro, and A. B. Capati, "Straw as a Source of Nutrients for Wetland Rice," *International Rice Research Newsletter* 4(6), IRRI, Manila, Philippines (Dec., 1979) 18–19.

"Process Design Manual for Land Treatment of Municipal Wastewater," U.S. Environmental Protection Agency, U.S. Army Corps of Engineers, and USDA, EPA 625/1-77-008 (COE EM 1110-1-501), 1977.

"Reclamation of Drastically Disturbed Lands," American Society of Agronomy, Crop Science Society of America, and Soil Science Society of America, Madison, Wis., 1978.

Robinson, D. W., and J. G. D. Lamb, (Eds.), *Peat in Horticulture,* New York: Academic Press, 1975.

"Soil Taxonomy: A Basic System of Soil Classification for Making and Interpreting Soil Surveys," USDA-Soil Conservation Service, Washington, D.C., *Agriculture Handbook 436,* Dec., 1975.

Springfield, H. W., "Using Mulches to Establish Woody Chenopods — An Arid Land Example." In *Special Readings on Conservation,* FAO Conservation Guide 4, Food and Agriculture Organization of the United Nations, Rome, Italy, 1978, pp. 51–64.

Tisdale, Samuel L., and Werner L. Nelson, *Soil Fertility and Fertilizers,* New York: Macmillan Pub. Co., 1975.

USDA, "Improving Soils with Organic Wastes," 1978. Wash., D.C.

USDA-SEA, "Utilization of Wastes on Land: Emphasis on Municipal Sewage," National Workshop, July 15–17, 1980 at University of Maryland; August 12–14, 1980 at Anaheim, California.

U.S.-EPA, "Manual for Composting Sewage Sludge by the Beltsville Aerated-Pile Method," EPA-600/8-80-OZZ, May, 1980. Municipal Environmental Research Laboratory, Office of Research and Development, U.S.-EPA, Cinncinnati, Ohio.

"Utilizing Municipal Sewage Wastewaters and Sludges on Land for Agricultural Production," North Central Regional Extension Publication No. 52, Nov., 1977.

CHAPTER ELEVEN

Economic and Efficient
Use of Fertilizers and Lime

11.1 INTRODUCTION

Selecting the proper analysis and amount of fertilizer and lime for a particular crop is an important farm management problem. The economic optimal amount and the analysis of fertilizer and lime to use depend on both economic and agronomic factors.

The commercial farmer is interested in maximum profits per hectare (acre). This is the point at which the last dollar spent to produce these yields returns just a dollar (Fig. 11–1). Economic factors that affect the optimum rate of fertilization are the prices paid for fertilizer, other production goods and services, and the prices received for the farm products. Other economic factors, such as organization of the farm business, availability of capital, and alternative opportunities for use of available capital may also influence fertilization practices. The grower realizes that he/she must spend money to make money. This is certainly true for fertilizers, lime, and (in some cases) other soil amendments.

Some agronomic factors to be considered are physical and chemical properties of the soil, past soil treatment, previous crops, and soil moisture levels. The effect of any one of these factors varies within and between seasons, thereby making yield response to fertilization both variable and complex in nature. The complexity is further increased by other factors such as cultivation; seedbed preparation; date and rate of seeding; stands; variety;

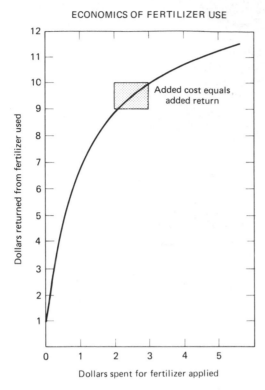

ECONOMICS OF FERTILIZER USE

Dollars returned from fertilizer used (y-axis)

Added cost equals added return

Dollars spent for fertilizer applied (x-axis)

Fig. 11-1. Illustration of a hypothetical yield response curve that shows dollars returned for dollars expended on fertilizer applied.

water control; fertilizer placement; weed, insect, and disease control; and harvesting practices. These factors, acting in combination, can bring about extreme variations in yield response from fertilizer during any one cropping season.

Weather is one of the most important and least predictable factors affecting crop response to fertilization. One of the main effects of climate is associated with the seasonal distribution of rainfall. However, the distribution of rainfall can be balanced on some farms with irrigation during dry periods.

It is interesting to compare the costs of certain farm inputs shown in Fig. 11-2. In the period 1967–1978 there was little change in fertilizer price. Fertilizer prices increased rapidly during the 1973–1975 period and decreased in the 1975–1978 period; they represent an input that has increased the least, pricewise, since 1967. Producers must maximize wise fertilizer use to take advantage of these relationships. They can substitute relatively low-cost fertilizers and improved technology for more expensive land, labor, and machinery. Although the price of fertilizer is expected to rise in the next few years, it is not expected to rise as fast as other farm inputs.

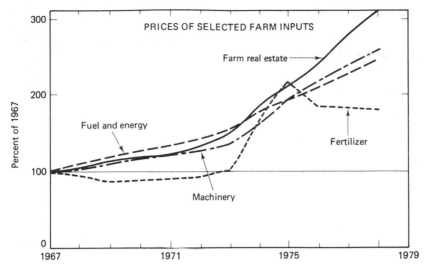

Fig. 11-2. Prices of selected farm inputs 1967–1978. Fertilizers are the best "bargain." (Source: *Handbook of Agricultural Charts*, USDA Special Supplement of AH561, Nov., 1979, unpaged.)

11.2 LAW OF DIMINISHING RETURNS

The *law of diminishing returns* recognizes that yield responses to a limiting growth factor are usually nonlinear. If all other factors remain the same, as the limiting growth factor (plant nutrient) is added in units of equal size, then each increment of yield response is smaller than the preceding increment. In 1909, this law was given mathematical expression by Mitscherlich.[1] This diminishing returns relationship results in a distorted sigmoidal curve similar to the one respresented in Fig. 11-3. The response curve shown in Fig. 11-3 extends from the zero level of nitrogen to an upper level well past that required for the maximum yield. In practice, soil nutrient levels are always greater than zero and commonly exceed *a*, the level corresponding to the inflection point on the curve. The maximum response occurs at *e* level of nitrogen. In the study of economically optimal levels (most profitable) of fertilization, the investigator is generally not concerned with the nature of the response function past the maximum yield. Thus, for finding the optimal level of fertilization, many investigators limit their range of interest between points *b* and *f* on the response curve. The optimum (most profitable) rate of nitrogen occurs at point *d* on the response curve. This is the point where the slope of the fertilizer cost line equals the slope of the response curve.

---

[1] E. A. Mitscherlich, "Das Gesetz des Minimums und das Gesetz des abnehmenden Bodenertrags," *Landwirtsch. Jahrb.*, 38:537–552, 1909.

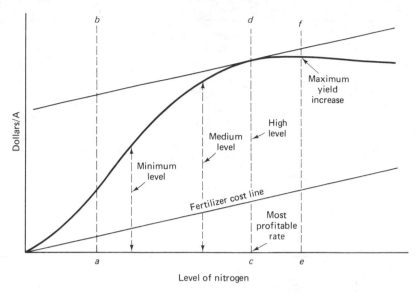

Fig. 11-3. A hypothetical response function showing the relationship between level of nitrogen applied and dollars per acre returned. There is an inflection point in the curve at *a* level of nitrogen and the maximum yield reponse occurs at *e* level of nitrogen. The optimum economic rate of nitrogen occurs at *c* level of nitrogen, and the optimum yield occurs at the point of tangency (*d*). Maximum yield is at point *f*.

Data obtained from field experiments look something like the curve shown in Fig. 11-3. From this curve, it is clear that a maximum yield increase is possible but that when the cost of fertilizer is considered, the maximum yield may not be the most profitable rate. It is important that economic aspects of fertilization be taken into account when fertilizer recommendations are beng made. When economics of fertilization are considered, other levels of fertilizer become logical. Beginning at the minimum level of fertilizer shown in Fig. 11-3, profits increase up to the point of the high level of fertilization. A choice of which level to use depends on the farmer's capital position, various agronomic factors, and the farmer's overall management ability.

The law of diminishing returns is illustrated by the yield response of corn to nitrogen fertilization using data from six site years of preplant nitrogen application in central Iowa (Fig. 11-4). Yields increased up to 180 lb/A (202 kg/ha) of nitrogen. Note that each additional 20 lb/A (22 kg/ha) of nitrogen produced a smaller yield increase than the previous 20-lb-increment. It is the characteristic of crop response to fertilizers that makes the calculation of economic optimum rates possible and their observation necessary. Decisions on fertilizer use are different from other agronomic inputs because there is a choice of rates. This can be contrasted to plowing where one can either plow or not plow, but not "half plow."

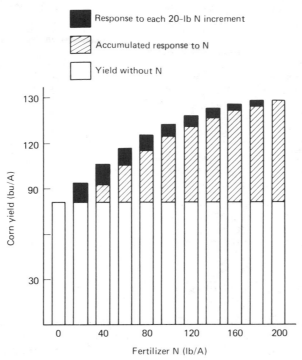

Fig. 11-4. Diminishing returns in yield response to fertilizer nitrogen. (Source: R. D. Voss, J. T. Pesek, and J. R. Webb, "Economics of Nitrogen Fertilizer for Corn," Iowa State Univ. Ext. Ser. Pm. 651, 1975, 12 pp.)

11.3 ECONOMIC RETURNS FROM FERTILIZER AND LIME

Crop profits are closely associated with high yields. Obviously, the most profitable crop years are experienced when yields are high and the selling price is high. Higher yields usually mean higher total production costs per acre, yet lower production costs per unit of yield, and (in the final analysis) more profit per acre. The data from Michigan (Table 11-1) shows that increasing the nitrogen rate increased net return and lowered the cost of production per unit of yield. Hence, the farm is better able to compete.

Fixed production costs such as land (including interest, taxes, and maintenance), machinery, and labor continue to increase. These costs are about the same regardless of yield. With a given set of management practices, fertilizer is the main input that will increase yields. However, when farmers look for a place to cut costs, they often decide to cut the fertilizer rate. This reduces cost but will probably reduce yields and profits and increase production cost per unit of yield.

Table 11-1. Response of Corn to Nitrogen Fertilization in Michigan

Rate of N (lb/A)	Yield (bu/A)	Gross Return ($/A)	Production[a] Cost ($/A)	Net Return ($/A)	Production Cost ($/bu)
0	83	$207.50	$275.00	− $67.50	$3.31
50	137	342.50	292.22	50.28	2.13
100	151	377.50	300.80	76.70	1.99
150	171	427.50	313.34	114.16	1.83

[a] Production cost based on $275/A plus 15¢/lb for N and 18¢/bu for the extra yield harvested. Corn price $2.50/bu. All other nutrients were assumed to be optimum. (W. L. Nelson, Planning for Maximum Economic Yields, 1977.)

NOTES:
kg/ha = lb/A × 1.121.
kg/ha = bu/A × 62.776.

11.3.1 Optimum Fertilizer Rate as Affected by Crop Prices

The optimum fertilizer rate is affected by both price of the crop and price of fertilizer. For example, the response to nitrogen for wheat in south-central Kansas is shown in Table 11–2. The optimum rate of nitrogen to apply as the price of wheat and nitrogen varies is shown in Table 11–3. For example, Table 11–3 indicates that the optimum rate of nitrogen would be 70 lb/A (78 kg/ha) if wheat were $2.00/bu and nitrogen were 10¢/lb. But if wheat were $4.00/bu and nitrogen were 40¢/lb, for example, 34 lb/A (38 kg/ha) of nitrogen would be the optimum rate of nitrogen. The data in Table 11–3 also suggests that sharp increases in nitrogen prices with little or no increase in wheat prices sharply changes the optimum fertilizer rate. In other words, higher

Table 11-2. Response of Wheat to Nitrogen Fertilization in South-Central Kansas

Rate of N (lb/A)	Yield (bu/A)	Yield Increase (bu/A)
0	32.2	0
30	48.7	16.5
60	50.8	2.1
90	52.2	1.4

NOTES:
kg/ha = lb/A × 1.121.
kg/ha = bu/A × 67.26.

SOURCE: Frank Orazem, L. S. Murphy, and D. A. Whitney, "Optimum Nitrogen Rates for Wheat," Kansas State University Ext. Ser. MF-356, 1974, 4 pp.

Table 11-3. Optimum Pounds of Nitrogen per Acre under Indicated Wheat and Nitrogen Prices in South-Central Kansas

Price per lb of N ($)	Wheat Price per Bushel ($)							
	2.00	2.50	3.00	3.50	4.00	4.50	5.00	5.50
	Optimum Pounds of Nitrogen per Acre							
0.10	70	74	77	80	83	85	88	90
0.15	40	50	62	73	76	78	81	83
0.20	34	39	43	57	70	73	76	80
0.25	32	34	37	40	48	60	70	76
0.30	30	31	34	36	38	46	52	60
0.35	28	30	32	34	36	39	41	47
0.40	25	28	30	32	34	37	39	41

NOTES:
kg/ha = lb/A × 1.121.
kg/ha = bu/A × 67.26.

SOURCE: D. A. Whitney and D. E. Kissel, "Optimum Nitrogen Rates for Wheat," Kansas State University Ext. Ser. MF-47, 1980, 4 pp.

rates of nitrogen application are more profitable under conditions of lower fertilizer prices or higher wheat prices.

The change in the optimum rate of fertilizer as the price of a crop drops is relatively small compared to the total amount of fertilizer applied. Field trials in Iowa show that the price of corn has little effect on the optimum rate of nitrogen and potassium fertilizer (Table 11-4). For example, as the price of corn dropped from $3.00 to $2.00/bu, the optimum rate of nitrogen at 15¢/lb decreased only 7 lb/A (8 kg/ha). Accordingly, the optimum rate of K_2O decreased only 20 lb/A (22 kg/ha) when crop price dropped from $3.00 to $2.00/bu. The higher nitrogen price also has a relatively minor effect on the optimum nitrogen application.

Table 11-4. How the Price of Corn Affects Fertilizer Rates (Iowa)

Corn Price ($/bu)	Optimum N Rate (lb/A)		Optimum K_2O Rate 9¢/lb K_2O-Low Soil Test (lb/A)
	15¢/lb N	25¢/lb N	
$1.50	166	146	100
2.00	173	159	125
2.50	178	166	135
3.00	180	171	145

NOTE: kg/ha = lb/A × 1.121.

SOURCE: Potash and Phosphate Institute.

11.3.2 Economics of Lime

The application of agricultural limestone, if needed, results in improved crop yields for an extended period. The cost of lime should not be charged against the first year's returns because a lime application lasts from 4 to 12 years. If is difficult to place a specific figure on the cost of lime because the source and transportation costs are highly variable. The cost at the quarry or source is obviously not the only factor involved. Spreading and transportation costs should be included when determining the economic response to lime.

In Table 11-5 are some calculated annual returns from lime at different application costs and yield increases. These data show that lime is a good investment even where the yield response is only 2 bu/A (135 kg/ha) of soybeans. When lime is needed, it can be profitable for both landowners and

Table 11-5. Annual per Acre Returns From Lime on Soybeans Under Specified Conditions (to Nearest Dollar)[a]

Total Cost of Lime per Acre Including Interest	Annual Yield Increase per Acre[b] (bu)																				
	2			3			4			5			6			8			10		
	Soybean Price per Bushel																				
	5	7	9	5	7	9	5	7	9	5	7	9	5	7	9	5	7	9	5	7	9
	Annual Returns ($/A)																				
$10	8	12	16	13	19	25	18	26	34	23	33	43	28	40	52	38	54	70	48	68	88
12	8	12	16	13	19	25	18	26	34	23	33	43	28	40	52	38	54	70	48	68	88
14	7	11	15	12	18	24	17	25	33	22	32	42	27	39	51	37	53	69	47	67	87
16	7	11	15	12	18	24	17	25	33	22	32	42	27	39	51	37	53	69	47	67	87
18	6	10	14	11	17	23	16	24	32	21	31	41	26	38	50	36	52	68	46	66	86
20	6	10	14	11	17	23	16	24	32	21	31	41	26	38	50	36	52	68	46	66	86
22	6	10	14	11	17	23	16	24	32	21	31	41	26	38	50	36	52	68	46	66	86
24	5	9	13	10	16	22	15	23	31	20	30	40	25	37	49	35	51	67	45	65	85
28	4	8	12	9	15	21	14	22	30	19	29	39	24	36	48	34	50	66	44	64	84
30	4	8	12	9	15	21	14	22	30	19	29	39	24	36	48	34	50	66	44	64	84
32	4	8	12	9	15	21	14	22	30	19	29	39	24	36	48	34	50	66	44	64	84
36	3	7	11	8	14	20	13	21	29	18	28	38	23	35	47	33	49	65	43	63	83
40	2	6	10	7	13	19	12	20	28	17	27	37	22	34	46	32	48	64	42	62	82
45	1	5	9	6	12	18	11	19	27	16	26	36	21	34	45	31	47	63	41	61	81

[a] This does not account for any increased harvesting or hauling costs associated with yield increases.

[b] Lime prorated over a 5-year period.

NOTES:
kg/ha = bu/A × 67.26.
$/ha = $/A × 2.471.

SOURCE: Woody N. Miley, "Liming Soybeans," University of Arkansas Ext. Ser. Leaflet 534, 1974.

renters. Too often each waits on the other, and both lose money as soil acidity increases through the years (see Chapter 8).

11.4 FERTILIZER-ENERGY RELATIONSHIPS

The production of satisfactory quantities of food to feed the world population is bound to the production and distribution of adequate quantities of chemical fertilizers. Likewise, the production and distribution of these chemical fertilizers is dependent on reliable sources of large quantities of fossil fuel energy (Fig. 11-5).

An estimated 0.8% of the total United States annual energy consumption is presently being used for fertilizer production. The production of anhydrous ammonia, which requires approximately 6300 Kcal/lb (13,780 Kcal/kg) of nitrogen represents nearly two-thirds of the total energy used for all fertilizers. Energy consumption values for phosphorus and potassium fertilizer are estimated at 1250 Kcal/lb (2755 Kcal/kg) of phosphate (P_2O_5) and 1000 Kcal/lb (2205 Kcal/kg) of potash (K_2O).[2]

Synthesis of ammonia is the major energy requirement for nitrogen fertilizers. A ton of ammonia typically requires about 38,130 ft^3 of natural gas. In addition, 9 gal of fuel oil and 54 kw of electricity are required (Table 11-6). The amount of natural gas used to produce 5 tons (4.5 metric tons) of ammonia would heat an average home in Iowa for 1 year. This same amount of natural gas, converted to nitrogen fertilizer, could result in enough extra corn production to satisfy the minimum protein and caloric requirements of 275 people for 1 year.

Because most nitrogen fertilizers are based on ammonia made from natural gas, a major problem looms on the horizon. Several nitrogen plants have faced shortages of natural gas. Shortages are expected to become more severe in a few years. Natural gas will become more expensive. The fertilizer industry will be forced to seek other sources of feedstock. Coal, our most abundant hydrocarbon, seems to be the best choice for the near future. The Tennessee Valley Authority (TVA) has the responsibility for fertilizer research and is giving top priority to the development of an alternative to natural gas as a source of hydrogen for ammonia production.[3]

Although the use of fuel oil, naphtha, and coal looks best for now, other ways to produce hydrogen could be attractive in the future. Researchers, for example, have reviewed existing and evolving technology for electrolysis of water. The large electrical energy requirement keeps this method from being feasible now, but new developments in power generation together with development of more efficient electrolytic cells may change that.

Energy use for the production of fertilizer, however, is not the entire story. Energy required for transporting fertilizers to the dealer and then to

[2] R. H. Follett, "Proper Fertilizer Management Saves Energy," Kansas State University Coop. Ext. Ser. MF-449, 1977.

[3] Annual Report, National Fertilizer Development Center, Tennessee Valley Authority, Muscle Shoals, Alabama, Circular Z-85, 1977.

Fig. 11–5. Fertilizer plants such as this one located in Kansas depend largely on fossil fuels. (Courtesy Farmland Industries.)

the farm is also a part of the energy budget. High-analysis fertilizers generally require more energy for production per ton of product but less energy for transportation. Because transportation costs may represent from 40 to 80% of the total farm cost of fertilizer, high-analysis fertilizers are usually more economical.

Table 11–6. Estimated Energy Required for Producing Nitrogen Fertilizer

Fertilizer Material	N (%)	Natural Gas (ft³/ton)			Fuel Oil (gal/ton)	Electricity (kwh/ton)
		For Hydrogen	For Heat	Total		
Anhydrous ammonia	82	22,220	15,910	38,180	9	54
Urea, solution	30	8,130	8,320	16,450	8	185
Urea, solid	45	12,200	13,080	27,280	12	156
Ammonium nitrate, solution	21	5,690	5,200	10,890	2	115
Ammonium nitrate, solid	34	9,210	10,940	20,150	8	217

NOTES:
ft³/ton × 0.031214 = m³/metric ton.
gal/ton × 4.172701 = *l*/metric ton.
kwh/ton × 0.907185 = kwh/metric ton.

SOURCE: W. C. White, "Fertilizer-Food-Energy Relationships," Illinois Fertilizer Conference Proceedings, 1974, 37–43.

Energy use for transportation varies greatly, depending on the fertilizer source being shipped, the methods of transportation, and the distance of travel. Approximately 403 Kcal are required to move 1 ton of fertilizer 1 mile by rail or barge and 1033 Kcals/ton-mi by truck. Pipelines are another means of transporting ammonia fertilizer and should be included in any energy-cost analysis.

Although fertilizers are expensive and consume large quantities of energy, they also help green plants utilize the sun's energy more efficiently. Green plants capture energy from the sun by the process of photosynthesis and store energy as carbohydrates, proteins, and oils, which eventually are available for human and animal consumption. Quite often the maximum absorption of energy by plants is not attained without the use of fertilizers.

11.5 EFFICIENT USE OF FERTILIZERS AND LIME

The fertilizer industry is the heaviest energy consumer of the agriculture and food systems. The energy needed to manufacture fertilizer is 610 trillion BTUs, compared to 391 trillion BTUs needed to operate field machinery, the second largest user of energy for agriculture. It has been estimated that by improving the efficiency of fertilizer use the energy conservation potential would exceed 10% (Fig. 11–6).

One of the most important ways the farmer or grower can conserve fertilizer energy is in making efficient use of fertilizer. Efficient use can be defined as maximum return per unit of fertilizer consumed. The greatest efficiency usually results from the first increment of added fertilizer. Additional increments of fertilizer usually result in a lower efficiency but may still be profitable. Fertilizer efficiency is improved by using lime, seeding rate, water management, and other management practices at optimum levels.

There are several ways to improve fertilizer efficiency. Efficient fertilizer use involves the input of information from many sources to predict conditions for fertilizer response. Soil tests, fertilizer costs, cultural and management factors, plus total production needs of the farm (cash grain versus livestock operation) must be taken into account.

11.5.1 Soil Testing

A good soil-testing program is essential to sound fertilizer use. The soil test value is the starting point. There are few soils being farmed today that do not require efficient fertilizer usage and good management for profitable crop production. On some soils, excess fertilizers are being applied and environmental concerns are being voiced. To determine the upper level of fertilization, economic considerations are important, especially the value of the expected crop increase in relation to the cost of the fertilizer. Once the upper level of fertilization for a crop has been decided, the actual amount of fertilizer needed to achieve that goal is the difference between the amount of available

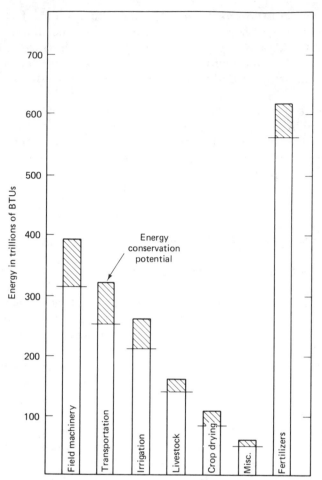

Fig. 11–6. Energy consumed by U.S. agricultural production inputs. (Source: R. H. Follett, "Proper Fertilizer Management Saves Energy," Kansas State University, Coop Ext. Ser. MF-499, 1977, 6 pp.)

nutrients present (soil test value) and the amounts needed to achieve that upper level.

Optimum economic fertilization can best be accomplished by a well-planned soil-testing program. There is no sense dumping fertilizer and lime on every acre of cropland if it does not need them. Soil testing is a means of evaluating the soil's ability to supply these nutrients. Some soils are naturally fertile or have been made more fertile by the use of fertilizers and lime. An assessment of what the soil can supply can be related to crop yields and is used extensively for making fertilizer and lime recommendations. Soils that test low in phosphorus (P) or potassium (K) or are strongly acid, for example, will need larger amounts of fertilizer phosphorus and potassium and lime than soils testing high in these two nutrients. Soil tests are also used to evaluate carryover levels of past fertilizer programs. In most cases a properly followed soil-testing program will result in more efficient use of energy and fertilizer (see Chapters 1 and 8).

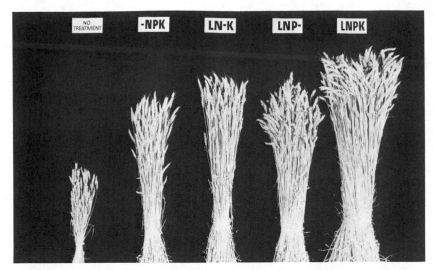

Fig. 11-7. Experiments have demonstrated that lime corrects soil acidity and provides better growing conditions for most crops. This helps to maximize the response to both fertilizer and lime as illustrated here with wheat (Brownstown Soil Experiment Field, Illinois). (Courtesy Potash and Phosphate Institute.)

11.5.2 Use of Lime

The proper use of lime can have a significant effect on fertilizer utilization and will improve the "efficiency" of the fertilizer applied (Fig. 11-7). Most of the essential plant nutrients increase in solubility and availability as the soil reaction is increased from pH 5.0 to 6.5 by liming. For example, the solubility of phosphorus compounds in the soil is acutely affected by soil pH. Under acid conditions (low pH), phosphorus is precipitated as iron and aluminum phosphates that are relatively insoluble and unavailable. With liming, phosphorus compounds in the soil become more soluble and could reduce the amount of phosphorus fertilizer needed.

The use of lime on acid soils encourages the growth of symbiotic nitrogen-fixing bacteria on legumes. The soil conditions are much more favorable for the growth of the nitrogen-fixing bacteria when the soil pH is in the neutral range than it is when the soil pH is 5.0.

Elements such as aluminum or manganese may be present in large quantities and become toxic to plant growth under acidic conditions. Proper liming decreases the solubility of these elements and, therefore, reduces the potential toxicity (see Chapter 8).

11.5.3 Fertilizer Placement

It has long been known that banding (placing fertilizer near the seed at planting) is frequently more efficient at supplying nutrients to the crop than broacast applications; yet many crop producers have moved away from

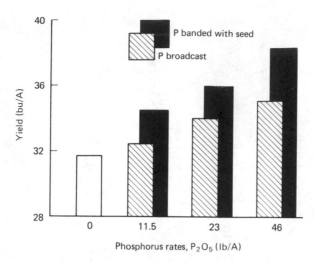

Fig. 11-8. Winter wheat yields related to phosphorus rates. Phosphorus was broadcast or banded with seed on soils testing 0-26 lb/A P (low), average of 13 sites. *NOTE:* (a) bu/A x 67.26 = kg/ha; (b) lb/A x 1.121 = kg/ha. (Source: C. A. Thompson, "Fertilizing Dryland Wheat on Upland Soils in the 20-26 Inch (51-66 cm) Rainfall Area in Kansas. Kansas State University Agr. Exp. Sta. Bul. 590, 1976, 27 pp.)

band applications. There have been good reasons for shifting from banding to broadcast applications. One reason for this shift is that many soils no longer respond to phosphorus and potassium fertilizers as they once did due to their accumulation in the soil. However, a response to band-placed phosphorus can still be seen in many locations, particularly under cool, wet conditions, even though many soils test medium to high in phosphorus.

In the dryland areas of Kansas, broadcast applications before planting have been popular because of the relatively low cost of phosphorus fertilizers, convenience of application, and time and labor saved at planting. The data in Fig. 11-8 indicates that band-placing phosphorus on neutral to alkaline soils with wheat can make phosphorus more efficient than broadcast applications.

The principal reason for the shift from band to broadcast fertilizer is low fertilizer cost in comparison to high labor cost and the importance of time. Many farming operations have expanded to the point where time and labor are the major factors affecting the productive capacity of the farm. Corn yields are often greatly affected by time of planting. Any operation that delays planting, such as filling fertilizer hoppers, can cause a reduction in yield, resulting in a significant economic loss.

Before a farmer switches from broadcasting to banding of certain nutrients, a thorough analysis of band application should be made. A farmer may not be ready to buy the equipment and machinery that is necessary to carry out the band application of fertilizer. However, banding usually means less fertilizer, fewer trips across the field, and quite often higher yields. Estimates of additional time involved in banding fertilizer are as low as 30 sec/A (74 sec/ha). As farmers become better equipped to handle bulk fertilizers for use in planters and as fertilizer prices rise, a reversal of this trend away from banding fertilizer is likely.

The dual application of both nitrogen and phosphorus together in the same soil zone has increased wheat yields in Kansas (see Chapter 7). Knifed application of nitrogen and phosphorus applied simultaneously has produced generally higher yields than either broadcast or band applications.[4]

This techique involves the injection of liquid nitrogen or anhydrous ammonia with liquid phosphorus simultaneously into the soil in the same retention zone. The concentration of large amounts of ammonium nitrogen in the same zone with phosphorus probably enhances the availability of phosphorus, in both alkaline and acid soils. Innovators are now taking this knowledge and finding equipment, formulations and techniques to make the concept practical and profitable to the grower.

11.5.4 Time of Application

Time of application is critical for efficient use of nitrogen. Although nitrogen can be applied in the fall, in the spring before planting, side-dressed after planting on row crops or topdressed on small grains, or applied in irrigation water, these times of application are not always equally effective. Nitrogen is most effective if applied when the plant is growing most rapidly. This indicates that delayed application of nitrogen as a side-dressing and/or application in irrigation water should be the most efficient means of applying nitrogen. Time of application is especially important on sandy soils where leaching losses may be high.

Side-dress applications of nitrogen after plant emergence may result in maximum efficiency. This is true particularly on shallow-rooted crops such as potatoes grown on sandy soils that are subject to leaching and on crops grown on fine-textured soils where denitrification is a problem. Agronomic research has shown that delaying the time of nitrogen application results in better usage of nitrogen. Corn and potatoes are good examples of crops having a high requirement of nitrogen late in the growing season. Side-dressed nitrogen will help to assure that nitrogen will be plentiful during the later stages of growth.

Applying nitrogen through the irrigation system is another means of improving nitrogen efficiency. Such a procedure requires little additional energy for application and assures that adequate nitrogen is available during the plant's greatest period of use. This practice is well adapted to sandy soils where leaching of nitrogen is a problem.

A model for efficient use of fertilizer nitrogen for corn growing on sandy soils would consist of applying a percentage of the total nitrogen requirement at four stages: (1) preplant, (2) 8-leaf, (3) 12–16 leaf, and (4) early tasseling (Fig. 11–9). Here is an example of one application scheme for supplying the 180 lb of fertilizer nitrogen (N) necessary for a 150 bu/A yield goal.

1. *Preplant* — One-sixth or 30 lb of the 180 lb of nitrogen (N) should be included with the starter fertilizer at planting time.

[4] D. R. Leikam, R. E. Lamond, P. J. Gallagher, and L. S. Murphy, "Improving N-P Application," *Agrichemical Age,* April, 1978.

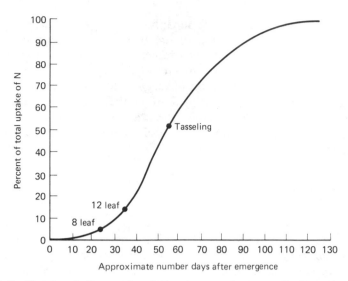

Fig. 11-9. Uptake of nitrogen in relation to corn plant growth. Note that about 50% of the total nitrogen has been taken up at time of tasseling. (Source: R. P. Schoper, A. C. Caldwell, J. B. Swan, and G. L. Malzer, "Corn Fertilization on Irrigation Sandy Soils," Soils Fact Sheet No. 31, 1978, 2 pp.)

2. *8-leaf stage* — The demand for nitrogen approximately 3 weeks after emergence is beginning to accelerate and the addition of 30 lb of fertilizer nitrogen (N) supplies plant needs for approximately the next 10 days.

3. *12–16 leaf stage* — Nitrogen requirement is at its peak during this period. Half or 90 of the 180 lb of nitrogen (N) required should be applied during this stage. Depending on the method of application, this amount of nitrogen can be applied in one or several applications over a 3-week period beginning at the 12-leaf stage. A nitrogen shortage in the plant during this growth period could reduce yield significantly. If nitrogen is to be applied several times exclusively through the irrigation system, a rain could eliminate an irrigation and therefore a nitrogen application.

4. *Early tassel* — Because nitrogen uptake continues until the corn plant matures, the application of 30 lb of nitrogen (N) at this growth stage will help to ensure adequate nitrogen for proper ear development.

11.5.5 Use of Legumes

An outstanding characteristic of legumes is their ability to obtain nitrogen from the air. Legumes, whose roots live in the soil with bacteria known as *Rhizobium,* are able to convert atmospheric nitrogen to a form that can be used by the plant. Therefore, cultivation of legumes has the double advantage of producing a valuable crop with the addition of little or no nitrogen fertilizer. Although this fact has been known for decades, it is now more economically feasible to utilize legumes because of the rising cost of nitrogen fertilizers (see Chapter 10).

11.5.6 Animal Manures

Animal manure has been highly valued as a fertilizer for centuries (see Chapter 10). The supplementary values of manure as a soil amendment should also be considered by those evaluating the worth of manure as a nitrogen-phosphorus-potassium fertilizer. Animal manure also contains an entire complement of plant nutrients that can alleviate or prevent crop micronutrient deficiencies. Manure has been shown to increase soil aggregation, permeability, and water-holding capacity. The value of manure in crop response is usually greater than its value in terms of plant nutrients. Long-term effects of manure on the fertility and physical condition of the soil account for the added value (Fig. 11–10).

The chemical composition of manure varies considerably depending on the source, the method of handling, and even within a given stockpile of manure. The manure should be analyzed as an aid in determining application rates and the amounts of the various nutrients added per ton of manure. Incorporation of the manure immediately after application will reduce volatilization losses of ammonia from manure and result in better nutrient recovery.

Nitrogen fertilizer can be saved by using manure and by plowing down legumes. Table 11–7 indicates the adjustments in nitrogen requirements for corn in Wisconsin when manure and/or legumes are plowed down. The first year corn without manure requires 130 lb/A (146 kg/ha) of N. If a previous crop, 60 to 100% legume, was plowed down, the N requirement would be lowered to 70 lb/A (78 kg/ha). If 10 tons/A (22 metric tons/ha) of manure was

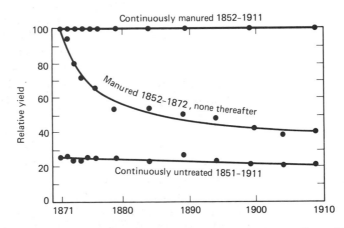

Fig. 11–10. Residual effects of 14 tons of manure applied annually on long-term barley yields at the Rothamsted Experiment Station in England. Yields are expressed relative to the continuously manured and continuously untreated barley. (Source: J. Azevedo, "Farm Animal Manures, An Overview of Their Role in the Agricultural Environment," California Agr. Exp. Sta. Manual 44, 1974, p. 64.)

Table 11-7. Nitrogen Fertilizer Recommendation for 120–145 Bu/A (7535–9100 kg/ha) Yield Goal for Corn in Wisconsin

Treatment	% Legume in Previous Crop Plowed Down		
	0–20	20–60	.60–100
	N Recommendation (lb/A)		
No manure	130	110	70
10 tons/A manure	90	70	30
20 tons/A manure	50	30	0

NOTES:
bu/A × 62.776 = kg/ha.
lb/A × 1.121 = kg/ha.
tons/acre × 2.24 = metric tons/ha.

SOURCE: R. D. Powell, "What to Do if You Can't Get Enough Fertilizer," University of Wisconsin Ext. A2693, 1975, 4 pp.

also added to the field, the fertilizer requirement would be lowered to 30 lb/A (34 kg/ha) of N.

11.5.7 Other Soil and Crop Management Factors

Selecting the best adapted variety or cultivar for your cropping area can result in the maximum potential production of the crop and a higher return per unit of fertilizer consumed. Good weed control is important because weeds compete with the crop for moisture, light, and nutrients. Proper use of insecticides and fungicides when needed can pay big dividends. Planting rates and planting dates are also important factors to consider. The time of application and amount of fertilizer may depend on the soil moisture situation. Using management systems that increase stored moisture at planting time is highly important to obtain the most efficient returns from fertilizer.

Practices that prevent or control wind and water erosion are other ways to help obtain efficient fertilizer use. A frozen soil should not be fertilized when surface runoff is a possibility. The majority of the fertilizer nutrient loss from surface runoff is associated with soil erosion. Unless there is soil particle movement, water passing over an unfrozen soil surface where fertilizer has recently been applied will carry off only a small amount of fertilizer nutrients.

Fertilizers often carry over from one year to another. More carryover can be expected with high application rates and following droughty years. Yield reduction due to drought, poor stand, insect, or disease problems often results in less nutrient uptake and removal, which can significantly increase the carryover of fertilizer. Nutrient carryover from manured soils also contributes to the fertility of the soil. Not all the nutrients in manure are released in the first year of application. Only half the nitrogen and phosphorus are considered available in the first year of application. All the potassium should be available during the first crop year.

Crop rotation is also important in using fertilizer efficiently. A lower nutrient-requiring crop such as soybeans following a heavily fertilized corn crop may require little or no fertilizer. Such practices are common and helpful in utilizing fertilizers efficiently.

11.6 WHAT ABOUT FERTILIZER RECOMMENDATIONS?

The value of a soil test lies in the making of efficient fertilizer recommendations to farmers. Soil test services have the primary goal of assessing the nutrient status of the soil and suggesting a fertilizer program that will result in a highest yield with the least fertilizer cost. This objective is not easily accomplished, but to achieve this goal certain steps must be followed: (1) soil samples must be collected from a field in such a manner to represent the soil nutrient levels of the field; (2) laboratory analyses must be precise, reproducible, and provide a meaningful index of nutrient levels; and (3) this index must be effectively utilized to predict the amounts of fertilizer that will maximize profit. A soil test is related to the soil's ability to provide a specific nutrient to a growing crop. It is essential that the soil test's numerical rating be correlated and calibrated with field nutrient responses. Without correlation and calibration, soil tests have no value and may be misleading.

It is impossible to predict with absolute certainty the ability of a soil to supply a given nutrient for current crop production because of the high degree of variability of the soil and climate. However, with proper calibration and correlation data, soil testing is a valuable tool in making fertilizer suggestions that efficiently utilize fertilizer resources.

A wide selection of soil test laboratory services is available to the producer. These services are provided by independent commercial and land-grant university laboratories. Numerous crop producers, as well as university agronomists, have observed large differences among fertilizer recommendations from various laboratories for the same year, field, and crop. Therefore, agronomists in the states of Kansas[5] and Nebraska[6] have conducted field studies in several locations in both states. The primary objective of these studies was to compare, under controlled conditions, recommendations of various soil testing laboratories and the net return to the farmer from those recommendations.

The procedures for conducting the studies were quite similar in both Kansas and Nebraska. A representative soil sample was taken from each field plot

[5] D. A. Whitney, "Comparison of Soil Testing Lab Fertilizer Recommendations and Resulting Yields on the Same Field," *Extension Agronomy,* Kansas State University (Mimeos for 1974 through 1979).

[6] Soil Fertility Staff, "A Comparison of Suggested Fertilizer Programs Obtained from Several Soil Testing Laboratories Services," University of Nebraska, Agronomy Department Report No. 26, 1978.

area, thoroughly mixed and split into subsamples to be mailed to the laboratories. The manner of handling and mailing the samples was such that no laboratory would know that their services were to be utilized. Each laboratory received a request for fertilizer recommendations to grow a given crop at a specific location and at a specified reasonable yield goal. This approach was used in order to reflect the normal service and normal fertilizer recommendations provided to any growers who would request soil test analyses and a suggested fertilizer program. This procedure eliminates dealer-producer contact, which might alter the suggested program in a producer situation. All nutrients suggested by the various laboratories were assumed to be needed and were applied. All fertilizers were broadcast and incorporated prior to planting. Fertilizer treatments were replicated several times at each experimental location. At each location during the first year, all laboratories were testing the same soil; however, in subsequent years, each laboratory received only soil samples from plots to which fertilizer had been applied according to their recommendations.

The data in Table 11-8 indicates the variation in fertilizer recommendations from five different soil-testing laboratories. The results indicate that the suggested fertilizer recommendation made by various laboratories on a given field may not be consistent. The cost of fertilizer ranged from a low of $15.41/A ($38.08/ha) for laboratory E to a high of $44.75/A ($110.58/ha). There were no significant differences in the yields as a result of the different fer-

Table 11-8. Comparison of Fertilizer Recommendations by Five Soil Testing Laboratories, Fertilizer Costs, and Grain Sorghum Yields, Bourbon County, Kansas

Nutrient	Laboratory					Control
	A	B	C	D	E	
	Fertilizer Recommendation (lb/acre)					
Nitrogen (N)	125	185	135	130	100	—
Phosphorus (P$_2$O$_5$)	110	70	75	40	30	—
Potassium (K$_2$O)	95	20	125	100	—	—
Magnesium (MgO)	15	—	—	—	—	—
Sulfur (S)	28	30	10	—	—	—
Zinc (Zn)	7	6	3	—	—	—
Boron (B)	1	—	0.5	—	—	—
Copper (Cu)	—	1	—	—	—	—
Fertilizer cost ($/A)	44.75	40.62	36.95	20.96	15.41	0
Yield, bu/A	74	70	69	70	72	49

NOTES:
1. Fertilizer cost based on spring of 1974 prices, $/ha = $/A × 2.471.
2. Yield is average of three replications, bu/A × 62.776 = kg/ha.
3. lb/A × 1.121 = kg/ha.

tilizer recommendations. The various fertilizer treatments all increased the yield significantly above the no-fertilizer treatment.

Table 11-9 shows a summary of 4 years of fertilizer recommendations data for the Bourbon County, Kansas, location. This is the same location as the data shown in Table 11-8. The 4 years of data still indicate no significant difference for grain sorghum yields among fertilizer recommendations from the five soil testing laboratories. However, there was a considerable difference in the total cost of fertilizer over the 4-year period of time.

The data accumulated by the University of Nebraska shows the same type of information as Kansas State University relative to fertilizer programs suggested by different laboratories. Table 11-10 shows a summary of 5 years of fertilizer recommendation data for the soil at the Mead, Nebraska, location. For this location all laboratories were asked to recommend nutrients for a 170 bu/A (10,672 kg/ha) yield goal. Although the yield differences were nonsignificant, it is interesting to note that laboratory C had the lowest total yield and the highest total fertilizer cost for the 5 years.

In addition to the data shown in Tables 11-9 and 11-10, agronomists at Kansas State University and the University of Nebraska have conducted similar tests at several different locations. No significant differences in yield

Table 11-9. Summary of Grain Sorghum Yields and Fertilizer Costs from Soil Test Recommendations by Five Laboratories on the Same Field, Bourbon County, Kansas

	Laboratory					
Year	A	B	C	D	E	Control
	Grain Yield (bu/A)					
1974	74	70	69	70	72	49
1975	50	56	56	51	54	37
1976	77	74	74	70	72	45
1977	81	79	80	81	82	31
4-year total	282	279	279	272	280	162
	Fertilizer Cost ($/A)					
1974	44.75	40.62	39.65	29.96	15.41	—
1975	62.21	57.51	52.12	43.02	22.98	—
1976	39.95	22.68	27.05	36.65	15.55	—
1977	44.99	26.24	30.54	42.02	16.82	—
4-year total	191.90	147.05	149.36	151.65	70.77	—
	Fertilizer Cost per Bushel Increase in Yield ($)					
	$1.60	$1.26	$1.28	$1.38	$0.60	

NOTES:
1. Fertilizer cost based on fertilizer prices for individual years. $/ha = $/A × 2.471.
2. Yield is the average of three replications, bu/A × 62.776 = kg/ha.

Table 11–10. Summary of Irrigated Corn Yields and Fertilizer Costs from Recommendations by Five Laboratories on the Same Field, Mead, Nebraska

	Laboratory				
Year	A	B	C	D	E
Yield of Corn (bu/A)					
1973	152	148	153	148	160
1974	139	131	131	137	133
1975	163	157	151	160	158
1976	143	143	129	143	137
1977	148	143	136	142	145
5-year total	745	722	702	730	733
Fertilizer Cost ($/A)					
1973	42.00	32.00	41.00	25.00	17.00
1974	96.00	69.00	90.00	57.00	39.00
1975	95.00	74.00	83.00	46.00	46.00
1976	51.15	31.75	60.85	34.80	31.60
1977	37.80	55.05	67.85	50.50	22.80
5-year total	321.95	261.80	342.70	213.30	156.40

NOTES:
1. Fertilizer cost based on individual year's fertilizer prices, $/ha = $/A × 2.471.
2. bu/A × 62.776 = kg/ha.

resulted from the varied fertilizer recommendations at any of the locations during the entire 4- to 5-year period of these studies. The small differences shown in Tables 11–8, 11–9 and 11–10 are due to random variation. However, there are definite differences in costs of fertilizer programs suggested by the different laboratories. These differences existed for individual years as well as the 4- or 5-year totals. Thus, the minimum fertilizer recommendation by soil testing laboratory E in all cases would have been the most economically advantageous to the farmer.

Laboratories that recommend less fertilizer than is needed for maximum profits do their growers a great disservice because of the favorable ratio between the cost of fertilizer and the price of most farm products. On the other hand, recommending more fertilizer than is scientifically and economically sound reduces grower profits and increases the amount of applied nutrients that may pollute surface water or groundwater. Growers using more fertilizer than needed may or may not continue to produce good yields depending on the crop involved and its sensitivity to nutrient excess or imbalance. In some instances, costly yield reductions result from nutrient imbalances that are hard to identify.

The results reported in Tables 11–8, 11–9, and 11–10 indicate that suggested fertilizer recommendations made by various soil testing laboratories on a given field can vary considerably. Farmers must be aware of this variability and evaluate suggested fertilizer recommendations in relation to fertilizer research results and their experience for the area. If the recommendations appear unusual for the area, it would be wise to evaluate their use on a limited area before investing heavily in fertilizer for the entire field.

Reliable fertilizer recommendations are the result of soil test calibrations developed from extensive field and greenhouse research. In order to interpret a soil test value, it is necessary to correlate the test value with known field response for various crops. In all soil-testing programs tables are prepared, or computer programs written, showing soil test results and suggested fertilizer use for various crops. Soil-testing laboratories must have access to both local and regional soil test-crop response correlation research in order to develop economical fertilizer recommendations.

11.7 UNIT COST CALCULATIONS

A very important part of determining profitable fertilizer rates is the cost of the fertilizer. The farmer is interested in the most economical source, but he may be accustomed to buying on the basis of cost per ton of fertilizer rather than on the cost per unit of plant nutrients. Comparisons of fertilizer costs and fertility values, in general, are most reliable when based on cost per unit of available plant nutrients such as price per pound (kilogram). Table 11–11

Table 11–11. Cost per Pound of Plant Nutrient (N, P_2O_5, K_2O) in Some Common Solid Fertilizer Materials

	Fertilizer Cost ($/ton)								
Fertilizer Source	80	100	140	180	220	260	300	340	380
	Cost per lb of Nutrient ($/lb)								
N Sources									
82-0-0	0.05	0.06	0.08	0.11	0.13	0.16	0.18	0.20	0.23
45-0-0	0.09	0.11	0.16	0.20	0.24	0.29	0.33	0.38	0.42
34-0-0	0.12	0.15	0.21	0.26	0.32	0.38	0.44	0.50	0.56
32-0-0	0.13	0.16	0.22	0.28	0.34	0.41	0.47	0.53	0.59
28-0-0	0.14	0.18	0.25	0.32	0.39	0.46	0.53	0.61	0.68
21-0-0	0.19	0.24	0.33	0.43	0.52	0.62	0.71	0.81	0.90
P_2O_5 Source									
0-46-0	0.09	0.11	0.15	0.19	0.24	0.28	0.32	0.37	0.41
K_2O Source									
0-0-60	0.07	0.08	0.12	0.15	0.18	0.22	0.25	0.28	0.32

NOTES:
$/metric ton = $/ton × 1.102.
$/kg = $/lb × 2.204.

gives the cost per pound of nutrient of fertilizer materials at various tonnage prices.

Cost of Single-Nutrient Solid Materials

The cost per pound for a straight material, such as those listed in Table 11–11, can be determined by dividing cost per ton by pounds of nutrient in 1 ton.

For example, the cost of anhydrous ammonia (82–0–0) is $164/ton. The cost per pound is determined as follows:

$$\frac{\text{price/ton}}{\% \text{ nutrient} \times 2000 \text{ lb/ton}} = \text{cost/lb of nutrient}$$

$$\frac{\$164}{0.82 \times 2000} = \frac{164}{1640} = \$0.10 \text{ or } 10¢/\text{lb}$$

If the price for anhydrous ammonia was $180/metric ton, then the cost per kilogram would be calculated as follows:

$$\frac{\text{price/metric ton}}{\% \text{ nutrient} \times 1000 \text{ kg/metric ton}} = \text{cost/kg of nutrient}$$

$$\frac{\$182}{0.82 \times 1000} = \frac{180}{820} = \$0.22 \text{ or } 22¢/\text{kg}$$

NOTE: The cost of $0.10/lb is equivalent to $0.22/kg ($/kg = $0.10/lb × 2.204 lb/kg = $0.22/kg).

Cost of Mixed (Formulated) Materials

For mixed materials having a common $N–P_2O_5–K_2O$ ratio, the buyer needs to calculate an average value for all nutrients before comparing prices.

For example, 5–20–20 costs $180/ton. The total nutrient content is 5% + 20% + 20% = 45%.

$$\frac{\$180}{2000 \times 0.45} = \frac{180}{900} = \$0.20 \text{ or } 20¢/\text{lb}$$

For mixed fertilizers that do not have the same $N–P_2O_5–K_2O$ ratio, the farmer should figure the price per pound before comparing prices of different analyses. The price per ton is not much help unless the farmer is comparing the same analysis. The cost per pound will vary with fluctuating price; therefore, the local figures should be used for N, P_2O_5, and K_2O.

For example, if a dry 10–20–10 fertilizer ($N–P_2O_5–K_2O$) costs $160/ton, is it a fair price? Assume the local price is 23¢/lb for nitrogen, 23¢/lb for phosphate (P_2O_5), and 10¢/lb for potash (K_2O). First, multiply the cost of N, P_2O_5, and K_2O by the percent of the nutrients to find cost per pound. Then add the three results and multiply by 2000 for price per ton.

$0.23 \times 0.10 = \$0.023$
$0.23 \times 0.20 = 0.046$
$0.10 \times 0.10 = \underline{0.010}$
$$\$0.079/lb$$

$0.079 \times 2000 = \$158/ton$

Therefore, according to the calculations, $160 is a reasonable price for the 10–20–10 fertilizer.

Cost of Liquid Fertilizers

The cost of liquid fertilizers is calculated as just shown for the 10–20–10 analysis. Sometimes liquid fertilizers are sold by the gallon. If the weight of the liquid fertilizer is 10.4 lb/gal, then there is about 192 gal/ton. To find out how much the fertilizer costs per gallon, divide the price by 192. If a 10–20–10 liquid fertilizer sold for $160/ton, it would cost about $0.84/gal ($0.22/l). Therefore, a comparable liquid fertilizer selling for $2.75/gal would be over-priced.

Other Factors

In addition to the actual cost of the material, the grower must also consider the cost of storage, transportation, and labor used in applying the fertilizer. These costs may be difficult to evaluate. However, if the actual price of the nutrients from one source is the same as another, farmers will probably buy the one requiring the least labor. The higher-analysis materials will require less labor in handling and less time because fewer stops are made in applying the material.

SUMMARY

The most profitable rate of plant nutrients is related to expected increase in yield from each additional increment of fertilizer according to the *law of diminishing returns*. Economic factors that affect the optimum rate of fertilization are the prices paid for fertilizer, the price of other production goods and services and the price received for the farm product.

The economic optimum or most profitable rate of fertilizer is determined by moving along the curve shown in Fig. 11–3. At each step the additional benefits of increasing fertilizer are weighed against the additional costs. If the additional benefits of another step on the response curve exceed the additional costs, then another step is taken until the optimum point has been gained. Farmers aim for maximum profits by fertilizing to the point where added costs equal added returns. However, there is a law of diminishing returns as evidenced by the response curve. Therefore, farmers should reduce the fertilizer input when they can no longer earn profits equal to costs plus a reasonable profit.

The production and distribution of chemical fertilizers are dependent on reliable sources of large quantities of fossil fuel energy. One of the most important ways the farmer or grower can conserve fertilizer energy is in making efficient use of fertilizer. Efficient use can be defined as maximum return per unit of fertilizer consumed. Agricultural producers have many opportunities to make efficient use of fertilizers. Management practices such as soil testing, liming, band placement of fertilizers, timing of nitrogen application to coincide with the crop's period of greatest use, use of manure, legumes, carryover fertilizer, and other crop management factors can all help to conserve fertilizer.

A very important part of determining profitable fertilizer rates is the cost of the fertilizer. Fertilizer prices vary greatly among products. Comparisons of fertilizer costs, in general, are most reliable when based on cost per unit of plant nutrient such as the price per pound (kilogram) of N, P_2O_5 or K_2O.

PROBLEMS

1. Discuss the factors that determine the most profitable rate of plant nutrients.

2. Explain the *law of diminishing returns.*

3. How can the efficient use of fertilizer conserve energy? Explain.

4. What factors would you consider in trying to improve fertilizer efficiency?

5. Is animal manure a valuable source of plant nutrients? Why or why not?

6. How should a farmer evaluate a fertilizer recommendation that appears to be unusual for his area?

7. Triple superphosphate, 0–45–0, costs $130/ton. (a) What is the cost of P_2O_5 per pound? (b) What is the cost of 0–45–0 per metric ton? (c) What is the cost of P_2O_5 per kilogram? (d) What is the cost of P per kilogram?

8. A farmer plans to apply 90 lb/A of 8–32–16 to a field, as indicated by a soil test. (a) How many kilograms per hectare does he apply? (b) How many pounds per acre (lb/A) and kilograms per hectare (kg/ha) of the following nutrients does he apply?

	lb/A	kg/ha
Nitrogen (N)	_____	_____
Phosphate (P_2O_5)	_____	_____
Potash (K_2O)	_____	_____

9. You are quoted the following prices: 3–12–12 for $125/ton and 10–10–10 for $230/ton. Which would you choose and why?

10. You are quoted a price of $3.00/gal for 9–18–9. (a) Is this fertilizer a good buy with regard to price? Explain. (b) How much does this fertilizer cost per ton? (c) How much does it cost per liter? (d) per gallon?

REFERENCES

Cady, F. G., and R. J. Laird, "Treatment Design for Fertilizer Use Experimentation," CIMMYT Research Bulletin No. 26, International Maize and Wheat Improvement Center (1973). Mexico 6, D. F.

Cooke, G. W., *Fertilizing for Maximum Yield,* London: Crosby Lockwood & Son Ltd., 1972.

Hoeft, R. G., and J. C. Siemens, "Do Fertilizers Waste Energy?" *Crops and Soils Magazine,* Nov., 1975, pp. 12–14.

Sanders, D. H., and R. A. Olson, "Use Nitrogen More Effectively," University of Nebraska, Neb. Guide G 74–111, undated.

Tisdale, S. L., and W. L. Nelson, *Soil Fertility and Fertilizers,* New York: Macmillan Publishing Co., 1975.

Vitosh, M. L., "Fertilizer Management to Save Energy," Michigan State University Ext. Bul. E-1136 (1977).

Vitosh, M. L., R. E. Lucas, and R. J. Black, "Effect of Nitrogen Fertilizer on Corn Yield," Michigan State University Ext. Bul. E-802 (1974).

Voss, R. D., J. T. Pesek and J. R. Webb, "Economics of Nitrogen Fertilizer for Corn," Iowa State University Coop. Ext. Serv., Pm 651 (1975).

APPENDIX

Conversion Factors

In 1866 the United States Congress authorized the use of *both* the United States customary and the metric units of weights and measures. On numerous occasions since then, the Congress took action that "inched up" on making mandatory the metric sytem. The last of such legislative action was U.S. Public Law 94–168, Metric Conversion Act of 1975. This Act authorized the establishment of a 17-member board to make specific plans for voluntary conversion to metric units by 1985.

The United States is a member of the International Bureau of Weights and Measures. All member countries have agreed on a uniform system of metric units that are known as *SI units*. These initials stand for the French term, *Le Système International d'Unités*.

Although agreeing to use all SI units, the United States has reserved the right to continue using these eight non-SI units *with SI units:* day, degree, hour, liter (litre), minute (of an arc), minute (of time), second, and tonne (metric ton). Furthermore, reservation has been retained for the United States to use 10 unrelated non-SI units. Five of these 10 units used in this book include Ångstrom, atmosphere, bar, gallon, and hectare.

The conversion factors in these tables include metric (SI) and other units essential for a more thorough understanding of the various weights and measures used throughout the world and in this book.

REFERENCES

"Agricultural Statistics, 1979," USDA, pp. IV–IX.

"Metric Practice Guide," Canadian Fertilizer Institute, Ottawa, Ontario, Canada, 1977, 119 pp.

Chester H. Page and Paul Vigoureux, (Ed.), "The International System of Units (SI)," National Bureau of Standards, U.S. Department of Commerce, NBS Special Publication 330, 1977.

Robert C. Weast, (Ed.), *Handbook of Chemistry and Physics*, The Chemical Rubber Co., 60th ed., 1979, pp. F-307–F-329.

NOTES:

1. According to international agreement, numbers are divided into groups of three, separated by a space and not a comma. In France, Greece, and several other countries, a comma is used with numbers in the same way as a period is used in the United States.
2. Centigrade is now known as Celsius.

CONVERSION FACTORS

To Convert from Column A	To Column B	Multiply Column A by
Acres	Hectares	0.404 686
	Square feet	43 560.000
	Square kilometers	0.004
	Square meters	4 046.856 422
	Square miles (statute)	0.001 562
	Square yards	4 840.000
Acre-feet	Acre-inches	12.000
	Cubic feet	43 560.000
	Cubic meters	1 233.482
	Cubic yards	1 613.333
	Gallons (U.S.)	325 851.560
	Hectare-centimeters	12.335
	Hectare-meters	0.123 35
Acre-feet/day	Cubic feet/second	0.504 17
	Cubic meters/second	0.014 28
Acre-inches	Acre-feet	0.083 33
	Cubic feet	3 630.000
	Cubic meters	102.790 33
	Gallons (U.S.)	27 154.286
	Hectare-centimeters	1.028
Ångstroms	Centimeters	1.000×10^{-8}
	Inches	3.937×10^{-9}
	Meters	1.000×10^{-10}
	Millimeters	1.000×10^{-7}

CONVERSION FACTORS (cont.)

To Convert from Column A	To Column B	Multiply Column A by
Atmospheres	Bars	1.013 25
	Centimeters of Hg (0°C)	76.000
	Centimeters of H_2O (4°C)	1 033.260
	Feet of H_2O (4°C)	33.900
	Grams/square centimeter	1 033.230
	Inches of Hg (0°C)	29.921 3
	Millimeters of Hg (0°C)	760.000
	Pounds/square inch	14.696 0
Barley (bushels/acre) (48 lb/bu)	Kilograms/hectare	53.808
Barrels (petroleum, U.S.)	Cubic feet	5.614 583
	Gallons (U.S.)	42.000
	Liters	158.982 84
Bars	Atmospheres	0.986 923
	Centimeters of Hg (0°C)	75.006 2
	Feet of H_2O (15°C)	33.488 3
	Grams/square centimeter	1 019.716
	Inches of Hg (0°C)	29.530 0
	Millibars	1 000.000
	Pounds/square inch	14.503 8
Board-feet	Cubic centimeters	2 359.737 2
	Cubic feet	0.083 333
	Cubic inches	144.000
British thermal units (Btu, mean)	Joules	1 055.870
	Kilowatt-hours	0.000 293
Bushels (U.S.) (level full, known as stricken or struck bushel. A heaping bushel is 1¼ stricken bushel)	Cubic centimeters	35 239.070
	Cubic feet	1.244 456
	Cubic inches	2 150.420
	Cubic meters	0.035 239
	Cubic yards	0.046 091
	Gallons (U.S., dry)	8.000
	Gallons (U.S., liquid)	9.309 177
	Liters	35.239 07
	Pecks (U.S.)	4.000
Calcium (Ca)	Calcium carbonate ($CaCO_3$)	2.497 255
	Calcium hydroxide [$Ca(OH)_2$]	1.848 802
	Calcium oxide (CaO)	1.399 202
Calcium carbonate ($CaCO_3$)	Calcium (Ca)	0.400 440
	Calcium hydroxide [$Ca(OH)_2$]	0.740 334
	Calcium oxide (CaO)	0.560 296
Calcium hydroxide [$Ca(OH)_2$]	Calcium (Ca)	0.540 891
	Calcium carbonate ($CaCO_3$)	1.350 742
	Calcium oxide (CaO)	0.756 815
Calcium oxide (CaO)	Calcium (Ca)	0.714 693 2
	Calcium carbonate ($CaCO_3$)	1.784 772
	Calcium hydroxide [$Ca(OH)_2$]	1.320 856 6

To Convert from Column A	To Column B	Multiply Column A by
Calories, gram (mean)	British thermal units (Btu)	0.003 974
	Calories, kg (mean)	0.001
	Foot-pounds	3.090 40
	Horsepower-hours	$1.560\,81 \times 10^{-6}$
	Joules	4.190 02
	Kilowatt-hours	$1.163\,90 \times 10^{-6}$
Centimeters	Ångstrom units	1×10^8
	Feet	0.032 808
	Inches	0.393 701
	Meters	0.010
	Microns	10 000.000
	Millimeters	10.000
	Millimicrons	1×10^7
Centimeters/second	Feet/minute	1.968 504
	Kilometers/hour	0.036
	Meters/minute	0.600
	Miles/hour	0.022 369
Chains (Gunter's)	Feet	66.000
	Furlongs	0.100
	Links (Gunter's)	100.000
	Meters	20.116 8
	Miles, statute	0.012 5
	Yards	22.000
Clover seed (bushels/acre) (60 lb/bu)	Kilograms/hectare	67.260
Cords, standard	Cubic feet	128.000
	Cubic meters	3.624 573
Corn (bushels/acre) (56 lb/bu)	Kilograms/hectare	62.776
Cotton, lint (bales/acre) (500 lb/bale)	Kilograms/hectare	560.500
Cubic centimeters	Cubic feet	3.531×10^{-5}
	Cubic inches	0.061 024
	Cubic meters	1×10^{-6}
	Cubic yards	$1.307\,950 \times 10^{-6}$
	Cups	0.004 227
	Gallons (U.S., liquid)	0.000 264
	Liters	0.001
	Ounces (U.S., fluid)	0.033 814
	Quarts (U.S., liquid)	0.001 057
Cubic feet	Acre-feet	2.296×10^{-5}
	Bushels (U.S.)	0.803 564
	Cords (of wood)	0.007 812
	Cubic centimeters	28 316.847
	Cubic inches	1 728.000

To Convert from Column A	To Column B	Multiply Column A by
Cubic feet (cont.)	Cubic meters	0.028 317
	Cubic yards	0.037 037
	Gallons (U.S., liquid)	7.480 520
	Liters	28.316 847
	Ounces (U.S., fluid)	957.506 49
	Pounds of water at 21°C, (avoirdupois)	62.366 3
Cubic feet/second	Acre-feet/day	1.983 333
	Cubic centimeters/second	28 316.847
	Cubic meters/second	0.028 317
	Hectare-centimeters/hour	1.019 4
	Liters/minute	1 698.963
	Liters/second	28.316 05
	Million gallons/day	0.646 412
Cubic inches	Board feet	0.006 944
	Bushels (U.S.)	0.000 465
	Cubic centimeters	16.387 064
	Cubic feet	0.000 579
	Cubic meters	$1.638\,706 \times 10^{-5}$
	Cubic yards	2.143×10^{-5}
	Gallons (U.S., liquid)	0.004 329
	Liters	0.016 387
	Milliliters	16.387 064
	Ounces (U.S., fluid)	0.554 113
	Quarts (U.S., liquid)	0.017 316
Cubic meters	Acre-feet	0.000 811
	Cords (of wood)	0.384
	Cubic centimeters	1×10^{6}
	Cubic feet	35.314 667
	Cubic inches	61 023.740
	Cubic yards	1.307 951
	Gallons (U.S., liquid)	264.172 05
	Liters	1 000.000
Cubic miles (U.S. statute)	Acre-feet	3.379×10^{6}
Cubic yards	Acre-feet	6.190×10^{-4}
	Cubic centimeters	764 554.860
	Cubic feet	27.000
	Cubic inches	46 656.000
	Cubic meters	0.764 555
	Gallons (U.S., liquid)	201.974 03
	Liters	764.555
Cups	Cubic centimeters	236.588
	Liters	0.236 588
	Ounces	8.000
Feet	Centimeters	30.480 37
	Inches	12.000

To Convert from Column A	To Column B	Multiply Column A by
Feet (cont.)	Kilometers	3.048×10^{-4}
	Meters	0.304 8
	Miles (statute)	$1.893\ 93 \times 10^{-4}$
	Yards	0.333 333
Feet/second	Kilometers/hour	1.097 28
	Meters/second	0.304 8
Flax seed (bushels/acre) (56 lb/bu)	Kilograms/hectare	62.776
Furlongs	Chains (Gunter's)	10.000
	Feet	660.000
	Meters	201.168
	Miles, statute	0.125
	Rods	40.000
	Yards	220.000
Gallons (U.S., liquid)	Acre-feet	$3.068\ 883 \times 10^{-6}$
	Cubic centimeters	3 785.411 8
	Cubic feet	0.133 681
	Cubic inches	231.000 23
	Cubic meters	0.003 785
	Cubic yards	0.004 951
	Gallons (British)	0.832 675
	Gallons (U.S., dry)	0.859 367
	Liters	3.785 412
	Ounces (U.S., fluid)	128.000
	Pints (U.S., liquid)	8.000
	Quarts (U.S., liquid)	4.000
Gallons (U.S.) of H_2O (15°C)	Pounds of water	8.328 23
Gallons/acre	Liters/hectare	9.353
Gallons/day	Liters/second	4.381×10^{-5}
Gallons (U.S. liquid)/minute	Acre-feet/day	0.004 419
	Cubic feet/second	0.002 228
	Cubic meters/second	0.631×10^{-4}
	Cubic meters/hour	0.227
	Hectare-centimeters/hour	0.002 27
	Liters/second	0.063 1
Grams	Kilograms	0.001
	Ounces (apothecary or troy)	0.032 151
	Ounces (avoirdupois)	0.035 274
	Pounds (apothecary or troy)	0.002 679
	Pounds (avoirdupois)	0.002 205
	Tons (metric)	1×10^{-6}
Grams/cubic centimeter	Kilograms/cubic meter	1 000.000
	Pounds/cubic foot	62.427 961

To Convert from Column A	To Column B	Multiply Column A by
Grans/cubic centimeter (cont.)	Pounds/cubic inch	0.036 127
	Pounds/gallon (U.S., liquid)	8.345 404
Grams/liter	Parts/million	1 000.000
	Pounds/cubic foot	0.062 426
Hectares	Acres	2.471 054
	Square feet	107 639.100
	Square kilometers	0.010
	Square meters	10 000.000
	Square miles	0.003 861
Hectare-centimeters	Acre-feet	0.081 08
	Acre-inches	0.972 76
	Cubic feet	3 531.844 8
	Gallons (U.S. liquid)	26 419.966
Hectare-centimeters/hour	Cubic-feet/second	0.981
	Gallons (U.S.)/minute	440.300
Hectare-meters	Acre-feet	8.108
	Acre-inches	97.290
	Cubic feet	353.184×10^3
	Gallons (U.S. liquid)	264.200×10^4
Hundredweights (British long)	Kilograms	50.802 345
	Pounds (avoirdupois)	112.000
	Tons (long)	0.050
	Tons (metric)	0.050 817
	Tons (short)	0.056
Hundredweights (U.S. short)	Kilograms	45.359 237
	Pounds (avoirdupois)	100.000
	Tons (long)	0.044 643
	Tons (metric)	0.045 359
	Tons (short)	0.050
Inches	Ångstrom units	2.540×10^8
	Centimeters	2.540
	Feet	0.083 333
	Meters	0.025 4
	Yards	0.027 778
Joules (SI)	BTU (mean)	9.472×10^{-4}
	Foot-pounds	0.737 684
	Kilowatt-hours	2.778×10^{-7}
	Watt-seconds (SI)	1.000
Kilograms	Ounces (apothecary or troy)	32.150 737
	Ounces (avoirdupois)	35.273 962
	Pounds (apothecary or troy)	2.679 229
	Pounds (avoirdupois)	2.204 623
	Quintals	0.010
	Tons (long)	9.842×10^{-4}
	Tons (short)	0.001 102

CONVERSION FACTORS (cont.)

To Convert from Column A	To Column B	Multiply Column A by
Kilograms/cubic meters	Grams/cubic centimeter	0.001
	Metric tons/cubic meter	0.001
	Pounds/cubic foot	0.062 428
	Pounds/gallon (U.S. liquid)	0.008 345
Kilograms/hectare	Pounds/acre	0.892
Kilograms/metric ton	Pounds/short ton	2.000
Kilometers	Centimeters	100 000.000
	Feet	3 280.839 9
	Meters	1 000.000
	Miles (statute)	0.621 371
	Yards	1 093.613 3
Kilometers/hour	Centimeters/second	27.777 778
	Feet/hour	3 280.839 9
	Meters/second	0.277 778
	Miles (statute)/hour	0.621 371
Liters	Bushels (U.S.)	0.028 378
	Cubic centimeters	1 000.000
	Cubic feet	0.035 315
	Cubic inches	61.023 744
	Cubic meters	0.001
	Cubic yards	0.001 308
	Gallons (U.S., liquid)	0.264 172
	Ounces (U.S., fluid)	33.814 023
	Quarts (U.S., liquid)	1.056 688
Meters	Ångstrom units	1×10^{10}
	Centimeters	100.000
	Feet	3.280 840
	Inches	39.370 079
	Kilometers	0.001
	Miles (statute)	6.214×10^{-4}
	Millimeters	1 000.00
	Rods	0.198 839
	Yards	1.093 613
Miles (statute)	Chains (Gunter's)	80.000
	Feet	5 280.000
	Inches	63 360.000
	Kilometers	1.609 344
	Meters	1 609.344
	Miles (nautical international)	0.868 976
	Rods	320.000
	Yards	1 760.000
Miles/hour	Feet/hour	5 280.000
	Feet/minute	88.000
	Feet/second	1.466 667
	Kilometers/hour	1.609 344

To Convert from Column A	To Column B	Multiply Column A by
Miles/hour (cont.)	Meters/minute	26.822 4
	Miles/minute	0.016 667
Millet (bushels/acre) (48 lb/bu)	Kilograms/hectare	53.808
Millet (bushels/acre) (50 lb/bu)	Kilograms/hectare	56.050
Milligrams/liter	Grains/gallon (U.S., liquid)	0.058 416
	Grams/liter	0.001
	Parts/million	1.000
	Pounds/cubic foot	6.243×10^{-5}
Millimeters	Centimeters	0.100
	Feet	0.003 281
	Inches	0.039 37
	Meters	0.001
Million gallons/day	Acre-inches/day	36.828
	Cubic feet/second	1.547
	Cubic meters/minute	2.629
Nitrogen (N)	Ammonia (NH_3)	1.216 274
	Crude protein	6.250
	Nitrate (NO_3)	4.426 124
Oats (bushels/acre) (32 lb/bu)	Kilograms/hectare	35.872
Ounces (avoirdupois)	Grams	28.349 523
	Ounces (apothecary or troy)	0.911 458 3
	Pounds (apothecary or troy)	0.075 955
	Pounds (avoirdupois)	0.062 5
Ounces (U.S., fluid)	Cubic centimeters	29.573 730
	Cubic inches	1.804 688
	Cubic meters	2.957×10^{-5}
	Cups	0.125
	Gallons (U.S., liquid)	7.812×10^{-3}
	Liters	0.029 573
	Quarts (U.S., liquid)	0.031 25
	Tablespoons	2.000
Parts/million	Grams/liter	0.001
	Milligrams/liter	1.000
Peanuts (bushels/acre) Virginia type (17 lb/bu)	Kilograms/hectare	19.057
Spanish type (25 lb/bu)	Kilograms/hectare	28.025
Phosphorus (P)	Phosphorus (P_2O_5)	2.291
Phosphorus (P_2O_5)	Phosphorus (P)	0.436
Pints (U.S., dry)	Bushels (U.S.)	0.015 625
	Cubic centimeters	550.610 47

To Convert from Column A	To Column B	Multiply Column A by
Pints (U.S., dry) (cont.)	Cubic inches	33.600 312
	Gallons (U.S., dry)	0.125
	Gallons (U.S., liquid)	0.145 456
	Liters	0.550 595
	Pecks (U.S.)	0.062 5
	Quarts (U.S., dry)	0.500
Pints (U.S., liquid)	Cubic centimeters	473.176 47
	Cubic feet	0.016 710
	Cubic inches	28.875
	Cubic yards	0.000 619
	Cups	2.000
	Gallons (U.S., liquid)	0.125
	Gills (U.S.)	4.000
	Liters	0.473 163
	Ounces (U.S., fluid)	16.000
	Quarts (U.S., liquid)	0.500
Potassium (K)	Potassium (K_2O)	1.205
Potassium (K_2O)	Potassium (K)	0.830
Potatoes, Irish (bushels/acre) (60 lb/bu)	Kilograms/hectare	67.260
Pounds (avoirdupois)	Grams	453.592
	Kilograms	0.453 592
	Ounces (apothecary or troy)	14.583 333
	Ounces (avoirdupois)	16.000
	Pounds (apothecary or troy)	1.215 277
	Quintals	0.004 536
Pounds of water at 15°C, (avoirdupois)	Cubic feet	0.016 034
	Cubic inches	27.737 013
	Gallons (U.S., liquid)	0.120 074
Pounds/acre	Kilograms/hectare	1.121
	Metric tons/hectare	0.001 121
	Quintals/hectare	0.011 21
Pounds/cubic foot	Grams/cubic centimeter	0.016 018
	Kilograms/cubic meter	16.018 463
	Pounds/cubic inch	5.787×10^{-4}
Pounds/gallon (U.S., liquid)	Kilograms/liter	0.119 826
Pounds/short ton	Kilograms/metric ton	0.500
Pounds/square inch	Atmospheres	0.068 046
	Bars	0.068 948
	Grams/square centimeter	70.306 958
Quarts (U.S., liquid)	Cubic centimeters	946.352 95
	Cubic feet	0.033 420
	Cubic inches	57.750
	Gallons (U.S., dry)	0.214 842

To Convert from Column A	To Column B	Multiply Column A by
Quarts (U.S., liquid) (cont.)	Gallons (U.S., liquid)	0.250
	Liters	0.946 326
	Ounces (U.S., liquid)	32.000
	Pints (U.S., liquid)	2.000
Quintals	Kilograms	100.000
	Pounds (avoirdupois)	220.462 26
	Metric tons	10.000
Quintals/hectare	Kilograms/hectare	100.000
	Metric tons/hectare	0.100
	Pounds/acre	89.206 07
Rice, rough (bushels/acre) (45 lb/bu)	Kilograms/hectare	50.445
Rods	Feet	16.500
	Feet (U.S. survey)	16.499 967
	Furlongs	0.025
	Inches	198.000
	Meters	5.029 2
	Miles (statute)	0.003 125
	Yards	5.500
Rye (bushels/acre) (56 lb/bu)	Kilograms/hectare	62.776
Sorgo seed (bushels/acre) (50 lb/bu)	Kilograms/hectare	56.050
Sorghum (bushels/acre) (56 lb/bu)	Kilograms/hectare	62.776
Soybeans (bushels/acre) (60 lb/bu)	Kilograms/hectare	67.260
Square centimeters	Square feet	1.076×10^{-3}
	Square inches	0.155 000
	Square meters	0.000 1
	Square yards	1.196×10^{-4}
Square chains (Gunter's)	Acres	0.100
	Square meters	404.686
	Square miles	0.000 156
	Square rods	16.000
	Square yards	484.000
Square feet	Acres	2.296×10^{-5}
	Hectares	$9.290\ 3 \times 10^{-6}$
	Square centimeters	929.030 4
	Square inches	144.000
	Square meters	0.092 903
	Square miles	3.587×10^{-8}
	Square yards	0.111 111
Square feet/acre	Square meters/hectare	0.022 957

To Convert from Column A	To Column B	Multiply Column A by
Square inches	Square centimeters	6.451 6
	Square feet	6.944×10^{-3}
	Square meters	6.452×10^{-4}
Square kilometers	Acres	247.105 38
	Hectares	100.000
	Square feet	1.076×10^{7}
	Square inches	1.550×10^{9}
	Square meters	1×10^{6}
	Square miles (statute)	0.386 102
	Square yards	1.196×10^{6}
Square meters	Acres	2.471×10^{-4}
	Hectares	0.000 1
	Square centimeters	10 000.000
	Square feet	10.763 910
	Square inches	1 550.003 1
	Square kilometers	1×10^{-6}
	Square miles	3.861×10^{-7}
	Square yards	1.195 990
Square miles (statute)	Acres	640.000
	Hectares	258.998 81
	Square feet	2.788×10^{7}
	Square kilometers	2.589 988
	Square meters	2 589 988.100
	Square yards	3 097 587.500
Square rods	Acres	0.006 25
	Hectares	0.002 529
	Square centimeters	252 928.526 4
	Square feet	272.250
	Square inches	39 204.000
	Square meters	25.292 853
	Square miles	9.766×10^{-6}
	Square yards	30.250
Square yards	Acres	2.066×10^{-4}
	Hectares	8.361×10^{-5}
	Square centimeters	8 361.273 6
	Square feet	9.000
	Square inches	1 296.000
	Square meters	0.836 127
	Square miles	3.228×10^{-7}
Sudangrass (bushels/acre) (40 lb/bu)	Kilograms/hectare	44.840
Sweet potatoes (bushels/acre) (55 lb/bu)	Kilograms/hectare	61.655
Tablespoons	Cubic centimeters	14.786 76
	Cups	0.062 5

To Convert from Column A	To Column B	Multiply Column A by
Tablespoons (cont.)	Ounces	0.500
	Teaspoons	3.000
Teaspoons	Cubic centimeters	4.928 922
	Cups	0.020 829
	Ounces	0.150 150
	Tablespoons	0.333 333
Tons (long)	Hundredweights (long)	20.000
	Hundredweights (short)	22.400
	Kilograms	1 016.046 9
	Ounces (avoirdupois)	35 840.000
	Pounds (apothecary or troy)	2 722.220
	Pounds (avoirdupois)	2 240.000
	Tons (metric)	1.016 047
	Tons (short)	1.120
Tons (metric)	Grams	1×10^6
	Hundredweights (short)	22.046 226
	Kilograms	1 000.000
	Ounces (avoirdupois)	35 273.962
	Pounds (apothecary or troy)	2 679.228 9
	Pounds (avoirdupois)	2 204.622 6
	Quintals	0.100
	Tons (long)	0.984 207
	Tons (short)	1.102 311
Tons (metric)/hectare	Kilograms/hectare	1 000.000
	Pounds (avoirdupois)/acre	892.180
	Tons (short)/acre	0.446
	Quintals/hectare	10.000
Tons (short)	Hundredweights (short)	20.000
	Kilograms	907.184 74
	Ounces (avoirdupois)	32 000.000
	Pounds (apothecary or troy)	2 430.555
	Pounds (avoirdupois)	2 000.000
	Tons (long)	0.892 857
	Tons (metric)	0.907 185
Tons (short)/acre	Tons (metric)/hectare	2.242
Vetch (bushels/acre) (60 lb/bu)	Kilograms/hectare	67.260
Wheat (bushels/acre) (60 lb/bu)	Kilograms/hectare	67.260
Yards	Centimeters	91.440
	Feet	3.000
	Inches	36.000
	Meters	0.914 4

Author Index

Miles, I. E., 392
Miley, W. N., 286
Miller, H. F., 213
Miller, J. R., 412
Miller, M. H., 126, 158, 166, 180, 189
Miller, R. H., 483, 505
Miller, R. J., 334
Miller, R. W., 4, 7, 90, 505
Milne, R. A., 430
Mitscherlich, E. A., 509
Monato, J. R., 18
Moore, W. A., 111, 439
Morey, D. D., 83
Morrill, L. G., 158
Morrison, F. B., 486
Morrow, G. E., 14
Mortvedt, J. J., 216, 232, 234, 244, 256, 257, 261, 263
Moschler, W. W., 380, 392, 471
Mugwira, L., 480
Mulder, E. G., 276
Muldoon, K., 18
Mullinier, H. R., 326
Mulvaney, D. L., 349
Murdock, J. T., 42
Murdock, L. W., 189, 416
Murphy, B. C., 291
Murphy, D. R., 371
Murphy, L. S., 31, 32, 38, 46, 54, 55, 72, 88, 96, 100, 129, 130, 132, 138, 145, 152, 155, 156, 207, 219, 221, 224, 232, 235, 251, 253, 313, 315, 320, 322, 323, 326, 336, 354, 374, 392, 418, 423, 452, 512, 521
Myers, F., 358, 359, 370
Myhre, D. L., 415

Namiki, M., 376
Nason, M. C., 159
Nebgen, J. W., 422
Nelson, G. S., 392
Nelson, W. L., 83, 159, 189, 506, 512, 533
Nesmith, J., 422
Nettles, W. C., 375
Newman, W. J., 141
Nicholas, D. J. D., 298
Nielsen, D. R., 90
Nilson, E. B., 460
Nishimoto, R. K., 146
Norman, A. G., 88
North, C. P., 336
Norvell, W. A., 216, 261

Nyborg, M., 422
Nye, P. H., 472

Obcemea, W. N., 74, 75
Ohki, K., 299
Ohlrogge, A. J., 126, 128
O'Leary, D., 16
Olsen, R. J., 219
Olsen, S. R., 97, 247
Olson, O. E., 422
Olson, R. A., 83, 97, 126, 219, 319, 348, 533
Oschwald, W. R., 506
Osmond, C. A., 392
Oster, J. D., 457
Overdahl, C. J., 213
Owensby, C. E., 33

Page, A. L., 83, 217
Parker, D. T., 36
Parks, W. L., 350
Parvin, P. L., 146
Patrick, W. H., Jr., 101
Paukstellis, J. V., 91
Pauli, A. W., 74
Paulsen, G. M., 91, 155, 207
Pearson, G. A., 435
Pearson, R. W., 85, 88, 415, 422
Peevy, W. J., 141, 142
Penas, E. J., 155, 234
Pendleton, J. W., 349, 350
Penney, D. C., 422
Pesek, J. T., 136, 341, 511, 533
Peterson, A., 490
Peterson, G., 147
Phillips, R. E., 391
Phillips, S. H., 379, 391
Phillips, W. M., 460
Place, G. A., 241
Poincelot, R. P., 495
Polijakova, G. V., 156
Polles, S. G., 238
Ponnamperuma, F. N., 244, 506
Poole, W. D., 376
Porter, H. C., 485
Porter, L. K., 204
Posler, G., 152
Potts, H. M., 133
Powell, R. D., 524
Powers, W. L., 71, 72, 83, 423, 439, 452
Prato, J. D., 13
Pratt, P. F., 457
Prokoscheva, M. A., 156

Qualset, C. O., 13

Racz, G. J., 92
Rader, L. F., Jr., 346
Ragland, J. L., 95, 471
Rai, R. K., 158
Raines, G. A., 111, 380
Randall, G. W., 258, 280, 376
Rapp, E., 430
Rauschkolb, R. S., 87, 333, 334
Read, D. W., 128
Reichman, G. A., 457
Reid, R. L., 213
Reisenauer, H. M., 248
Reitemeier, R. F., 391
Reitz, H., 267
Reuss, J. O., 87, 333
Revle, C. A., 454
Reynolds, E. B., 149
Rhoades, J. D., 437, 454, 457
Richards, G. E., 348, 351
Richardson, B. Z., 471
Richenderfer, J. L., 493
Richey, C. B., 470
Riley, D., 158, 319
Robbins, J. M., 485
Robertson, L. S., 72, 206, 219, 225, 263, 264, 274, 284
Robinson, D. W., 506
Robinson, F. E., 74, 76
Roemig, I. J., 213
Rolston, D. E., 83, 87, 333, 334
Rosell, R. A., 247
Rosenstein, L., 325
Rozycka, T., 219
Rubins, E. J., 262
Rule, C. A., 443
Russ, O. G., 374, 460
Russell, H. L., 486
Russell, J. C., 469

Sabbe, W. E., 83
Sabey, B. R., 34, 35
Saladaga, F. A., 74, 75
Salako, E. A., 232
Sanchez, P. A., 21, 78, 423
Sander, D. H., 328, 348, 533
Sandoval, F. M., 457
Sauchelli, V., 158
Savant, N. K., 92
Sawyer, E. W., 418
Schield, S. J., 100
Schmidt, B. L., 169
Schmidt, W. H., 392

Subject Index

Subjec